Springer Tracts in Mechanical Engineering

About this Series

Springer Tracts in Mechanical Engineering (STME) publishes the latest developments in Mechanical Engineering - quickly, informally and with high quality. The intent is to cover all the main branches of mechanical engineering, both theoretical and applied, including:

- Engineering Design
- Machinery and Machine Elements
- Mechanical structures and Stress Analysis
- Automotive Engineering
- Engine Technology
- Aerospace Technology and Astronautics
- Nanotechnology and Microengineering
- Control, Robotics, Mechatronics
- MEMS
- Theoretical and Applied Mechanics
- Dynamical Systems, Control
- Fluids mechanics
- Engineering Thermodynamics, Heat and Mass Transfer
- Manufacturing
- Precision engineering, Instrumentation, Measurement
- Materials Engineering
- Tribology and surface technology

Within the scopes of the series are monographs, professional books or graduate textbooks, edited volumes as well as outstanding PhD theses and books purposely devoted to support education in mechanical engineering at graduate and post-graduate levels.

Indexed by SCOPUS and Springerlink.

To submit a proposal or request further information, please contact: Dr. Leontina Di Cecco Leontina.dicecco@springer.com or Li Shen Li.shen@springer.com.

Please check our Lecture Notes in Mechanical Engineering at http://www.springer.com/series/11236 if you are interested in conference proceedings. To submit a proposal, please contact Leontina.dicecco@springer.com and Li.shen@springer.com.

More information about this series at http://www.springer.com/series/11693

Michael Okereke • Simeon Keates

Finite Element Applications

A Practical Guide to the FEM Process

Springer

Michael Okereke
Department of Engineering Science
University of Greenwich
Chatham Maritime
Chatham, Kent, UK

Simeon Keates
Faculty of Engineering & Science
University of Greenwich
Chatham Maritime
Chatham, Kent, UK

ISSN 2195-9862 ISSN 2195-9870 (electronic)
Springer Tracts in Mechanical Engineering
ISBN 978-3-319-88382-3 ISBN 978-3-319-67125-3 (eBook)
https://doi.org/10.1007/978-3-319-67125-3

Softcover re-print of the Hardcover 1st edition 2018

Printed on acid-free paper

This Springer imprint is published by Springer Nature
The registered company is Springer International Publishing AG
The registered company address is: Gewerbestrasse 11, 6330 Cham, Switzerland

This work is dedicated to our lovely families…for their love, devotion and dedication.

Foreword

The design of complex structures in modern engineering has been made possible by the routine use of computer-based numerical simulation techniques such as the finite element method (FEM). These techniques allow the designer to explore with relative ease previously unthinkable geometric shapes in combination with the use of a variety of modern materials and construction techniques. Visionary architects and engineers that are unwilling to restrict themselves to simple geometries, like Gaudi in the previous century, are now freed from the complexities and cost of building physical models of their projects and can rely on accurate computer predictions of how their designs will bear the loads imposed on them. Hence, FEM is now used routinely in structural, civil, aerospace, mechanical and manufacturing engineering industries and in many other fields of applied science.

Having a good understanding of the principles underpinning finite element techniques and the practical applications of FEM is therefore a key element of the education of modern engineers. This book aims to provide its readers with a thorough grounding in FEM from the point of view of its application to structural problems and solid mechanics. It covers most of the topics of importance in this field, from linear elasticity to large strains and plastic or viscoelastic behaviour of more advanced materials. This book provides the reader with an overview of MATLAB, a simple and user-friendly coding platform to implement the principles of FEM, but this is complemented with the description of use of commercial software such as ABAQUS which will be often required to handle large-scale realistic applications. It pays attention to issues of critical practical importance such as boundary conditions and the generation of good-quality finite element meshes. This book also provides a comprehensive set of examples which will illustrate the power of FEM and will help the reader fully understand its application to real problems.

I strongly believe that the authors have put together an excellent and comprehensive textbook that will become a much-used educational tool for modern engineers.

Greenwich, UK Javier Bonet

Preface

The text demonstrates the application of the finite element modelling principles to the solution of real-life problems. The presentation tracks the implementation from generation of virtual domains for the finite element modelling (FEM) process to derivation of numerical solutions of practical problems drawn from a wide class of industrial sectors. This textbook is aimed at final year undergraduate and postgraduate students as well as graduate-level industry staff.

The text will equip the reader with the ability to develop models of real-life problems using industry-standard finite element packages. The text will show the reader how to assess the validity of the FEM solution. The authors presuppose that the reader will have a working knowledge of common FEM solvers. If this is not the case, the reader is encouraged to get familiar with the working principles of some FEM solvers such as ABAQUS, ANSYS, LS-DYNA and COSMOL. These FEM solver packages have software documentations that will serve as a good start point.

As a result, the book is not a toolkit for setting up finite element modelling (FEM) solutions within FEM solvers. Rather, whichever the FEM solver package the reader is familiar with, the contents of this textbook will help the reader understand the outputs and probe what is really *under the hood*. This aspect is not always well articulated in the software documentations. Understanding what is *under the hood* is essential for users to deploy the FEM techniques in developing robust solutions to real-life problems.

The key benefits of the textbook include the following:

- The text is a hands-on textbook aimed at helping the reader quickly master the FEM process and start developing numerical solutions to real-life problems. The text cuts out a lot of the technical jargon associated with the FEM process, thus making the material accessible to both postgraduate students and graduate-level industry staff.
- The textbook has several case studies of real-life problems – in which the authors have demonstrated the successful application of the FEM process to challenging problems.

- The textbook opens up the *black box* of FEM solver design – thus helping students learn the key steps in the development of their versions of FEM solver packages. This skill, the authors believe, is central to developing newer and better FEM solvers, which will be needed in future uses of the FEM process.
- The textbook is supported by interactive additional resources, which will enhance the student's learning. These resources are published on the textbook's website and include user-defined material model (as Fortran code), MATLAB™ scripts, model generation software, etc.
- The textbook is written in easy to follow language with the emphasis on students being able to get started immediately with developing their FEM models right from the first chapter.

The textbook is divided into two parts:

- ***Part I: Introduction to the FEM Process*** This part comprises a technical introduction to FEM by initially describing the place of FEM within computational mechanics. Also, this part describes the *direct stiffness method* – a key principle at the centre of FEM solver simulation engine. It also presents a brief introduction to MATLAB™. Finally, to illustrate the process of creating a bespoke FEM solver, this part also guides the reader through steps in developing their first FEM solver using MATLAB™. This part consists of *four* chapters.
- ***Part II: FEM Principles*** The FEM process comprises many analysis pillars. This part is the core of the textbook and guides the reader through the different pillars or principles of the FEM process. These principles range from domain generation, boundary conditions, material models, to contact mechanics. It represents a handy reference material for the reader as they begin to create, understand and adapt FEM solutions to real problems. This part concludes with a reflection on the future of the FEM technique and suggestions of streams of research that can be undertaken to enhance the FEM process. There are *seven* chapters in this part.

We appreciate the contributions of colleagues, mentors and students during the different stages of development of this book. We offer this text as a resource to help anyone interested in the FEM process to get started.

Greenwich, London — Michael Okereke
Greenwich, London — Simeon Keates
August, 2017

Acknowledgements

Firstly, we acknowledge all of our students, whose project works inspired many sections of this book, as well as the examples and problems. In particular, the contributions of Mr. Ambrose Akpoyomare towards the periodic boundary conditions and UMAT development sections are gratefully acknowledged. It was rewarding supervising all of you, and the many illuminating conversations we had will be forever treasured.

We acknowledge contributions of Emeritus Prof. C. Paul Buckley and Prof. Nik Petrinic, both of Oxford University. You both got Michael excited about computational mechanics, and the many discussions he had about the principles of finite element modelling have remained evergreen in his mind. This work will not be possible without you. Thank you for being both great academic mentors and friends.

All colleagues that we have met along the academic journey are gratefully acknowledged. You have both excited and challenged us. The brilliance of your ideas and the company we shared are special, and we have been formed on the basis of those interactions.

We would like to also acknowledge our young families for their patience during the many weekends and evenings that were spent away from them as the book was written. It is a huge sacrifice that you have made, and now that this project is over, you will have us back to yourselves completely.

Finally, we acknowledge our general family and friends, who stood as great bulwarks of support all through the journey. You saw our many frustrations, listened to our endless prattle and endured many times away from your company, and yet you remained faithful. You are the champions of this work and we acknowledge you unreservedly. Thank you!

Contents

About the Authors

Dr. Michael Okereke is a Senior Lecturer in Engineering Mechanics at University of Greenwich, UK where he leads research in computational mechanics, impact engineering and constitutive model development for different engineering materials. He obtained his Ph.D. from University of Oxford and has researched extensively on the impact behaviour and micromechanics of composite materials. He teaches both undergraduate and post-graduate engineering mechanics courses, with a specific focus on finite element modelling. He also collaborates with commercial organizations on use of computational mechanics to improve their product design, material modelling and manufacturing processes.

Professor Simeon Keates is Deputy Pro Vice Chancellor of the Faculty of Engineering and Science and Head of Engineering at the University of Greenwich. He was formerly Chair of HCI and Head of the School of Engineering, Computing and Applied Mathematics at Abertay University and Associate Professor at the IT University of Copenhagen. He obtained his MA and Ph.D. from the Departing of Engineering at the University of Cambridge, where he later worked as an Industrial Research Fellow in the Engineering Design Centre. After leaving Cambridge, he moved to the US and joined the IBM TJ Watson Research Centre for a number of years before moving to Boston and working at ITA Software (now part of Google).

Acronyms

1-D one-dimensional.
2-D two-dimensional.
3-D three-dimensional.
AEM Applied Element Method.
BC Boundary Condition.
BEM Boundary Element Method.
BVP Boundary Value Problem.
CAD Computer Aided Analysis.
CFD Computational Fluid Dynamics.
DSM Direct Stiffness Method.
FDM Finite Difference Method.
FEA Finite Element Analysis.
FEM Finite Element Modelling.
FE Finite Element.
FVM Finite Volume Method.
GFEM Generalized Finite Element Method.
GPS Global Positioning System.
MATFESE™ MATLAB™ Finite Element Simulation Engine.
MFC Multi-freedom Constraint.
MFM Meshfree Method.
MSM Matrix Stiffness Method.
PBC Periodic Boundary Condition.
PRAC Particle Reinforced Aluminium alloy Composite.
PUM Partition of Unity Method.
RVE Representative Volume Element.
SM Spectral Method.
STIM Solder Thermal Interface Material.

SmFEM Smoothed Finite Element Method.
UD Unidirectional (used especially for composites).
UMAT User-defined Material model.
XFEM Extended Finite Element Method.

Part I
Introduction to the Finite Element Method

This part of the *Finite Element Applications* textbook focuses on developing an understanding of the finite element method. The principles that guide the development of a finite element solution to practical problems will be presented here and readers will be guided through a structured way of developing and implementing a simple finite element solver within MATLAB™. The objective is to equip readers with the ability to analyze from first principles the outputs that result from the **FE** process.

Chapter 1
Computational Mechanics and the Finite Element Method

Abstract An essential part of an engineer's training is the development of necessary skills to analyse and predict the behaviour of engineering systems under different loading conditions. Only a small proportion of real engineering problems can be solved analytically; hence, there arises the need to use numerical methods capable of accurately simulating real phenomena. The finite element method is one such widely used numerical method. This chapter introduces the principles of computational mechanics within the field of engineering mechanics. The finite element method is a key pillar in computational mechanics and this chapter explores the FE method within computational mechanics. Here, the necessity for the FE method and the limitations to its use in solving practical problems are established. The chapter concludes by discussing currently available FE solvers both as open source and proprietary versions.

Keywords Computational mechanics • Finite element method • Numerical methods • Finite element solver

1.1 Chapter Objectives

This chapter presents introductory facts about the Finite Element Method (**FEM**) by establishing the principles underpinning the method. It is expected that readers should be conversant with some basic principles of *mechanics of materials*, as well as *numerical methods*. However, it is the view that as much details as possible will be included in this chapter to breach any gaps in knowledge of the reader.

The following objectives will be addressed in this chapter:

(a) Introduce readers to computational mechanics.
(b) Establish the FEM approach, highlighting the need and limitations to their use in solving practical problems.
(c) Explore the different software for carrying out the finite element analysis.

M. Okereke, S. Keates, *Finite Element Applications*, Springer Tracts in Mechanical Engineering, https://doi.org/10.1007/978-3-319-67125-3_1

1.2 Introduction to Computational Mechanics

This section presents an overview to computational mechanics as a solution modality for engineering problems. The section concludes by presenting the different numerical methods that fall under computational mechanics, of which the finite element method is one of them. This section presents to the reader the place of FEM within computational mechanics.

1.2.1 Reflections About Engineering Mechanics

Engineering Mechanics is a branch of engineering dedicated to understanding and predicting the effects of forces and motions on engineering systems. Such systems can include particles, rigid bodies, and deformable bodies. To understand the stress state i.e. assess the impact of externally applied loads on the internal forces on a deformable body, certain equations need to be understood and solved. Engineering mechanics equips engineers with a wide range of equations for making informed decisions on the effects of external loads on the internal structure of the materials. These equations help provide analytical solutions. Once a material's response cannot be solved by simple equations, as deduced from the approach stated above, then more robust, alternative solution techniques must be considered to obtain a solution. There is the necessity to broaden the experience of the reader to those fields that cannot readily yield to analytical solutions. The introduction of the finite element method is a way of exposing readers to another (complex) tool for investigating, predicting and assessing the effect of forces on loaded members. This is of particular interest in complex systems where the solution is nonlinear.

1.2.1.1 Divisions of Engineering Mechanics

To solve a practical problem using the principles of engineering mechanics, there are several considerations that the engineer has to make. Consider a typical deformable body as shown in Fig. 1.1, subjected to a distributed external loading, $\mathbf{F}_{ext}$. Typical examples of such bodies can be range from two sub-atomic particles, macro-molecular polymeric chains, a car's axle shaft, truss structure of a bridge or even planetary bodies under the influence of gravitational forces.

The deformable body of Fig. 1.1 is bounded by a domain that encloses a set of *internal forces*, $\mathbf{F}_{int}$, which create the **constitutive behaviour** of the body. It is the objective of engineering mechanics to find the relationship between external and internal forces acting on such a deformable body. There are different strategies that can be used in understanding the interaction between external and internal forces. Some of these approaches can include:

(a) Design practical experiments that investigate the response of the deformable body under the effect of external forces. Load cells, strain gauges, video

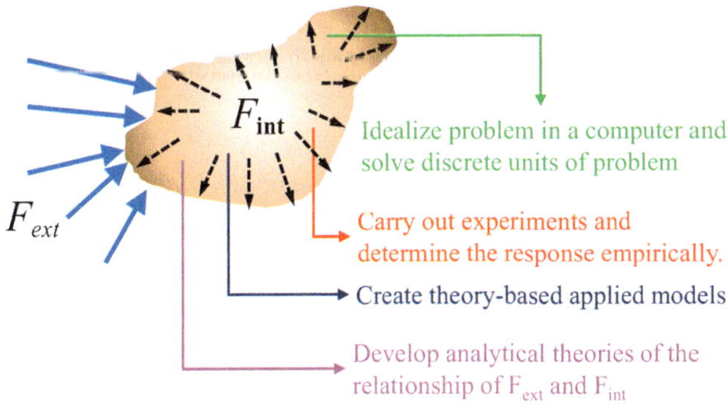

Fig. 1.1 An illustration of the different approaches for solving an engineering mechanics problem

extensometers, force and temperature transducers can be used to measure the external loads on the body and mechanics of materials theories used similarly to relate external forces to internal loads. This is the principle of *Experimental Mechanics*.

(b) Create theory-based models that use known **constitutive behaviours** to automate the understanding and prediction of behaviour of such bodies under diverse external loads. This approach falls within the field of *Applied Mechanics*.
(c) Develop complex analytical models based on the physics of deformation of the body such that a set of equations that link external and internal forces can be established. This approach falls within the area of *Theoretical mechanics*.
(d) Idealize the problem space (domain) into a virtual domain, which can be split up into tiny/finite solution units with simpler boundaries and loading conditions. Each solution unit can then be solved 'easily' and such solutions can then be assembled into a global solution representative of the original problem domain. This is the principle embedded in *Computational Mechanics*.

Based on the above approaches, engineering mechanics as a branch of (mechanical) engineering can be divided into four classes as described in more details below.

Theoretical Mechanics deals with the development and understanding of fundamental laws and principles that underpin the behaviour of deformable bodies. Engineers working in this field typically spend time deducing equations for describing the behaviour of materials. Often the experts in this field are called *modellers* and understand the physics of material behaviour that define the response of a given material under different load cases. A speciality of *Engineering Mechanics* concerned with developing better material models for describing/modelling the behaviour of materials under use is described as *constitutive modelling*.

Applied Mechanics utilizes tools established by theoretical mechanics experts to develop applications that can be useful for the end user. They specialize in understanding equations that have already been derived and deploy them into different applications. For example, an engineer at work who uses fatigue equations to undertake a fatigue investigation software/tool works in the field of Applied Mechanics.

Experimental Mechanics relates to the use of empirical tests to assess the validity, or not, of theories established in Theoretical Mechanics. They are often proponents of the most reliable set of data about the relationship between external loading and internal forces within a material. The data derived from carefully designed experiments cannot be refuted and demonstrate the most reliable set of data about the mechanics of such material/structures.

Unfortunately, what is realizable within a laboratory is often restricted by practical limitations. This means, for example, that not all load cases may be experimentally investigated. The environmental setup, the length scale and time scale of material behaviour may not be easily replicated in experiments due to limitations of test equipment, etc. Additionally, the costs associated with experiments are often too excessive. Consequently, irrespective of the dependable nature of data sets from experiments, the above limitations open the door for another approach that may not be so constrained, time consuming or expensive.

Computational Mechanics provides solutions to difficult problems by model-based simulations through the use of different numerical methods. The science of numerical analysis has built a deeper understanding of robust approaches to solving complicated problems by breaking them into small, manageable units. At the core of this approach is the implementation of *numerical methods* that 'divides' the problem domain into *finite* units. Each unit is then solved comprehensively before another numerical scheme is used to 'assembly' the solution units such that a holistic picture of the mechanics of such systems can be built. Computational mechanics has grown tremendously over the last two decades, due essentially to the advances in computing resources now available to engineers. This is the approach under which the finite element method is classed.

1.2.1.2 Impact of the Different Engineering Mechanics Branches

The comparative importance of the different branches of *Engineering Mechanics* can be assessed by considering typical Internet search results of both general and scholarly publications involving the different branches of engineering mechanics. The result, shown in Fig. 1.2 shows that although nearly the same volume of research is ongoing in all four branches, there is a higher interest in *computational mechanics* and *applied mechanics* in the general Internet community. These two branches contribute nearly 90% of the general web search results.

The theoretical and experimental mechanics are important and have seen nearly the same volume of research publications as the Computational and Applied Mechanics branches. However, it seems there is a higher importance placed on Computational and Applied mechanics. It is known that Computational and Applied

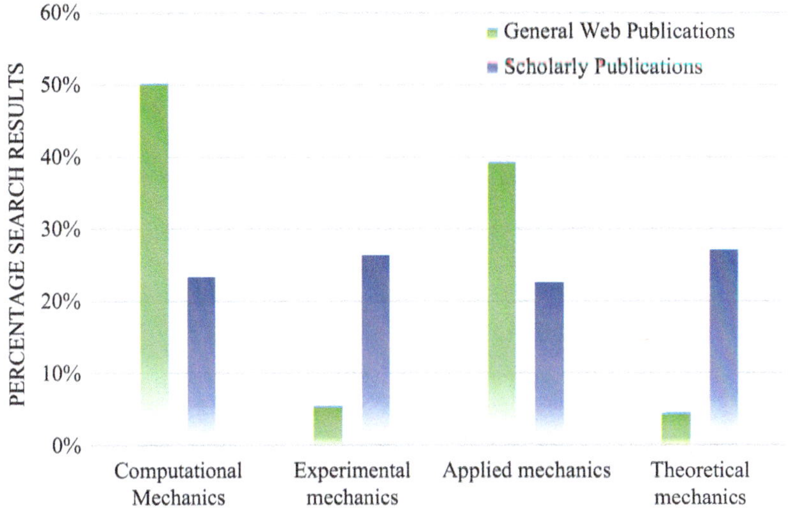

Fig. 1.2 Internet search results showing the relative impacts of the four classifications of *Engineering Mechanics*

Mechanics are better suited for solving practical challenging problems of both today and the future by creating powerful tools that engineers and scientists find essential for tackling real problems. Applied and Computational Mechanics provide a bridge between physical theories and their applications to technology. They offer practical insights to problems and create *solution platforms*, which many users can use without a deep understanding of the underlining theories driving the solution platforms. This is not the case with the Theoretical or Experimental Mechanics approaches, which generally require expert knowledge. This is why the Computational and Applied Mechanics branches attract the most significant Internet interest today, even though there are nearly equal research interest across all four branches. This trend is expected to continue into the future, especially as the capacity of computers continues to improve.

1.2.2 Aspects of Computational Mechanics

A computational mechanics problem can be differentiated according to either length scale or time scale of the problem. The length scale is related to the virtual domain (volume in space) in which the computational work is done. The time scale element relates to the speed of the deforming material or its externally applied loads for the problem under investigation. These two indices determine the type of computational mechanics procedure one has to adopt in the analysis of a typical problem. The following describe these aspects in more details.

1.2.2.1 Differentiation by Length Scale

In order to use the computational mechanics approach in solving problems, the domain of the problem has to be understood. This, therefore, introduces sub- and super-divisions to the computational mechanics virtual domain. If the method of distinction between domains is based on the length scale (i.e. the refinement of the microstructure), one can identify computational mechanics approaches for solving a given problem to be sub-divided into such length scale analysis types as: ***macro-molecular*** *mechanics,* ***nanomechanics***, ***micromechanics***, *continuum mechanics* and finally *structural* or *system mechanics*. This is illustrated in Fig. 1.3 and in the following, each of the analysis types will be discussed briefly.

Molecular mechanics: A computational mechanics problem can be set in the molecular mechanics length scale, which represents the scale of molecules that make up the material. For example, a finite element analysis of polypropylene can be set at molecular level in which the behaviour of **macromolecular** chains can be investigated. Also, analysis of nanocomposites usually begins at the molecular (or, in other cases, crystal lattice) length scale.

Nanomechanics involves analysis at molecular and atomic scales. It can be used interchangeably with the molecular mechanics scale, depending on the characteristic size of molecules under consideration. Length scale measurements within this branch are usually 10^{-9} m.

Micromechanics involves differentiation of the microscale components of the material domain under investigation. Micromechanics varies from one material type

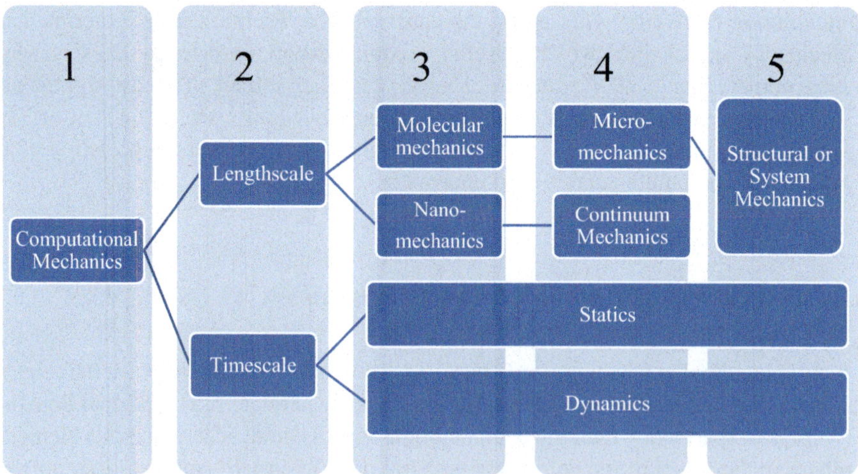

Fig. 1.3 The differentiation of a typical computational mechanics problem can be done in consideration of either length scale (and its sub-scales) and timescale domains. *Numbers 1 to 5 indicate different levels of analysis of the problem*

to another. For metals, it can represent the crystallographic and granular levels of matter that make up the metallic material under investigation. For composites, micromechanics can represent a scale in which it is easy to distinguish between the fibre and matrix/polymer/ceramic material that were used to manufacture the composite. Within a biological system, such as bone, micromechanics can represent the trabecular or cancellous bone. Typically measurements within this scale are in the range of 10^{-6} m. Entities at this scale are often visualized using some type of microscope.

Continuum mechanics is a scale where the test domain is visible to the human eye. It is usually in measurements of 10^{-3} m. Some textbooks differentiate this scale to sub-scales of *mesoscale* and *macroscale* but it is common in most literature to identify the continuum mechanics to operate at the macroscopic scale.

Mesoscopic length scales are usually introduced where the transition from microscale to macroscale will mean some underlining features of the test material are not represented. For example, in woven composites – whilst at microscale we identify the microstructure to consist of fibre and polymeric matrix, the arrangement of the fibre bundles within the matrix medium is captured by identifying a mesoscale. Consequently, it becomes possible, having represented the underlining features, to treat at the macroscale the test composite panel as homogeneous. At the macroscale, continuum models developed by theoretical mechanics experts can be applied directly to the material. Also, test data from experiments are easily compared with predictions from the macroscale without the need to scale up or scale down the data to suite predictions at sub- or super- length-scales.

System or structural mechanics involves analysis of the test domain where measurements are usually of the order of 100 m. This can represent an entire F1 car made from carbon fibre reinforced composites, the turbine blade of an airplane made from titanium material, or a human femur. This scale identifies mechanical, biomedical or other objects with clearly differentiable functions. The order of arrangement and typical dimensions of all these length scales differentiations is given in Fig. 1.4.

1.2.2.2 Differentiation by Timescale

With respect to differentiating computational mechanics approaches based on the timescale, we identify two types of simulations under this category: *statics* and *dynamics*. Statics relate to continuum mechanics simulations requiring the neglect of inertial forces. As shown in Fig. 1.5, if a block of weight, W is pulled by force, P under the effect of frictional force, F_f, the effect of the timescale on its mechanics can be distinguished by assessing its static or dynamic responses. This is distinguished by the value of its acceleration, a at a given time. If $a \to 0$, then the system is considered a static system, whilst if $a \to \infty$, such system is a dynamic system.

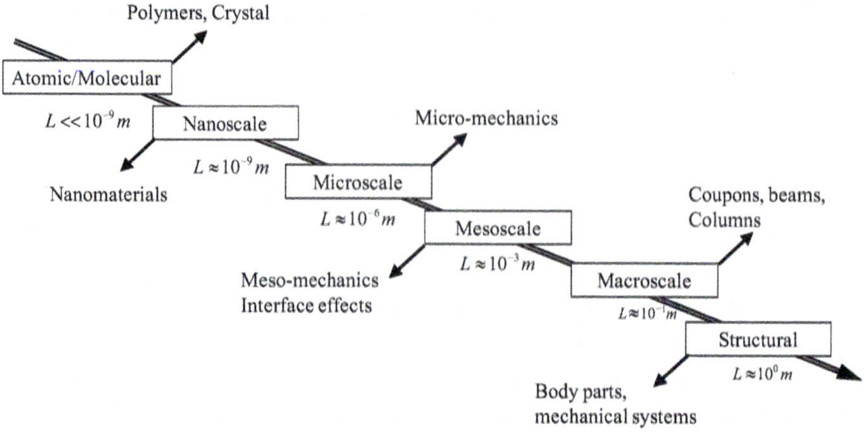

Fig. 1.4 The ordering of the typical length scales used in computational mechanics. The different scales are shown in order on increasing dimensions (Image reproduced with kind permission of Okereke and Akpoyomare [9])

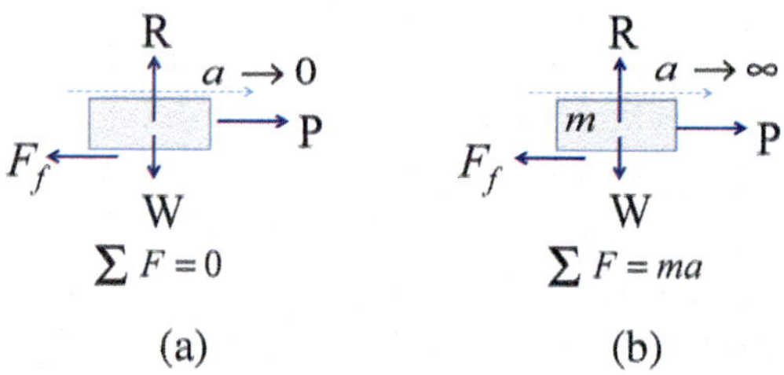

Fig. 1.5 A free body diagram of a body under (**a**) *static* and (**b**) *dynamic* loading

As well as acceleration, another marker of timescale effects in computational mechanics is ***rate-dependence***. Most statics are time-invariant (or rate-independent) in which case it is not important to the behaviour of such system to consider the effect of changing time.

Consider a typical example such as the tensile testing of a material. If the stress-strain response of the material is invariant with test speed (a measure of time) then the constitutive behaviour is said to be time-invariant. Most metals behave like this. However, if the speed of testing (in other words changing time) leads to changes in the mechanical response of the material then it is said to be time-dependent. In this latter case, it is important that we consider the effect of timescale to the continuum mechanics of the system.

A typical time-dependent material behaviour is shown in Fig. 1.6. The graph indicates how varying the timescale of loading for a given experiment can cause change, for example, of the yield stress of polypropylene (as in Fig. 1.6) from 40 MPa at $10^{-4}\ s^{-1}$ strain rate to 150 MPa at $10^{4}\ s^{-1}$ strain rate.

Dynamics on the other hand must recognize the effect of time in calculating the inertial forces. Without recognizing the time-dependence, the calculation will not be correct. Consequently, a computational mechanics solution may approach a problem

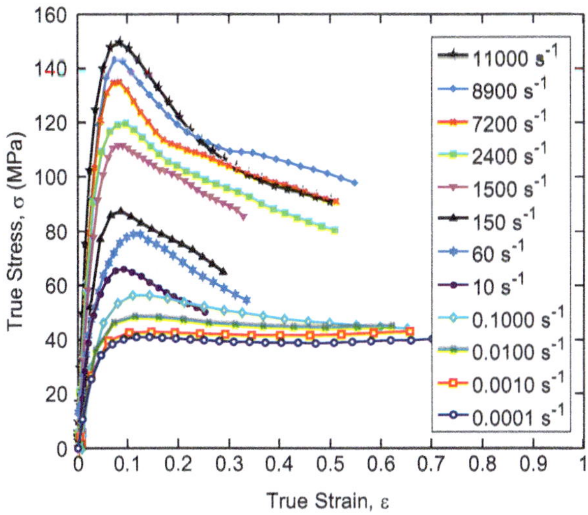

Fig. 1.6 An illustration of the effect of timescale on the rate-dependent compressive response of normal grade of polypropylene homopolymer – tested at a temperature of 25 °C. The observed response was seen following tests across a wide range of strain rates (Image reproduced with kind permission of Okereke et al. [10])

using either statics or dynamics as a basis. These two approaches can either neglect or account for associated inertial effects on the body during the loading process.

1.2.3 Numerical Methods for Computational Mechanics

As established previously, computational mechanics rely heavily on numerical methods. This section details some of the existing numerical methods that have been used for dealing with computational mechanics problems. Some of the commonly used numerical methods in computational mechanics are shown in Fig. 1.7.

Numerical methods in computational mechanics involve developing numerical approximations of the mathematical equations that define the behaviour of a problem domain. In order to make the numerical approximations work, the problem domain is often *discretized* into solvable units. Discretization is the process of converting a continuous body, functions, models or equations into discrete units that can are amenable to numerical evaluation within, say, digital computers. The resultant discretization of the physical variables leads to *discrete equations*.

Numerical methods are needed to simplify the solution of partial differential equations (PDEs). Numerical methods are tools that help in obtaining *discrete equations* that can be deployed in computers to solve many of the problem domains that characterize real life. This process is essential if computers are to solve computational problems. This leads us to assess the various discretization

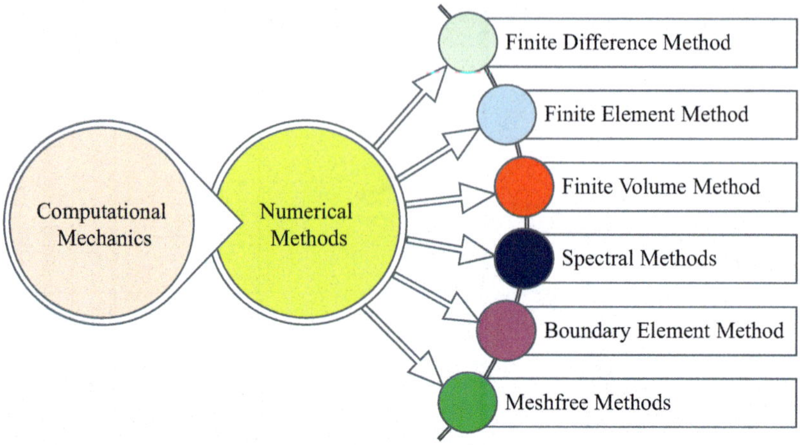

Fig. 1.7 Examples of commonly used numerical methods used in computational mechanics

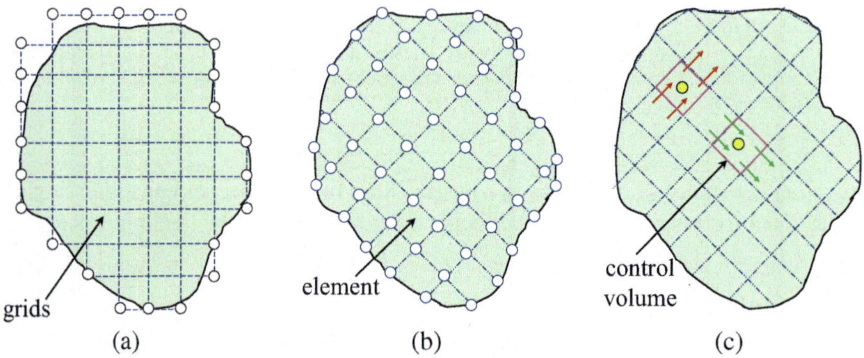

Fig. 1.8 An implementation of numerical discretization on an arbitrary domain using (**a**) finite difference methods (FDM) discretized through equally spaced grids that approximately describe the boundary profile of the arbitrary domain; (**b**) finite element methods (FEM) showing elements with nodes (unfilled circles) and (**c**) finite volume methods (FVM) – discretized using quadrilaterals, and showing two control volumes with inflow and outflow fluxes

modalities used in computational mechanics, highlighting their differences as numerical methods. Figure 1.8 shows two commonly used discretization modalities for converting continuum models into small counterparts for use in simulation studies within a digital computer. These discretization modalities have to be assessed in consideration of them being tools for obtaining a solution to PDEs.

Finite difference method (FDM) This is the earliest of the numerical methods for solving PDEs. It was made popular in the 1950s. This discretization method relies on the *difference* between the exact function (model, equation, body) and an approximate solution. The FDM approach involves the application of local ***Taylor's series*** *expansion* to represent the PDEs. The closer the match between the exact and approximate solution, the lesser the difference. The FDM approach requires that the problem domain has to be discretized into a topographic square assembly of lines usually of same grid sizes.

Finite element method (FEM) This numerical method became increasingly popular as the limitations of the FDM approach became apparent. It was developed in the 1960s. It offers also a numerical approximation of the problem domain, and its accuracy depends on the ***smoothness*** of the problem data. However, even for *non-smooth* data, the finite element method can still be used, although convergence to the desired solution may be delayed. Since most problem domains have non-smoothed properties, the finite element method can be adapted to obtain solutions to most practical problems.

The FDM and FEM approaches rely on a *meshed* geometric discretization of the problem domain. The mesh typically consists of *nodes* and *elements*. The FEM approach is commonly used in both structural, thermal and computational fluid dynamics studies. The FEM is the main focus of this book and more attention will be paid to it in subsequent sections and chapters.

Finite volume method (FVM) This is similar to the FDM and FEM except that in the FVM approach, volume integrals of a given '*finite volume*' are converted to a surface integral using **Gauss' theorem**. The smallest unit/element of an FVM problem is called a *control volume*. These control volumes are *finite sized* and of significantly higher sub-domain sizes than corresponding FEM elements. *Diffusion fluxes* – resulting from the implementation of the Gauss' theorem on volume integrals – are derived at discrete spatial locations that make up the meshed geometry. A key criteria for FVM solutions is that the *diffusion fluxes* entering and exiting the control volumes must be *conserved*, i.e. equal and opposite to themselves. The FVM approach is used especially in computational fluids mechanics, heat transfer and mass diffusivity problems.

Spectral method (SM) Following the success of the FDM and FEM approaches, further research in the 1970s led to the discovery of the Spectral method. It is widely regarded that FDM, FEM and SM are the "big three" discretization technologies for solution of PDEs [14]. The Spectral method gives the highest order of accuracy among the "big three" technologies for solution of PDEs. The SM can achieve up to ten digits of accuracy in comparison with the FEM and FDM approaches that can only achieve two or three digits of accuracy [14].

Spectral methods develop solutions to PDEs as a sum of what is called *basis functions* over an entire problem domain. Such basis functions are usually **Fourier series**, hence both SM and FEM involve the use of Fast Fourier Transforms (FFTs) in obtaining solutions to PDEs. Although FEM and SM are quite similar, the literature states that SM uses a *global* approach whilst FEM uses *local* approach to the solution of PDEs. Also, SM discretization require that the defining data for the problem domain is *smooth*, else spectral methods cannot be used.

Boundary element method (BEM) This is a numerical method for solving linear PDEs by using only the boundary integral equations of the problem domain. The

BEM approach requires the definition of surface integrals formulated according to **Green's functions**. If the behaviour at the boundaries of a physical problem is known, one can iteratively use this information to determine the behaviour of the media within the domains. The BEM approach is best used if the physical problem has linear homogeneous characteristics: an essential condition for using the Green's function. Due to this requirement, the BEM approach has limited applicability and in its purest forms cannot be used to solve nonlinear problems. Although, adaptations of the BEM can be used for solving nonlinear problems but such adaptations complicate the solution by introducing volume integrals to the solution [1].

Meshfree method (MFM) As the name implies, this type of numerical method for domain discretization does not require a mesh. Rather, the simulation domain is discretized into a collection of point clouds or nodes. Each node is enforced with a predefined set of equations that define its interaction with its neighbours. The **constitutive behaviour** of the material domain discretized using the MFM can be established using the connectivity equations that link the set of nodes. Model parameters such as mass, kinetic energy and displacement vectors are applied to the nodal points, unlike in FEM where mass and energy are applied to the elements. The meshfree method has been shown to significantly reduce computational cost [4].

1.3 The Finite Element Method

This section presents the classic finite element method, highlighting its necessity in the solution of engineering problems as well as its limitations. The section concludes by describing the current adaptations of classic FEM in order to address the limitations identified.

1.3.1 An Introduction to the Finite Element Process

There are many approaches for deploying a computational mechanics based solution to a typical problem. The focus of this work is on the finite element method. The implementation of a computational mechanics solution using the finite element method consist of three key steps. These are:

(a) *idealization of solution domain*;
(b) *application of governing equations for the solution*; and finally,
(c) *implementation of numerical methods* to obtain the desired solution.

Consider a body as shown in Fig. 1.9 simply supported as shown and experiencing an off-axis pull of force, **P**. The objective is to determine the mechanical response of the system under the effect of **P**. This statement describes the *physical problem*. Using the computational mechanics process, the physical domain of the

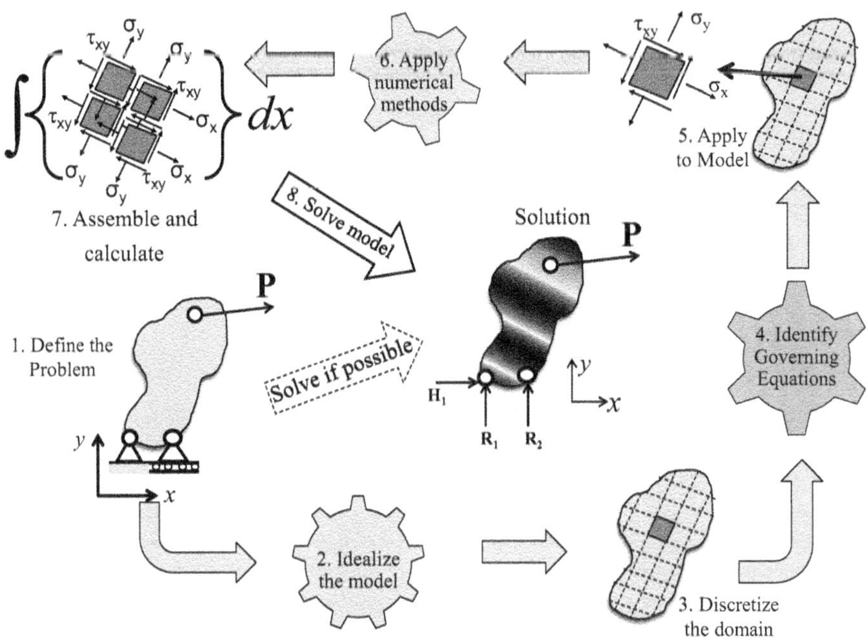

Fig. 1.9 A schematic representation of the key processes that comprise the finite element method

problem has to be idealized within a computational platform as a first step. As part of the idealization process, it is essential to discretize the system into 'small solvable units' – called *finite elements* as will be described in detail later.

Based on one of the finite elements, the next stage involves the application of principles of engineering mechanics to solve for any unknown parameters associated with the mechanical response of the physical system. The principles of mechanics also refers to the *governing equations* for a given problem. If it is a structural problem, the governing equations will be drawn from *structural mechanics* whilst *fluid mechanics* formulations will be used in obtaining solutions to a problem involving fluid flow, for example.

The final step involves the application of numerical methods to determine a solution to the physical problem. Numerical methods give approximate solutions and herein lies the strength of the finite element process. Although the solutions are approximate, the discretization of the problem domain into finite elements allows for simplification of the solution domain into solvable units using the numerical process. Part of the numerical step is the *assembly* of the solution units into a cohesive domain with applicable compatibility and equilibrium criteria. The *assembly* step involves integration of solutions from each finite element into a global output which defines the desired solution to the physical problem. All these processes are at the core of any computational mechanics solution involving the finite element method.

The solution from the FE process depends on the size of the finite elements. More resolution or predictive fidelity can be achieved by adding more nodes/elements to the discretizaion of the geometric domain. This leads to greater accuracy/precision but at a greater computational effort i.e. more calculations required as well as more effort to build such model. The aim should always be to find a computational solution that is *accurate enough* to meet the user's needs.

However, how 'accurate' is *accurate enough*? Of course, it depends on the problem and the context in which the user is developing the numerical solution. Getting a computational answer that is not accurate enough risks failure of what is being designed. Being too accurate suggests wasted computational effort in generating unnecessary detail. One therefore needs to find a satisfactory balance between these.

Galerkin Method and FEM: One of the key objectives of a structural or solid mechanics FEM problem is to solve for the displacement field of the solution. The displacement field is then coupled with a standard strain-displacement relationship (e.g. normal or shear strain or even small-strain small-stress relations) to define the strain field. Since FEM is a numerical analysis method for obtaining approximate solutions to the problem domain, a classic Galerkin method is commonly used in FEM for the numerical solution. The Galerkin method was invented by a Russian mathematician called Boris Galerkin and its formulation is often referred to as a *Galerkin method of weighted residuals*. A similar formulation that is also applicable in obtaining numerical approximate solutions in FEM is described as the *Ritz variation method*. For further reading of the Galerkin method in FEM, refer to [3, 5].

1.3.2 The Need for the Finite Element Method

As mathematical modelling of physical, chemical and biological phenomena becomes more and more common place, so does the need to be able to solve partial differential equations (PDEs). Different problems from fluid dynamics, electromagnetism, financial analyses, economy, material science, can now be approached using numerical methods based on partial differential equations. The formulation of the PDEs is so complicated that obtaining a closed form solution to them by purely analytical means is very improbable, if not impossible. The impetus, therefore, is to obtain good enough approximate solutions for such problems. Numerical approaches, of which the finite element method is one, provides a viable route towards creating numerical approximations to an unknown analytical solution.

According to Jacques Hadamard, a French mathematician who made contributions to development of PDEs, the mathematical modelling of physical phenomena requires that the problem should be *well-posed*. This implies that such PDEs must obey the following criteria, otherwise referred to as *Hadamard criteria*:

(a) *The PDE should have a solution that exists;*
(b) *The solution must be unique; and finally,*
(c) *The solution of the PDE should change continuously with the initial (boundary) conditions defined for the PDE.*

There are many examples of well-posed problems, one of which is the heat equation. Well-posed problems can be solved easily using computers where numerically stable algorithms can be designed for obtaining such solutions. The converse of well-posed problems are *ill-posed* problems. These problems violate some if not all of the Hadamard criteria. To solve these problems in a computer, the user will often face the challenge of numerical instability and hence the problem has to be reformulated before numerical analysis principles are implemented on them.

Finite element methods are particularly suitable for obtaining approximate solutions to PDEs for many physical models. Originally, FEM was designed for solving structural and solid mechanics problems. However, in recent times, they can be applied to solving a wide range of physical phenomena such as dynamics, thermal analyses, wave propagation, acoustic studies, biomedical and biomaterial studies, fracture mechanics and problems of discontinuum mechanics. In many cases, the problems can be ill-posed, but the finite element solver (the computational engine of FEM tools) will find ways of re-formulating the problem to make it solvable using common numerical methods. At the core of this process is the discretization of the problem domain allowing for such discrete/finite units to be solved.

The above explains why the finite element method has continued to attract great interest as a computational mechanics method for tackling a large range of physical problems. Herein lies the need for the FEM, as it opens the possibility for problems that cannot always be solved analytically to be solved numerically following acceptable discretization of the problem domain. Although the solutions have inherent discretization errors, there is often a close enough agreement between physical experimental results and numerical solutions of the problem.

1.3.3 Limitations of the Finite Element Method

In spite of the importance of the finite element method, it is fraught with a lot of challenges which will have to be detailed at this early stage of the book. It is important for the reader to understand these challenges and how they have led to adaptations of classic FEM to create better, more robust FEM tools. The reader is encouraged to explore more widely the implications of these limitations. The objective here is simply to highlight them to provide a complete picture of the classic FEM. Adaptations of classic FEM implementations will be presented in Sect. 1.3.4.

The overly stiff/locking problem: This is the main problem encountered with the classic FEM. Classic FEM is implemented based on assumed compatibility of equilibrium across the problem domain. This implies that all modes of deformation must exist within the problem domain. This is not always the case in practical problems, leading to what is commonly reported in standard FEM software as *shear locking* or *hourglassing* of elements. This problem is most pronounced for triangular (linear) elements. This is an inherent limitation to classic FEM, and traditionally artificial numerical fixes are introduced to FEM schemes to eliminate this problem.

The stress accuracy problem: The presence of overly stiff elements in an FEM solution will result in incorrect stresses for such elements. This in itself will create stress accuracy issues. This is particularly a problem with classic FEM where a change from one element type to another can introduce significant stress variance that casts doubts on the reliability of the numerical solutions.

Mesh distortion problem: FEM relies on high quality mesh generation and where this is not possible, significant errors can arise. One particularly challenging case is modelling finite deforming structures, for example a hyperelastic material which experiences a strain of, say, 400%. Some of the finite elements might become highly distorted. These elements are likely to fail to retain the high mesh quality required by classic FEM and so introduce numerical errors. The mesh distortion issue is related to the inherent requirement in FEM for a *Jacobian matrix* to be evaluated for isoparametric elements.

Element shape problem: It is widely known that FEM solutions are more accurate if quadrilateral or hexahedral elements shapes are used to solve 2D or 3D problems respectively. However, meshing of complex geometries are easiest with triangular (for 2D problems) or tetrahedral elements (for 3D problems). The mesh efforts required for generating triangular or tetrahedral elements is quite minimal and can be automated (especially where remeshing algorithms are incorporated with a solution algorithm). Any gains in mesh generation efforts as in this case will lead to numerical errors resulting from the use of such elements to obtain numerical solutions to classic FEM problems. An ideal situation will be to create an FEM scheme where such triangular or tetrahedral elements are used with minimal element shape errors.

Discontinuity conundrum: Classic FEM requires that the discretization domain has to be continuous over the problem domain. This continuity requirement must be preserved all throughout the deformation stages of the solution. This is not always the case where fracture or material losses occur during the deformation. In these instances, discontinuities are introduced to the solution domain, which means classic implementations of FEM have to be reformulated to deal with this problem. In an FEM solution, the element's integrity has to be retained and cannot be allowed to 'split' – as when this happens, the philosophy of classic FEM is broken.

Real structures experience discontinuities, especially as cracks evolve through the material. The crack path might have to pass through an element, but classic FEM is not designed to tackle this problem. This introduces what is herein referred to as the *discontinuity conundrum*. There are approaches in the existing classic FEM methods to deal with discontinuities, however these are limited and expensive in terms of computational time. The good news is that emerging adaptations of classic FEM, some of which are reported in Sect. 1.3.4, are increasingly capable of dealing with these problems.

1.3.4 Adaptations of the Classic Finite Element Method

There are different types of FEM implementations that have arisen over the years. The diversity of these FEM implementations attest to the attraction of FEM as a numerical technique for solving many practical challenges. Although the following are regarded as types of FEM, they actually represent adaptations of the classic structure of FEM.

Applied element method (AEM): This is a recent adaptation of the classic FEM method, made popular by the seminal paper of Meguro and Tagel-Din [6]. It was developed essentially for progressive structural collapse. It is said to be the main numerical tool for analyzing structural collapse through all stages of loading: small and large displacements, collision and collapse. AEM is similar to classic FEM but the problem domain is usually discretized into quadrilaterals (2D analyses) or parallelepiped (3D analyses) elements. These elements are considered as rigid bodies and connected through the whole surface to other elements via three springs: one normal and two shear deformation springs. Deformation occurs strictly on element surfaces and not within the elements. AEM is particularly powerful in that it predicts reliably both continuum and discrete behaviours of structures.

Generalized finite element method (GFEM): This method is an extension of the classic/standard FEM method aimed at tackling complex physical problems that cannot be tackled using normal FEM approach [13]. Such problems can include: domains with re-entrant corners, propagating cracks in elastic continua, and domains involving composite materials with wide variation in material properties, etc. Also, GFEM has been applied to material interfaces problems, such as a typical lap joint between two materials of different material coefficients. GFEM was made popular in 2000 by Strouboulis et al. [12]. This method combines the principles of classic FEM with what is described as the *Partition of Unity Method (**PUM**)*.[1] The distinguishing thing about the GFEM is that it introduces special functions that are proven to approximate the exact solution of the PDEs that define the physical problem. Local and global accuracy of classic FEM solutions can be significantly increased by introducing such special functions.

[1] Partition of Unity Method (**PUM**) is a generalization of classic FEM. It involves the discretization of the geometric domain using random local (enrichment) function spaces. A mesh is required in classic FEM studies but such is not used here. Such local function spaces can represent intersecting circles, squares, ellipses. This method is useful in adaptive refinement processes. For more information on PUM, consider the work of Melenk and Babûska [7].

Extended finite element method (XFEM): This method was invented by Ted Belytschko and colleagues [8] to address problems of evolving discontinuities or crack propagation without the requirement for remeshing. The method is similar to the GFEM approach, incorporating classic FEM with the PUM. However, in this approach, a set of discontinuous fields are incorporated across the crack faces away from the crack tip. New crack fronts are accounted for by redefining the crack tip location and adding new crack segments. The method depends on the PUM properties of FEM problems and the following enrichment functions are typically used: a **Haar function**, and an **asymptotic near-tip field** function. It is widely used in the solid mechanics community for problems involving crack modelling, cohesive crack growth, modelling holes and inclusions.

Smoothed finite element method (SmFEM): Traditional FEM is fraught with many limitations some of which are already identified in Sect. 1.3.3. The SmFEM is another numerical method that attempts to solve a few of the limitations that face classic FEM. This method incorporates the traditional FEM implementation with meshfree methods.

Other methods: Under this category, we can include various combinations of the numerical methods with the FEM method. Examples include: the *spectral element method* which brings aspects of spectral methods into the FEM process; the *stretched grid method* uses aspects of the FDM approach within a typical FEM solution; and the *FEM-meshfree method* which borrows aspects of **MFMs** within an FEM scheme. It is expected that future numerical algorithms that model complex physical problems will increasingly be based on different combinations of the numerical methods discussed in Sect. 1.2.3.

1.4 The Finite Element Modelling Software

At this introductory stage of this book, it is important to explore the FEM software that is available within the computational modelling community. This will give the reader a quick review of these FEM software platforms, highlighting their history, evolution and suitability for different engineering problems. The section also presents the ancillary software that can be used to drive these standard FEM solution platforms.

1.4.1 The Finite Element Modelling Solver

The solution of the partial differential equations for temperature, stress, fluid flow that are associated with a physical problem is often complicated as the mesh size of the discretized domain starts to grow. It becomes very difficult to track the solutions of these equations in the numerical scheme using simple analytical tools. Hence, computational methods presented earlier are crucial to dealing with the wide range of data associated with the FEM process. Solving them often require use of **finite element modelling solvers** designed for this.

A *finite element modelling solver* is a software package developed specifically to implement the solution of partial differential equations. There are many solvers on the market and more are continuously being introduced.

The following sections will highlight some solvers, with a brief description of their special features.

(a) **NASTRAN**: This is the earliest general purpose finite element program – developed initially through a NASA funded project in 1965 by a Californian research group led by Dick MacNeal. It was later released into public domain as a commercial FE solver after many bugs were corrected in the initial version. As of 1990, many industries used NASTRAN as the mainstay of their FE work. It is now called MSc Nastran – named after the MacNeal-Schwendler Corporation (MSc) that currently markets it.
(b) **ANSYS**: This software was the next to be developed, after NASTRAN. John Swanson developed this program for Westinghouse Electric Corporation for the analysis of nuclear reactors. This software could analyse both linear and non-linear problems and became widely adopted by companies in 1996. As of 2017, ANSYS Inc has a total market value of $10.7 billion [11].
(c) **LS-DYNA**: This is another non-linear software package developed initially by John Hallquist for Lawrence Livermore National Laboratory. When John left the company in 1989, he founded a company called Livermore Software and Technology Corporation (LSTC) which now markets the program. The specific attraction of this software to the industry was its nonlinear dynamic capabilities, which were developed specifically for crashworthiness analysis, sheet metal forming and prototype simulations, such as drop tests. The software has, however, been expanded to included static analysis.
(d) **ABAQUS**: This software is one of the newest players in the range of commercial/proprietary FE solvers. It was developed in 1978 by a company called Hibbitt, Karlsson & Sorensen, Inc., (HKS) – named after the developers of the software. ABAQUS was initially focussed on nonlinear applications, but gradually linear capabilities were added. This was the earliest software where gateways for the addition of new material models and elements were possible. The current marketers of the program, Dassault Systemes Simulia Corporation (DSSC), bought the company from HKS in 2005 for $413 million.
(e) **COSMOL Multiphysics**: This software was initially called FEMLAB and was developed in 2005. It is a different type of FE solver compared with the traditional commercially available ones listed above. It is specialized in

dealing with multi-physics problems and so contains modules designed for specific physical phenomena from solid/structural mechanics, fluid mechanics, thermodynamics, electromagnetics, acoustic, semiconductor, to particle tracing, etc. Each of these modules can be added to a solution for a realistic simulation of a given problem. It is designed to interface with MATLAB™. COSMOL was initially started by graduate students Swedish university.

(f) **Mimics Innovation Suite**: This is a dedicated set of tools for biomedical professionals that enables them to perform a diverse range of engineering operations on biological systems. It was developed by researchers in Katholieke Universiteit Leuven, but currently marketed by Materialise NV, an additive manufacturing company specialising in medical, dental, and additive manufacturing problems. The capability of the Mimics software ranges from image acquisition using 3D Tomography modalities to manufacture of prototypes. The key capabilities of the Suite include: (a) Mimics (Medical Image segmentation and 3D creation); (b) 3-matic – A **CAD** tool that includes mesh generation; (c) Engineering services; (d) **FEA/CFD** modelling; and finally (e) Anatomical models.

(g) **ANSA**: This is a pre-processor software with advanced capabilities for mesh generation. ANSA stands for **A**utomatic **N**et generation for **S**tructural **A**nalysis. It is one of the few FE software packages that retains the connection between geometry and mesh so that if the geometry is changed, only the changed part is meshed whilst the original mesh is retained. It is widely used in automotive industries and that is where it originated from. ANSA is developed by BETA CAE Systems S. A. (Greece), but marketed by BETA CAE Systems, USA.

(h) **Digimat**: This is a nonlinear multi-scale material and structure modelling platform. It is developed by e-Xtreme Engineering, an MSc Company. It represents the future of computational modelling with integration of different modules to solve holistically several mechanics problem. It is a multi-scale platform, which implies that it tracks the computational solution across multiple length and time scales.

1.4.2 Other Ancillary Software

The FE solvers listed above are specific for solving the partial differential equations required for the **FEM** process. However, there is a plethora of ancillary software that support the process. These are developed to help either in pre- or post-processor stages of FEM. They are usually numerical tools for developing microenvironment applications that can be bolted onto the backbone of the FE solvers for solving specific non-standard tasks. Such tasks can include: creation of user-defined material models; implementation of periodic boundary conditions; batch-processing of files; and the automated manipulation of FE outputs. In most cases, these ancillary software are developed using the same underlining programming language that drives the original FEM solver. For example, ABAQUS is developed using Python programming language. Hence, ancillary software to support ABAQUS model development is written in Python.

Some examples of ancillary software that drive the FE solver can include:

(a) **FORTRAN**: This is general purpose programming language widely used for development of user-defined materials for the FEM process. Prior to 1990, most FEM textbooks contain code based on FORTRAN; however, in recent times, FORTRAN is gradually being phased out, especially in engineering courses. It is particularly powerful for scientific computing.

(b) **C and C++**: C is also a general purpose programming language. It allows for structured programming and C has influenced the development of a wide variety of other programming languages and development of operating systems, such as UNIX and LINUX. Subsequently, C++ was developed based on C, but with imperative, object-oriented and generic programming features. *Imperative programming* suggests that the programming language can be called to change the sequence of events thus changing the *ease of execution* of the code.

Object-oriented code development is a modern feature of programming where *objects* are created and can be called upon to execute specific tasks. Consider objects to be tools/blocks that are required and pulled up when needed for execution of a construction task. Finally, generic programming requires that codes are developed and stored away – only to be called up when needed. This is evidenced by the availability of sub-routines in C programming. The C++ programming language is essentially C language except with additional bells and whistles to make for easy use. User-defined material models can now be easily developed in a C++ environment using, for example Microsoft Visual Studio: an integrated development environment (IDE) from Microsoft. These are used to create programs for the windows environment.

(c) **MATLAB™**: This is one of the most modern and fastest growing programming languages. It is marketed by Mathworks – a US company. MATLAB™ is a high-level language and interactive environment for numerical computation, visualization, and programming. It is fondly referred by the developers as the *language of technical computing* although the original ideal for low level programming has now been expanded by more demanding object-oriented programming using its Simulink module. In fact, MATLAB™ now sits astride a wide range of engineering application areas. Some research groups are developing FE solvers using the MATLAB™ platform.[2]

(d) **Python**: This is a high-level general purpose programming language, which is simple to use, highly robust and can be used to create software in many application areas. Code readability is a major advantage for codes written in Python. Python was developed in 1989 by Guido Van Rossum. Python can be

[2]For those interested in use of MATLAB™ in FE-code development, please refer to the textbook by Ferreira [2].

interfaced with or embedded within existing programming applications, hence it is widely used in the ABAQUS FE community. Programming instructions to ABAQUS jobs are passed using a Python scripting interface in ABAQUS so knowing a little more about Python will always be useful to students.

(e) **Shell scripting languages**: These are programming languages designed to run with a UNIX shell/environment. The tasks executed by shell scripts are quite basic, such as file manipulation, running programs and writing text to files. Batch jobs can be submitted and automated using Shell scripts too.

1.5 Conclusions

Computational mechanics is an increasingly popular solution strategy within engineering mechanics and other application areas. This chapter has presented the principles of computational mechanics and the numerical methods for obtaining numerical solutions to practical problems. One such numerical method is the finite element method and this has also been introduced in this chapter. The need for the **FEM** was presented as well as current limitations and adaptations of the classic FEM approach. FE solvers are at the core of the FEM method, so this chapter has introduced the existing FEM solvers as well as ancillary software that drive the **FEM** process.

At the end of this chapter, the reader should be able to:

- Appreciate the principles of computational mechanics and how the finite element method relates to the solution of mechanics problems,
- Understand the different aspects of the finite element method and finally
- Appreciate the range of different software that can be used as part of the FEM process.

References

1. Beer, G., Smith, I., Duenser, C.: The Boundary Element Method with Programming: for Engineers and Scientists. Springer, England (2008). https://books.google.co.uk/books?id=FAun2DgME20C
2. Ferreira, A.: MATLAB Codes for Finite Element Analysis: Solids and Structures. Solid Mechanics and Its Applications. Springer, Netherlands (2008). https://books.google.co.uk/books?id=skdFXpM0XycC
3. Huebner, K.: The Finite Element Method for Engineers. A Wiley-Interscience publication, Wiley, New York (2001). https://books.google.co.uk/books?id=f3MZE1BYq3AC

4. Liu, G.: Meshfree Methods: Moving Beyond the Finite Element Method, 2nd edn. CRC Press, Hoboken (2009). https://books.google.co.uk/books?id=JWqE-LzjvfEC
5. Logan, D.: First Course in the Finite Element Method. Thomson, Canada (2007). https://books.google.co.uk/books?id=wjr3ArdvAc4C
6. Meguro, K., Tagel-Din, H.: Applied element method for structural analysis: theory and application for linear materials. Struct. Eng. Earthq. Eng. **17**(1), 21s–35s (2000)
7. Melenk, J., Babuska, I.: The partition of unity finite element method: basic theory and applications. Comput. Methods Appl. Mech. Eng. **139**(1–4), 289–314 (1996). http://dx.doi.org/10.1016/S0045-7825(96)01087-0. http://www.sciencedirect.com/science/article/pii/S0045782596010870
8. Moës, N., Dolbow, J., Belytschko, T.: A finite element method for crack growth without remeshing. Int. J. Numer. Methods Eng. **46**(1), 131–150 (1999). 10.1002/(SICI)1097-0207(19990910)46:1<131::AID-NME726>3.0.CO;2-J, http://dx.doi.org/10.1002/(SICI)1097-0207(19990910)46:1<131::AID-NME726>3.0.CO;2-J
9. Okereke, M., Akpoyomare, A.: A virtual framework for prediction of full-field elastic response of unidirectional composites. Comput. Mater. Sci. **70**, 82–99 (2013). http://dx.doi.org/10.1016/j.commatsci.2012.12.036, http://www.sciencedirect.com/science/article/pii/S0927025612007744
10. Okereke, M.I., Buckley, C.P., Siviour, C.R.: Compression of polypropylene across a wide range of strain rates. Mech. Time-Depend. Mater. **16**(4), 361–379 (2012). 10.1007/s11043-012-9167-z, http://dx.doi.org/10.1007/s11043-012-9167-z
11. Orbis: Orbis Company Report: ANSYS Inc. [Online Database]. Assessed 20 Jul 2017. (2017)
12. Strouboulis, T., Babuska, I., Copps, K.: The design and analysis of the generalized finite element method. Comput. Methods Appl. Mech. Eng. **181**(1–3), 43–69 (2000). http://dx.doi.org/10.1016/S0045-7825(99)00072-9, http://www.sciencedirect.com/science/article/pii/S0045782599000729
13. Strouboulis, T., Copps, K., Babuska, I.: The generalized finite element method. Comput. Methods Appl. Mech. Eng. **190**(32–33), 4081–4193 (2001). http://dx.doi.org/10.1016/S0045-7825(01)00188-8, http://www.sciencedirect.com/science/article/pii/S0045782501001888
14. Trefethen, L.: Spectral Methods in MATLAB. Software, Environments, and Tools. Society for Industrial and Applied Mathematics, Philadelphia (2000). https://books.google.co.uk/books?id=9Zu4YqPQKocC

Chapter 2
A Brief Introduction to MATLAB™

Abstract MATLAB™ is a programming language. It is widely used for technical computation, programming and visualization of data. The scientists of this generation are taught the use of MATLAB™ in tackling diverse problems. MATLAB will be used to demonstrate the **FEM** process in this book and therefore it is important to offer a brief introduction to MATLAB. This chapter presents an elementary introduction to MATLAB™. For experienced users of MATLAB™, it is okay to skip this chapter. However, if one is not conversant with MATLAB™ then this brief introduction will equip you with skills to understand MATLAB™ and appreciate its in subsequent chapters. The brief introduction will cover manipulation of data, visualization of data as two- and three-dimensional profiles and finally the principles of programming (scripts) development.

Keywords MATLAB™ • Matrices • Technical computation • Data visualization • Linear equations • Programming scripts

2.1 Chapter Objectives

The following objectives will be addressed in this chapter:

(a) Use MATLAB™ to define matrices, write basic MATLAB™ functions and scripts to accomplish simple tasks, etc.
(b) Start developing MATLAB™ scripts for different computing problems
(c) Manipulate matrices within MATLAB™ for solving different problems.
(d) Plot **2-D** and **3-D** graphs within MATLAB™ from the command line and using student-developed programming scripts.

M. Okereke, S. Keates, *Finite Element Applications*, Springer Tracts in Mechanical Engineering, https://doi.org/10.1007/978-3-319-67125-3_2

2.2 Introduction to MATLAB Matrices

This section will highlight a few of some basic notions needed for understanding, interpreting and developing MATLAB™ scripts. For detailed understanding of MATLAB™, there is a plethora of resources on the MathWorks website or the documentation that comes with the MATLAB™ installation. There are also several textbooks on MATLAB™ that provide a more extensive introduction into the use of MATLAB™.

2.2.1 Generation of Matrices

Matrices are at the centre of the MATLAB™ Technical computing language. In fact, **MATLAB** essentially means **MAT**rix **LAB**oratory. Both scalar and vector quantities are defined as matrices within MATLAB™. For example, if one intends to input the value of a scalar quantity like potential energy, $PE = 2000$ J, into MATLAB™, this can be written as:

```
>>  PE = 2000

    PE =
              2000
```

The above is a 1×1 matrix that defines the scalar value of the potential energy in a given system. Also, one may define a three-dimensional concentrated force system acting at a given point in *ijk*-vectorial representation, as: $\mathbf{F} = 2000\mathbf{i} + 3000\mathbf{j} - 5000\pi\mathbf{k}\ N$. Within MATLAB, this vector system can be represented by a 1×3 column vector and inputted into the MATLAB™ command, with results as shown below:

```
>>  F = [2000 3000 -5000$*pi$]
    F =
       1.0e+004 *
              0.2000    0.3000    -1.5708
```

Notice that the above expression displays the result in what MATLAB™ likes to describe as an *engineering format* where numbers are shown in exponentials of 10. Note that `2000` becomes `1.0e+004*0.2000` in order words 0.2×10^4. We can change this display format to standard fixed or floating point format with five digits displayed by typing into the command window: `format short g` as demonstrated below. If you want to display the long display format, then use the `format long g` command instead.

```
>> format short g
>> F = [2000 3000 -5000*pi]
   F =
           2000          3000          -15708
```

When writing the matrix-style representation of $\mathbf{F} = 2000\mathbf{i} + 3000\mathbf{j} - 5000\pi\mathbf{k}$ N, only the coefficients of the vectors, enclosed in square brackets, are specified in the matrix equation.

Similarly, consider the stress state of a given point in a composite material. It is usually a 3×3 matrix. This stress state can be defined as a stress tensor, **S**, which can be defined in MATLAB™ as also a 3×3 matrix. The typical example shown below is the stress tensor of a fully anisotropic composite material. In specifying the matrix, each individual row of the matrix is separated by a semi-colon.

```
>> S = [20 30 -25;  -40 80 100;  5 -30 50]
   S =
        20     30    -25
       -40     80    100
         5    -30     50
```

It is also possible to generate matrices without typing out the terms of the matrix out as above by using pre-defined functions within MATLAB™. Here are some examples of four functions used to create a 3×3 matrix.

```
>> format short
>> A = rand(3)
A =
        0.39223         0.70605        0.046171
        0.65548        0.031833        0.097132
        0.17119         0.27692         0.82346

>> B = hilb(3)
B =
     1.0000     0.5000     0.3333
     0.5000     0.3333     0.2500
     0.3333     0.2500     0.2000

>> C = magic(3)
C =
       8     1     6
       3     5     7
       4     9     2

>> D = eye(3)
D =
       1     0     0
       0     1     0
       0     0     1
```

Note that the `format short` command was introduced to switch the display format to a scaled fixed point display of five digits. Brief notes on meaning of these in-built MATLAB functions are given below:

- `rand(N)` is used to create uniformly distributed **pseudorandom** numbers where N is number and returns an N-by-N matrix. It can also be used to create a row or column matrix by specifying `rand(1,N)` or `rand(N,1)` respectively.
- `magic(N)` is used to create a square matrix with the numbers taken from 1 to N^2. The matrix has equal rows, columns and diagonal sums.

- `hilb(N)` produces a *Hilbert Matrix* which is a badly conditioned matrix.
- `eye(N)` produces an *identity matrix* i.e. a matrix with diagonal terms all 1 and every other term being zeros.

2.2.2 *Manipulation of Matrices in MATLAB*

Matrices are manipulated similarly as scalar numbers. We can add, subtract, multiply and transpose matrices. Let us define a random 4 × 4 matrix, A, and a magic 4 × 4 square matrix, B, as shown below:

```
>> A = rand(4)
A =
    0.8147    0.6324    0.9575    0.9572
    0.9058    0.0975    0.9649    0.4854
    0.1270    0.2785    0.1576    0.8003
    0.9134    0.5469    0.9706    0.1419

>> B = magic(4)
B =
    16     2     3    13
     5    11    10     8
     9     7     6    12
     4    14    15     1
```

Operations on matrices: The matrices A and B can be manipulated to create new matrices: $C = A+B, D = A-B, E = A*B, F = A \div B, G = \sqrt{A*B}$ and $H = A^B$ as shown below. In all these cases, all the terms within the matrix are operated upon by either the $+, -, *$ or $\div$ signs as well as the square room and exponent signs. The reader is encouraged to check out if these MATLAB™ outputs are correct by undertaking their own long-hand calculations.

```
>> C = A + B
C =
   16.8147    2.6324    3.9575   13.9572
    5.9058   11.0975   10.9649    8.4854
    9.1270    7.2785    6.1576   12.8003
    4.9134   14.5469   15.9706    1.1419


>> D = A - B
D =
  -15.1853   -1.3676   -2.0425  -12.0428
   -4.0942  -10.9025   -9.0351   -7.5146
   -8.8730   -6.7215   -5.8424  -11.1997
   -3.0866  -13.4531  -14.0294   -0.8581

>> E = A*B
```

```
E =
   28.6436   28.6883   28.8703   28.0975
   25.6059   16.4340   16.7627   24.6197
    8.0439   15.6247   16.1158    6.5705
   26.6513   16.6230   16.1608   28.0379


>> F = A\B
F =
  -58.3257  241.3247  217.8700   12.0384
    4.2511   12.7466   15.2165   -3.1587
   55.4284 -221.2590 -199.1475  -10.9061
    8.1052    9.5946    6.8524   16.3317


>> G = sqrt(A*B)
G =
    5.3520    5.3561    5.3731    5.3007
    5.0602    4.0539    4.0942    4.9618
    2.8362    3.9528    4.0145    2.5633
    5.1625    4.0771    4.0200    5.2951


>> H = A.^B
H =
    0.0377    0.3999    0.8779    0.5660
    0.6097    0.0000    0.6995    0.0031
    0.0000    0.0001    0.0000    0.0690
    0.6960    0.0002    0.6391    0.1419
```

Note that when dividing MATLAB™ matrices say, $A \div B$, one uses the backslash, \ sign. One can also use the forward slash, /, as a matrix operator however, there are subtle differences in the use of / and \.

If the matrices A and B have the same rows, then use the backslash \. However if the two matrices have the same column, then use \. Also, note that $A\backslash B$ is not equal to $A \setminus B$. They are related by the equation: $B/A = (A'\backslash B')'$ where the apostrophe represents the transpose of A (i.e. A') and B' is the transpose of matrix B.

Multiplication of matrices by a scalar: One can also use a scalar quantity to multiply a matrix. For example, given the equation, $B = 3A$ where A is a magic square 3×3 matrix, MATLAB™ multiplies all terms in A by 3 as shown.

```
>> A = magic(3)
A =
     8     1     6
     3     5     7
     4     9     2
```

```
>> B = 3*A
B =
    24     3    18
     9    15    21
    12    27     6
```

Multiplication of individual terms of two matrices: There may be instances where we want only the terms of two matrices to be operated upon against one another. To do this, we include a full-stop prefix before the multiplication sign (.$*$) for example. Example outputs of the resultant matrices based on A and B above are given below.

```
>> C = A.*B
C =
   192     3   108
    27    75   147
    48   243    12
```

```
>> F = A.^2
F =
    64     1    36
     9    25    49
    16    81     4
```

Transpose of a matrix: This is also another important matrix operation and will be used extensively in this book. For example, given a column matrix, A, defined below, one can determine its vector form by taking the transpose, B = A'.

```
>> A = [2 3 6]
A =
     2     3     6
```

```
>> B = A'
B =
     2
     3
     6
```

Trigonometric functions in MATLAB™: The common trigonometric functions: cosine, sine, tangent, etc., exist in MATLAB™. They are initiated using `cos`, `sin` and `tan` for cosine, sine and tangent functions respectively. In the examples above, `cosd(A)` is used to obtain cosine of angle A in degrees and `cos(A)` is cosine of angle A in radians.

```
>> A = 30
A =
    30
```

```
>> B = cos(A)
B =
    0.1543
```

```
>>C = cosd(A)
C =
    0.8660
```

```
>> D = tan(A)
D =
   -6.4053
```

MATLAB™ distinguishes between radian and degree values of trigonometric functions. For degrees, you attach a 'd' to the trigonometric function.

Also, MATLAB™ distinguishes the *hyperbolic* functions from other trigonometric functions by attaching an h to the parent trigonometric function. For example, the hyperbolic sine, cosine and tangent are evaluated by using the commands: `cosh`, `sinh` and `tanh` respectively.

Sub-matrix: When manipulating a matrix, it may be necessary to isolate only some of the terms in the matrix for further analysis. A *sub-matrix* refers to a sub-section of the parent matrix. There are also commands in MATLAB for such analysis. For example, let us define the matrix, A.

```
>> A = [10 20 30 40 50 60 70 80 90]
A =
    10    20    30    40    50    60    70    80    90
```

Assuming, one wants to isolate the first three terms of matrix, or the third to sixth, we use the following commands.

```
>> B = A(1:3)
B =
    10    20    30
```

```
>> C = A(3:6)
C =
    30    40    50    60
```

Further matrix manipulations of the sub-matrices are given below.

```
>> D = 3*(A(7:9))
D =
   210   240   270
```

```
>> E = cosd(A([5,7]))
E =
    0.6428    0.3420
```

The E matrix above represents the cosine (in degrees) of terms 5 and 7 of the A matrix. As part of creating a sub-matrix, we can choose a set of terms and assemble them within a square bracket before operating on them. This is the case for the E matrix above, in which a trigonometric function was used to operate on the square-bracketed terms.

2.3 Plotting Functions

In MATLAB™, one can plot two-dimensional as well as three-dimensional graphs. Most graphs used in this book are usually 2D involving a variable on the $x-$axis and another variable on the $y-$axis. For 3D plots, we have a third set of variables representing a third axis, for example the $z-$axis. We will consider briefly plotting these types of graphs. MATLAB™ supports other graphical representations of data such as pie charts, histograms, quiver plots, scatter plots, etc. We will not be describing these in this book but interested users should refer to the MATLAB™ documentation.

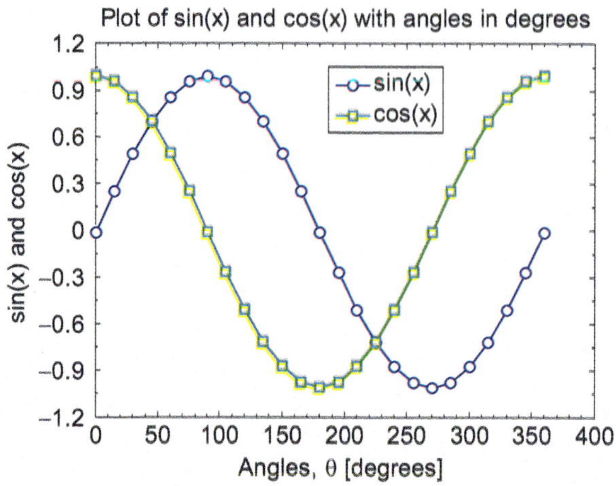

Fig. 2.1 An example of a 2D plot of `sin(x)` and `cos(x)`

Plotting 2D graphs: The command `plot` is used in MATLAB™ to plot a 2D graph inside a `figure` command using the $x-$ and $y-$coordinates. For example, let us plot the graph of `sin(X)` and `cos(X)` with values from 0 to 360°, with increments of 15°. The set of command line instructions that will generate the 2D plot is are given below. Figure 2.1 is the result of these commands.

```
>> X = 0:15:360;
>> S = sind(X);
>> C = cosd(X);
>> figure(1)
>> plot(X,S,'-o')
>> hold all
>> plot(X,C,'-s')
>> legend('sin(X)','cos(X)')
>> xlabel('Angles, \theta [degrees]')
>> ylabel('sin(X) and cos(X)')
>> title('Plot of sin(X) and cos(X) with angles in degrees')
```

Plotting 3D graphs: For 3D plots, it is best illustrated using an example. Let us plot a graph of $C = ae^{-a^2-b^2}$, with values of a and b taken from -2 to $+2$ with incremental steps of 0.1. The `meshgrid` function transforms a 2D domain space of variables a and b into set of vectors that can be used to create a 3D plot based on function C. Thereafter, the `surf` function is used to create a 3D surface plot of the function, C.

There are many 3D plotting commands in MATLAB™ and these are divided into *surface/mesh plots* and *volume visualization* representation. For example, surface/mesh plot commands include: `surf`, `mesh` and `surface` while representation of gridded volume data commands are: `cylinder`, `ellipsoid` and `sphere`.

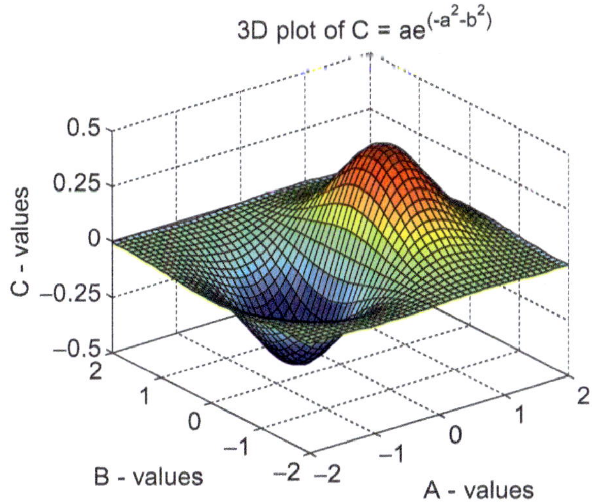

Fig. 2.2 An example of a 3D surface plot of $C = ae^{-a^2-b^2}$

Here, we will use `surf` command to create the 3D plot of C. The set of command-line instructions for creating the 3D plot is given below and the resulting plot is shown in Fig. 2.2. The `colormap` command enforces a gray contour mapping on the 3D plot.

```
>> [A, B] = meshgrid(-2:0.1:2, -2:0.1:2);
>> C = A.*exp(-A.^2 - B.^2);
>> surf(A,B,C)
>> xlabel('A - values')
>> ylabel('B - values')
>> zlabel('C - values')
>> title('3D plot of C = ae^{(-a^2-b^2)}')
>> colormap gray
```

2.4 Solution of Linear Equations

As part of the brief introduction to MATLAB, this section explains how MATLAB can be used to solve a set of linear equations. This is similar to the solution of *simultaneous equations*.

For example, consider the set of linear equations given below:

$$a + 3b - c + 2d = 2$$

$$10a + 5b + c - 6d = 32$$

$$a + 2b + c + 2d = 29$$

$$-5a + b + c + 5d = -12$$

The above equation can be re-written into a matrix format on a algebraic linear equation like this: **AX = b** thus:

$$\mathbf{AX} = \mathbf{b} \quad \Longrightarrow \quad \begin{bmatrix} 1 & 3 & -1 & 2 \\ 10 & 5 & 1 & -6 \\ 1 & 2 & 1 & 2 \\ -5 & 1 & 1 & 5 \end{bmatrix} \begin{bmatrix} a \\ b \\ c \\ d \end{bmatrix} = \begin{bmatrix} 2 \\ 32 \\ 29 \\ -12 \end{bmatrix}$$

Inputting these known matrices into MATLAB becomes:

```
>> A = [1 3 -1 2; 10 -5 3 -6; 1 2 5 2; -5 1 1 5]
A =
     1     3    -1     2
    10    -5     3    -6
     1     2     5     2
    -5     1     1     5
```

```
>> b = [2 32 29 -12]'
b =
     2
    32
    29
   -12
```

```
>> X = A\b
X =
     2
     3
     5
    -2
```

The b matrix was transposed from a row vector to a column vector. Using MATLAB, the solution of the above system of linear equations can be obtained by evaluating $A \setminus b$ directly. The above implies that the solution of the linear equation is the matrix, **X** with values defined as: $a = 2, b = 3, c = 5$ and $d = -2$. In Sect. 3.5.3, the *Augumented matrix* approach has been shown and this can be used to determine the unknown values of a system of linear equations.

2.5 Programming in MATLAB

Apart from a powerful computational interface, MATLAB™ is also a very useful programming language. Most of the common uses of MATLAB™ are in the form of a programming language. Due to the intuitive nature of MATLAB™, it is quite easy to start programming in MATLAB™. There are two type of program files available in MATLAB™ namely:

(a) *M-files* or *Scripts* which do not accept input files or return outputs;
(b) *Functions*, which accept input files and return specified output files.

2.5.1 M-Files or Scripts

This type of program file contains a list of commands/instructions that MATLAB™ executes to accomplish a set of tasks. The script uses input files in the workspace of

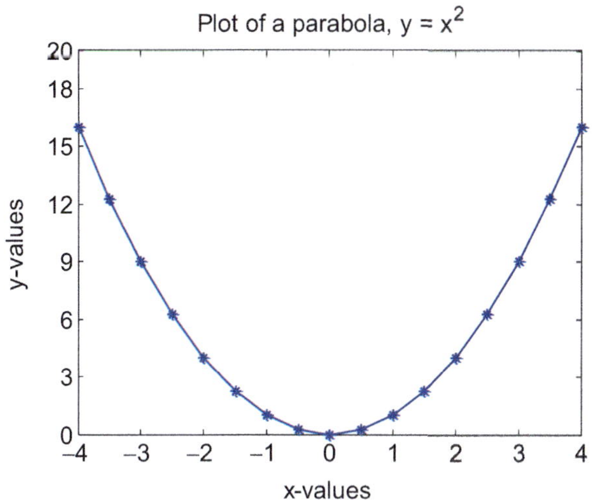

Fig. 2.3 Graph of $y = x^2$

the active MATLAB session and also writes its outputs to the same workspace. All programming in MATLAB is carried out in the MATLAB text editor interface. The approach to take in developing the M files or scripts are given in the example below.

Example 2.1 Write a script that plots the graph of $f(x) = x^2$ taking values of x from -4 to 4 for an incremental step of 0.5.

Solution

- Type into the MATLAB text-editor the commands given below.

```
%% Plot Parabola Script: plotParabolaScript
%Author:     Dr. Michael I. Okereke
%Date:       6th June, 2016
%About:      A simple Function script that plots
%            a parabola  of type y=x^2

%Specify the x-variable data range
    x = -4:0.5:4;    %Range of X-variable

%Write expression for f(x) = y = x^2
    y = x.^2;

%Plot Function
    plot(x, y, '*-')

%Label Axis and provide title
    xlabel('x-values')
    ylabel('y-values')
    title('Plot of a parabola, y = x^2')
```

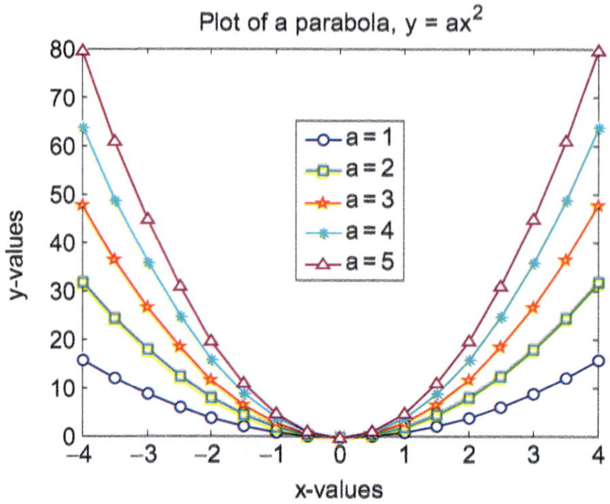

Fig. 2.4 Graph of $y = ax^2$

- It is good practice to begin every script with a little bit of information about it. Prefix the comment with a percentage sign (%).
- Save the script to a known location in your computer. For the purpose of this demonstration, the file name that we will use is: `plotParabolaScript.m`
- To run the script, click the `Run` button in the MATLAB text editor window.

Example 2.2 We may be interested in plotting a generic form of the parabolic function, $y = ax^2$ where $a = 1, 2, \cdots, 5$. For the case when $a = 1$, then we recover the expression of Example 2.1, i.e. $y = x^2$. Modify the original M-file script to automate the plotting for the different values of a.

Solution

One of the common tools for automatic MATLAB operations are the conditional statements such as `if ... else` and `for ... end`. We will use the `for ... end` conditional statement to automate the plotting of graphs for the chosen values of the coefficient, a. A typical M-file for automating this process is given below and the graph shown in Fig. 2.4.

```
%% Plot Parabola Script: plotParabolaScriptImproved
% Author:     Dr. Michael I. Okereke
% Date:       6th June, 2016
% About:      A simple Function script that plots
%             a parabola  of type y=ax^2 for different
%             values of 'a'
```

```
%Specify the x-variable data range
    x = -4:0.5:4;       %Range of X-variable
    y = cell(1);        %Cell array for storing values of y
    l = 'osp*^';        %plot marker characters

 for k =1:5
    %Write expression for y = ax^2
        y{k} = k.*x.^2; %individual value of y computed

    %Plot Function
        plot(x, y{k}, [l(k),'-'])
        hold all
 end
 legend('a=1','a=2','a=3','a=4','a=5')

%Label Axis and provide title
    xlabel('x-values')
    ylabel('y-values')
    title('Plot of a parabola, y = ax^2')
```

2.5.2 Functions

The *functions* programming approach is a slightly different approach than that explained in Sect. 2.5.1. In fact, most of the commands in MATLAB are simply in-built functions. It is good practice for the reader to master how to develop function scripts as it makes programming neater and quicker, and the codes are better re-used even inside other bigger codes.

Code for a function is enclosed inside a `function ... end` loop. Two things are important in writing functions in MATLAB™ and these are:

(a) *Always save the file containing the function by the function name; and*
(b) *Specify a set of input variables. As an option, you may specify a set of output variables too. These output variables are the only ones stored in the workspace, from the function, after execution of the script.*

We will now illustrate the writing a MATLAB™ function with an example.

Example 2.3 The script developed in Example 2.2 was for plotting the parabolic function, $y = ax^2$ where $a = 1, 2, \cdots, 5$. To illustrate the steps for writing a function script, re-write the original M-file script as a function for any range of x-values, increments and any coefficients, a.

Solution The following details the steps and essential notes for creating the function scripts:

- Re-save the original M-file script with a new name: `plotParabola.m`
- At the top (before the first line) of the re-saved M-file script, add the commands: `function [y]=plotParabola(xStart, xInc, xEnd, aStart, aEnd)`
- At the end of the script, add the word: end to complete the function loop
- Inside the script, change `x=-4:0.5:4` to `=xStart:xInc:xEnd`.
- Within the `for-end` loop, we will change `k = 1:5` to `k = aStart:aEnd`

There are sections inside the script which you have to modify to ensure the legend displays the correct updated legend results. Figure handles are introduced to store legend inputs so that the correct labelling of plotted profiles can be correctly displayed.

Below is the updated function script. Run the script to recover the original plot using the command: `plotParabola(-4,0.5,4, 1, 5)`. This will give us Fig. 2.4.

```
function [y] = plotParabola(xStart, xInc, xEnd,aStart, aEnd)

%% Plot Parabola Function: plotParabola
%Author:     Dr. Michael I. Okereke
%Date:       8th June, 2016
%About:      A simple Function script that plots a parabola
%            of type y=ax^2 by requesting:
%            (a) start, increment and end values of x-variable
    and
%            (b) the start and end values of the a-constant.

%Specify the x-variable data range
    x = xStart:xInc:xEnd;     %Range of X-variable
    y = cell(1);
    l = 'osp*^+-h';        %plot marker characters
    m = 0;                 %initialize legend counter

%Create a Figure handle to put the plots
figure(1)
    [b,c,d,e]= legend; %#ok<ASGLU>

%Iterate through values of a as required
for k = aStart:aEnd
    %Write expression for f(x) = y = kx^2
    y{k} = k.*x.^2;

    %Define legend marker
    m = m+1;
    %Plot Function
    plot(x, y{k}, [l(m),'-'])
    hold all

    %Create the Legend Input
```

```
    a             =   ['a = ',num2str(k)];
    [b,c,d,e]     =   legend([e a]); %#ok<ASGLU>
end

%Label Axis and provide title
    xlabel('x-values')
    ylabel('y-values')
    title(['y=ax^2 for a=',num2str(aStart),' to ',num2str(aEnd)
    ])
clc

end
```

Notice the new function script is enclosed inside a `function ... end` loop. This is typical of functions. The output from the script is `y` while the input is: `xStart, xInc, xEnd, aStart` and `aEnd`. The only variable stored in the workspace after execution of the script is `y`. The results of the plot of $y = ax^2$ obtained for an x-range of $-5 \leq x \leq 5$ for $a = 5, 6, \ldots, 9$ is shown in Fig. 2.5.

2.6 Conclusions

This chapter has provided a brief introduction into MATLAB™. This software is designed as a tool for technical computing, visualization of data and development of programming codes. The contents of this chapter are primarily for the reader who does not have much experience of using MATLAB™ and the presentation here is to help bridge any gap in knowledge. In subsequent chapters, the principles developed in this chapter will be used to develop MATLAB™ code that will help with understanding the FEM process.

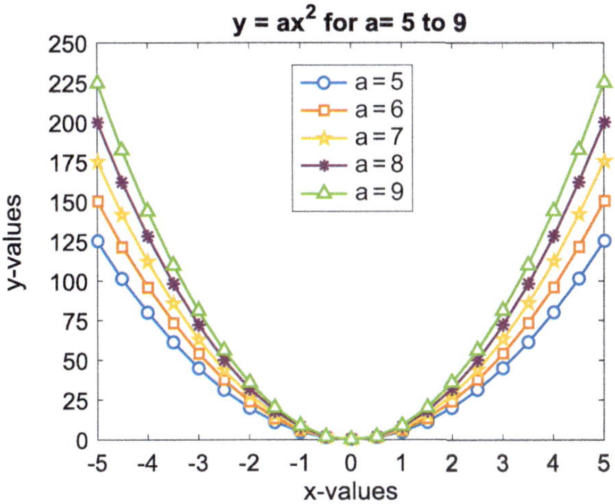

Fig. 2.5 Graph of $y = ax^2$ for $a = 5, 6, \ldots, 9$

At the end of this chapter, the reader should be able to:

(a) Understand different methods for manipulating matrices in MATLAB™.
(b) Visualize numerical data using simplistic plotting functions for two- and three-dimensional profiles.
(c) Create MATLAB™ M-files and functions for automatic simple processes as in this case plotting of graphs.

2.7 Problems: A Brief Introduction to MATLAB™

Problem 2.1 The distances travelled by three long-distance lorry drivers are tabulated as follows (Table 2.1):

(a) Create a matrix of distances travelled by each driver.
(b) Using the in-built MATLAB™ function for `sum`, find the total distance travelled by each driver.
(c) Using the in-built MATLAB™ function, find the average distance travelled by each driver.
(d) What is the total distance travelled, per trip, by all the drivers?
(e) What is the absolute difference, per trip, between the distances travelled by David and Stephen?
(f) How different is the combined distances, per trip, travelled by David and Stephen, when compared with Ian?

Problem 2.2 Using the following in-built MATLAB™ functions: `rand(N)`, `hilb(N)`, `eye(N)`, `ones(N)`, `zeros(N)`, and `magic(N)`, obtain 4×4 matrices and label them **A**, **B**, **C**, **D**, **E**, and **F** respectively.

Use the right MATLAB™ commands in the command window to evaluate the outputs for:

(a) $\mathbf{A} + \mathbf{B} + \mathbf{C} + \mathbf{D} + \mathbf{E} + \mathbf{F}$
(b) $\mathbf{AB} + \mathbf{CD} + \mathbf{EF}$ and $2\mathbf{A} - 10\mathbf{F}$
(c) $\mathbf{AF}$, $\mathbf{F(D + E)}$ and $\mathbf{U} - \mathbf{V}$
(d) Show that $\mathbf{F(D + E)} = \mathbf{F}$ and $\mathbf{FE} = \mathbf{D}$
(e) Evaluate $\mathbf{U} = \sin(\mathbf{F})$ in degrees and $\mathbf{V} = \sin(\mathbf{F})$ in radians

Table 2.1 Travel log for three long-distance lorry drivers

Driver	Distances covered on 8 trips (miles)							
David	207	154	51	89	21	500	127	63
Stephen	45	81	116	56	341	192	54	35
Ian	233	97	28	88	94	658	521	208

Problem 2.3 Based on the angles, θ, defined in the range $0° \leq \theta \leq 360°$ with increments of $\Delta\theta = 10°$:

(a) Plot the graphs of $\tan\theta$, $\sin\theta$ and $\cos\theta$ all plotted on the same graph
(b) Label the graphs accordingly using the right legend, $x-$ and $y-$ axes and the title information
(c) On the graphs, identify the $x-$ and $y-$ values of the points before angle 50° where $\tan\theta$ intersects with $\cos\theta$ as well as the points after angle 200° where $\sin\theta$ and $\cos\theta$ meet.
(d) Re-plot the graphs again (on a different figure handle) with the range of $\theta = 0°$ to 2π but evaluating the sine and cosine plots in radian mode. Note increments should be $2\pi/36$.
(e) Do you notice any difference between this graph and the previous [degrees-based] graphs for sine and cosine? Comment on $\tan\theta$ in radian mode.

Problem 2.4 Using MATLAB™, construct a 3D plot of the following expressions:

(a) $c = a^2 + b^2$ and $c = a^{-2} + b^{-2}$ on the same coordinate system
(b) $c = a^{-2}\sinh(a) + b^{-2}\sinh(b)$
(c) $c = e^{-(a^2+b^2)}$ and $c = -e^{-(a^2+b^2)}$ on the same coordinate system
(d) $c = \sinh(a)\cosh(b)\tanh(a)$
(e) $c = \sqrt[3]{\dfrac{a^3 + b^4}{\cos^2(2a) + \sin^2(2b)}}$

In all cases, assume the range of values for both a and b starts from -4 to $+4$, with increments of 0.1.

Problem 2.5 Use MATLAB to solve for the unknown constants (a, b, c, d and e) in the following system of linear equations:

$$\begin{aligned} a + b + c - d + e &= 3 \\ 2a - 3b + 6c + 5d - 2e &= -12 \\ 3a + 2b - 3c - d - 5e &= 26 \\ 4a - 5b + c + d - e &= -7 \\ -2a - 2b - 5c - 6d - 8e &= 0 \end{aligned}$$

Problem 2.6 Write a simple M-file for plotting the graph of $y = ae^x$. The script should be dynamic enough for any range of values of x taken from a given x_{start} to x_{end} with incremental steps, x_{inc}.

(a) Also, allow the coefficient, a to vary from 1 to 10.
(b) Show example profiles from the M-file showing plots for $y = 2e^x$ and $y = 10e^x$ for the range of $5 \leq x \leq 10$.

Problem 2.7 Given a polynomial function: $y = ax^3 + bx^2 + cx + d$.

(a) Write the function file such that you will require inputs for a, b, c and d as well as the x-range and increment values.
(b) Illustrate the working of your function by outputting results for $a = 1, b = -3, c = 10, d = -5$ and $-5 \leq x \leq 5$.
(c) Re-use your function script to plot the graphs of $y = ax^3 - 3x^2 + 10x - 5$, where $a = 1, 2, \cdots, 6$.

Problem 2.8 During material model development for the finite element modelling process, visualizing the 3D plots of functions is essential in understanding the spatial variation of material parameters, say a and b with c. Assume the range of values of a and b to vary from -3 to $+3$ using incremental steps of 0.1.

(a) Use MATLAB™ to construct a 3D plot of the equations: $c = e^{-(a^2+b^2)}$ and $c = -e^{-(a^2+b^2)}$ with both graphs plotted on the same coordinate axis.
(b) Document your solution procedure within a MATLAB™ M-file.
(c) In order to better understand the spatial variation of the above equations, the student decides to consider the following variants of the original equation:

   (i) $c_3 = c_1 + c_2$
   (ii) $c_4 = c_1 - c_2$
   (iii) $c_5 = c_1 c_2$
   (iv) $c_6 = c_1^{c_2}$

   Construct the 3D plots for these different scenarios and comment on the shapes of the graphs and their implications for the modelling process.

Problem 2.9 In order to develop a user-defined material model for a typical engineering material, five material constants (i.e. a, b, c, d and e) were found to be related in five simultaneous equations as given below.

$$\begin{aligned}
2a + 5b + 3c - 8d + 6e &= -20 \\
2a - 5b + 3c - 8d - 6e &= -10 \\
2a + 5b + 3c - 8d - 6e &= 15 \\
2a - 5b + 3c + 8d - 6e &= 16 \\
2a - 5b - 3c - 8d + 6e &= 25
\end{aligned}$$

Use in-built MATLAB™ functions to solve for the unknown material parameters. Document your solution procedure within a MATLAB™ M-file.

Problem 2.10 Write a MATLAB™ function script that calculates the area of particular *regular* geometry.

Make your script as robust as possible so that it can cater for a diverse range of geometric shapes, for example, it should be able to calculate areas of circles, triangles, squares, rectangles, ellipses, pentagon.

(a) Illustrate how your script works by running a few test scenarios.
(b) Prove the outputs are correct by providing hand calculations of the areas.
(c) Create a short report to document the outputs thereby demonstrating the performance of the code.

Hint: Consider the case where all the different areas are circumscribed within a circle, so you only require the radius of the circle to calculate these areas.

You may have an 'Area selector' option, which you need to specify to determine which area type you want to calculate.

You may also specify a matrix of all area types and use a `for ... end` loop to calculate all the areas for all shapes you are interested in.

Refer to Sect. 8.6.4 of the further ideas about tackling this problem.

Chapter 3
Direct Stiffness Method

Abstract At the core of the finite element modelling process are a diverse possible range of solution approaches for any particular problem. Each of these approaches are adapted for the type of problem that one is interested in, for example structural, fluid, thermal or acoustic problems. The commonest type of problems that FEM addresses are the *structural and solid mechanics* problems and the direct stiffness method is the heart of the solution strategy. This chapter describes the principles of the direct stiffness method. Simple truss elements are introduced as the crudest finite elements for demonstrating the direct stiffness method, although other more advanced discretization finite elements can also be used. The mechanics of the direct stiffness method will be explained. In particular, the discussion highlights the use of nodal properties for the truss elements to determine displacements, velocities, internal and external forces, etc. for a given truss system. The chapter concludes with practical example problems.

Keywords Direct stiffness method • Stiffness matrix • Discretization • Planar truss • Nodal forces • Nodal displacements • Internal stress

3.1 Chapter Objectives

Having established the principles of computational mechanics and the FEM process in Chap. 1, this chapter explores the philosophy of the FEM process as a strategy for solving structural problems, using the *direct stiffness method* (**DSM**). The DSM is at the core implementation for majority of the FEM solver discussed in Sect. 1.4.1 hence it is important that the reader understands its implementation within the FEM process. This chapter will explain this method and illustrate how it can be used in obtaining a solution of a finite element problem.

The following objectives will be addressed in this chapter:

(a) Understand what the DSM is and how it fits within FEM.
(b) Use analytical equations to illustrate the DSM.
(c) Derive a DSM solution of a problem by hand calculations.

M. Okereke, S. Keates, *Finite Element Applications*, Springer Tracts in Mechanical Engineering, https://doi.org/10.1007/978-3-319-67125-3_3

3.2 Introduction to Structural Analysis

Finite element modelling was initially solely designed to solve structural analysis problems. In this section, we will introduce the principles of structural analysis and the different approaches used for undertaking such analysis.

3.2.1 *Structural Analysis*

Structural analysis aims to determine the effect of forces/loads on structures and their components. Examples of such structures can include buildings, bridges, cars, cranes, bone structures, tissue scaffolds, etc. Consider two typical structures shown in Fig. 3.1, consisting of complex arrangement and geometry of trusses.

These structures are designed through careful assessment of the forces, stresses and strains each truss will have to bear. The range of investigation required for understanding the suitability of the design to bear loads and not fail fall within the area of structural analysis – a subset of engineering mechanics.

The structural engineer would normally determine such parameters as: *internal* and *external loads* on the structures; *geometry* of structural components; *nature of supports* and their *reaction values*; and, finally, the required *material properties* to sustain any load type and geometry configurations. Outputs of the structural analysis are stresses, displacements (and, at times, strains) as well as support reactions. The outputs are compared against known constitutive behaviour of the test material to ensure that the structure does not fail.

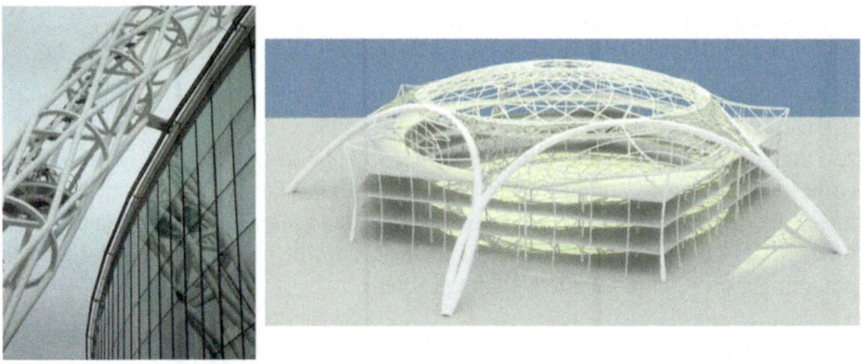

Fig. 3.1 Typical engineering structures showing *Left*: section of Wembley Arch (Image source: Flickr) and *Right*: concept design for Arena Dubai (Image source: Grasshopper algorithmic modelling rhino)

3.2.2 Structural Analysis Approaches

To identify all the internal parameters of the structure, as well as reactions, there are numerous methods for solving structural analysis problems. These methods will be highlighted briefly, but detailed discussion in the next session will be dedicated to the DSM, which is better suited for a finite element analysis.

The main methods for solving structural analysis problems are:

Mechanics of materials approach: This is the simplest approach, also called the *classical mechanics approach*. It is used in determination of stresses, displacements and reactions for simple structural arrangements such as beams, cantilever, etc. Details of this approach are usually available in *Mechanics of Materials* undergraduate textbooks. The implementation helps determine displacement of axially loaded beams, deflections of beams and stresses of members/structures carrying diverse types of loads.

The limitation of this method is that it is cumbersome to use for complex structures as shown in Fig. 3.1. When structures have simple truss arrangements, the ***Methods of joints and sections*** is useful in solving a structural problem. However, these methods become increasing onerous when complex arrangements of structures are to be analyzed, hence a more robust approach has to be used for such complex truss structures.

The elasticity approach: In the previous method, we were not concerned about the material response of the material used in the design of the structural members. The calculation process (using the above approach) for determining the displacements of structural members will remain the same irrespective of whether the structural member is made of plastic, composite or metallic parts. However, the magnitude of the displacements and stresses will be different depending on material type. It is therefore essential that the structural engineer should also explore the effect of changing material type on the calculated displacements, stresses and reactions at the supports.

The elasticity method therefore introduces to the analysis the extra feature of constitutive behaviour of the structural members. The analysis is, however, limited to linear elastic behaviour, in which case **Hookean** material response will become applicable in the structural analysis. This method is often integrated with the *mechanics of materials approach* and it is quite common not to treat them as two distinct and separate approaches.

The numerical or finite element approach: The use of numerical methods is commonplace in structural analysis. It involves the use of approximate solutions of assembled partial differential equations of structural members. Different numerical methods are then used to determine the displacements, stresses and reactions of the complex structure.

The commonest of the numerical methods is the finite element method (FEM) – the principle of which has already been presented in Chap. 1, where the complex structure is discretized into small elements (units). Principles of Mechanics of Materials are therefore applied to that element to determine its individual displacements, stresses and support reactions. These unit outputs are assembled into a body of elements and solved numerically to obtain desired outputs of the whole structure. This approach depends largely on the application of matrix algebra through the *direct stiffness method* (DSM). Further considerations will be given to this in subsequent sections of this chapter.

Energy method: This approach requires the application of balance of work and energy within a system to determine the forces, stresses and strains within the structure. In other words, this method establishes the relationship between internal structural parameters (such as stresses, strains, deformations, displacements etc.) and externally applied structural parameters. This relationship is formulated by considering the form of energy or work done by the external forces to alter the energy state of the internal variables. One of the particularly promising approaches under this method is the *virtual work principle.*

Matrix force method: This is similar to the DSM, but in this case the forces acting on individual members that make up the structure are assembled into a *flexibility matrix* or *force matrix*. This matrix defines the overall behaviour of the structure and can be manipulated to determine other unknown structural response of the structure such as nodal displacements. The flexibility matrix is often the inverse of a stiffness matrix, hence both this method and the direct stiffness method are used interchangeably.

3.3 Introduction to Direct Stiffness Method

The *Direct Stiffness Method* is at the centre of the finite element method for numerically determining the forces and displacements of units of structural members that make up a structure. Each structure has an associated stiffness and these stiffnesses are summed up *directly* in the DSM to determine the global behaviour of the structure.

DSM is also called the *Matrix Stiffness Method* (**MSM**) since it requires assembling a matrix of stiffnesses of all the components of the whole structure. This singular matrix assembly (of all component members of the structure) is then used as a standard tool for describing the overall behaviour of the structure. For further reading, the reader should consult the following references: [2, 5, 6]. It is a relevant method for any computer-assisted analysis for determining the forces and displacements of the whole structure.

To illustrate the principle behind the DSM, let us consider the metallic support structure of a bridge as shown in Fig. 3.2a. This structure can be analysed by any of the methods mentioned previously, however, for our purposes here, we will use the DSM. To do so, we will convert the support structure into an idealized truss arrangement as shown in Fig. 3.2b. In this case, the truss has been extrapolated beyond the image to create a consistent arrangement. For the structural analysis, and

(a) (b)

Fig. 3.2 An ilustration of DSM: (**a**) Physical system: Metallic support structure of a bridge (Image source: Flickr); (**b**) Idealized system: the metallic support is idealized into a simply supported truss with distributed loading

to ensure equilibrium and compatibility conditions are obeyed, the necessary simply supported arrangement at the base has been specified as well as a distributed load at the top to represent the weight of the bridge resting on the trusses. We are now ready to undertake the structural analysis of this simply supported truss arrangement shown in Fig. 3.2b.

Now that a free body diagram of the structure has been developed, as shown in Fig. 3.2b, these are the key steps that have to be completed.

(a) *discretization of the problem;*
(b) *formulation of the stiffness matrix of each truss;*
(c) *assembly of the stiffness matrix for all trusses; and,*
(d) *solution for displacements and internal forces of the structure.*

Consequently, the structural analysis of the bridge using the **DSM** will be focussed on addressing each of the above steps. The remainder of this chapter will be dedicated to each of these steps to formulate a holistic understanding of the DSM, with a view to applying the same in the finite element method.

3.3.1 Discretization of the Problem

The idealization of the physical structure of Fig. 3.2a into a truss structure of Fig. 3.2b is key step of the DSM. Idealization can also be described as discretization in finite element methods nomenclature. This requires the conversion of physical bodies (in this case, the truss) into discrete component parts, so that they can be solved by a computer.

For the truss system of Fig. 3.2b, one can discretize it into smaller *unit length* pin-jointed trusses for a start, as shown in Fig. 3.3b. Another structural engineer may decide to discretize the physical structure into much smaller units made up of

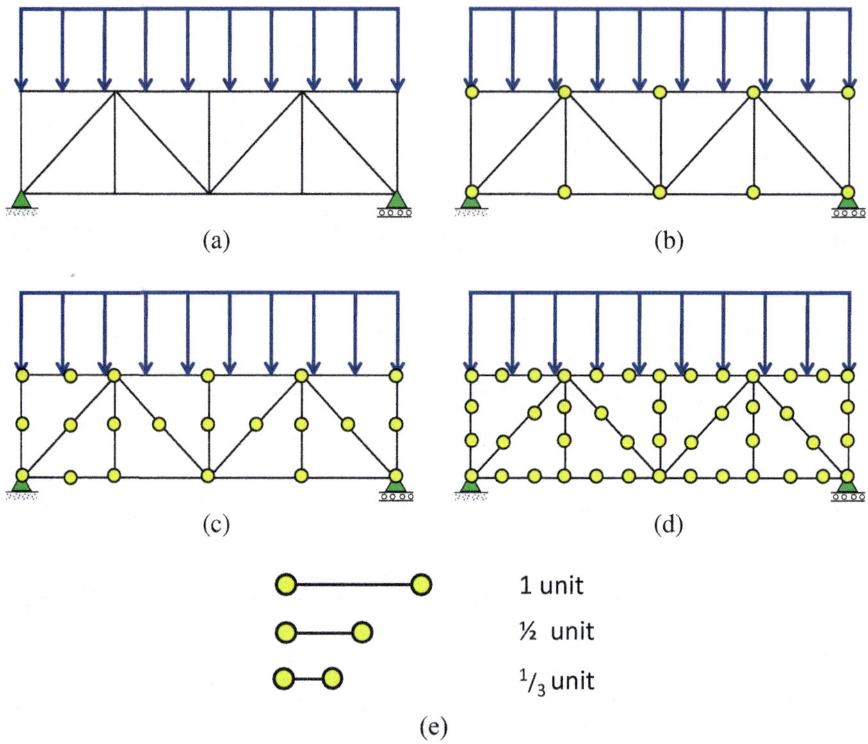

Fig. 3.3 Discretization of: (**a**) physical truss system using (**b**) a unit length, (**c**) half-a-unit length, (**d**) one-third-of-a-unit length, and (**e**) comparison of the three discretization lengths

half-unit or *one-third unit lengths* of pin-jointed members, as shown in Fig. 3.3c, d respectively. One can continue to discretize the physical structure into smaller and smaller units as one chooses. The more discretizations there are to a problem, the larger the number of equations that need solving for that problem. This will improve the solution obtained at the end, but for greater computational effort.

The smallest unit for each discretization consists of a joint (symbolized by a circle) and a bar (symbolized with dark line) as shown in Fig. 3.3e. In finite element nomenclature, the joint is equivalent to a *node* while the bar is equivalent to an *element*. In subsequent sections, these terms will be defined in more detail and their relevance to the FEM explained.

3.3.2 Discretization Error

The discretization of the problem domain shown in Fig. 3.3 will always create some form of error, which is called *truncation* or *discretization error*. This error originates

Fig. 3.4 A bridge arch represented by a parabolic function, $f(x) = -x^2$

from the fact that a continuous physical model (or in mathematical terms – *function*) is idealized for numerical solution by breaking that continuity into manageable smaller units (in mathematics, called *discrete functions*) which approximates the continuous function [1, 3].

Bigger discretized units yield comparatively larger discretization errors. We illustrate with Example 3.1, the impact of discretization errors on the calculation of area under the arch of a bridge.

Consequently, the best solution will be to continuously reduce the unit size of the discrete parts of the continuous function until a good enough solution can be obtained. Such reduction in size will lead to more equations that need to be solved hence the computational cost will be consequently higher. The structural engineer will have to balance the need for accuracy of solutions with computational cost of obtaining the solutions.

Example 3.1 Consider an arch support of a bridge to have the shape of a half parabola described by mathematical function, $f(x) = -x^2$. To calculate the area under the arch, the engineer decides to discretize the arch into smaller rectangular units.

(a) Demonstrate how you can determine the discretization error of the area under the arch of the bridge.
(b) Compare your calculated areas with the area determined from integration.

Solution

Consider the arch function, $f(x) = -x^2$ to be described within a cartesian reference frame such that the centre of the arch (topmost part) is located at $x = 0$. We assume a limit to the arch to range from $x_{start} = -6$ to $x_{end} = +6$.

The true value of the area under the arch can be calculated by obtaining the integral of the continuous function, $f(x) = -x^2$ for the limits: $-6 \leq x \leq +6$.

We can do this because the arch function is continuous over the limits specified so a simple integration can give us the true area, A_{true} under the function.

$$A_{true} = \int_{-6}^{+6} \left(-x^2\right) dx \quad \Longrightarrow \quad \left[-\frac{x^3}{3}\right]_{-6}^{+6} = 144$$

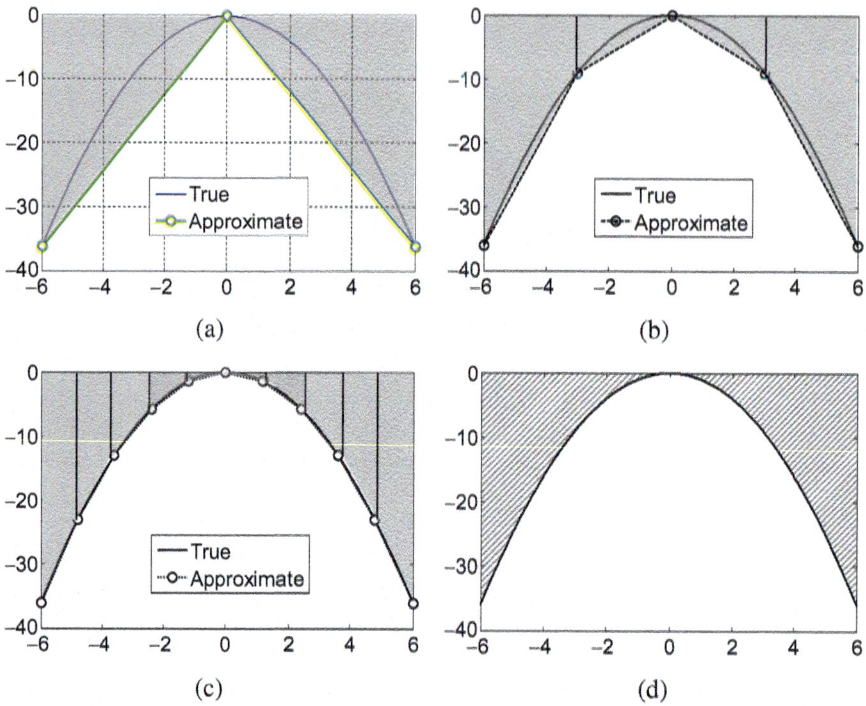

Fig. 3.5 An illustration of the discretization errors associated with a bridge arch function, $f(x) = -x^2$ where the function has been discretized int the following number of bins: (**a**) 2 (**b**) 4 (**c**) 10 and (**d**) infinite number of bins (integrated value)

Table 3.1 Discretization error associated with a parabolic bridge arch function

Number of bins	$\|A_{approx}\|$	$\|A_{true}\|$	Discretization error, (%)
2	216.00	144.00	50.0 %
4	162.00	144.00	12.5 %
10	146.88	144.00	2.0 %
∞	144.00	144.00	0.0 %

In discretizing the arch function, we convert the arch into linear segments depending on the *number of bins* of discretization chosen. For an illustration, the function has been discretized into 2, 4 and 10 bins as shown in Fig. 3.5. For the case of two bins, the area under the function approximates to area of two right-angled triangles. For the 4 and 10 bins cases, the total area under the graph is equal to the sum of the triangles and trapezia obtained from the discretization linear functions.

We can observe that, increasing number of bins, leads to increasing accuracy of area calculations, as illustrated in Table 3.1. The case shown in Fig. 3.5d represents an infinite number of discretization bins, whose area should converge to the true area determined previously.

3.3.3 Formulation of Displacement and Force Matrix Equations

Once the discretization exercise is complete, the next challenge facing the structural engineer is to formulate the matrix equations applicable to the physical problem. This matrix equation will be used in the direct stiffness method. To reduce the complexity of the solution, we will use only a simple planar triangular-shaped truss arrangement that is simply supported as shown in Fig. 3.6. It is described as planar, because the truss is allowed displacement only in the $x-$ and $y-$ directions. All loads on the members are transmitted axially hence the truss cannot sustain bending.

Consider that the planar triangular truss of Fig. 3.6 is subject to an angular load, $\boldsymbol{F}$, acting at angle, θ, as shown in Fig. 3.7a. The structural engineer's challenge here will be to determine the forces and displacements in each of the three joints (or nodes) of the triangular truss. To do so, we will determine the reaction forces of the truss and resolve the prescribed load into its $x-$ and $y-$ axis values. This is illustrated in Fig. 3.7b. These boundary reaction forces will be determined using classical mechanics methods through demanding analytical calculations.

However, using the DSM there is no need to obtain these boundary forces as they are determined using the numerical scheme. Once a *Stiffness Matrix* is obtained, the forces can be calculated individually. We will define the stiffness matrix in the next section. Consistent with the DSM, let us represent the joint or nodal forces and displacements in both $x-$ and $y-$ axes as shown in Fig. 3.7c.

It is also important that we obtain the material and geometric properties of the three members, B_1, B_2 and B_3. The geometric parameters are cross-sectional area, A_i of the members and their lengths, L_i where $i = 1, 2, 3$. The material parameters are **Young's Modulus**, E_i, and density, ρ_i, of the members, as shown in Fig. 3.7d.

With respect to Fig. 3.7c, the joints (or nodes) have nodal forces and nodal displacements. For a given node, i where $i = 1, 2, 3$ the nodal forces are represented as: f_{xi} and f_{yi} in the $x-$ and $y-$ axes respectively. Similarly, the nodal displacement for the i-node becomes: u_{xi} and u_{yi} for $x-$ and $y-$ axes respectively.

Let us assemble the set of nodal forces, $\mathbf{f}$, and nodal displacements, $\mathbf{u}$, both represented as vectors, into a single column matrix given in Eq. 3.1. Clearly, some of the nodal forces and displacements are zero for support reactions and prescribed

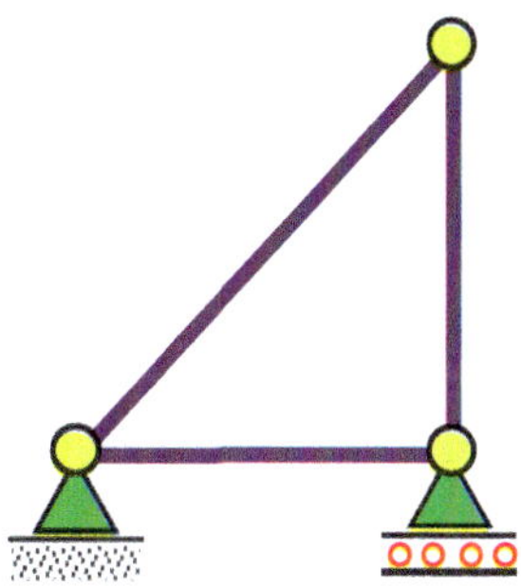

Fig. 3.6 A triangular truss which is the simplest part of bridge truss of Fig. 3.3a

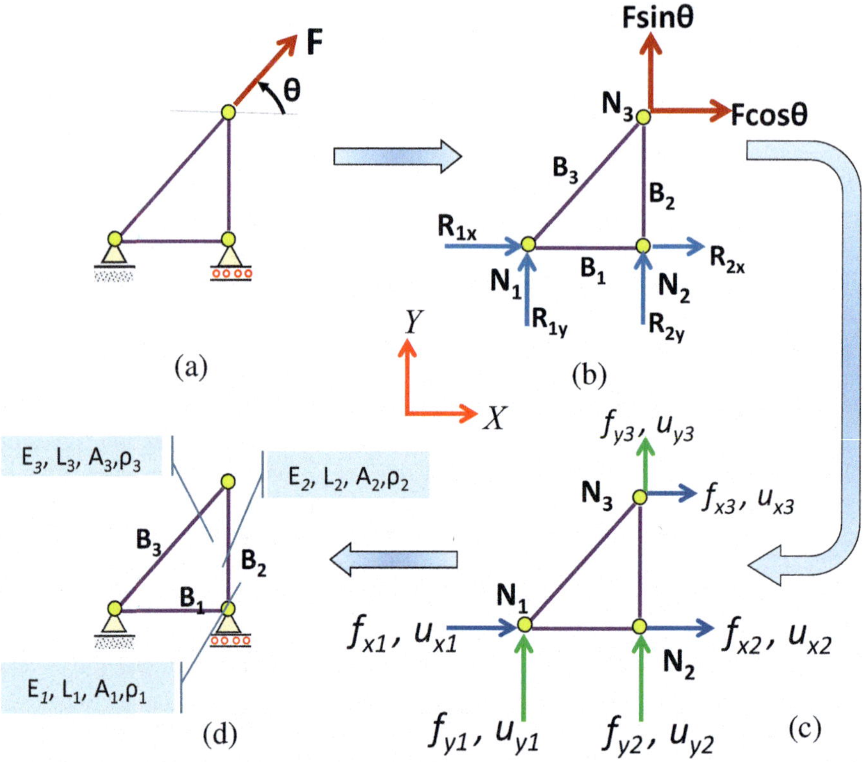

Fig. 3.7 (**a**) A typical triangular simply supported pin-jointed truss subjected to an angular load, **F** (**b**) Free body diagram of the pin-jointed truss showing reaction forces, **R** prescribed load, **F**, node numbers, N_i and Bar number, B_i where $i = 1, 2, 3$ (**c**) nodal forces and displacements (determined using the DSM) and (**d**) geometric and material parameters of the truss

loads. For example, applying the equations of equilibrium according to Fig. 3.7b, c, we see that:

$$
\begin{aligned}
f_{x1} &= R_{x1} \neq 0 & \qquad f_{y1} &= R_{y1} \neq 0 \\
f_{x2} &= R_{x2} = 0 & \qquad f_{y2} &= R_{y2} \neq 0 \\
f_{x3} &= F\cos\theta & \qquad f_{y3} &= F\sin\theta.
\end{aligned}
$$

We will not concern ourselves now with substituting these into the terms of Eq. 3.1. The computer-assisted analysis will deal with these once the *Stiffness Matrix* is obtained.

$$\text{Nodal forces, } \mathbf{f} = \begin{bmatrix} f_{x1} \\ f_{y1} \\ f_{x2} \\ f_{y2} \\ f_{x3} \\ f_{y3} \end{bmatrix} \qquad \text{Nodal displacements, } \mathbf{u} = \begin{bmatrix} u_{x1} \\ u_{y1} \\ u_{x2} \\ u_{y2} \\ u_{x3} \\ u_{y3} \end{bmatrix} \tag{3.1}$$

From the above, it is becoming clear how important the *Stiffness Matrix* is in obtaining the required nodal forces (and hence displacements) using the DSM. In the remainder of this chapter, we will focus on establishing the steps required in obtaining the *Stiffness Matrix* for any given structure.

3.4 Determination of the Structural Stiffness Matrix

In discussing the *Stiffness Matrix* for a DSM, we distinguish two types of stiffness matrices: *structural* and *member* stiffness matrices. The *structural stiffness matrix* is a singular stiffness matrix that contains all the individual stiffnesses of every member/truss that comprise the structure. It is also the matrix that is used for calculating (within a finite element method) the nodal forces (or the nodal displacements) of the structure. On the other hands, the *member stiffness matrix* is also a stiffness matrix, but only for a single member/bar/truss that comprises the structure.

For example, let us consider the planar triangular truss of Fig. 3.6. Let us assume that the *member stiffness matrix* for bars/members B_1, B_2 and B_3 are given as: $\mathbf{K}_1$, $\mathbf{K}_2$ and $\mathbf{K}_3$ respectively. The *structural stiffness matrix*, $\mathbf{K}$, will thus consist of the sum of all three member stiffness matrix, i.e. $\mathbf{K} = \mathbf{K}_1 + \mathbf{K}_2 + \mathbf{K}_3$.

According to finite element method nomenclature, the member stiffness matrix is essentially the *element stiffness matrix* (for a chosen element) while the structural stiffness matrix is an assembly of all the element stiffness matrices for the physical model under analysis. As we discretize a structure into much smaller parts, the structural stiffness matrix begins to grow into a very large matrix and therefore requires *substantial* computational power to store the matrix elements and to solve for the resulting nodal forces or displacements of the structures. This is one reason why the higher the mesh density of a model, the longer the simulation time taken to solve it, as much computational resources are diverted to dealing with the 'huge' resultant stiffness matrix.

The steps required for determining the structural stiffness matrix consist of the following:

(a) Determination of the *transformation matrix*;
(b) Determination of all the *member stiffness matrices*;

(continued)

(c) Rotation of each *member stiffness matrix* from local to global axes; and,
(d) Assembly of the *member stiffness matrices* into the *structural stiffness matrix*.

In the following sections, we will address the above steps individually, so that we can develop the structural stiffness matrix. The triangular planar truss, and other combinations of it, will be our structure of choice throughout the presentation. We will also introduce the coordinate systems needed in describing the orientation of members, and determining the member stiffness matrices, of the planar trusses. Finally, we will show how the coordinate systems can be manipulated during the *assembly* operation of compiling the structural stiffness matrices.

3.4.1 Analysis of Planar Trusses

In this section, we will introduce the concept of coordinate systems required in describing the orientation of the individual members of a given structure. There is a need to distinguish between local and global coordinate systems.

3.4.1.1 Local and Global Coordinate Systems

It is important at this point to distinguish between *local* and *global coordinates systems*. In both systems, we will use the rectangular, or Cartesian, coordinate system of $x-$ and $y-$ axes. The local coordinates are attached directly to the members (irrespective of their orientation) in space, while the global coordinate is attached to the overall spatial arrangement of the structure (with respect to a universal spatial reference frame). Whereas the local coordinate system always changes in orientation as each member's spatial arrangement changes, the global coordinate system remains fixed.

Conceptually, consider a typical **GPS** in a car navigation system. Such systems usually offer two display modes:

(a) Where the car is always pointing towards the top of the screen and thus north rotates around it; and,
(b) Where north is always at the top of the screen and the representation of the car changes orientation as it is being driven.

The first of these models is based on a local coordinate system since it is always displayed relative to the car. The second model is based on a global coordinate system where the car moves in a defined space that never changes irrespective of the car's movement.

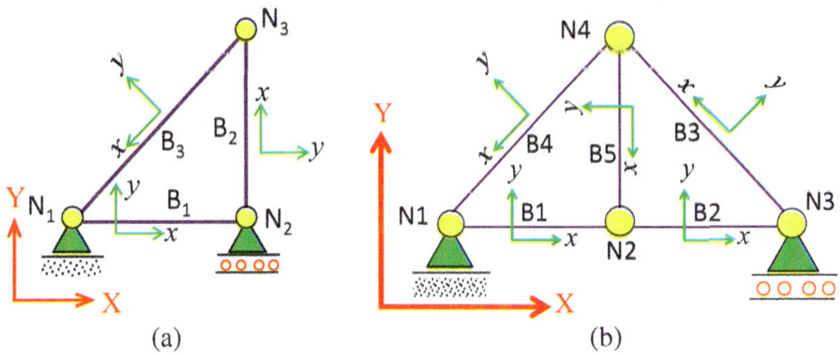

Fig. 3.8 An illustration of *global* and *local* Cartesian coordinates for (**a**) one-triangle and (**b**) two-triangle planar trusses. Note: The *global* coordinate system is fixed whilst the *local* coordinate system is attached to each bar and changes its direction with change of orientation of that bar

Also, the local coordinate systems will always have the $x-$ axis always acting along the length of the member while the $y-$ axis acts perpendicular to the length of the member. Both coordinate systems are illustrated using two structures in Fig. 3.8. Note that the local coordinates are given in italicized lowercase x- and y-symbols while the global coordinates are given as uppercase X- and Y-symbols.

The orientation of the local coordinates goes in an anticlockwise direction from the lower node number to the higher node number. For example, for left truss of Fig. 3.8, the nodes/joints are labelled in a counter-clockwise direction thus: $N_1 \rightarrow N_2 \rightarrow N_3$. Similarly, the members or elements are labelled in same counter-clockwise direction thus: $B_1 \rightarrow B_2 \rightarrow B_3$. Therefore, the orientation of the local coordinate systems should reflect this with origin of bar, B_1 starting at node, N_1 and origin of bar, B_2 starting at node, N_2.

3.4.1.2 Disassembly of the Planar Trusses

To derive the *member stiffness matrix*, each member/element of the truss structure in Fig. 3.8a, has to be analysed independently. This will involve disassembly of the structure into individual members. Once they are dissociated, each member will have to be rotated to align with the global reference frame as shown in Fig. 3.9. The act of rotation will be explained later. In this instance, the local coordinate system ($x-$ and $y-$ axes) is transformed to a *rotated reference frame* ($x'-$ and $y'-$axes).

3.4.1.3 Force-Displacement Formulation for a Given Bar

With respect to Fig. 3.9, the solution for the *member stiffness matrix* is obtained by assembling the individual stiffness of the three members that make up the structure. Here, we establish the stiffness for member, B_1 bounded by nodes, N_1 and N_2, of length, L. Let us make the assumption that the truss/member behaves according to a linear elastic (or **hookean**) material response. To determine the force-displacement

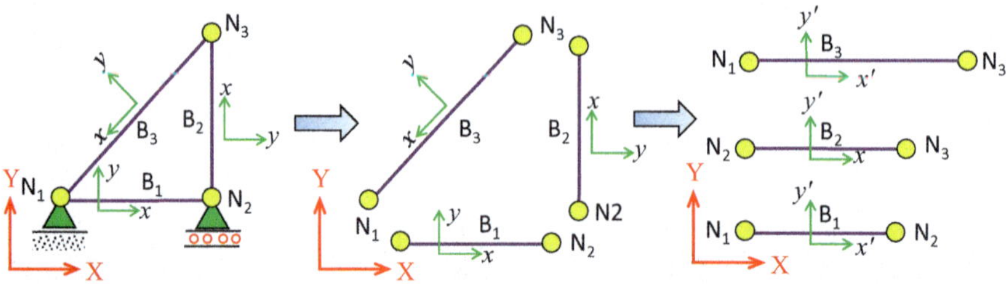

Fig. 3.9 A dissociation of the truss structure into individual members, and subsequent rotation of member's local coordinate system to align with the global coordinate system. Note that the rotated frame is x'− and y'−axis

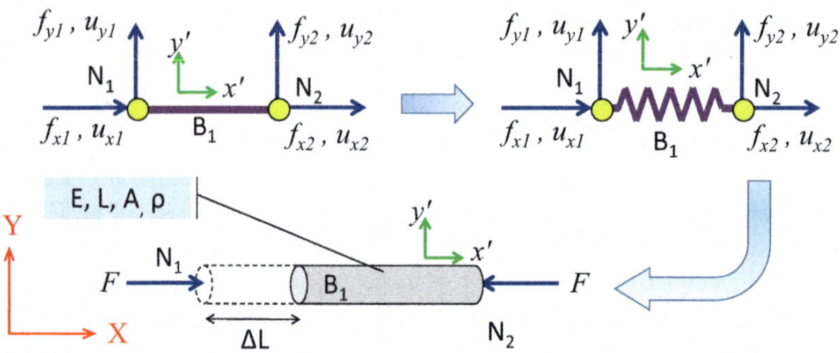

Fig. 3.10 Each truss member is equivalent to a linear elastic spring or (linear elastic) axially loaded member

relationship for such a bar, we assume that the behaviour of the bars can be idealized to be linear springs and so is equivalent to the spring arrangement of Fig. 3.10.

Treating the truss as a linear elastic element, we can deduce the force-displacement relationship associated with its axial deformation by first considering the truss member to be say a prismatic bar. If the bar/truss member is homogeneous with respect to **Young's Modulus**, E, cross-sectional area, A, density, ρ, and length, L, and subject to externally applied load such that it experiences an axial deformation, ΔL, as shown in Fig. 3.10, we can establish the force-displacement relationship for such axially-loaded member subjected to force, **F**, as in Eq. 3.2.

$$\Delta L = \frac{FL}{EA} \quad \Longrightarrow \quad F = \frac{EA}{L}\Delta L \quad \Longrightarrow \quad F = ku \tag{3.2}$$

where spring constant, $k = EA/L$, and displacement of bar, $u = \Delta L$. The spring constant, k, is a measure of the stiffness (i.e. resistance to axial deformation) of the truss member. The use of this force-displacement relationship enforces axial deformation response on the discretized element/member. Consequently, this will become the mechanical response of the truss element under consideration.

Using the nodal forces of Fig. 3.10, the nodal force of node, N_1, can be expressed in vectorial form as: $\mathbf{F_1} = f_{x1}\mathbf{i} + f_{y1}\mathbf{j}$. Also, equation: $\mathbf{F_2} = f_{x2}\mathbf{i} + f_{y2}\mathbf{j}$ will apply at node, N_2. Similarly, the nodal displacements at nodes, N_1 and N_2 are $\mathbf{u_1} = u_{x1}\mathbf{i}+u_{y1}\mathbf{j}$ and $\mathbf{u_2} = u_{x2}\mathbf{i} + u_{y2}\mathbf{j}$ respectively.

These nodal forces and displacements will have to be determined as part of the finite element modelling process. The stiffness of Eq. 3.2 represents the proportionality constant that links nodal forces to nodal displacements. The vectorial representation of the force-displacement equation becomes: $\mathbf{F}_j = k\mathbf{u}_j$ where $j = 1, 2$ for the case of Fig. 3.10.

3.4.2 Derivation of Transformation Matrix

The orientation of the member bars in the local coordinate systems is not always coincident with the global coordinate systems. For the DSM to work, it is important that stiffness parameters in either local or global coordinates need to be added together. This leads us to define a transformation matrix, which ensures that nodal displacement or matrix in one coordinate system can be transformed to values in another coordinate system.

To derive the required transformation matrix, we deal with two case studies:

(a) Where the local and global coordinate systems coincident in orientation; and,
(b) Where there is an arbitrary angle between the global and local coordinate systems.

The following sections present the derivation of the transformation matrices for displacement and forces for the planar truss. The interdependence of the transformation matrices will also be established.

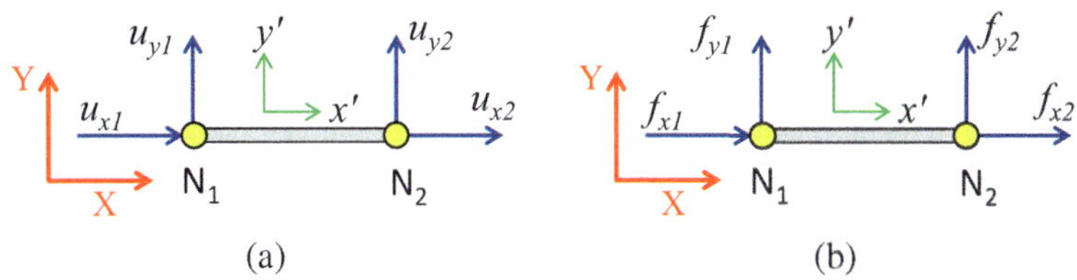

Fig. 3.11 A representation in rotated reference frame of the nodal (**a**) displacements and (**b**) forces for nodes, N_1 and N_2. The global coordinate system (X- and Y-axes) is shown to be coincident in orientation with the transformed reference frame (x'— and y'—axes)

3.4.2.1 Case I: Bar Oriented with Coincident Local and Global Axes

With respect to Eq. 3.2, the relationship between nodal forces and displacement can be re-written as:

$$F = \frac{EA}{L}\Delta L \implies f_{x1} = \frac{EA}{L}\left[u_{x1} - u_{x2}\right] \tag{3.3}$$

Similarly, through equilibrium, it can be shown that the following is also true:

$$f_{x1} = -f_{x2} = \frac{EA}{L}\left[u_{x1} - u_{x2}\right] \implies f_{x2} = \frac{EA}{L}\left[-u_{x1} + u_{x2}\right] \tag{3.4}$$

Since the trusses are axially loaded members, they only carry load along their main axis and so will experience axial deformation only.

As a result, there is no associated y-axis displacement, hence these conclusions are valid:

$$f_{y1} = 0 \implies f_{y1} = \frac{EA}{L}\left[0u_{y1} - 0u_{y2}\right] \tag{3.5}$$

$$f_{y2} = 0 \implies f_{y2} = \frac{EA}{L}\left[-0u_{y1} + 0u_{y2}\right] \tag{3.6}$$

Combining Eq. 3.1 with Eqs. 3.3, 3.4, 3.5, and 3.6, we can now convert the later equations into a matrix-format such that the resulting nodal force-displacement matrix equation becomes:

$$\begin{bmatrix} f_{x1} \\ f_{y1} \\ f_{x2} \\ f_{y2} \end{bmatrix} = \frac{EA}{L}\begin{bmatrix} 1 & 0 & -1 & 0 \\ 0 & 0 & 0 & 0 \\ -1 & 0 & 1 & 0 \\ 0 & 0 & 0 & 0 \end{bmatrix}\begin{bmatrix} u_{x1} \\ u_{y1} \\ u_{x2} \\ u_{y2} \end{bmatrix} \implies \mathbf{f} = \mathbf{ku} \tag{3.7}$$

In the above equation, we can observe that:

$$\text{Member stiffness matrix (in local axes)} : \mathbf{k} = \frac{EA}{L}\begin{bmatrix} 1 & 0 & -1 & 0 \\ 0 & 0 & 0 & 0 \\ -1 & 0 & 1 & 0 \\ 0 & 0 & 0 & 0 \end{bmatrix} \tag{3.8}$$

The above demonstrates how to determine the stiffness matrix in a local coordinate system for an instance where the global coordinate axis is *coincident* or *parallel* to the local coordinate system. However, this is a very restrictive case and in the next case we will widen the derivation to a case where the orientation of the truss member is at an arbitrary angle, θ. For now, let us consider an example of

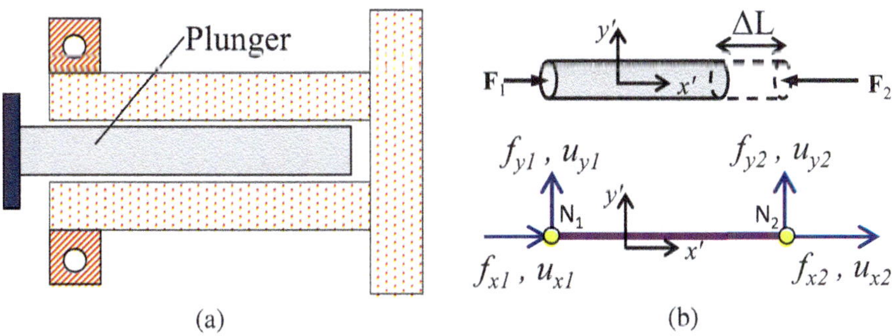

Fig. 3.12 (**a**) *Physical Problem* showing a plunger-cylinder assembly and (**b**) *Idealized System* showing *Top:* Free-body diagram of the piston and *Bottom:* discretization element with applicable nodal forces and displacements at nodes, N_1 and N_2

this restrictive case to illustrate how one can use the stiffness matrix to calculate the nodal forces or displacement of a structure.

Example 3.2 A plunger system, shown in Fig. 3.12, consists of a steel plunger of diameter, $d = 25$ mm and of length, $L = 150$ mm. The piston is designed to ensure full compression at a plunger force, $F = 500$ kN. The plunger is retained within another cylinder such that it can only experience axial deformation along the length of the plunger. As a result of fatigue, the piston end was observed to have worn off by 1 mm such that more force is required to ensure full compression of the piston. Assume that the Young's modulus of steel, $E_{steel} = 210$ GPa.

(a) Determine the new piston force.
(b) What is the percentage reduction in efficiency of the mechanism?

Solution

Guiding Assumptions

- We will focus only on the plunger and assume the casing of the plunger is rigid such that it does not experience any deformation.
- The plunger is assumed to be linear elastic with the fatigue-caused damage of the plunger end to represent nodal displacement of node, N_2.
- The global and local coordinates are coincident so we can treat this problem as a *Case I* bar with the relevant stiffness matrix defined in Eq. 3.8.

Calculation of Geometric and Material Properties

- Cross-sectional area, A of plunger:
$$A = \frac{\pi d^2}{4} \quad \Longrightarrow \quad A = 0.25 \times \pi \times 0.025^2 = 4.9087 \times 10^{-4}\,\text{m}^2$$
- Calculate the stiffness coefficient, k:
$$k = \frac{EA}{L} \quad \Longrightarrow \quad k = \frac{210 \times 10^9 \times 4.9087 \times 10^{-4}}{0.150} = 6.8722 \times 10^8\,\text{Nm}^{-1}$$

Determination of nodal displacement and forces

- According to Eq. 3.7, we can conclude that the following are the applicable boundary conditions for the plunger system of Fig. 3.12b:

$$
\begin{aligned}
f_{x1} &= F_1 && \Longrightarrow F_1 = F = 500\,\text{kN known external load} \\
f_{x2} &= F_2 && \Longrightarrow F_2 \text{ unknown reaction load at node, } N_2 \\
u_{x1} &= 0 && \Longrightarrow \text{no movement of node, } N_1 \\
u_{x2} &= 1 \times 10^{-3}\,\text{m} && \Longrightarrow \text{worn-off end, i.e. node, } N_2 \\
u_{y1} &= u_{y2} = 0 && \Longrightarrow \text{no y-axis displacement}
\end{aligned}
$$

- Equation 3.7 can be re-written for the plunger-cylinder assembly problem thus:

$$
\mathbf{f} = \mathbf{ku} \quad \Longrightarrow \quad \begin{bmatrix} F_1 \\ f_{y1} \\ f_{x2} \\ f_{y2} \end{bmatrix} = \frac{EA}{L} \begin{bmatrix} 1 & 0 & -1 & 0 \\ 0 & 0 & 0 & 0 \\ -1 & 0 & 1 & 0 \\ 0 & 0 & 0 & 0 \end{bmatrix} \begin{bmatrix} 0 \\ 0 \\ 1 \times 10^{-3} \\ 0 \end{bmatrix}
$$

- The expansion of the third term in the force-displacement matrix equation above gives:

$$
\begin{aligned}
f_{31} &= \frac{EA}{L}\left\{ -1 \times 0 + 0 \times 0 + 1 \times (1 \times 10^{-3}) + 0 \times 0 \right\} \\
&= (1 \times 10^{-3})\frac{EA}{L} \\
&= 1 \times 10^{-3} \times 6.8722 \times 10^{8} \\
&= 687.22\,\text{kN}
\end{aligned}
$$

- The magnitude of the compressive force with worn ends piston is thus: $F_{new} = 687\,\text{kN}$. This implies that an extra force: $\Delta F = F_{new} - F = 187.22\,\text{kN}$ has to be applied to ensure full compression of the plunger.

Calculate percentage reduction in efficiency of plunger system

In view of the requirement for extra force of 187 kN that needs to be imposed on the plunger to achieve full (optimal) compression, the system is therefore not operating at the desired efficiency. The reduction in efficiency can be calculated thus:

$$
\begin{aligned}
\%\ \text{Redunction in efficiency} &= \frac{F_{new} - F}{F} \\
&= \frac{187}{500} * 100\% \\
&= 37.4\%
\end{aligned}
$$

The above illustrates how knowledge of a stiffness matrix for a structure (in this instance a plunger) can be used to determine unknown nodal forces. We can also go ahead and use the full Eq. 3.7 to determine any list unknown displacements in this deformed/worn configuration of the plunger.

3.4.2.2 Case II: Bar Oriented at an Arbitrary Angle, θ

Structural members are often oriented at an arbitrary angle from the global reference frame. This case study investigates the determination of member stiffness for such bars whose local reference frames are oriented at an arbitrary angle, θ to the global reference frame as shown in Fig. 3.13.

The local reference frame for the bars is the $x-$ and $y-$ axes, whilst the reference frame (transformed from local reference frame) is the $x'-$ and $y'-$axes. The global reference frame is the X- and Y-axes. The transformed reference frame is coincident with the local reference frame and both of them make angle θ with the global reference frame, X- and Y-axes.

In order to relate the nodal displacements of global coordinates systems (similar to Case I) to the nodal displacements of the local reference frame, we isolate node, N_1 as shown in Fig. 3.14. Here, the nodal displacements (of the global reference frame) along X- and Y-axes represented as u_{X1} and u_{Y1} respectively. The nodal displacements of the local reference frame for the $x-$ and $y-$ axes are u_{x1} and u_{y1} respectively. Figure 3.14 indicates resolution of displacements of the global coordinates into their components on the local reference frame.

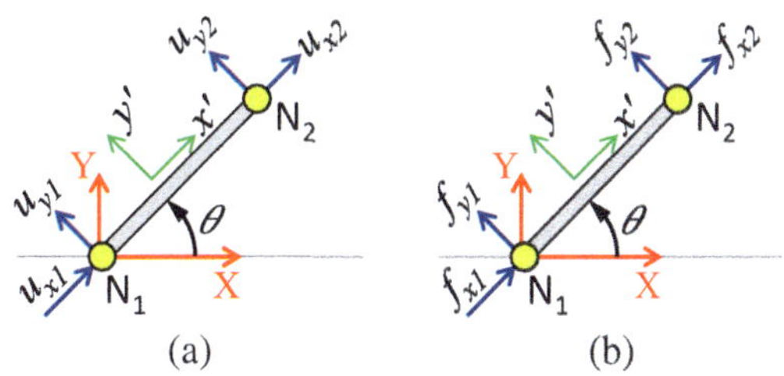

Fig. 3.13 A representation of a two-node bar oriented at an arbitrary angle, θ showing the nodal (**a**) displacements and (**b**) forces for nodes, N_1 and N_2

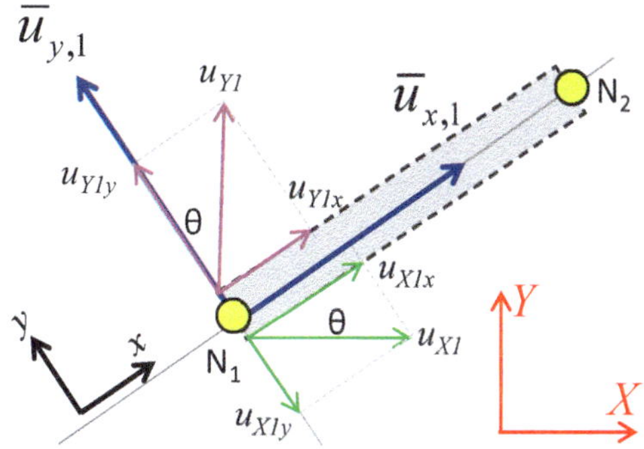

Fig. 3.14 Resolution of nodal displacements at node N_1 of a two-node bar showing transformation of the nodal displacements from global reference frame to local reference frame

Based on Fig. 3.14, let us apply equations of equilibrium along x- and y-axes gives (where $c = \cos\theta$ and $s = \sin\theta$), taking the sum of displacements along x- and y- axes results in:

$$\sum U_x = 0 \quad \Rightarrow \quad \begin{aligned} \bar{u}_{x,1} &= U_{X1x} + U_{Y1x} \\ &= U_{X1}\cos\theta + U_{Y1}\sin\theta \\ &= U_{X1}c + U_{Y1}s \end{aligned} \tag{3.9}$$

and

$$\sum U_y = 0 \quad \Rightarrow \quad \begin{aligned} \bar{u}_{y,1} &= U_{Y1y} - U_{X1y} \\ &= -U_{X1}\sin\theta + U_{Y1}\cos\theta \\ &= -U_{X1}s + U_{Y1}c \end{aligned} \tag{3.10}$$

respectively. Similarly, with respect to node N_2, we write as follows for $x-$ and $y-$ axes:

$$\begin{aligned} \bar{u}_{x,2} &= U_{X2}c + U_{Y2}s \\ \bar{u}_{y,2} &= -U_{X2}s + U_{Y2}c \end{aligned} \tag{3.11}$$

If we assemble Eqs. 3.9, 3.10, and 3.11 into a single matrix results:

$$\begin{bmatrix} \bar{u}_{x1} \\ \bar{u}_{y1} \\ \bar{u}_{x2} \\ \bar{u}_{y2} \end{bmatrix} = \begin{bmatrix} c & s & 0 & 0 \\ -s & c & 0 & 0 \\ 0 & 0 & c & s \\ 0 & 0 & -s & c \end{bmatrix} \begin{bmatrix} U_{X1} \\ U_{Y1} \\ U_{X2} \\ U_{Y2} \end{bmatrix} \quad \Longrightarrow \quad \bar{\mathbf{u}} = \mathbf{T_d U} \tag{3.12}$$

The $\mathbf{T}_d$-matrix is called a *Displacement transformation matrix.* It is the matrix that ensures that nodal displacements (expressed in the global coordinates – $U_{X,i}, U_{Y,i}$ for $i = 1, 2$) can be transformed into their local coordinates values such that member stiffness matrix can be determined for such a member. In Eq. 3.12, $\bar{\mathbf{u}}$ is a column matrix of nodal displacements in the local coordinates system while $\mathbf{U}$ is the column matrix of nodal displacements in a global coordinate system.

Similarly, the same process applies for the nodal forces, but this time around, we resolve the member forces from the local coordinates system to their global coordinate system values. We begin by considering the resolution of nodal forces as shown in Fig. 3.15.

Applying equations of equilibrium to Fig. 3.15, we have:

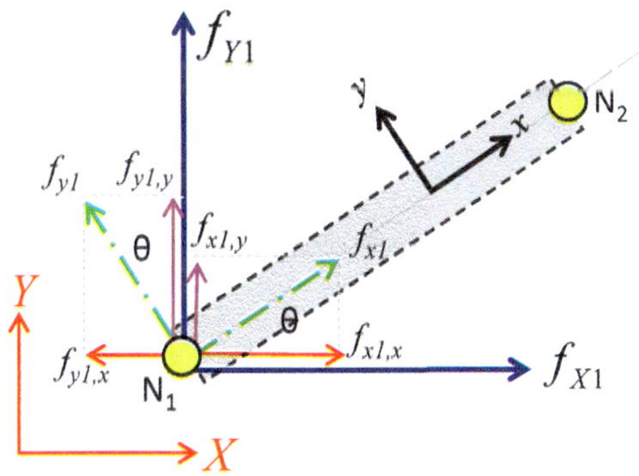

Fig. 3.15 Resolution of nodal forces at node N_1 of a two-node bar showing transformation of the nodal forces from local reference frame to global reference frame

$$\sum F_x = 0 \quad \Rightarrow \quad \bar{f}_{X1} = f_{x1,x} - f_{y1,x} = f_{x1}\cos\theta - f_{y1}\sin\theta = f_{x1}c - U_{y1}s \tag{3.13}$$

and

$$\sum F_y = 0 \quad \Rightarrow \quad \bar{f}_{Y1} = f_{x1,y} + f_{y1,y} = f_{x1}\, sin\theta + f_{y1}\cos\theta = f_{x1}s + f_{y1}c \tag{3.14}$$

Similarly, with respect to node N_2, we write the global coordinates nodal forces as follows:

$$\begin{aligned} \bar{f}_{X2} &= f_{x1}c + f_{y2}s \\ \bar{f}_{Y2} &= -f_{x1}s + f_{y2}c \end{aligned} \tag{3.15}$$

Now, let us assemble Eqs. 3.13, 3.14, and 3.15 into a single matrix results:

$$\begin{bmatrix} \bar{f}_{X1} \\ \bar{f}_{Y1} \\ \bar{f}_{X2} \\ \bar{f}_{Y2} \end{bmatrix} = \begin{bmatrix} c & -s & 0 & 0 \\ s & c & 0 & 0 \\ 0 & 0 & c & -s \\ 0 & 0 & s & c \end{bmatrix} \begin{bmatrix} f_{x1} \\ f_{y1} \\ f_{x2} \\ f_{y2} \end{bmatrix} \quad \Longrightarrow \quad \bar{\mathbf{F}} = \mathbf{T_f f} \tag{3.16}$$

The $\mathbf{T_f}$-matrix is called a *force transformation matrix*. This matrix is used to transform the nodal forces from one co-ordinate system (in this case the local one) into another coordinate system (in the above case, the global coordinate system). In Eq. 3.16, $\mathbf{f}$ is column vector of forces in local coordinates while $\bar{\mathbf{F}}$ is the column

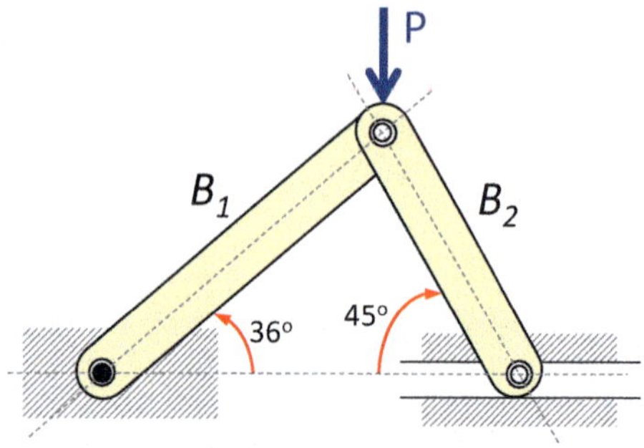

Fig. 3.16 A pin-jointed structure supporting a concentrated force, **P**

vector of nodal forces in the global coordinate system. Notice from Eqs. 3.12 and 3.16 that the force transformation matrix is the transpose of the displacement transformation matrix: $\mathbf{T}_f = \mathbf{T}_d^T$.

Example 3.3 A pin-jointed structure consists of two members of lengths L_1 and L_2 and supports a point load, **P**, as shown in Fig. 3.16. Assume both members are made of the same material and have the same cross-sectional areas.

(a) Determine the member force and displacement transformation matrices.
(b) Comment on whether both matrices are the transpose of one another.

Solution

All angles, θ, are measured anti-clockwise from the positive global X-axis.

For the structure of Fig. 3.16, there exists two bars/members, which we have labelled B_1 and B_2. The joints/nodes are labelled $N_1 \rightarrow N_3$ in an anti-clockwise manner. The angles the members make with the positive global $x-$ axis are given as θ_1 and θ_2 respectively for bars B_1 and B_2.

Calculation of Displacement and Force Transformation Matrices for Bar, B_1

$$\theta_1 = 36^\circ \qquad c = \cos\theta_1 = \cos 36^\circ = 0.8090$$
$$s = \sin\theta_1 = \sin 36^\circ = 0.5878$$

Recall from Eq. 3.12, the expression for the displacement transformation matrix is:

$$\mathbf{T}_d = \begin{bmatrix} c & s & 0 & 0 \\ -s & c & 0 & 0 \\ 0 & 0 & c & s \\ 0 & 0 & -s & c \end{bmatrix} \Longrightarrow \mathbf{T}_{d,B_1} = \begin{bmatrix} 0.8090 & 0.5878 & 0 & 0 \\ -0.5878 & 0.8090 & 0 & 0 \\ 0 & 0 & 0.8090 & 0.5878 \\ 0 & 0 & -0.5878 & 0.8090 \end{bmatrix} \tag{3.17}$$

Similarly, the force transformation matrix (based on Eq. 3.16) for bar B_1 is given thus:

$$\mathbf{T}_f = \begin{bmatrix} c & -s & 0 & 0 \\ s & c & 0 & 0 \\ 0 & 0 & c & -s \\ 0 & 0 & s & c \end{bmatrix} \Longrightarrow \mathbf{T}_{f,B_1} = \begin{bmatrix} 0.8090 & -0.5878 & 0 & 0 \\ 0.5878 & 0.8090 & 0 & 0 \\ 0 & 0 & 0.8090 & -0.5878 \\ 0 & 0 & 0.5878 & 0.8090 \end{bmatrix} \tag{3.18}$$

Calculation of Displacement and Force Transformation Matrices for Bar, B_2

$$\theta_2 = 135^\circ \qquad c = \cos\theta_2 = \cos 135^\circ = -0.7071$$
$$s = \sin\theta_2 = \sin 135^\circ = 0.7071$$

The expressions for the displacement and force transformation matrices are given thus:

$$\mathbf{T}_d = \begin{bmatrix} c & s & 0 & 0 \\ -s & c & 0 & 0 \\ 0 & 0 & c & s \\ 0 & 0 & -s & c \end{bmatrix} \Longrightarrow \mathbf{T}_{d,B_2} = \begin{bmatrix} -0.7071 & 0.7071 & 0 & 0 \\ -0.7071 & -0.7071 & 0 & 0 \\ 0 & 0 & -0.7071 & 0.7071 \\ 0 & 0 & -0.7071 & -0.7071 \end{bmatrix} \tag{3.19}$$

and

$$\mathbf{T}_f = \begin{bmatrix} c & -s & 0 & 0 \\ s & c & 0 & 0 \\ 0 & 0 & c & -s \\ 0 & 0 & s & c \end{bmatrix} \Longrightarrow \mathbf{T}_{f,B_2} = \begin{bmatrix} -0.7071 & -0.7071 & 0 & 0 \\ 0.7071 & -0.7071 & 0 & 0 \\ 0 & 0 & -0.7071 & -0.7071 \\ 0 & 0 & 0.7071 & -0.7071 \end{bmatrix} \tag{3.20}$$

Notice that for both members: $\mathbf{T}_{f,B_1} = \mathbf{T}'_{d,B_1}$ and $\mathbf{T}_{f,B_2} = \mathbf{T}'_{d,B_2}$, which confirms the expectation that the force transformation matrix is the transpose of the displacement transformation matrix (Fig. 3.17).

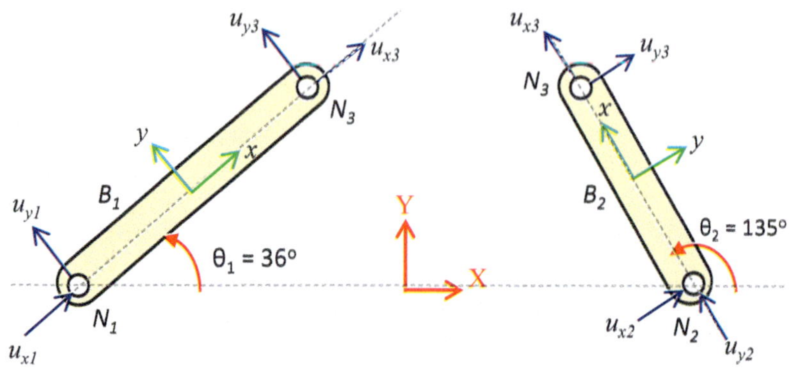

Fig. 3.17 An idealized system of structure of Fig. 3.16 showing node and bar numbers, angles of inclination of bars as well as nodal displacements. Also, the same applies to the nodal forces

3.4.3 *Formulation of Member Stiffness Matrix for the Global Coordinate System*

With respect to Eqs. 3.12 and 3.16, we can determine the global coordinate system representation of the nodal displacement, **U**, and force, **F**, as shown in Eq. 3.21.

$$\bar{\mathbf{u}} = \mathbf{T}_d\mathbf{U} \quad \Longrightarrow \quad \mathbf{U} = \mathbf{T}'_d\bar{\mathbf{u}} \quad \text{and} \quad \bar{\mathbf{F}} = \mathbf{T}_f\mathbf{f} \tag{3.21}$$

where $\mathbf{T}'_d = \mathbf{T}_f$, and **U** and **F** are nodal displacement and force column vectors respectively in a *global coordinate system*. On the other hand, $\bar{\mathbf{u}}$ and **f** are nodal displacements and force column vectors in a *local coordinate system*.

To formulate the *member stiffness matrix* that combines the nodal forces and displacements, we have to determine the member stiffness matrix using either global or local coordinates. For this instance, let us determine the desired member stiffness matrix first using the global coordinate system. This is essential as it makes for easy addition of the different stiffness matrices that make up a body since they are oriented *similarly* in global coordinates. In a global coordinate system, the force-displacement matrix for a bar/element is written thus:

$$\mathbf{F}^e = \mathbf{K}^e\mathbf{U}^e \tag{3.22}$$

where $\mathbf{F}^e$ is the bar- or element-level nodal force vector (in global coordinate system) while $\mathbf{U}^e$ is the bar- or element-level nodal displacement vector (also in global coordinate system). Note, with respect to Eqs. 3.21 and 3.22: $\mathbf{F}^e = \bar{\mathbf{F}}^e$ and $\mathbf{U}^e = \mathbf{U}$.

Therefore, combining Eqs. 3.12, 3.21, and 3.22 into a single equation gives:

$$\mathbf{F}^e = \mathbf{K}^e\mathbf{U}^e \quad \Longrightarrow \quad \mathbf{F}^e = \mathbf{T}'_d\mathbf{k}^e\mathbf{T}_d\mathbf{U}^e \tag{3.23}$$

The above implies therefore that:

$$\mathbf{K}^e = \mathbf{T}_d'\mathbf{k}^e\mathbf{T}_d \qquad \text{or} \qquad \mathbf{K}^e = \mathbf{T}_f'\mathbf{k}^e\mathbf{T}_f \tag{3.24}$$

Executing the matrix manipulation that relates to Eq. 3.24 (using inputs from Eqs. 3.12, 3.21, and 3.22) results in the following *member stiffness matrix* – in a global coordinate system representation:

$$\mathbf{K}^e = \mathbf{T}_d'\mathbf{k}^e\mathbf{T}_d \Longrightarrow$$

$$\mathbf{K}^e = \begin{bmatrix} c & -s & 0 & 0 \\ s & c & 0 & 0 \\ 0 & 0 & c & -s \\ 0 & 0 & s & c \end{bmatrix} \frac{E^eA^e}{L^e} \begin{bmatrix} 1 & 0 & -1 & 0 \\ 0 & 0 & 0 & 0 \\ -1 & 0 & 1 & 0 \\ 0 & 0 & 0 & 0 \end{bmatrix} \begin{bmatrix} c & s & 0 & 0 \\ -s & c & 0 & 0 \\ 0 & 0 & c & s \\ 0 & 0 & -s & c \end{bmatrix} \tag{3.25}$$

where $\mathbf{T}_d'$ is the transpose of the *displacement transformation matrix*. Also, in the above expression, E^e = **Young's Modulus** of the element, A_e = cross-sectional area of the element and L_e = length of the element. Recall also that $c = \cos\theta$ and $s = \sin\theta$ where θ is the angle the element/bar/member makes with the positive $x-$ axis, with the angles measured in an anti-clockwise fashion.

When the matrix equation of Eq. 3.25 is evaluated, we obtain the follow *transformed global coordinate member stiffness matrix*:

$$\text{Global member stiffness matrix:} \quad \mathbf{K}^e = \frac{E^eA^e}{L^e} \begin{bmatrix} c^2 & sc & -c^2 & -sc \\ sc & s^2 & -sc & -s^2 \\ -c^2 & -sc & c^2 & sc \\ -sc & -s^2 & sc & s^2 \end{bmatrix} \tag{3.26}$$

where $c^2 = \cos^2\theta, s^2 = \sin^2\theta$, and $sc = \sin\theta\cos\theta$, where θ is specified in degrees. In the following, we illustrate this implementation using an example.

Example 3.4 A loaded simply supported triangular planar truss shown in Fig. 3.18, consists of three members labelled, B_1, B_2 and B_3. The diameters of B_1, B_2 and B_3 are 3, 5 and 7 mm and of lengths 150, 98 and 132 mm. B_1, B_2 and B_3 were made of steel, titanium and brass respectively. Assume the Young's Modulus of steel, E_s = 210 GPa, brass, E_b = 125 GPa and titanium, E_t = 110 GPa.

Determine the global coordinate member stiffness matrices of the three bars.

Solution

The discretization of the structure in *global* (i.e. X- and Y-axes) and *local* (i.e. *x− and y−* axes) *coordinate systems* is given in Fig. 3.19.

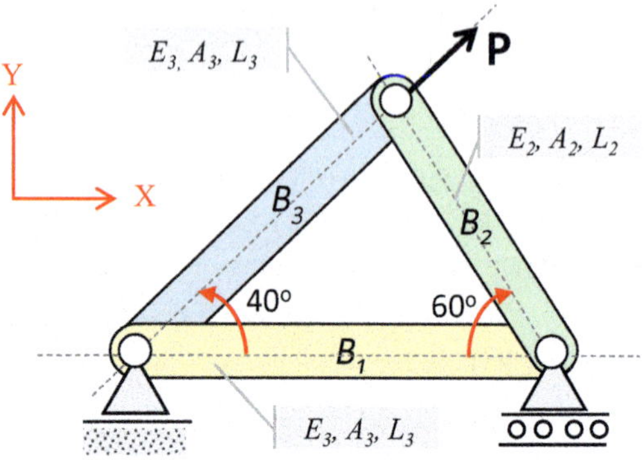

Fig. 3.18 A plane triangular truss subjected to a load, P made from three materials namely, steel (B_1), brass (B_2) and titanium (B_3)

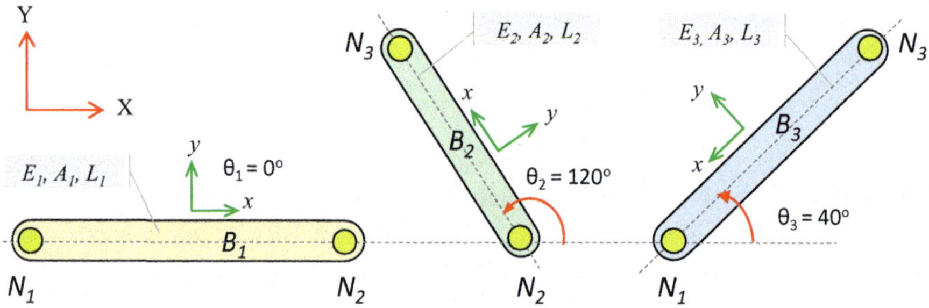

Fig. 3.19 An idealized (discretized) model of structure of Fig. 3.18 showing the correct angles of inclination of the bars, measured from the positive x-axis

Calculation of the member stiffness matrices (global)

For Bar, B$_1$: Steel

$$\text{Angle: } \theta_1 = 0^\circ \quad \Rightarrow c = \cos\theta_1 = \cos 0^\circ = 1.0 \longrightarrow c^2 = 1.0$$
$$s = \sin\theta_1 = \sin 0^\circ = 0.0 \longrightarrow s^2 = 0.0 \text{ and } cs = 0.0$$

$$\text{Area: } A_1 = \frac{\pi d_1^2}{4} \Rightarrow A_1 = \frac{\pi \times 0.003^2}{4}$$

$$A_1 = 7.0686 \times 10^{-6}\,\text{m}^2,$$

$$\text{Axial rigidity: } E_1A_1 \Rightarrow E_sA_1 \Rightarrow = (210 \times 10^9\,\text{Nm}^{-2})(7.0686 \times 10^{-6}\,\text{m}^2)$$

$$= 1.4844 \times 10^6\,\text{N}$$

$$\text{Stiffness constant: } \frac{E_1A_1}{L_1} \Rightarrow \frac{E_sA_1}{L_1} \Rightarrow \frac{E_sA_1}{L_1} = \frac{1.4844 \times 10^6\,\text{N}}{0.150\,\text{m}}$$

$$= 9.8960 \times 10^6 \text{Nm}^{-1}.$$

Global member stiffness matrix, $\mathbf{K}_1^e$ for bar B_1, (using Eq. 3.26) becomes:

$$\mathbf{K}_1^e = \frac{E_1A_1}{L_1}\begin{bmatrix} c^2 & sc & -c^2 & -sc \\ sc & s^2 & -sc & -s^2 \\ -c^2 & -sc & c^2 & sc \\ -sc & -s^2 & sc & s^2 \end{bmatrix} \Longrightarrow \mathbf{K}_1^e = 9.8960 \times 10^6 \begin{bmatrix} 1 & 0 & -1 & 0 \\ 0 & 0 & 0 & 0 \\ -1 & 0 & 1 & 0 \\ 0 & 0 & 0 & 0 \end{bmatrix} \text{Nm}^{-1}$$

For Bar, B$_2$: Titanium

$$\text{Angle: } \theta_2 = 120° \quad \Rightarrow \quad \begin{aligned} c &= \cos\theta_2 = \cos 120° = -0.5000 \to c^2 = 0.2500 \\ s &= \sin\theta_2 = \sin 120° = 0.8660 \to s^2 = 0.7500 \\ cs &= (-0.5000)(0.8660) = -0.4330 \end{aligned}$$

$$\text{Area: } A_2 = \frac{\pi d_2^2}{4} \Rightarrow A_2 = \frac{\pi \times 0.005^2}{4}$$

$$A_2 = 1.9635 \times 10^{-5}\,\text{m}^2$$

$$\text{Axial rigidity: } E_2A_2 \Rightarrow E_tA_2 \Rightarrow = (110 \times 10^9\,\text{Nm}^{-2})(1.9635 \times 10^{-5}\,\text{m}^2)$$

$$= 2.1598 \times 10^6\,\text{N}$$

$$\text{Stiffness constant: } \frac{E_2A_2}{L_2} \Rightarrow \frac{E_tA_2}{L_2} \Rightarrow \frac{E_tA_2}{L_2} = \frac{2.1598 \times 10^6\,\text{N}}{0.098\,\text{m}}$$

$$= 22.04 \times 10^6 \text{Nm}^{-1}.$$

Global member stiffness matrix, $\mathbf{K}_2^e$ for bar B_2 becomes:

$$\mathbf{K}_2^e = 22.04 \times 10^6 \begin{bmatrix} 0.2500 & -0.4330 & -0.2500 & 0.4330 \\ -0.4330 & 0.7500 & 0.4330 & -0.7500 \\ -0.2500 & 0.4330 & 0.2500 & -0.4330 \\ 0.4330 & -0.7500 & -0.7500 & 0.7500 \end{bmatrix} \text{Nm}^{-1}$$

For Bar, B_3: Brass

$$\text{Angle: } \theta_3 = 40^\circ \Rightarrow \begin{array}{l} c = \cos\theta_3 = \cos 4^\circ = 0.7660 \quad \longrightarrow \quad c^2 = 0.5868 \\ s = \sin\theta_3 = \sin 40^\circ = 0.6428 \quad \longrightarrow \quad s^2 = 0.4132 \\ cs = (0.7660)(0.6428) = -0.4924 \end{array}$$

$$\text{Area: } A_3 = \frac{\pi d_3^2}{4} \Rightarrow A_3 = \frac{\pi \times 0.007^2}{4}$$

$$A_3 = 3.8485 \times 10^{-5}\,\text{m}^2$$

$$\text{Axial rigidity: } E_3A_3 \Rightarrow E_tA_3 \Rightarrow = (125 \times 10^9\,\text{Nm}^{-2})(3.8485 \times 10^{-5}\,\text{m}^2)$$

$$= 4.8106 \times 10^6\,\text{N}$$

$$\text{Stiffness constant: } \frac{E_3A_3}{L_3} \Rightarrow \frac{E_bA_3}{L_3} \Rightarrow \frac{E_bA_3}{L_3} = \frac{4.8106 \times 10^6\,\text{N}}{0.132\,\text{m}}$$

$$= 36.44 \times 10^6\text{Nm}^{-1}$$

Global member stiffness matrix, $\mathbf{K}_3^e$ for bar B_3 becomes:

$$\mathbf{K}_3^e = 36.44 \times 10^6 \begin{bmatrix} 0.5868 & 0.4924 & -0.2500 & -0.4924 \\ 0.4924 & 0.4132 & -0.4924 & -0.4132 \\ -0.5868 & -0.4924 & 0.5868 & 0.4924 \\ -0.4924 & -0.4132 & -0.4132 & 0.4132 \end{bmatrix} \text{Nm}^{-1}$$

3.4.4 *Formulation of Structural Stiffness Matrix*

The formulation of a *structural stiffness matrix* essentially involves the assembly of all member stiffness matrices provided that the member stiffness matrices are in exactly the same global reference frames. Two processes must be obeyed during the assembly process and these are:

(a) ***Compatibility criterion***: This criterion is imposed on the nodal displacements and requires that the displacement of all members that meet at a node/joint must be the same. As a result, the mechanics of that joint will show a compatible deformation: all members at that joint will displace equally, due to this criterion.

(b) ***Equilibrium criterion:*** This establishes that at any joint/node, all the member/bar (internal) forces acting at that point must balance with any externally applied force acting on same joint. This means that the externally imposed loads on the joint/node must cause and equal and opposite internal response in the member.

To illustrate this process, let us consider the triangular planar truss of Fig. 3.8a, originally disassembled as part of the process for determining the member stiffness matrices of the component members/bars. Here, to determine the structural stiffness matrix of the same truss, an assembly process of all members needs to be carried out. This is often referred to as *merging* in finite element solvers, for example within ABAQUS CAE, in the *Assembly Module*. In the FEM process, the merging process is essential to ensure compatibility of displacements and equilibria of both internal and external forces acting on the structure. Without merging, the solution will not proceed correctly and the solutions will be spurious.

For the plane truss shown in Fig. 3.20, consider the nodal forces at node, N_3 (in a local coordinate system). For example, the force identified as $f_{x3}^{[3]}$, with a square-bracketed 3 superscript, this refers to the local coordinate nodal force acting on node, N_3 along the $x-$ direction, with the square-bracketed number indicating the bar/member associated with the force (in this case, B_3).

Also, notice that at node N_3, there is another force, $f_{x3}^{[2]}$, acting there. This is also a local coordinate nodal force, acting along the $x-$ axis of the B_2 bar/member, with a square-bracketed number 2 superscript: the later signifying that the nodal force is associated with the B_2 bar. The same argument applies when identifying the nodal displacements, for similar directions as the nodal forces identified here.

Based on the merged structure of Fig. 3.20, and considering only the *local coordinate system*, we can write the two equations that address the *compatibility* and *equilibrium* criteria thus:

$$\text{Compatibility criterion:} \quad u_{x1}^{[1]} = u_{x1}^{[3]} \qquad u_{x2}^{[1]} = u_{x2}^{[2]} \qquad u_{x3}^{[2]} = u_{x3}^{[3]}$$

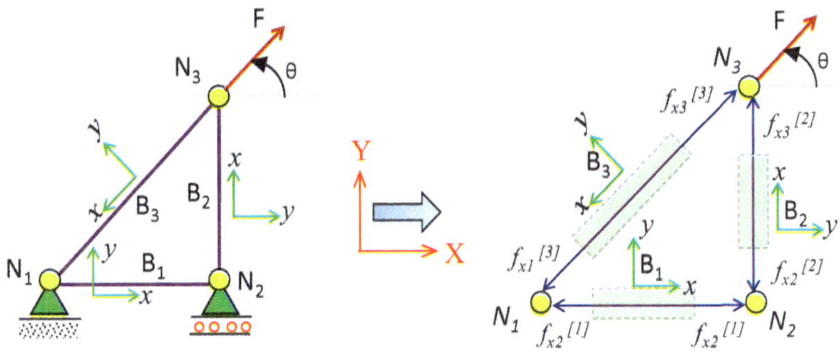

Fig. 3.20 An illustration of the *merging* of members/elements into a structural unit by enforcing that the *compatibility* and *equilibrium* criteria are met for that structure

and

$$\text{Equilibrium criterion:} \quad F_{x3} = f_{x3}^{[2]} + f_{x3}^{[3]} \quad \equiv \quad F_{x3} = f_{x3}^{[1]} + f_{x3}^{[2]} + f_{x3}^{[3]}$$

$$F_{y3} = f_{y3}^{[2]} + f_{y3}^{[3]} \quad \equiv \quad F_{y3} = f_{y3}^{[1]} + f_{y3}^{[2]} + f_{y3}^{[3]}$$

In terms of vector format, the expression for the forces at node N_3 above becomes: $\mathbf{F}_3 = \mathbf{f}^{[1]} + \mathbf{f}^{[2]} + \mathbf{f}^{[3]}$, where $\mathbf{F}$ = vector of external forces and $\mathbf{f}_i$ = vector of internal forces, for $i = 1, 2, 3$. The is the comprehensive expression for the equilibrium of forces between internal and external forces at node, N_3. However, to be more precise, the contribution of internal forces from bar B_1 at this point is redundant since it is not part of the forces acting on node N_3. Therefore, during the execution, it is okay to simply neglect it and evaluate only the contributions from bars B_2 and B_3.

The assembly of member stiffness matrices to form the structural matrices can be done by two processes:

(a) ***Manual process***: Here, the structural engineer does the analysis by analytical hand-calculations, which involves matrix manipulations; and,
(b) ***Computer process***: In this process, a computer program is written to automate the process of matrix manipulation required for the assembly process.

The computer process is amenable to the FEM process and will be used subsequently in this textbook (see Chap. 4), but for the purpose of understanding the steps involved in developing the *Structural Stiffness Matrix*, a manual process is illustrated in the next section.

3.4.4.1 Illustration of Manual Method for Assembling a Structural Stiffness Matrix

We will aim to determine the structural stiffness matrix of the triangular planar truss shown in Fig. 3.16 of Example 3.3. Assuming a global coordinate system is used, we can isolate the nodal forces and displacements acting on the structure as shown in Fig. 3.21.

We will use the member stiffness matrices (global coordinates) of Example 3.3 for bars B_1, B_2 and B_3. Note that the compatibility and equilibrium criteria have to be implemented here such that the force-displacement relationship for the three bars (in global coordinates) can be written thus:

<u>For Bar, B_1</u>

$$\mathbf{F}^e = \mathbf{K}_1^e \mathbf{U}^e \Longrightarrow \begin{bmatrix} \mathbf{F}_{X1}^{[1]} \\ \mathbf{F}_{Y1}^{[1]} \\ \mathbf{F}_{X1}^{[2]} \\ \mathbf{F}_{Y1}^{[2]} \end{bmatrix} = 9.8960 \times 10^6 \begin{bmatrix} 1 & 0 & -1 & 0 \\ 0 & 0 & 0 & 0 \\ -1 & 0 & 1 & 0 \\ 0 & 0 & 0 & 0 \end{bmatrix} \begin{bmatrix} \mathbf{U}_{X1}^{[1]} \\ \mathbf{U}_{Y1}^{[1]} \\ \mathbf{U}_{X1}^{[2]} \\ \mathbf{U}_{Y1}^{[2]} \end{bmatrix} \tag{3.27}$$

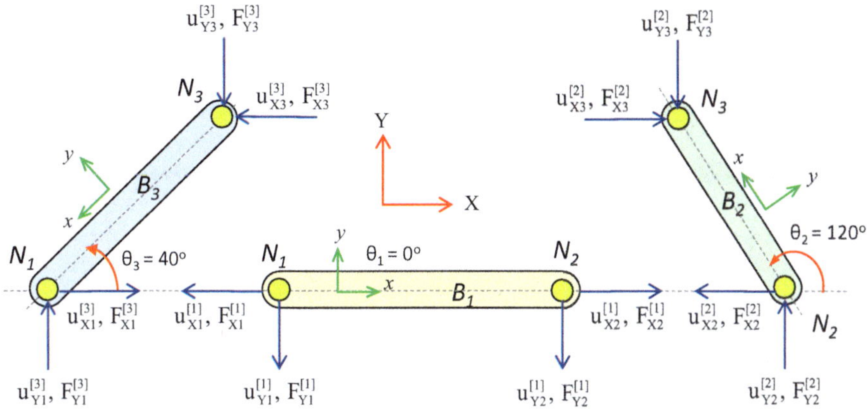

Fig. 3.21 An illustration of applicable nodal displacements and forces specified in terms of bar number and global coordinates systems representation

For Bar, B_2

$$\mathbf{F}^e = \mathbf{K}_2^e\mathbf{U}^e \Longrightarrow \begin{bmatrix} \mathbf{F}_{X2}^{[2]} \\ \mathbf{F}_{Y2}^{[2]} \\ \mathbf{F}_{X3}^{[3]} \\ \mathbf{F}_{Y3}^{[3]} \end{bmatrix}$$

$$= 22.04 \times 10^6 \begin{bmatrix} 0.2500 & -0.4330 & -0.2500 & 0.4330 \\ -0.4330 & 0.7500 & 0.4330 & -0.7500 \\ -0.2500 & 0.4330 & 0.2500 & -0.4330 \\ 0.4330 & -0.7500 & -0.7500 & 0.7500 \end{bmatrix} \begin{bmatrix} \mathbf{U}_{X2}^{[2]} \\ \mathbf{U}_{Y2}^{[2]} \\ \mathbf{U}_{X3}^{[3]} \\ \mathbf{U}_{Y3}^{[3]} \end{bmatrix} \tag{3.28}$$

For Bar, B_3

$$\mathbf{F}^e = \mathbf{K}_3^e\mathbf{U}^e \Longrightarrow \begin{bmatrix} \mathbf{F}_{X1}^{[1]} \\ \mathbf{F}_{Y1}^{[1]} \\ \mathbf{F}_{X3}^{[3]} \\ \mathbf{F}_{Y3}^{[3]} \end{bmatrix}$$

$$= 36.44 \times 10^6 \begin{bmatrix} 0.5868 & 0.4924 & -0.2500 & -0.4924 \\ 0.4924 & 0.4132 & -0.4924 & -0.4132 \\ -0.5868 & -0.4924 & 0.5868 & 0.4924 \\ -0.4924 & -0.4132 & -0.4132 & 0.4132 \end{bmatrix} \begin{bmatrix} \mathbf{U}_{X1}^{[1]} \\ \mathbf{U}_{Y1}^{[1]} \\ \mathbf{U}_{X3}^{[3]} \\ \mathbf{U}_{Y3}^{[3]} \end{bmatrix} \tag{3.29}$$

Given a truss-based structure, the structural stiffness matrices, $\mathbf{K}^s$, is determined as the summation of all corresponding bar member stiffness matrices, $\mathbf{K}_i^s$. It is essential that the member stiffness matrix is represented in a global coordinate system with all nodal displacement and force terms of the structure represented in a holistic force-displacement matrix equation.

To assemble the global coordinate systems structural stiffness matrix, it is customary to do so by summing the individual member stiffness matrices according to Eq. 3.30,

$$\mathbf{K}^s = \sum_{i=1}^{n} \mathbf{K}_i^s \tag{3.30}$$

where $\mathbf{K}^s$ is the global coordinate system (structural) stiffness matrix, and $\mathbf{K}_i^s$ is the global coordinate system (member) stiffness matrix of bar i. You must distinguish $\mathbf{K}_i^s$ from the $\mathbf{K}_i^e$, where the later is the local coordinate system member stiffness matrix (see Sect. 3.4.3). According to Eq. 3.30, the summation of the individual (global coordinate) member stiffness matrices ranges from bar 1 to bar n for an n-bar structure.

Equation 3.30 demands that the left sign of the equation is equal to the right hand side of the equation. For the equality condition to be established, the (local coordinate system) member stiffness matrix originally developed in Eq. 3.25 as $\mathbf{K}^e$ has to be represented in global coordinate system as $\mathbf{K}_i^s$. To do this, we have to ensure that $\mathbf{K}_i^e$ is expanded so that all the nodes associated with the structure have terms represented in $\mathbf{K}_i^s$.

To apply Eq. 3.30 for the structure of Fig. 3.19, all the force-displacement matrix equations (3.27, 3.28, and 3.29) should have the nodal forces and displacements written as a comprehensive list of all applicable forces and displacements on the three nodes – $N_1, N_2,$ and N_3. Unfortunately, these equations are written only in terms of the applicable nodal forces and displacement for a given bar. For example, nodes N_1 and N_2 appear only in bar, B_1, hence Eq. 3.27 contains only nodal forces and displacements for these two nodes. Notice that the rows and columns of bar B_1 relate to nodes N_1 and N_2, as shown in Fig. 3.22.

We have to expand the force-displacement equation of say Fig. 3.22, to include the terms that do not relate to the specific bar, but are part of the complete structure. This is important as equilibrium and compatibility criteria consider the structure holistically. Hence, a change in behaviour at one point should result in a corresponding change at another part of the structure.

Let us illustrate this using the force-displacement equation of bar B_1 of structure shown in Fig. 3.21. We will have to rewrite the above equations, but this time around replacing values in the columns with zero where nodes do not belong to the member of interest. This is illustrated for the three bars in Figs. 3.23, 3.24, and 3.25. The

$$\begin{matrix} \\ N_1 \\ \\ N_2 \end{matrix} \begin{bmatrix} F_{X1}^{[1]} \\ F_{Y1}^{[1]} \\ F_{X1}^{[2]} \\ F_{Y1}^{[2]} \end{bmatrix} = 9.8960 \times 10^6 \overset{\qquad N_1 \qquad\qquad N_2}{\begin{bmatrix} 1 & 0 & -1 & 0 \\ 0 & 0 & 0 & 0 \\ -1 & 0 & 1 & 0 \\ 0 & 0 & 0 & 0 \end{bmatrix}} \begin{bmatrix} U_{X1}^{[1]} \\ U_{Y1}^{[1]} \\ U_{X1}^{[2]} \\ U_{Y1}^{[2]} \end{bmatrix}$$

Fig. 3.22 An illustration of the separation of terms of the Force-Displacement $(F - u)$ equation for bar, B_1 into terms associated with nodes N_1 and N_2

$$\begin{matrix} \\ N_1 \\ \\ N_2 \\ \\ N_3 \end{matrix} \begin{bmatrix} F_{X1}^{[1]} \\ F_{Y1}^{[1]} \\ F_{X1}^{[2]} \\ F_{Y1}^{[2]} \\ F_{X1}^{[3]} \\ F_{Y1}^{[3]} \end{bmatrix} = 9.8960 \times 10^6 \overset{\qquad N_1 \qquad\quad N_2 \qquad\quad N_3}{\begin{bmatrix} 1 & 0 & -1 & 0 & 0 & 0 \\ 0 & 0 & 0 & 0 & 0 & 0 \\ -1 & 0 & 1 & 0 & 0 & 0 \\ 0 & 0 & 0 & 0 & 0 & 0 \\ 0 & 0 & 0 & 0 & 0 & 0 \\ 0 & 0 & 0 & 0 & 0 & 0 \end{bmatrix}} \begin{bmatrix} U_{X1}^{[1]} \\ U_{Y1}^{[1]} \\ U_{X1}^{[2]} \\ U_{Y1}^{[2]} \\ U_{X1}^{[3]} \\ U_{Y1}^{[3]} \end{bmatrix}$$

Fig. 3.23 Extension of $F - u$ equation of bar, B_1 to include zero-terms for node N_3

$$\begin{matrix} \\ N_1 \\ \\ N_2 \\ \\ N_3 \end{matrix} \begin{bmatrix} F_{X1}^{[1]} \\ F_{Y1}^{[1]} \\ F_{X1}^{[2]} \\ F_{Y1}^{[2]} \\ F_{X1}^{[3]} \\ F_{Y1}^{[3]} \end{bmatrix} = 22.04 \times 10^6 \overset{\quad N_1 \qquad\qquad N_2 \qquad\qquad\qquad N_3}{\begin{bmatrix} 0 & 0 & 0 & 0 & 0 & 0 \\ 0 & 0 & 0 & 0 & 0 & 0 \\ 0 & 0 & 0.2500 & -0.4330 & -0.2500 & 0.4330 \\ 0 & 0 & -0.4330 & 0.7500 & 0.4330 & -0.7500 \\ 0 & 0 & -0.2500 & 0.4330 & 0.2500 & -0.4330 \\ 0 & 0 & 0.4330 & -0.7500 & -0.7500 & 0.7500 \end{bmatrix}} \begin{bmatrix} U_{X1}^{[1]} \\ U_{Y1}^{[1]} \\ U_{X1}^{[2]} \\ U_{Y1}^{[2]} \\ U_{X1}^{[3]} \\ U_{Y1}^{[3]} \end{bmatrix}$$

Fig. 3.24 Extension of $F - u$ equation of bar, B_2 to include zero-terms for node N_1

columns and rows that are highlighted are the extra zero-terms rows and columns introduced to complete the three-nodal structure.

As a consequence, the global stiffness matrices become 6×6 matrices. Similarly, a structure with n-nodes will have a $2n \times 2n$ structural stiffness matrix. The increasing discretization of a structure into increasing node numbers will lead to

$$\begin{bmatrix} F_{X1}^{[1]} \\ F_{Y1}^{[1]} \\ F_{X1}^{[2]} \\ F_{Y1}^{[2]} \\ F_{X1}^{[3]} \\ F_{Y1}^{[3]} \end{bmatrix} = 36.44\times10^{6} \begin{bmatrix} 0.5868 & 0.4924 & 0 & 0 & -0.2500 & -0.4924 \\ 0.4924 & 0.4132 & 0 & 0 & -0.4924 & -0.4132 \\ 0 & 0 & 0 & 0 & 0 & 0 \\ 0 & 0 & 0 & 0 & 0 & 0 \\ -0.5868 & -0.4924 & 0 & 0 & 0.5868 & 0.4924 \\ -0.4924 & -0.4132 & 0 & 0 & -0.4132 & 0.4132 \end{bmatrix} \begin{bmatrix} U_{X1}^{[1]} \\ U_{Y1}^{[1]} \\ U_{X1}^{[2]} \\ U_{Y1}^{[2]} \\ U_{X1}^{[3]} \\ U_{Y1}^{[3]} \end{bmatrix}$$

Fig. 3.25 Extension of $F-u$ equation of bar, B_3 to include zero-terms for node N_2

a larger stiffness matrix, hence requiring more computational memory to store and manipulate it. This is why highly discretized/meshed structures tend to take longer to solve as the computational resources are in greater demand for obtaining a solution to such problems.

The structural stiffness matrix for all three members is calculated by summing up the individual member stiffness matrices (expanded to include non-applicable nodal terms). For the test case investigated in Figs. 3.23, 3.24, and 3.25, the resultant global coordinate structural stiffness matrix for the three-node triangular planar truss structure becomes:

$$\mathbf{K}^s = \mathbf{K}_1^s + \mathbf{K}_2^s + \mathbf{K}_3^s$$

$$\mathbf{K}^s = \begin{bmatrix} 1.2024 & 0.1809 & -0.9896 & 0 & -0.2128 & -0.1809 \\ 0.1809 & 0.1537 & 0 & 0 & -0.1809 & -0.1537 \\ -0.9896 & 0 & 1.0459 & -0.0957 & 0.0563 & 0.0957 \\ 0 & 0 & -0.0957 & 0.1627 & 0.0957 & -0.1627 \\ -0.2128 & -0.1809 & -0.0563 & 0.0957 & 0.2691 & 0.0852 \\ -0.1809 & -0.1537 & 0.0957 & -0.1627 & 0.0852 & 0.3165 \end{bmatrix} \times 10^8\,\mathrm{Nm}^{-1} \tag{3.31}$$

This $\mathbf{K}^s$ matrix is a structural parameter needed for every design involving this three-node truss. We will use a another example to illustrate this process more.

Example 3.5 A simple right-angled planar triangular truss has all its members made from steel of diameter, $d = 2.5\,\mathrm{mm}$. The cartesian global coordinate nodal positions of ends of the trusses are specified in Fig. 3.26. Assume that the Young's Modulus of steel, $E_s = 210\,\mathrm{GPa}$ and all coordinate positions are in units of meters.

(a) Determine the global coordinate member stiffness matrices of the bars.
(b) Determine the structural stiffness matrix of the truss.

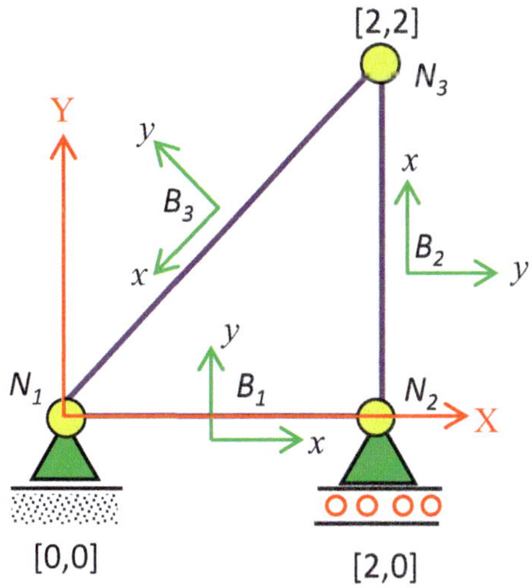

Fig. 3.26 A right-angled triangular planar truss, with all bars made of the same material and same bar diameters

Solution

Calculation of geometric and material parameters

Area: $$A = \frac{\pi d^2}{4} \quad \Rightarrow \quad A = \frac{\pi \times 0.0025^2}{4} = 4.9087 \times 10^{-6}\,\mathrm{m}^2$$

Lengths: Based on the node numbers of structure of Fig. 3.26:

$$[x_1, y_1] = [0, 0] \qquad [x_2, y_2] = [2, 0] \qquad [x_3, y_3] = [2, 2]$$

$$L_1 = \sqrt{(x_2 - x_1)^2 + (y_2 - y_1)^2} \quad \Rightarrow L_1 = \sqrt{(2-0)^2 + (0-0)^2}$$
$$\Rightarrow L_1 = 2\,\mathrm{m}$$

$$L_2 = \sqrt{(x_3 - x_2)^2 + (y_3 - y_2)^2} \quad \Rightarrow L_2 = \sqrt{(2-2)^2 + (2-0)^2}$$
$$\Rightarrow L_2 = 2\,\mathrm{m}$$

$$L_3 = \sqrt{(x_3 - x_1)^2 + (y_3 - y_1)^2} \quad \Rightarrow L_3 = \sqrt{(2-0)^2 + (2-0)^2}$$
$$\Rightarrow L_3 = 2.8284\,\mathrm{m}$$

Consider Fig. 3.27 which is a disassembly of the structure of Fig. 3.26. The angles of inclination of the constituent bars that make up the structure are calculated, with respect to the global coordinate system, as follows:

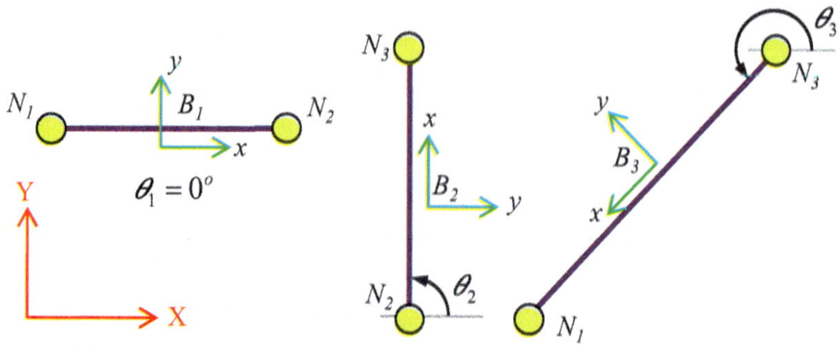

Fig. 3.27 A disassembled representation of structure of Fig. 3.26 showing angles of inclination of the constituent bars

Angles:

$$\theta_1 = 0° \qquad \theta_2 = 90° \quad \text{and}$$

$$\theta_2 = 180° + \tan^{-1}\left(\frac{y_3 - y_1}{x_3 - x_1}\right) \implies \theta_3 = 180° + \tan^{-1}\left(\frac{2-0}{2-2}\right) = 225°$$

Stiffness constants:

$$k_1 = k_2 = \frac{E_s A}{L_1} = \frac{210 \times 10^9 * 4.9087 \times 10^{-6}}{2} = 5.1541 \times 10^5\,\text{Nm}^{-1}$$

$$k_3 = \frac{E_s A}{L_3} = \frac{210 \times 10^9 * 4.9087 \times 10^{-6}}{2.8284} = 3.6446 \times 10^5\,\text{Nm}^{-1}$$

Calculation of member stiffness matrix in local coordinates

Using the *transformed global coordinates member stiffness matrix*, $\mathbf{K}^e$, shown in Eq. 3.26 obtained above, we can find the individual member stiffness matrices for the three bars as follows (where $\mathbf{T}'_{d_i}$ is the displacement transformation matrix for bar B_i for $i = 1, 2, 3$):

$$\mathbf{K}_1^e = \mathbf{T}'_{d_1} k_1 \mathbf{T}_d \implies \mathbf{K}_1^e = 5.1541 \times 10^5 \begin{bmatrix} 1 & 0 & -1 & 0 \\ 0 & 0 & 0 & 0 \\ -1 & 0 & 1 & 0 \\ 0 & 0 & 0 & 0 \end{bmatrix} \text{Nm}^{-1}$$

$$\mathbf{K}_2^e = \mathbf{T}'_{d_2} k_2 \mathbf{T}_d \quad \longrightarrow \quad \mathbf{K}_2^e = 5.1541 \times 10^5 \begin{bmatrix} 0 & 0 & 0 & 0 \\ 0 & 1 & 0 & -1 \\ 0 & 0 & 0 & 0 \\ 0 & -1 & 0 & 1 \end{bmatrix} \text{Nm}^{-1}$$

$$\mathbf{K}_3^e = \mathbf{T}'_{d_3} k_3 \mathbf{T}_d \Longrightarrow$$

$$\mathbf{K}_3^e = 3.6446 \times 10^5 \begin{bmatrix} 0.5000 & 0.5000 & -0.5000 & -0.5000 \\ 0.5000 & 0.5000 & -0.5000 & -0.5000 \\ -0.5000 & -0.5000 & 0.5000 & 0.5000 \\ -0.5000 & -0.5000 & 0.5000 & 0.5000 \end{bmatrix} \text{Nm}^{-1}$$

Determination of member stiffness matrix in global coordinates

The above member stiffness matrices were specified for the only applicable nodal coordinates per given bar, i.e. in all cases only two nodes out of three are reported (in the local coordinate system). We have to now expand them to include those nodes that are not necessarily attached to the given member. This action will convert the member stiffness matrices from the local coordinate system representation to a global coordinate system representation.

For Bar, B_1: $$\mathbf{K}_1^s = 5.1541 \times 10^5 \begin{bmatrix} 1 & 0 & -1 & 0 & 0 & 0 \\ 0 & 0 & 0 & 0 & 0 & 0 \\ -1 & 0 & 1 & 0 & 0 & 0 \\ 0 & 0 & 0 & 0 & 0 & 0 \\ 0 & 0 & 0 & 0 & 0 & 0 \\ 0 & 0 & 0 & 0 & 0 & 0 \end{bmatrix}$$

For Bar, B_2: $$\mathbf{K}_2^s = 5.1541 \times 10^5 \begin{bmatrix} 0 & 0 & 0 & 0 & 0 & 0 \\ 0 & 0 & 0 & 0 & 0 & 0 \\ 0 & 0 & 0 & 0 & 0 & 0 \\ 0 & 0 & 0 & 1 & 0 & -1 \\ 0 & 0 & 0 & 0 & 0 & 0 \\ 0 & 0 & 0 & -1 & 0 & 1 \end{bmatrix}$$

For Bar, B_3: $$\mathbf{K}_3^s = 3.6446 \times 10^5 \begin{bmatrix} 0.5000 & 0.5000 & 0 & 0 & -0.5000 & -0.5000 \\ 0.5000 & 0.5000 & 0 & 0 & -0.5000 & -0.5000 \\ 0 & 0 & 0 & 0 & 0 & 0 \\ 0 & 0 & 0 & 0 & 0 & 0 \\ -0.5000 & -0.5000 & 0 & 0 & 0.5000 & 0.5000 \\ -0.5000 & -0.5000 & 0 & 0 & 0.5000 & 0.5000 \end{bmatrix}$$

Assembly of the global structural stiffness matrix

The *global structural stiffness matrix* of the structure is the summation of all three (global) member stiffness matrices of the bars that make up the structure which is given thus:

$$\mathbf{K}^s = \mathbf{K}_1^s + \mathbf{K}_2^s + \mathbf{K}_3^s$$

$$\mathbf{K}^s = \begin{bmatrix} 1.3953 & 0.3645 & -1.0308 & 0 & -0.3645 & -0.3645 \\ 0.3645 & 0.3645 & 0 & 0 & -0.3645 & -0.3645 \\ -1.0308 & 0 & 1.0308 & 0 & 0 & 0 \\ 0 & 0 & 0 & 1.0308 & 0 & -1.0308 \\ -0.3645 & -0.3645 & 0 & 0 & 0.3645 & 0.3645 \\ -0.3645 & -0.3645 & 0 & -1.0308 & 0.3645 & 1.3953 \end{bmatrix} \times 10^7\ \text{N/m} \tag{3.32}$$

3.5 Determination of Global Nodal Reaction Forces and Displacements

Up until now, the solution process has neglected the effect of boundary conditions i.e. boundary nodal forces and displacements on the structure. The focus had been on simply determining the structural stiffness matrix (which is a property) for the structure and will remain constant irrespective of the nature of the imposed boundary conditions.

However, the structural engineer's objective is always to determine the global reaction forces and displacements of the structure. This knowledge will aid the design process as well as help determine the stresses that will be sustained by the structural members upon loading. It will also help determine the load limit for such a structure, and where necessary, the required modification to sustain more loads. This brings us full circle back to the problem statement of most finite element problems which is: *how do we determine the global reaction forces and displacements?*

The improvement we have in this analysis procedure is that we now have a global stiffness matrix, $\mathbf{K}^s$ as established in Sect. 3.4. The $\mathbf{K}^s$ obtained gets us closer to the solution than ever before. In the following, we are going to establish matrix manipulations that will help us determine the nodal forces and displacements for any structure solved using the *direct stiffness matrix* (**DSM**).

3.5.1 Expressions for Nodal Forces and Displacements

For a typical three-node plane structure as shown in Fig. 3.18, the force-displacement equation (in vector format) in the global cartesian coordinate system representation for bar members B_1, B_2 and B_3 is given as:

$$\mathbf{F}^s = \mathbf{K}^s\mathbf{U}^s \quad \Longrightarrow \quad \begin{bmatrix} \mathbf{F}_{X1}^{[1]} \\ \mathbf{F}_{Y1}^{[1]} \\ \mathbf{F}_{X2}^{[2]} \\ \mathbf{F}_{Y2}^{[2]} \\ \mathbf{F}_{X3}^{[3]} \\ \mathbf{F}_{Y3}^{[3]} \end{bmatrix} = \begin{bmatrix} K_{11}^s & K_{12}^s & K_{13}^s & K_{14}^s & K_{15}^s & K_{16}^s \\ K_{21}^s & K_{22}^s & K_{23}^s & K_{24}^s & K_{25}^s & K_{26}^s \\ K_{31}^s & K_{32}^s & K_{33}^s & K_{34}^s & K_{35}^s & K_{36}^s \\ K_{41}^s & K_{42}^s & K_{43}^s & K_{44}^s & K_{45}^s & K_{46}^s \\ K_{51}^s & K_{52}^s & K_{53}^s & K_{54}^s & K_{55}^s & K_{56}^s \\ K_{61}^s & K_{62}^s & K_{63}^s & K_{64}^s & K_{65}^s & K_{66}^s \end{bmatrix} \begin{bmatrix} \mathbf{U}_{X1}^{[1]} \\ \mathbf{U}_{Y1}^{[1]} \\ \mathbf{U}_{X2}^{[2]} \\ \mathbf{U}_{Y2}^{[2]} \\ \mathbf{U}_{X3}^{[3]} \\ \mathbf{U}_{Y3}^{[3]} \end{bmatrix} \tag{3.33}$$

where $\mathbf{F}^s$ is a column vector of all nodal forces in the global coordinate system, $\mathbf{K}^s$ is a global 6×6 stiffness matrix of the structure and, finally, $\mathbf{U}^s$ is a column vector of nodal displacements (in global coordinates) of the structure.

The structure of Fig. 3.18 is subjected to an externally applied force, **P**, at node N_3. It is important that we determine the nodal displacements of not only node N_3 but also nodes N_1 and N_2 in order to understand the deformation of the structure under the effect of **P**. To do so, we have to determine the nodal displacement column vector, $\mathbf{U}^s$, of Eq. 3.33, which defines the planar motion of the three nodes. The solution of Eq. 3.33 involves obtaining the *inverse of the global stiffness matrix*, written as $[\mathbf{K}^s]^{-1}$. The new equation that determines $\mathbf{U}^s$ is given thus:

$$\mathbf{U}^s = [\mathbf{K}^s]^{-1}\mathbf{F}^s \tag{3.34}$$

The matrix of $\mathbf{F}^s$ would be given for a given problem hence the right hand side of the equation would be known. The global structure stiffness matrix, however, would have to be inverted and this challenge will be addressed in the next section.

3.5.2 The Challenge of Inverting a Global Stiffness Matrix

Equation 3.34 presents yet another challenging question and that is: *how do we determine the inverse of a global stiffness matrix?* The problem with most stiffness matrices obtained in the format shown above (by assembly of member stiffness matrix) is that they are always ***singular*** even though they may be ***square***. These matrices are usually *ill-conditioned* and the severity of their ill-conditioning is demonstrated by considering their ***condition number***. This is a measure of the effect of changes in output value of a function with a slight change in its input values. It gives an indication of how sensitive model predictions can be due to slight errors in input to such models.

For example, consider the displacement-force linear equation: $u = k^{-1}f$, specified in a local coordinate system representation. A large condition number for k will imply that any rounding-off approximates from the exact solution for the displacement, u, will significantly affect the forces, f, acting on the structure.

The solution in a computer of the linear system of equations that represents a truss involves providing approximate floating values to the terms of matrix, **K**. A high condition number for K implies that this will significantly influence the reliability of outputs that can be derived from the numerical procedure. The condition number is a property of the given problem under investigation. As result, conclusions about the problem can be drawn simply by understanding the condition number. In fact, Kannan et al. [4] have published a methodology for detecting the ill-conditioning of a structural FEM model by simply detecting its condition number and also identifying which parts of the model the ill-conditioning arises from.

A global structural stiffness matrix, $\mathbf{K}^s$, is nearly always ***symmetric***, arising essentially from how it is assembled as illustrated previously. However, it is common for it to be ill-conditioned. $\mathbf{K}^s$ is ill-conditioned when:

(a) There is not sufficient connectivity of a few set of nodal pairs with the rest of the model. This is often manifest as a *'floating'* node without any structural support that links it to other parts of the structure. It may also manifest as *over-* or *under-constrained* nodal positions which allows for minimal or excessive deformation of any affected nodal position.
(b) Some elements/members of the model have stiffnesses that are disproportionate to the rest of the model. Such elements might be very stiff in comparison with neighbouring elements. The force and displacement continuity between such members and the rest of the model is broken hence causing ill-conditioning of the stiffness matrix.

Given a stiffness matrix, $\mathbf{K}^s$, the inverse of the matrix can be defined in terms of its ***adjugate*** and its *determinant* as shown in Eq. 3.35.

$$[\mathbf{K}^s]^{-1} = \frac{\mathrm{adj}(\mathbf{K}^s)}{\det(\mathbf{K}^s)} \tag{3.35}$$

where $\mathrm{adj}(\mathbf{K}^s)$ is the adjugate and $\det(\mathbf{K}^s)$ is the determinant of the global stiffness matrix, $\mathbf{K}^s$.

It is a requirement that for any matrix to be *invertible* i.e. its inverse exists, such a matrix must be *square* and *non-singular* or *non-degenerate*. *Singular matrices* often have a zero determinant and are inherently *non-invertible*. Thus, if such matrices are inverted, they tend to '*blow up*' which within a numerical scheme implies non-convergence to a desired numerical solution. Consequently, although we have determined a stiffness matrix, it is plagued by this inherent problem of inversion and hence strategies to obtain their inversions have to be found.

3.5.3 The Requirement for Suppressing Rigid Body Motions of Structures

The singularity of the traditional global stiffness matrix results from the fact that the matrix includes within it, columns and rows for boundary forces and displacements, that are *non-zero*. If this format is kept the way it is, it leads to a *floating* structure which cannot be ever subjected to a load. The term for describing this is *rigid body modes*, and its defined as the free translation and rotation of a body without any significant internal deformation of the structure. It is important that structures to be analyzed using the DSM are imposed with some representative, realistic boundary conditions that will cause at least one of the nodes to experience zero displacement and forces, whilst other parts of the model experience internal deformation. This is termed *suppressing rigid body motions*.

In practice, the *suppression of rigid body motions* is already established with most structures since the presence of supports on such structure will enforce this suppression of rigid body motions. However, the assembly process for the global structural stiffness matrix has neglected any effects arising from the boundary conditions. As a result, the information from inherent support reactions of real structures has not been reflected in $\mathbf{K}^s$.

Let us now illustrate the rigid body motions suppression using the examples shown in Fig. 3.28. In particular, the body shown in Fig. 3.28b, is simply supported with a fixed left support and a roller right support. This structure already has three zero displacements values namely: $u_{x1} = u_{y1} = 0$ at node N_1 and $u_{y2} = 0$ at node N_2. With respect to Fig. 3.28c, suppression of rigid body motions is achieved again

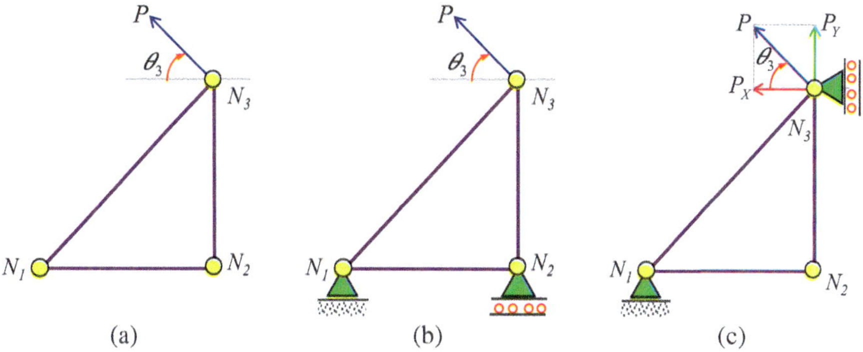

Fig. 3.28 Examples of: (**a**) floating structure, (**b**) simply-supported structure with load P, and (**c**) another fixed structure with load, P, decomposed into P_X and P_Y

by the presence of three zerodisplacements values namely: $u_{x1} = u_{y1} = 0$ at node N_1 and $u_{x3} = 0$ at node N_3. In terms of nodal forces for structure of Fig. 3.28b, $f_{x2} = 0$ at node N_2 whilst for structure of Fig. 3.28c, $f_{x3} = P_X$ and $f_{y3} = P_Y$ at node N_3.

Having established the requirement of suppression of rigid body motions for achieving the inversion of a structural stiffness matrix, the information of nodal displacement and forces of the nodal supports will have to be transferred into Eq. 3.34. In effect, the degrees of freedom that show zero displacements (supported nodes) will be replaced with zero in Eq. 3.34. If we want to manually solve the overall stiffness equations (using the force-displacement equation of Eq. 3.34), then it is simplest to account for support conditions by removing from the said equation any of the rows/columns with known zero joint displacements. This is illustrated in more detail using the following two examples.

Example 3.6 Using the same truss arrangement of Example 3.5 given in Fig. 3.26, let us consider it subject to a force, $\mathbf{F} = 2\mathbf{i} + 3\mathbf{j}$ kN acting at node N_3, as shown in Fig. 3.29. Note that distances are in units of metre.

Determine the unknown nodal displacements and forces of the whole structure.

Solution

The structure here is the same as Fig. 3.26, hence the same global stiffness matrix derived in Example 3.5 and reported as Eq. 3.36 will be used here. Recall:

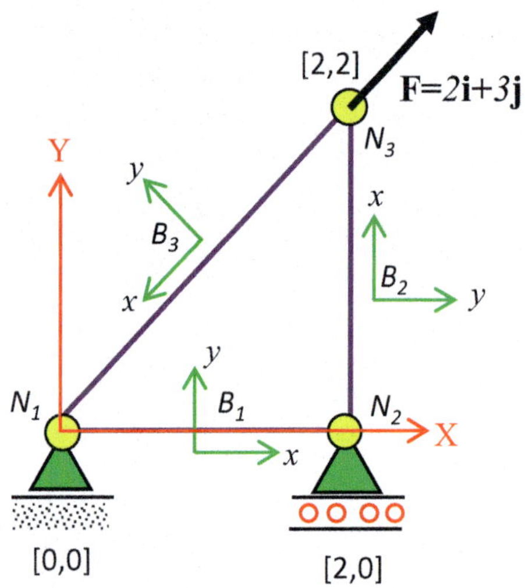

Fig. 3.29 A right-angled planar triangular truss subjected to force, **F**

$$\mathbf{K}^s = \begin{bmatrix} 1.3953 & 0.3645 & -1.0308 & 0 & -0.3645 & -0.3645 \\ 0.3645 & 0.3645 & 0 & 0 & -0.3645 & -0.3645 \\ -1.0308 & 0 & 1.0308 & 0 & 0 & 0 \\ 0 & 0 & 0 & 1.0308 & 0 & -1.0308 \\ -0.3645 & -0.3645 & 0 & 0 & 0.3645 & 0.3645 \\ -0.3645 & -0.3645 & 0 & -1.0308 & 0.3645 & 1.3953 \end{bmatrix} \times 10^7 \, \text{N/m} \quad (3.36)$$

We will illustrate in the following steps the determination of nodal forces and displacements of the structure of Fig. 3.29 by exploiting the *suppression of rigid body motions* condition of the structure. This will allow us to obtain the inverse of the reduced structural stiffness matrix.

Step 1: Boundary Conditions (BC) of the problem

BC_1 at node N_1: $u_{X1} = 0$ $u_{Y1} = 0$ $f_{X1} \neq 0$ $f_{Y1} \neq 0$

BC_2 at node N_2: $u_{X2} \neq 0$ $u_{Y2} = 0$ $f_{X2} = 0$ $f_{Y2} \neq 0$

BC_3 at node N_3: $u_{X3} \neq 0$ $u_{Y3} \neq 0$ $f_{X3} = 2000\,\text{N}$ $f_{Y3} = 3000\,\text{N}$

Step 2: Substitute BC values into Force-displacement equation

Based on the above boundary conditions and the global stiffness matrix of Eq. 3.36, the next solution step is to create a modified Force-displacement equation. This is done by incorporating these pieces of information into Eq. 3.33 (dropping superscript references to bar/member numbers). The resulting equation becomes:

$$\begin{bmatrix} f_{X1} \\ f_{Y1} \\ 0 \\ f_{Y2} \\ 2000 \\ 3000 \end{bmatrix} = \begin{bmatrix} 1.3953 & 0.3645 & -1.0308 & 0 & -0.3645 & -0.3645 \\ 0.3645 & 0.3645 & 0 & 0 & -0.3645 & -0.3645 \\ -1.0308 & 0 & 1.0308 & 0 & 0 & 0 \\ 0 & 0 & 0 & 1.0308 & 0 & -1.0308 \\ -0.3645 & -0.3645 & 0 & 0 & 0.3645 & 0.3645 \\ -0.3645 & -0.3645 & 0 & -1.0308 & 0.3645 & 1.3953 \end{bmatrix} \begin{bmatrix} 0 \\ 0 \\ u_{X1} \\ 0 \\ u_{X3} \\ u_{Y3} \end{bmatrix} \times 10^7 \, \text{N}. \quad (3.37)$$

Step 3: Remove the rows and columns that relate to zero nodal displacements

The zero nodal displacements will have to be removed to simplify the above equation for the solution to be possible and enforce *suppression of rigid body motions* of the structure. This will mean all rows and columns that relate to the zero displacements namely x- and y-axes of node N_1 and y-axis of node N_2. This is illustrated as shown in Fig. 3.30.

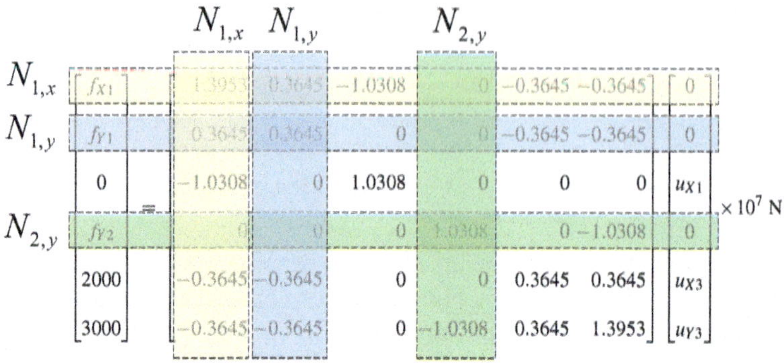

Fig. 3.30 An illustration of the isolation process required to remove effect of zero nodal displacement terms in the force-displacement equation

Step 4: Isolate the modified force-displacement equation

After the isolation process of Fig. 3.30, the resulting modified force-displacement equation becomes:

$$\begin{bmatrix} 0 \\ 2000 \\ 3000 \end{bmatrix} = \begin{bmatrix} 1.0308 & 0 & 0 \\ 0 & 0.3645 & 0.3645 \\ 0 & 0.3645 & 1.3953 \end{bmatrix} \begin{bmatrix} u_{X2} \\ u_{X3} \\ u_{Y3} \end{bmatrix} \times 10^7\,\mathrm{N}$$
$$\Longrightarrow \begin{bmatrix} 0 \\ 2000 \\ 3000 \end{bmatrix} = \mathbf{K}^s_{**} \begin{bmatrix} u_{X2} \\ u_{X3} \\ u_{Y3} \end{bmatrix} \tag{3.38}$$

The above equation has resulted in a much smaller global stiffness matrix, $\mathbf{K}^s_{**}$, which is now a 3 × 3 matrix having initially been a 6 × 6 matrix. It is called a *reduced stiffness matrix*. This matrix will be the new stiffness matrix to be used in determining the relevant nodal displacements namely: u_{x2}, u_{x3}, and u_{y3}.

Step 5: Obtain the reduced stiffness matrix displacement-force equation

To solve for the remaining displacements, we will have to re-write the force-displacement equation into a displacement-force equation. To do this, we need to compute the inverse of the reduced stiffness matrix. The following step describes an approach for determining the inverse of the reduced global stiffness matrix.

$$\begin{bmatrix} 0 \\ 2000 \\ 3000 \end{bmatrix} = \mathbf{K}^s_{**} \begin{bmatrix} u_{X2} \\ u_{X3} \\ u_{Y3} \end{bmatrix} \quad \Longrightarrow \quad \begin{bmatrix} u_{X2} \\ u_{X3} \\ u_{Y3} \end{bmatrix} = [\mathbf{K}^s_{**}]^{-1} \begin{bmatrix} 0 \\ 2000 \\ 3000 \end{bmatrix} \tag{3.39}$$

Step 6: Compute the inverse of the reduced stiffness matrix

To obtain the inverse of $\mathbf{K}^s_{**}$, you need to check if this matrix is invertible. This implies that it should be non-singular and square. Undergraduate mathematics textbooks can guide the reader through the process of obtaining an inverse of a matrix. One such approach is the *Gauss elimination method.* The method does not require that $\mathbf{K}^s_{**}$ is always invertible.

The key steps for the method are as follows:

(a) Create a matrix, **A** consisting of the coefficients of the unknown variables in the linear equation of $\mathbf{Ax} = \mathbf{b}$.
(b) Create another matrix consisting of the *right hand side* values of the linear equations, where this matrix consists of values '**b**' in $\mathbf{Ax} = \mathbf{b}$.
(c) Create an ***augmented matrix*** which consists of **A** and **b** values.
(d) Based on the augmented matrix, start applying the *Gauss-Jordan elimination method*.

For the problem under consideration here, the system of linear equations becomes:

$$\mathbf{Ax} = \mathbf{b} \quad \Longrightarrow \quad \mathbf{K}^s_{**}\begin{bmatrix} u_{X2} \\ u_{X3} \\ u_{Y3} \end{bmatrix} = \begin{bmatrix} 0 \\ 2000 \\ 3000 \end{bmatrix} \tag{3.40}$$

which means that:

$$\mathbf{A} = \mathbf{K}^s_{**} = \begin{bmatrix} 1.0308 & 0 & 0 \\ 0 & 0.3645 & 0.3645 \\ 0 & 0.3645 & 1.3953 \end{bmatrix} \times 10^7\,\text{N} \quad \text{and} \quad \mathbf{b} = \begin{bmatrix} 0 \\ 2000 \\ 3000 \end{bmatrix} \tag{3.41}$$

Let us define the ***augmented matrix***, **Q**, of the above linear system as:

$$\mathbf{Q} = [\mathbf{A}|\mathbf{b}] \quad \Longrightarrow \quad \mathbf{Q} = \left[\begin{array}{ccc|c} 1.0308 & 0 & 0 & 0 \\ 0 & 0.3645 & 0.3645 & 0.0002000 \\ 0 & 0.3645 & 1.3953 & 0.0003000 \end{array}\right] \times 10^7\,\text{N} \tag{3.42}$$

The augmented matrix, **Q** will have to be solved using the *Gauss-Jordan elimination method*. For our purpose here, we will use a MATLAB™ command that helps determine the final format of the augmented matrix after the *Gaussian elimination*. The command is `rref`.

The MATLAB™ command **rref** is short for *reduced row echelon form*. The MATLAB™ documentation states that if the command: $R = \texttt{rref(K)}$ is executed, it "*produces a reduced row echelon form of the matrix, K, using Gauss-Jordan elimination with partial pivoting.*"

The resulting reduced row echelon matrix essentially is an augmented version of an *identity matrix*, **I**, and the column matrix of the *desired solution*. Therefore, the last column from the resulting matrix following the use of the **rref** MATLAB™ command gives us the solution of the matrix. Hence, to continue with the solution, use of the **rref** command gives:

$$\mathbf{R} = \texttt{rref}(\mathbf{Q}) \quad \Longrightarrow \quad \mathbf{R} = \begin{bmatrix} 1 & 0 & 0 & 0 \\ 0 & 1 & 0 & 0.4518 \\ 0 & 0 & 1 & 0.0970 \end{bmatrix} \tag{3.43}$$

Using the last column as our results column, we obtain therefore the expected result thus:

$$\begin{bmatrix} u_{X2} \\ u_{X3} \\ u_{Y3} \end{bmatrix} = \begin{bmatrix} 0 \\ 0.4518 \\ 0.0970 \end{bmatrix} \Longrightarrow \begin{aligned} u_{X2} &= 0 \\ u_{X3} &= 0.452 \times 10^{-3}\,\text{m} = 450\,\text{mm} \\ u_{Y3} &= 0.097 \times 10^{-3}\,\text{m} = 97\,\text{mm} \end{aligned} \tag{3.44}$$

Step 7: Calculation of Nodal Forces

To calculate the nodal forces, we will need to recall the global stiffness matrix and replace the unknown nodal displacements with the values of Eq. 3.44.

The resulting force-displacement equation becomes:

$$\begin{bmatrix} f_{X1} \\ f_{Y1} \\ 0 \\ f_{Y2} \\ 2000 \\ 3000 \end{bmatrix} = 10^7 \begin{bmatrix} 1.3953 & 0.3645 & -1.0308 & 0 & -0.3645 & -0.3645 \\ 0.3645 & 0.3645 & 0 & 0 & -0.3645 & -0.3645 \\ -1.0308 & 0 & 1.0308 & 0 & 0 & 0 \\ 0 & 0 & 0 & 1.0308 & 0 & -1.0308 \\ -0.3645 & -0.3645 & 0 & 0 & 0.3645 & 0.3645 \\ -0.3645 & -0.3645 & 0 & -1.0308 & 0.3645 & 1.3953 \end{bmatrix} \begin{bmatrix} 0 \\ 0 \\ 0 \\ 0 \\ 0.451 \\ 0.097 \end{bmatrix} \tag{3.45}$$

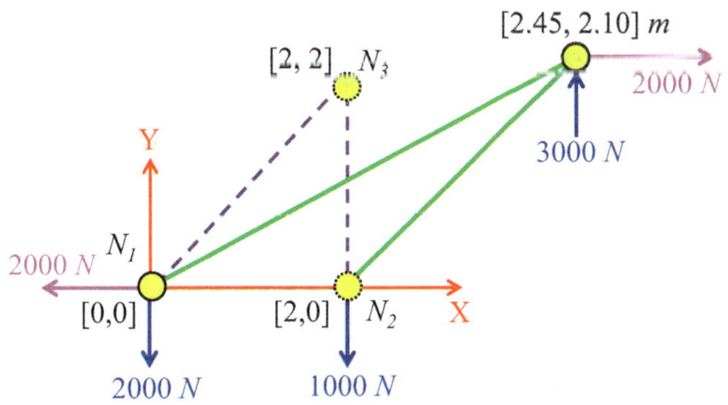

Fig. 3.31 A graphical representation of the deformed profile of the triangular truss of Example 3.6. Solid lines indicate the deformed profile and broken lines are the undeformed profile. *The sketch is not drawn to scale*

Expanding the above matrix gives column matrices for the nodal forces and displacements thus:

$$\begin{bmatrix} f_{X1} \\ f_{Y1} \\ f_{X2} \\ f_{Y2} \\ f_{X3} \\ f_{Y3} \end{bmatrix} = \begin{bmatrix} -2000 \\ -2000 \\ 0 \\ 0 \\ 2000 \\ 3000 \end{bmatrix} \text{N} \qquad \begin{bmatrix} u_{X1} \\ u_{Y1} \\ u_{X2} \\ u_{Y2} \\ u_{X3} \\ u_{Y3} \end{bmatrix} = \begin{bmatrix} 0 \\ 0 \\ 0 \\ 0 \\ 0.451 \\ 0.097 \end{bmatrix} \text{m} \quad (3.46)$$

The sketch for the deformed triangular truss/body in comparison with the original (undeformed structure) is given in Fig. 3.31.

Example 3.7 Using the Direct Stiffness Method, determine all nodal displacements and forces for the truss shown in Fig. 3.32. It is subjected to an external force, $\mathbf{F} = -1500\mathbf{i} + 2500\mathbf{j}\,\text{N}$. Sketch the shape of the deformed structure. Assume all the bars are made of steel of diameter, $d = 3$ mm and Young's Modulus, $E_s = 210$ GPa.

Solution

<u>*Calculation of Geometric and Material Parameters*</u>

First we obtain the cross-sectional area of bars using the diameter, $d = 5$ m.

$$A = \frac{\pi d^2}{4} \qquad \Rightarrow \qquad A = \frac{\pi \times 0.003^2}{4} = 7.0686 \times 10^{-6}\,\text{m}^2$$

In order to determine the lengths of the bars, based on the node numbers of structure of Fig. 3.32, we identify the nodal coordinates as:

$$[x_1, y_1] = [0, 0] \qquad [x_2, y_2] = [4, 0] \qquad [x_3, y_3] = [2, 2]$$

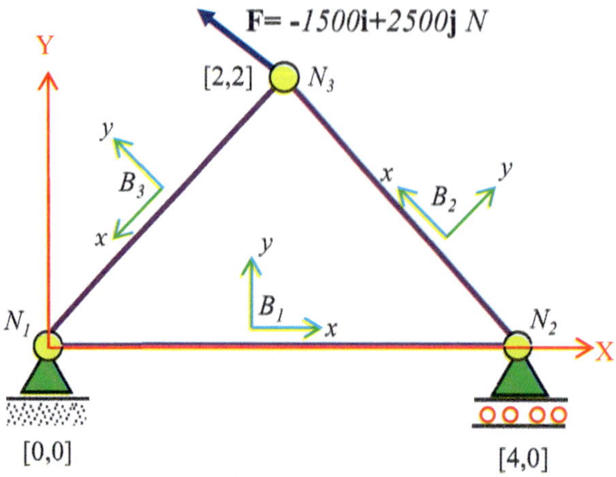

Fig. 3.32 A planar truss subjected to a force, **F**

The calculations for these lengths are:

$$L_1 = \sqrt{(x_2 - x_1)^2 + (y_2 - y_1)^2} \quad \Rightarrow L_1 = \sqrt{(4-0)^2 + (0-0)^2}$$
$$\Rightarrow L_1 = 4\,\text{m}$$

$$L_2 = \sqrt{(x_3 - x_2)^2 + (y_3 - y_2)^2} \quad \Rightarrow L_2 = \sqrt{(2-0)^2 + (4-2)^2}$$
$$\Rightarrow L_2 = 2.8284\,\text{m}$$

$$L_3 = \sqrt{(x_3 - x_1)^2 + (y_3 - y_1)^2} \quad \Rightarrow L_2 = \sqrt{(2-0)^2 + (2-0)^2}$$
$$\Rightarrow L_3 = 2.8284\,\text{m}$$

By observation of Fig. 3.32, we notice that the angles for the three bars are:

$$\theta_1 = 0°$$
$$\theta_2 = 180° - \tan^{-1}\left(\frac{y_3 - y_2}{x_3 - x_2}\right) \quad \Longrightarrow \theta_2 = 180° - \tan^{-1}\left(\frac{2-0}{4-2}\right) = 135°$$

$$\theta_3 = 180° - \tan^{-1}\left(\frac{y_3 - y_1}{x_3 - x_1}\right) \quad \Longrightarrow \theta_3 = 180° + \tan^{-1}\left(\frac{2-0}{2-0}\right) = 225°$$

Stiffness constants for the three bars can be calculated as shown below:

$$k_1 = \frac{E_s A}{L_1} = \frac{210 \times 10^9 \times 7.0686 \times 10^{-6}}{2} = 3.7110 \times 10^5\,\text{N/m}$$

$$k_2 = \frac{E_s A}{L_2} = \frac{210 \times 10^9 \times 7.0686 \times 10^{-6}}{2.8284} = 5.2482 \times 10^5 \text{ N/m}$$

$k_3 = k_2 = 5.2482 \times 10^5$ N/m (both bars are made of same geometry and material)

Calculation of member stiffness matrix (global coordinates)

Now we obtain $\mathbf{K}_i^s$ for the three bars where $i = 1, 2, 3$. Applying the transformation matrix as well as including the zero terms associated with nodes not directly attached to the bar of interest, we can prove that the following member stiffness matrices (in global coordinates) are the right ones for bars, B_1, B_2 and B_3.

$$\mathbf{K}_1^s = 7.4220 \times 10^5 \begin{bmatrix} 1 & 0 & -1 & 0 & 0 & 0 \\ 0 & 0 & 0 & 0 & 0 & 0 \\ -1 & 0 & 1 & 0 & 0 & 0 \\ 0 & 0 & 0 & 0 & 0 & 0 \\ 0 & 0 & 0 & 0 & 0 & 0 \\ 0 & 0 & 0 & 0 & 0 & 0 \end{bmatrix}$$

$$\mathbf{K}_2^s = 5.2482 \times 10^5 \begin{bmatrix} 0 & 0 & 0 & 0 & 0 & 0 \\ 0 & 0 & 0 & 0 & 0 & 0 \\ 0 & 0 & 0.5000 & -0.5000 & -0.5000 & 0.5000 \\ 0 & 0 & -0.5000 & 0.5000 & 0.5000 & -0.5000 \\ 0 & 0 & -0.5000 & 0.5000 & 0.5000 & -0.5000 \\ 0 & 0 & 0.5000 & -0.5000 & -0.5000 & 0.5000 \end{bmatrix}$$

$$\mathbf{K}_3^s = 5.2482 \times 10^5 \begin{bmatrix} 0.5000 & 0.5000 & 0 & 0 & -0.5000 & -0.5000 \\ 0.5000 & 0.5000 & 0 & 0 & -0.5000 & -0.5000 \\ 0 & 0 & 0 & 0 & 0 & 0 \\ 0 & 0 & 0 & 0 & 0 & 0 \\ -0.5000 & -0.5000 & 0 & 0 & 0.5000 & 0.5000 \\ -0.5000 & -0.5000 & 0 & 0 & 0.5000 & 0.5000 \end{bmatrix}$$

Calculation of Global Structural Stiffness matrix

The next step is to obtain the *global structural stiffness matrix*, $\mathbf{K}_i^s$ by taking the summation of the individual member stiffness matrices:

$$\mathbf{K}^s = \begin{bmatrix} 1.2670 & 0.5248 & -0.7422 & 0 & -0.5248 & -0.5248 \\ 0.5284 & 0.5248 & 0 & 0 & -0.5248 & -0.5248 \\ -0.7422 & 0 & 1.2670 & -0.5248 & -0.5248 & 0.5248 \\ 0 & 0 & -0.5248 & 0.5248 & 0.5248 & -0.5248 \\ -0.5248 & -0.5248 & -0.5248 & 0.5248 & 1.0496 & 0 \\ -0.5248 & -0.5248 & 0.5248 & -0.5248 & 0 & 1.0496 \end{bmatrix} \times 10^6\ \text{N/m}$$

Calculate the nodal displacements

- **Boundary conditions**: Here are the applicable boundary conditions for this structure.

$$\text{BC}_1 \text{ at node, } N_1{:}\ u_{X1} = 0 \quad u_{Y1} = 0 \quad f_{X1} \neq 0 \quad f_{Y1} \neq 0$$

$$\text{BC}_2 \text{ at node, } N_2{:}\ u_{X2} \neq 0 \quad u_{Y2} = 0 \quad f_{X2} = 0 \quad f_{Y2} \neq 0$$

$$\text{BC}_3 \text{ at node, } N_3{:}\ u_{X3} \neq 0 \quad u_{Y2} \neq 0 \quad f_{X3} = -1500\,\text{N} \quad f_{Y3} = 2500\,\text{N}$$

- **Modified Force-displacement equation** comprising the specified boundary conditions becomes:

$$\begin{bmatrix} f_{X1} \\ f_{Y1} \\ 0 \\ f_{Y2} \\ -1500 \\ 2500 \end{bmatrix} = \begin{bmatrix} 1.2670 & 0.5248 & -0.7422 & 0 & -0.5248 & -0.5248 \\ 0.5284 & 0.5248 & 0 & 0 & -0.5248 & -0.5248 \\ -0.7422 & 0 & 1.2670 & -0.5248 & -0.5248 & 0.5248 \\ 0 & 0 & -0.5248 & 0.5248 & 0.5248 & -0.5248 \\ -0.5248 & -0.5248 & -0.5248 & 0.5248 & 1.0496 & 0 \\ -0.5248 & -0.5248 & 0.5248 & -0.5248 & 0 & 1.0496 \end{bmatrix} \begin{bmatrix} 0 \\ 0 \\ u_{X1} \\ 0 \\ u_{X3} \\ u_{Y3} \end{bmatrix} \times 10^6\ \text{N} \tag{3.47}$$

- **Reduced global structure stiffness matrix:** After identifying and removing the rows and columns associated with zero displacements from Eq. 3.47, we obtain:

$$\mathbf{K}^s_{**} = \begin{bmatrix} 1.2670 & -0.5248 & 0.5248 \\ -0.5248 & 1.0496 & 0 \\ 0.5248 & 0 & 1.0496 \end{bmatrix} \times 10^6\ \text{N/m} \tag{3.48}$$

- **Augumented matrix:** Combining the $\mathbf{K}^s_{**}$ matrix with the force-matrix equation i.e. $\mathbf{b} = [0 \quad -0.001500 \quad 0.002500]' \times 10^6$ N, we obtain the augumented matrix of the problem:

$$\mathbf{Q} = [\mathbf{A}|\mathbf{b}] \quad \Longrightarrow \quad \mathbf{Q} = \begin{bmatrix} 1.2670 & -0.5248 & 0.5248 & | & 0 \\ -0.5248 & 1.0496 & 0 & | & -0.001500 \\ 0.5248 & 0 & 1.0496 & | & 0.002500 \end{bmatrix} \times 10^7\,\text{N} \tag{3.49}$$

- **Reduced row echolon form matrix, R**: Using the **`rref`** command in MATLAB, the **R** matrix becomes:

$$\mathbf{R} = \texttt{rref}(\mathbf{Q}) \quad \Longrightarrow \quad \mathbf{R} = \begin{bmatrix} 1 & 0 & 0 & -0.0027 \\ 0 & 1 & 0 & -0.0028 \\ 0 & 0 & 1 & 0.0037 \end{bmatrix} \tag{3.50}$$

- **Derived nodal displacements**: Based on the last column of the **R**-matrix above, we obtain the nodal displacements for the structure under consideration as:

$$\begin{bmatrix} u_{X2} \\ u_{X3} \\ u_{Y3} \end{bmatrix} = \begin{bmatrix} -0.0027 \\ -0.0028 \\ 0.0037 \end{bmatrix} \text{m}$$

$$\Longrightarrow \quad \begin{aligned} u_{X2} &= -0.0027 \times 10^{-3}\,\text{m} = -27\,\text{mm} \\ u_{X3} &= -0.0028 \times 10^{-3}\,\text{m} = -28\,\text{mm} \\ u_{Y3} &= 0.0037 \times 10^{-3}\,\text{m} = 37\,\text{mm} \end{aligned} \tag{3.51}$$

Calculation of Nodal forces

We now have a complete set of values for all the nodal displacements for the 3 nodes that make up the structure. Substituting these into Eq. 3.47, we obtain the following:

$$\begin{bmatrix} f_{X1} \\ f_{Y1} \\ 0 \\ f_{Y2} \\ -1500 \\ 2500 \end{bmatrix} = 10^6 \begin{bmatrix} 1.2670 & 0.5248 & -0.7422 & 0 & -0.5248 & -0.5248 \\ 0.5284 & 0.5248 & 0 & 0 & -0.5248 & -0.5248 \\ -0.7422 & 0 & 1.2670 & -0.5248 & -0.5248 & 0.5248 \\ 0 & 0 & -0.5248 & 0.5248 & 0.5248 & -0.5248 \\ -0.5248 & -0.5248 & -0.5248 & 0.5248 & 1.0496 & 0 \\ -0.5248 & -0.5248 & 0.5248 & -0.5248 & 0 & 1.0496 \end{bmatrix} \begin{bmatrix} 0 \\ 0 \\ -0.027 \\ 0 \\ -0.028 \\ 0.037 \end{bmatrix}. \tag{3.52}$$

Fig. 3.33 A graphical representation of the deformed profile of the triangular truss of Example 3.32. Solid lines indicate the deformed profile and broken lines are the undeformed profile. *The sketch is not drawn to scale*

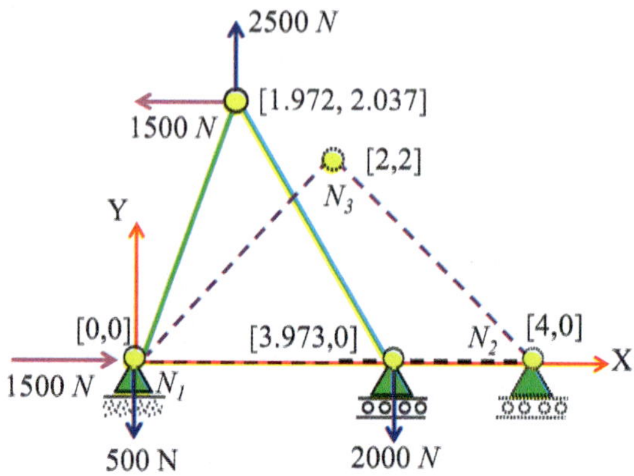

Expanding the above matrix gives column matrices for the nodal forces and displacements thus:

$$\begin{bmatrix} f_{X1} \\ f_{Y1} \\ f_{X2} \\ f_{Y2} \\ f_{X3} \\ f_{Y3} \end{bmatrix} = \begin{bmatrix} +1500 \\ -500 \\ 0 \\ 2000 \\ -1500 \\ 2500 \end{bmatrix} \text{N} \qquad \begin{bmatrix} u_{X1} \\ u_{Y1} \\ u_{X2} \\ u_{Y2} \\ u_{X3} \\ u_{Y3} \end{bmatrix} = \begin{bmatrix} 0 \\ 0 \\ -0.027 \\ 0 \\ -0.028 \\ 0.037 \end{bmatrix} \text{m} \tag{3.53}$$

Sketch of deformed structure

The sketch for the deformed body with respect to the original structure is shown in Fig. 3.33 above.

3.6 Determination of Internal Forces and Stresses on the Members/Bars

As well as determining the externally applied forces as shown in the previous section, we will also have to determine the internal forces within the bars. These internal forces are crucial for determining the stresses. A structural engineer is often more interested in the stresses within the structural members rather than the displacements of the nodes. This is because the size of the stresses will guide the choice of material for design of the members of the structure. The objective of this section is of particular interest to the structural engineer.

To determine the internal forces, let us consider the same right-angled planar triangular truss element of Fig. 3.20, but this time we section through each of the members to release the internal forces acting in them.

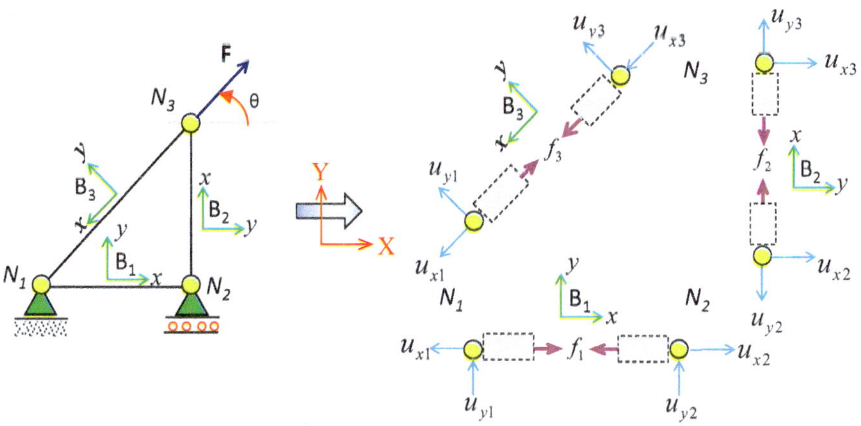

Fig. 3.34 An illustration of internal forces (on each member) following its sectioning

3.6.1 Calculation of Internal Forces

The *internal forces* and *displacements* would have to be related to global forces and displacements. Using the values of displacements specified in Fig. 3.34 and for each of the members, we will now relate the global displacements, $\mathbf{U}_X = [U_{X1} \quad U_{Y1} \quad U_{X2} \quad U_{Y2}]'$ to the local coordinate nodal displacements is defined as $\mathbf{u}_{xi} = [u_{x1} \quad u_{y1} \quad u_{x2} \quad u_{y2}]'$, according to Eq. 3.12. The resulting transformation matrix equation is given thus:

$$\begin{bmatrix} \bar{u}_{x1} \\ \bar{u}_{y1} \\ \bar{u}_{x2} \\ \bar{u}_{y2} \end{bmatrix} = \begin{bmatrix} c & s & 0 & 0 \\ -s & c & 0 & 0 \\ 0 & 0 & c & s \\ 0 & 0 & -s & c \end{bmatrix} \begin{bmatrix} U_{X1} \\ U_{Y1} \\ U_{X2} \\ U_{Y2} \end{bmatrix} \quad \Longrightarrow \quad \bar{\mathbf{u}} = \mathbf{T_d U} \tag{3.54}$$

In Eq. 3.54, local coordinate nodal displacements, $\mathbf{u}_{ji}$, for $j = x, y$ and $i = 1, 2$ can be determined based on the global nodal displacements, $\mathbf{U}_{ji}$, for $j = X, Y$ and $i = 1, 2$ obtained in Sect. 3.5. In order to determine the *internal forces* on the members shown in Fig. 3.34, we have to treat the members as undergoing axial deformation such that the governing axial-deformation linear elastic equation becomes

$$f_i = \frac{E_i A_i}{L_i} \Delta L \quad \Longrightarrow \quad \mathbf{f}_i = \frac{E_i A_i}{L_i} \mathbf{u}_{ji} \tag{3.55}$$

The above expression can be interpreted for example for the members/bars in Fig. 3.34, according to these equations:

$$f_1 = \frac{E_i A_i}{L_i}\left[u_{x2} - u_{x1}\right] \qquad f_2 = \frac{E_i A_i}{L_i}\left[u_{x3} - u_{x2}\right] \qquad f_3 = \frac{E_i A_i}{L_i}\left[u_{x3} - u_{x1}\right] \tag{3.56}$$

Since we are dealing with linear elastic axial deformation of the members, the internal forces calculated above depend only on the axial or x-axis deformations. The deformation in the y-axis (i.e. u_{y1} for node, N_1) will not appear in the above equations. For a different problem, say bending, we will include the $y-$ deformations, yielding different element/member behaviour.

The expansion of the matrix of Eq. 3.54 will yield the local coordinates' deformation terms and will subsequently lead to the calculation of the internal forces.

3.6.2 Calculation of Internal Stresses

Once the internal forces are determined, the calculation of internal stresses on each member becomes easy. For our purposes here, we define *stress* as the *intensity of an internal force over a given area.* In other words, it refers to the internal force acting per unit area. Knowing the internal force on a given structural member will make it easier to define the internal stress acting on that member. We can use this definition to obtain the internal stresses acting on the three members that make up the structure shown in Fig. 3.34. The guiding equation for internal stresses, σ_i, on an i–th bar becomes:

$$\sigma_i = \frac{f_i}{A_i} \quad \Longrightarrow \quad \text{For bar } B_1 \quad \sigma_1 = \frac{f_1}{A_1} \tag{3.57}$$

where f_1 and A_1 are internal force and cross-sectional area respectively of bar B_1. Example 3.8 will show the implementations of Eqs. 3.56 and 3.57 for obtaining all internal forces and stresses on the members that make up a given structure.

Example 3.8 Based on the structure shown in Fig. 3.29, determine the internal forces and internal stresses in all members of the system. Note that you can use the displacement values obtained in Example 3.6. Assume the bars are made of steel of diameter, $\phi = 3\,\text{mm}$ and Young's Modulus, $E_s = 210\,\text{GPa}$.

Solution

Calculation of internal forces

Recall from Eq. 3.53 that the global nodal displacement for the structure is given as:

$$\begin{bmatrix} u_{x1} & u_{y1} & u_{x2} & u_{y2} & u_{x3} & u_{y3} \end{bmatrix}' = \begin{bmatrix} 0 & 0 & -0.0027 & 0 & -0.0028 & 0.0037 \end{bmatrix}'$$

Using this displacement information, we can determine the local nodal displacement values from bar to bar;

<u>*For Bar B_1 consisting nodes, N_1 and N_2:*</u>

- Angle, $\theta = 0°$ hence we obtain $c = \cos\theta = \cos 0° = 1$ and $s = \sin\theta = \sin 0° = 0$

$$\begin{bmatrix} u_{x1} \\ u_{y1} \\ u_{x2} \\ u_{y2} \end{bmatrix} = \begin{bmatrix} 1 & 0 & 0 & 0 \\ 0 & 1 & 0 & 0 \\ 0 & 0 & 1 & 0 \\ 0 & 0 & 0 & 1 \end{bmatrix} \begin{bmatrix} 0 \\ 0 \\ -0.0027 \\ 0 \end{bmatrix} = \begin{bmatrix} 0 \\ 0 \\ -0.0027 \\ 0 \end{bmatrix} \text{m}$$

- Internal forces

$$f_1 = \frac{EA}{L}\left[u_{x2} - u_{x1}\right] \quad \Longrightarrow f_1 = \frac{210 \times 10^9 \times 0.25 \times \pi \times 0.003^2}{4} \times \left[-0.0027 - 0\right]$$

$$\Longrightarrow f_1 = -1002\,\text{N}(\text{compressive})$$

<u>*For Bar B_2 consisting nodes, N_2 and N_3:*</u>

- Angle, $\theta = 135°$ hence we obtain $c = \cos\theta = \cos 135° = -0.7071$ and $s = \sin\theta = \sin 135° = 0.7071$

$$\begin{bmatrix} u_{x2} \\ u_{y2} \\ u_{x3} \\ u_{y3} \end{bmatrix} = \begin{bmatrix} -0.7071 & 0.7071 & 0 & 0 \\ -0.7071 & -0.7071 & 0 & 0 \\ 0 & 0 & -0.7071 & 0.7071 \\ 0 & 0 & -0.7071 & -0.7071 \end{bmatrix} \begin{bmatrix} -0.0027 \\ 0 \\ -0.0028 \\ 0.0037 \end{bmatrix}$$

$$= \begin{bmatrix} 0.0019 \\ 0.0019 \\ 0.0046 \\ -0.0006 \end{bmatrix} \text{m}$$

- Internal forces

$$f_2 = \frac{EA}{L}\left[u_{x3} - u_{x2}\right] \quad \Longrightarrow f_2 = \frac{210 \times 10^9 \times 0.25 \times \pi \times 0.003^2}{2.8284} \times [0.0046 - 0.0019]$$

$$\Longrightarrow f_2 = +1417\,\text{N}(\text{tensile})$$

For Bar B_3 consisting nodes, N_3 and N_1:

- Angle, $\theta = 45°$ hence we obtain $c = \cos\theta = \cos 45° = 0.7071$ and $s = \sin\theta = \sin 45° = 0.7071$

$$\begin{bmatrix} u_{x1} \\ u_{y1} \\ u_{x3} \\ u_{y3} \end{bmatrix} = \begin{bmatrix} 0.7071 & 0.7071 & 0 & 0 \\ -0.7071 & 0.7071 & 0 & 0 \\ 0 & 0 & 0.7071 & 0.7071 \\ 0 & 0 & -0.7071 & 0.7071 \end{bmatrix} \begin{bmatrix} 0 \\ 0 \\ -0.0028 \\ 0.0037 \end{bmatrix} = \begin{bmatrix} -0.0006 \\ -0.0046 \\ 0 \\ 0 \end{bmatrix} \text{m}$$

- Internal forces

$$f_3 = \frac{EA}{L}\left[u_{x3} - u_{x1}\right] \quad \Longrightarrow f_3 = \frac{210 \times 10^9 \times 0.25 \times \pi \times 0.003^2}{2.8284} \times \left[0 + 0.0006\right]$$

$$\Longrightarrow \; f_3 = +315\,\text{N(tensile)}$$

Calculation of internal stresses

Based on Eq. 3.57, we can then calculate the relevant internal stresses on each of bars B_1, B_2 and B_3 as follows, considering that the cross-sectional area, A_i for all three bars are the same:

$$\text{For Bar } B_1\text{:}\ \sigma_1 = \frac{f_1}{A_1} \Longrightarrow \sigma_1 = \frac{-1002}{0.25 \times \pi \times 0.003^2} = -142\,\text{MPa}$$

$$\text{For Bar } B_2\text{:}\ \sigma_1 = \frac{f_2}{A_2} \Longrightarrow \sigma_2 = \frac{+1417}{0.25 \times \pi \times 0.003^2} = +200\,\text{MPa}$$

$$\text{For Bar } B_3\text{:}\ \sigma_1 = \frac{f_3}{A_3} \Longrightarrow \sigma_3 = \frac{+325}{0.25 \times \pi \times 0.003^2} = +50\,\text{MPa}$$

3.7 Conclusions

This chapter has explored the *direct stiffness method* (**DSM**). It has shown the necessity of the method as a vital tool for solving finite element problems especially the structural mechanics types. A triangular planar truss has been used to demonstrate a discretization methodology involving an axial-deforming bar.

The steps required in assembling a global structural stiffness matrix have been shown as well as how such a matrix can be used in determining global nodal displacements and forces. The chapter concludes by illustrating how internal forces and stresses in each of the constituent bars that make up a structure can be determined. These outputs are always the key parameters that a structural mechanics problem, and vice versa, a finite element solution of such problems seeks to answer.

The challenges of the DSM when solved by hand calculations are enormous especially for a *large* structure consisting of hundreds of constituent bars. In the next chapter, we demonstrate the use of computers to solve structural analysis problem using the DSM. The students will be exposed to computational strategies needed to undertake such analysis using MATLAB™ as a programming package.

At the end of this chapter, the reader should be able to:

(a) Understand the place of the DSM within the finite element modelling process;
(b) Undertake a hand calculation of structural analysis problems based on the DSM;
(c) Assembly of a global structural matrix for a simple planar truss structure; and finally,
(d) Determine structural outputs as global nodal displacements and forces; and internal forces and stresses in each of the constituent bars that make up the structure.

3.8 Problems: Direct Stiffness Method

Problem 3.1 A hydraulic hammer, as shown in Fig. 3.35, consists of a housing chamber, a piston and a hardened chisel tool (pointed end). It is used for breaking rocks during road construction. Assume the hammer is designed to sustain a constant vibratory force, $F = 20 \sin \omega t$ kN. The tool is cylindrical of diameter, $d = 8$ mm and has a working length of 500 mm. After a long period of use, the hammer was subjected to its designed maximum vibratory force and experienced a micro-indentation of 950 μm.

(a) Determine the limiting **Young's Modulus** of the material that was used for the above design.
(b) After many cycles, the tool was badly worn and so was wrongly replaced with a different grade of material of Young's Modulus, $E = 190$ GPa, determine the percentage severity to similar micro-indentation.

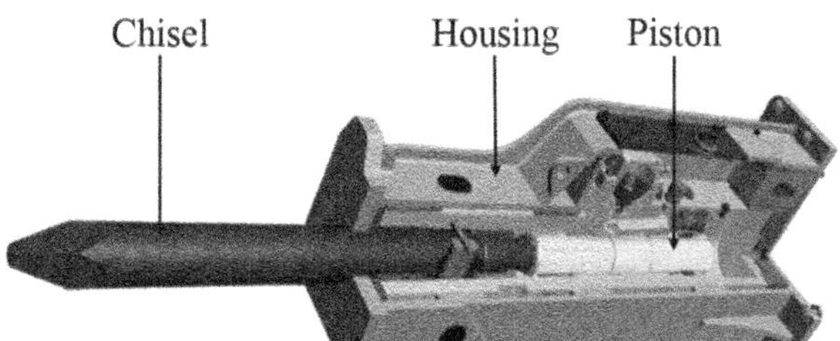

Fig. 3.35 A hydraulic hammer showing a hardened chisel tool (Image source: NPK Europe)

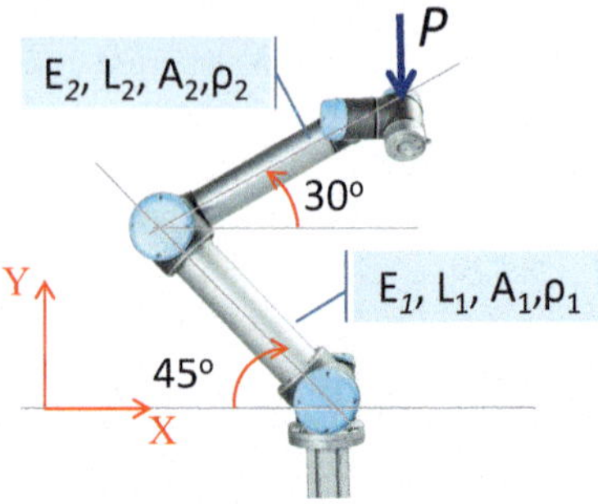

Fig. 3.36 A two-bar robotic arm that supports a concentrated load, P

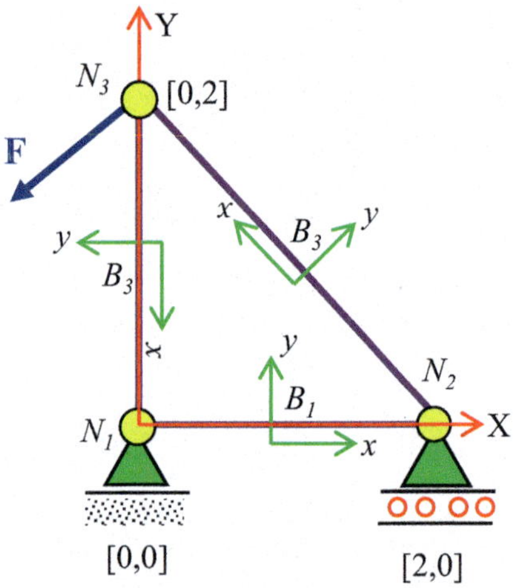

Fig. 3.37 A right-angled triangular truss subjected to a load, P

Hint: Neglect the contribution to the area of the conical chiselled ends of the tool.

Problem 3.2 A two-bar robotic arm consists of bars of lengths L_1 and L_2, as shown in Fig. 3.36. The robot transports a concentrated load, P, across a production floor. The bars have different Young's Moduli, E, and cross-sectional areas, A. In your analysis, treat the bar as a planar system.

(a) As a first step in the structural analysis of this robotic arm, determine the force and displacement transformation matrices for both members.
(b) Comment on whether both matrices are the transpose of one another.

Problem 3.3 A triangular planar truss shown in Fig. 3.37 supports a load, $\mathbf{F} = -2000\mathbf{i} - 3000\mathbf{j}$ N on node N_3. Each of the bars are made from steel of cylindrical cross-section of diameter, $d = 3$ mm and Young's Modulus of steel, $E_s = 210$ GPa.

(a) Determine the local coordinate member stiffness matrices for the 3 bars.
(b) Determine the global coordinate member stiffness matrix of the 3 bars.
(c) Assemble the matrices to derive the global structural stiffness matrix.

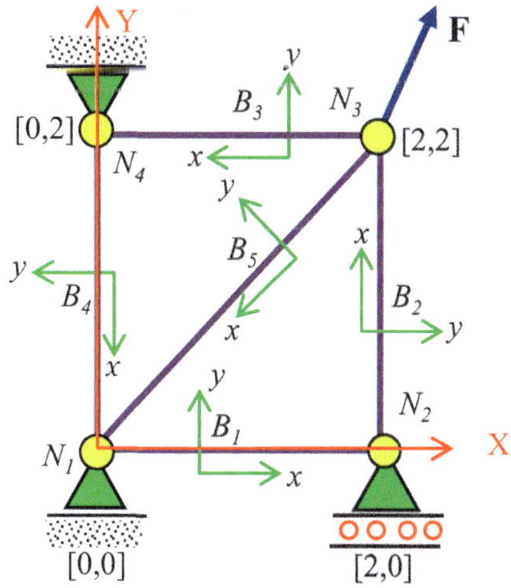

Fig. 3.38 A square truss subjected to a load, P

Problem 3.4 Based on the same structure of Problem 3.3, use the assembled global structure stiffness matrix for the following further analysis of the triangular (planar) truss of Fig. 3.37. Assume Young's Modulus of steel, $E_s = 210\,\text{GPa}$.

(a) Determine the global nodal displacement of the structure.
(b) Determine the global nodal forces (external) of the structure.
(c) Determine the global internal forces and stresses on the three members.
(d) Sketch the resultant deformed profile of the structure.

Problem 3.5 A square arrangement of a truss structure made from steel is shown in Fig. 3.38. Assume all the component bars are made of steel of diameter, $d = 3\,\text{mm}$ and Young's Modulus, $E_s = 210\,\text{GPa}$. This truss is subjected to an angular load $\mathbf{F} = 2000\mathbf{i} + 3000\mathbf{j}\,\text{N}$ at node N_3. Using the *direct stiffness method*:

(a) Determine all nodal (external) displacements and forces on the structure.
(b) Determine the internal forces and stresses in all member.
(c) Sketch the shape of the deformed structure.

Problem 3.6 A truss-based stick man shown in Fig. 3.39 is a concept design for simulating the weight-bearing and balancing capacity of a robot. The structure is fixed securely at nodes N_1 and N_2 and carries different weights, F_1 and F_2 at nodes N_4 and N_6 respectively. The structure is made from cylindrical solid steel bars of diameter, $d = 10\,\text{mm}$. The Young's Modulus of steel is $E_s = 210\,\text{GPa}$. Consider the short robotic neck and round head as rigid structures.

(a) Determine the global structural stiffness matrix of the robotic stick man.
(b) Determine the displacements of nodes, N_4 and N_6 if the robot carries loads $F_1 = F_2 = 400\,\text{N}$.
(c) Consider the robots is redesigned using annular steel tubes diameters of $d_1 = 8$ and $d_2 = 10\,\text{mm}$. What are the displacements of nodes N_4 and N_6 for loads $F_1 = F_2 = 400\,\text{N}$?

Fig. 3.39 A concept design of a robotic stick man

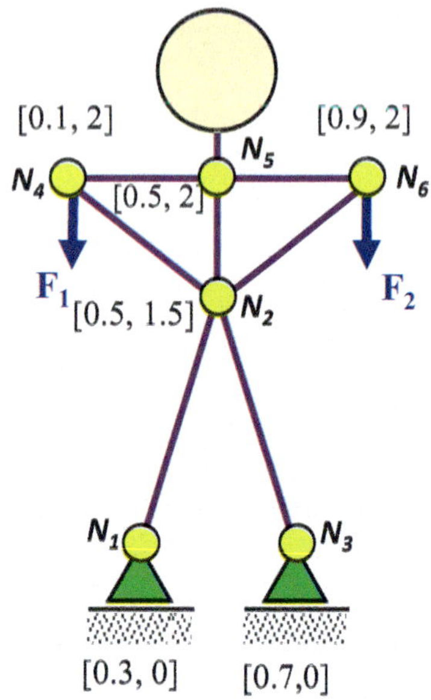

(d) What are the lateral displacements at nodes N_3 and N_5 if the robot is subjected to loads: $F_1 = 200\,\text{N}$ and $F_2 = 350\,\text{N}$?

(e) In order to increase the load-bearing capacity of the robotic stick man, it was suggested that the steel material be replaced with a tungsten of Young's Modulus, $E_t = 410\,\text{GPa}$. What new loads can the re-designed robot carry to achieve the same nodal displacements as (b) above? Assume the robot carries equal loads.

References

1. Ainsworth, M., Oden, J.: A Posteriori Error Estimation in Finite Element Analysis. Pure and Applied Mathematics: A Wiley Series of Texts, Monographs and Tracts. Wiley, New York (2011). https://books.google.co.uk/books?id=-V8ZLkvjaX0C
2. Ferreira, A.: MATLAB Codes for Finite Element Analysis: Solids and Structures. Solid Mechanics and Its Applications. Springer, Dordrecht (2008). https://books.google.co.uk/books?id=skdFXpM0XycC
3. Grätsch, T., Bathe, K.J.: A posteriori error estimation techniques in practical finite element analysis. Comput. Struct. **83**(4), 235–265 (2005)
4. Kannan, R., Hendry, S., Higham, N.J., Tisseur, F.: Detecting the causes of ill-conditioning in structural finite element models. Comput. Struct. **133**, 79–89 (2014)
5. Khennane, A.: Introduction to Finite Element Analysis Using MATLAB® and Abaqus. Taylor & Francis, Boca Raton (2013). https://books.google.co.uk/books?id=IHXfSNYXRD8C
6. Logan, D.: First Course in the Finite Element Method. Thomson, Canada (2007). https://books.google.co.uk/books?id=wjr3ArdvAc4C

Chapter 4
Design of Simple Finite Element Modelling Solver

Abstract This chapter presents a methodology for designing a simple finite element solver. The implementation is carried out inside MATLAB™ and the simulation engine is driven by the *Direct Stiffness Method (DSM)* presented in Chap. 3. In order to illustrate this, the authors have presented an in-house finite element solver they created called: **M**ATLAB **F**inite **E**lement **S**imulation **E**ngine (**MATFESE**™). This chapter presents the structure, implementation and execution of **MATFESE**™. The chapter concludes by illustrating the robustness of **MATFESE**™ in tackling a range of structural mechanics problems. It is expected that after reading through this chapter, the reader can start creating simple finite element solver by utilizing the principles of the DSM. ***MATFESE***™ *can be downloaded from the textbook website.*

Keywords **MATFESE**™ • Finite element modelling solver • Simulation engine • Direct stiffness matrix • Post-processor • Truss solver

4.1 Chapter Objectives

Having discussed the *direct stiffness method* (**DSM**) in Chap. 3, it was concluded that for increasing complexity of structures to be analyzed using the DSM, a computer-aided implementation of DSM is necessary. It is therefore the aim of this chapter to illustrate the processes involved in developing a numerical framework for deploying the DSM in solution of structural problems.

At the end of this chapter, the reader should be able to:

(a) Extend their understanding of the *direct stiffness method* towards its implementation as a finite element solver
(b) Develop a simple MATLAB™ program/software for solving *direct stiffness method* problems to obtain desired model outputs.
(c) Understand the use of the **MATFESE**™. The reader should be able to use the solver to solve simple 2D truss problems.

(continued)

M. Okereke, S. Keates, *Finite Element Applications*, Springer Tracts in Mechanical Engineering, https://doi.org/10.1007/978-3-319-67125-3_4

(d) Query results from **MATFESE**™ to determine their predictive accuracy.
(e) Begin developing their version of **MATFESE**™ using the same implementation principle to create their versions of the code.

4.2 Structure of a Finite Element Modelling Tool

A finite element modelling tool consists of three key components: *pre-processor*, *simulation engine* and a *post-processor*, as shown in Fig. 4.1. According to the figure, the FEM process starts from the *pre-processor* stage which creates what is often described as a *keyword* or *input file*. This file is also called the *model file* and consists of all instructions for the FEM simulation that is to be carried out.

The keyword file is then passed onto the *simulation engine* of the FEM solver. The simulation engine is the heart of the FEM tool and does the number crunching required for the determination of the stiffness matrix for the whole structure. The output from the simulation engine is an output data file. This data file contains solutions of displacements, forces, stresses, etc and are passed onto the *post-processor* analysis stage. Of course, in practical applications, these components may not easily be distinguishable and often dovetail into one another. Frequently, activities at the pre-processor stage may be required again during the post-processor stage depending on the nature of the analysis under investigation.

Further investigation of the output files is then carried out in *parametric studies*. There can also exist what is described here as *interfacing modules*, which establish linkages between the main FEM backbone and other tools that are introduced to aid the FEM analyses. In the next sections, we will be exploring these different stages in much detail.

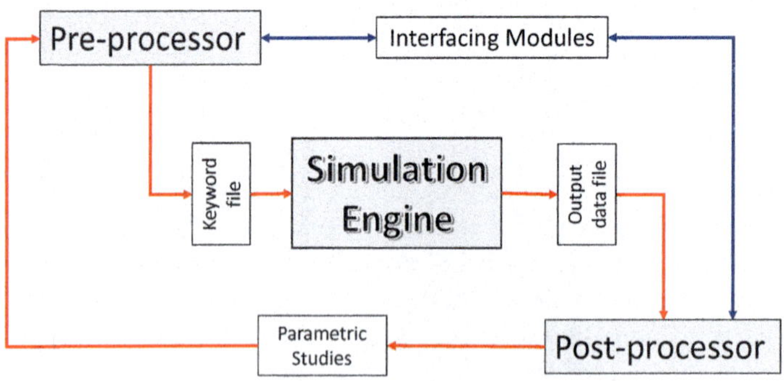

Fig. 4.1 Typical components that make up a finite element method

4.2.1 Pre-processor

The pre-processor program in the FEM consists of sets of instructions, activities, or programs that run to create a set of inputs required for executing the next step in a finite element analysis. These sets of inputs/commands are stored in a keyword file. The pre-processor stage always produces inputs required for another program to run. In fact, for most users of finite element solvers, a lot of effort is invested in developing the pre-processor components of the FEM problem. This aspect of the program is the one that most users interact with. Ultimately, after the pre-processor stage creates a set of keyword/command in structures which will have to be executed at the simulation stage of the problem.

For a finite element analysis, the pre-processor actions can include the following:

(a) *Model generation*: representation of physical problem as a computer model.
(b) *Discretization* or *idealization* of the physical problem into finite elements required for subsequent analysis. This is same as *meshing* and creates *elements* and *nodes*.
(c) Specification of *material model and their parameters* for different components of the model.
(d) *Association* of material components of the model to the elements and nodes.
(e) *Assembly* and *merging* of elements/nodes to create a global structure for analysis.
(f) Design and specification of relevant *boundary conditions*.
(g) Incorporation of necessary *contact mechanics* required in the model.
(h) Choice of required *model outputs* and the *regularity* of their outputting into workspace.
(i) Linking all the above steps into a *keyword/model file* for subsequent simulation exercise.

There is a wide range of pre-processors in commercially available finite element modelling environments. Some of the well known examples include:

(a) **ABAQUS/CAE**: This is the *Complete ABAQUS Environment* (CAE) for creating models using the ABAQUS solver.
(b) **ANSYS Workbench**: This is a comprehensive environment for the ANSYS solver with the key preprocessor modules namely *ANSYS DesignModeler* and *ANSYS Meshing*.
(c) **LS-DYNA LS-Prepost**: This serves as both pre- and post-processor environment for the LS-DYNA solver and is offered free to users who hold a valid LS-DYNA license.
(d) **COSMOL Desktop Graphical User Interface**: This is the development environment for the COSMOL FE solver.
(e) **Altair Hypermesh**: A pre-and post-processor tool by Altair as part of their HyperWorks engineering simulation suite.

Algorithms 4.1 and 4.2 show examples of typical pre-processor stage keyword/input files for the *ABAQUS* and the *LS-DYNA* FE solvers.

Algorithm 4.1 A snippet of an ABAQUS keyword input file

```
*Heading
** Job name: UDComposite Model name: Model-1
** Generated by: Abaqus/CAE 6.13-2
*Preprint, echo=NO, model=NO, history=NO, contact=NO
**
** PARTS
** -----------------------------------------------------
*Part, name=UDComposite
*Node
      1,           50.,    32.2917976
      2,           50.,    45.7082024
** -----------------------------------------------------
*Element, type=CPS3
  1, 385, 379, 368
  2, 296, 254, 253
**  -----------------------------------------------------
** STEP: Step-1
*Step, name=Step-1, nlgeom=NO
*Static
1., 1., 1e-05, 1.
*End Step
```

Algorithm 4.2 A snippet of an LS-DYNA keyword/input file

```
*KEYWORD
*TITLE
  TENSILE TEST ON UD COMPOSITE
*CONTROL_BULK_VISCOSITY
       1.5      0.06
*DATABASE_BINARY_D3PLOT
     1e-06
$**************************************************************
*MAT_COMPOSITE_DAMAGE
$   Mat.ID   density      Ea     Eb     Ec     Pr.ba     Pr.ca  Pr.
    cb
         3   1.6e-09  1.4e04 6.6e3  6.6e3 1.6e-02  1.6e-02
             0.14
$       Gab       Gbc       Gca KFail   AOPT      MACF
       3332      2897      3332     0      0         1
$        XP        YP        ZP    A1     A2        A3
          0         0         0     0      0         0
$        V1        V2        V3    D1     D2        D3       BETA
          0         0         0     0      0         0
$***************************************************************
*NODE
        1                0                0                 0
        2                0            0.125                 0
*END
```

4.2.2 *Simulation Engine*

This is the computational engine of any finite element application. It is essentially a unique numerical implementation for the determination of either forces or displacements resulting from loaded structures. This can take the form of determining the applicable global stiffness matrix for the structure under investigation and directly solving for the forces and/or displacements.

The simulation engine numerical implementation is where the robustness of one FE application is distinguished from another. Some developers have excellent ways of managing and allocating storage space for the calculation of the stiffness matrix, while others are not so elegant. The stiffness matrix in this case may even be matrices of order 500,000 × 500,000, but 10 × 10 in another case. It is therefore important that any finite element development must cater for a wide range of stiffness matrix sizes and structural complexities.

The key objective of the finite element process, for structural analysis and solid mechanics, is to determine the external forces that result following a set of displacement vectors imposed on the structure. There are many ways the solution can be approached, which were presented previously for a 2D truss system.

Consider for example the equation of the global stiffness matrix thus: $\mathbf{F}^s = \mathbf{K}^s\mathbf{U}^s$. The solution for the force vector will be the product of a stiffness matrix and displacement column vector. Regardless of the type of solver one chooses, it is always the challenge to solve for the force-displacement equation so that equality/equilibrium is ensured. To solve this equation, matrix manipulation of the stiffness matrix is a necessity. Different finite element packages have different ways of solving for the external and internal loads. Below are some typical examples of solution procedures that are commonplace in the FEM community.

4.2.2.1 Direct Solution (or Direct Sparse Method)

This approach is similar to the direct stiffness method presented previously where the solution procedure includes determining the global stiffness matrix for the structure under investigation. Also, the method includes determining the inverse of the stiffness matrix. The inverse matrix solution is memory/resource intensive and often slows down the solution process. The resource intensity results from the fact that this method requires a very large number of floating point operations for factorization. Once the stiffness matrices are determined, the solution for nodal displacement or nodal forces can be obtained by multiplication of the stiffness matrices with the displacement column vector (to determine the nodal forces), and vice versa. The method is applicable to problems where the equations of the system (and consequently the stiffness matrix) have a sparse structure, i.e. most of the elements that make up the matrix are zeros. Sparsity arise from the fact that the constituent parts that make up the structure are loosely coupled, as demonstrated in the previous chapter.

4.2.2.2 Iterative Solution (or FFEPlus Method)

This method is very different from the *direct sparse method* in that, instead of determining the stiffness matrix, the solution process makes an educated guess of the deformation based on the boundary conditions being imposed. Then, iterative processes will have to be carried out from the initial guessed displacement to determine the appropriate matrix equations. The iteration continues until an appropriate solution is obtained. This method requires that the iteration must converge to a solution. The iterative approach is highly sensitive to model geometries. If the model is blocky (cubes, plates, cuboids, etc.), they are amenable to an iterative solution. Such geometries (because of their lack of connectivity) have high sparsity and so are best solved using an iterative approach.

As computing resources are not limited, and the *direct solution* is resource intensive. Consequently, it is limited in scope by the types of FEM problems this method can be used for. For example, if a job has 10,000 nodes based on a 2D mesh, there exist 20,000 degrees of freedom, for such the job. It is best suited for a *direct sparse method.* On the other hand, for problems with many degrees of freedom in other words, high mesh density, (say 200,000 degrees of freedom) using the direct solution approach will be memory intensive. Hence, an the iterative method becomes the right approach for solving the FEM problem.

4.2.2.3 Implicit Methods

This is a direct solution method where the behaviour of the system (for example displacement) at a current time, t, is known as well as at a future time (i.e. $t + \delta t$). It requires that, for the implicit method to be used, the system of linear equations that represent the system has to be solved completely. With respect to the Direct Stiffness Method, implicit methods imply that the global stiffness matrix has to be determined, which will serve as a basis for analysis of the whole structure. Implicit solutions are usually applicable to static analyses.

4.2.2.4 Explicit Methods

This type of solution requires that the user is able to predict the behaviour of the system, e.g. the displacement, at a future point in time, based on understanding the behaviour of the system at a current time (for example, now). The explicit method is suitable for the iterative solution described previously. It is also best suited for problems that involve dynamic behaviour as well as damage and impact problems, which often involve interface discontinuities.

Within most FEM software, the *solution engines* are designed and differentiated along the lines of *implicit* and *explicit* methods. For example, in ABAQUS, we have ABAQUS/Standard which is designed to solve problems using the implicit route (often requiring the direct solution). There is also ABAQUS/Explicit, which is designed to solve problems using explicit methods. ANSYS has both Explicit and Implicit versions of its FEM solver. The ANSYS Explicit dynamics engineering simulation are well suited for solving problems that occur in short period of times and may involve material damage/failure as well as contact mechanics.

4.2.3 Post-processor

Most finite element software packages/applications have a post-processor section, which uses results from the *simulation engine* to further analyse the structure. There is a wide gamut of activities that apply to the post-processor stage. In many modern solvers, the post-processor environment is usually the same as the pre-processor, for example in ABAQUS and LS-DYNA: both use the same application environment for pre- and post-processor tasks.

Post-processing can include the calculation of internal forces and stresses acting on the structure. They can also include determining energy dissipation within the model as well as values of displacements and boundary forces that result following the load(s) imposed on the model. The post-processing can include display of the results in contour plots so that a visual solution can be observed. The post-processing activities require skill to be able to interpret the finite element solution. Most times, beginners to the FEM process will simply observe the FEM solution and immediately adopt it without resorting to proving that the results are convincing. The aspect of the finite element process that deals with justifying and/or verifying the solution as representative of true physical response is called *model validation*. Every finite element work should have some element of validation studies in order for the model outputs to be deemed acceptable.

A further aspect of post-processing activities is *parametric studies*. This is very important as it assesses the suitability of the finite element model in meeting wider practical scenarios. Also, parametric studies quantify the effect of model parameters on model outputs. The studies link back to the pre-processor stage of the FEM process by altering for example mesh density, element choice, boundary condition type in order to assess the solution of the physical problem. Where these studies result in meaningful results that agree with experimental data sets, the finite element model can be accepted as a suitable design tool for further investigation of the physical problem.

Readers should always take a doubtful stance when considering finite element results, because intrinsic to the finite element process are lots of errors that follow the solution. Some of such errors are *discretization errors* (due to the idealization of the physical problem for use within an FE scheme), and *solution errors* (following the numerical solution of mathematical models/matrix equations that describe the physical problem).

Also, as a result of the conversion of a physical problem into a mathematical model, there is an intrinsic *material modelling error*. For example, considering the behaviour of an elastic band subjected to a tensile load to be **hookean** in material response, is now accepted as true behaviour. However, that choice comes with modelling errors. The behaviour is true only at small strains. Once large (in excess of 400%) strains are subjected to the body, hyperelastic material response (see Sect. 10.7.2) has to be imposed on the model. So, for the latter case, unless hyperelasticity mechanics are included to the analysis, there will arise a *material modelling error* carried through with any numerical solution.

Consequently, it is important for readers to always carry out validation and parametric studies to assess the suitability of the finite element solution to different scenarios of the structural problem. The more one investigates the problem, the better one understands the computational model and the 'black box' solution generated from the FEM process. With these in place, the user would be better suited to use the computational model for subsequent design analysis of the physical problem under investigation.

4.2.4 Interfacing Modules

One common trend in modern FEM solvers is the incorporation of *interfacing modules* which are add-on options for including other third party software to the FEM process. This has become commonplace in the commercial FEM solver community where providers of one software are looking to allow users of their software communicate with other competitor software so that proper integration for enhanced productivity is ensured. Below are some examples of these popular *interfacing modules*.

4.2.4.1 ABAQUS CAD Associative Interfaces

This associative interface enables geometry information transfer between a **CAD** system and an ABAQUS/CAE framework. Establishing the connection between the CAD Software and the ABAQUS/CAE system ensures that geometry modifications within the CAD software can be updated easily in the CAE-equivalent model using the associative import option. Some of the CAD systems operational within ABAQUS are: CATIA V5, Pro/ENGINEER and SolidWorks. Without the associative interface, the model can be directly imported from CAD system to CAE without establishing the links, as such geometry updates in the CAD System cannot be effected in the ABAQUS/CAE system.

4.2.4.2 COSMOL LiveLinks

COSMOL Multiphysics operates on a modular basis where solver modules are integrated within an FE model to incorporate one or more multiphysics behaviour associated with the physical problem. This development philosophy has also been extended to other third party software as shown in Fig. 4.2.

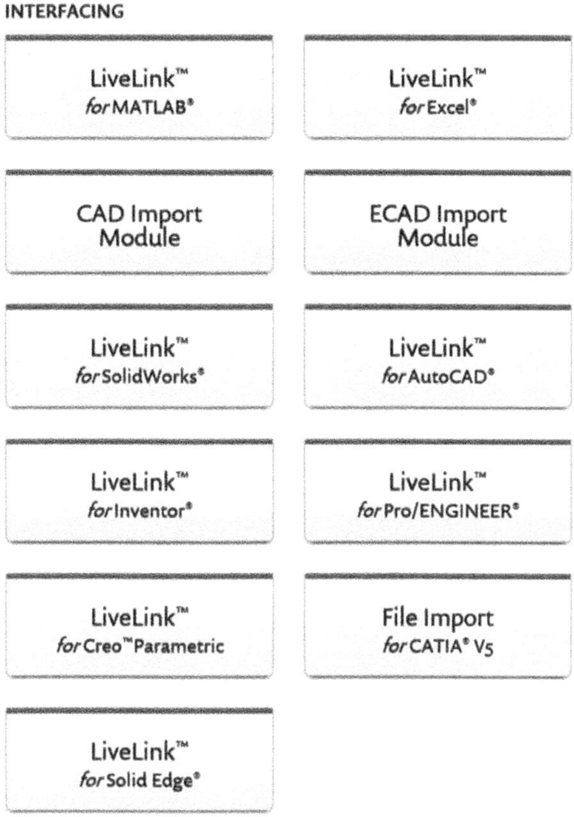

Fig. 4.2 A set of interfacing module for COSMOL Multiphysics. The figure shows that these different interfacing modules are aimed at establishing links with common CAD softwares like CATIA, ProEngineer, AutoCAD, etc as well as computational/analytical tools like MATLAB and Microsoft Excel (Image Source: COSMOL Multiphysics Software)

4.2.4.3 ANSYS LS-DYNA

ANSYS and LS-DYNA are both FEM solver developers, but have come together to offer this add-on module that offers users the best of both worlds. It is possible for subscribers to incorporate within their FE models the traditional ANSYS FE solution engine as well as explicit solution mode of the LS-DYNA system. This helps convert simple static models to dynamic models with capacity for crashworthiness analysis. This mode of generating FEM solutions is expected to be more commonplace in future as the FEM methodology becomes increasingly a design model for companies.

4.2.4.4 DIGIMAT Interfaces

Digimat is designed with several interfaces with the following as examples: (a) *Interfaces to **FEA*** which links with major FE solvers (b) *Interfaces to Injection moulding* – links to an injection moulding cycle (c) *Interfaces to draping* – takes information from a draping software (e.g. Simulayt) (d) *Interfaces to fatigue* – this couples FEM solutions with life assessment for high cycle fatigue (e.g. LMS Virtual.Lab Durability software) (e) *Interfaces to micro-CT* – here Digimat is configured to obtain model geometry using experimental microstructure information of the test material. It is a similar model framework as COSMOL and the whole FEM scheme is built around interfaces that can be added or not to a given solution, according to the problem needs of the user.

4.2.4.5 Scripting Interfaces, User-Defined Sub-Routines

These are special kinds of interface, but mainly serve the same purpose of extending the operational reach of the finite element software. In ABAQUS, a dedicated Python programming interface exists for creating automatic procedures for analysis of the jobs. Material libraries in FEM solvers need to be extended with modern, enhanced predictive fidelity models. As a result, the FEM user often undertakes the design of **user-defined material subroutines** (**UMAT**) to achieve this objective. They are optional interfaces that improve the usefulness of the FEM process in both academic and industrial sectors.

4.3 Development of a Simple FEM Solver for 2D Truss Problems

In this section, we will develop a simple FEM solver for solving 2D truss problems to illustrate the principles underpinning the development of an FEM solver. This is important as it will help readers understand how the finite element framework is designed. For interested readers, they can even start investigating the principles of developing their own finite element solvers along the lines presented here.

The mindset of developing software applications is at the core of engineering aptitude and is encouraged. This exercise, as illustrated using the 2D Truss problem FEM solver, will help hone the students understanding of the different aspects of an FE solver framework. It should also serve as motivation to explore developing bespoke FEM solvers without undue dependence on existing commercial FEM solvers.

4.3.1 Structure of the Proposed FEM Solver

The proposed FEM Solver is called **MATFESE**™ (**MAT**LAB **F**inite **E**lement **S**imulation **E**ngine). It is developed in MATLAB™ as series of M-files and function scripts which are executed from the MATLAB™ command line. The Solver is designed to deliver the following model outputs:

(a) *Global stiffness matrix* for the structure under investigation.
(b) *Nodal displacement and forces* (as column vectors).
(c) *Member stiffness matrices* for all component members/elements both in local and global coordinate systems.
(d) *Internal forces and internal stresses* on each member that make up the structure.
(e) *Deformed plot* of the 2D truss illustrated both as deformed and undeformed configurations.

In developing **MATFESE**™, it is important to mirror the structure of a typical FEM tool discussed in Sect. 4.2. This implies developing environments for *the pre-processor tasks, the solution engine* and finally *the post-processor tasks*. Figure 4.3 shows the typical structure of the proposed FEM solver, which demonstrates the alignment of the three key compartments of any FEM tool.

MATFESE™ runs from a front-end script, which is a MATLAB™ M-file where the major running of the solver is initiated. Algorithm 4.3 shows the full script of this front-end file of **MATFESE**™. This particular file shows the layout of **MATFESE**™ highlighting how it aligns with the FE traditional framework. Note the different sections of the script that are attributable to: *pre-processing section*, *simulation engine section* and the *post-processing section*. Different scripts are written for each of these different sections, but connected using the **MATFESE**™ front-end script. We will consider in the next section these different scripts that apply for the different FE-framework sections.

The home drive containing the **MATFESE**™ contains the different scripts that make up the system as shown in Fig. 4.4. The **memStiffnessData.mat** is an output data file that results from the execution of **MATFESE**™. It stores the member and global stiffness matrices of the structure and its component members/elements.

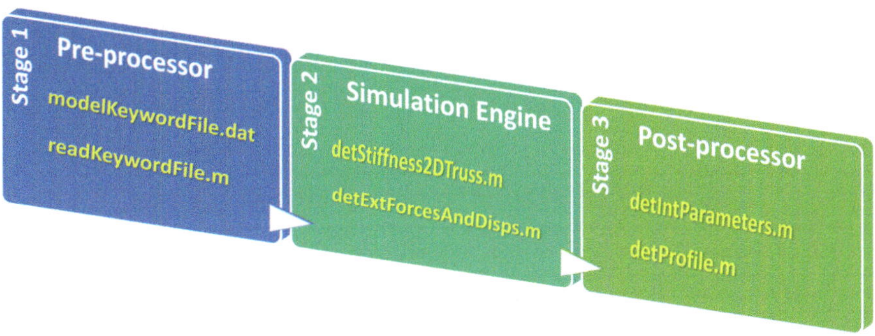

Fig. 4.3 The three-part compartmentalization of **MATFESE**™ in line with the framework of common FEM tools

Algorithm 4.3 Front-end script of **MATFESE**™

```
%% MATFESE:   MATLAB Finite Element Simulation Engine
%Date:   14th July, 2016
%Author: Dr. Michael I. Okereke
%About:  This MATLAB script is the front end of a set of
    scripts
```

```
%         dedicated to determining Finite-Element style
    solutions
%         for 2D Truss problems. This Solver outputs these:
%         [a] Structural Stiffness Matrix;
%         [b] External [Nodal] Forces and [Nodal] Displacements;
%         [c] Internal Forces and
%         [d] Internal Stresses
%Note:    This Solver requires a *.dat file to run. This file is
%         the Keyword File for the FE solution to be undertaken.

%% Read keyword/input data specification file (*.dat)
    readKeywordFile

%% Evaluate model data obtained from the keyword (*.dat) files
for evaluateData = 1:1
    numberElements          =   length(elementNodes);
    numberNodes             =   length(nodeCoordinates);
    xx                      =   nodeCoordinates(:,1);
    yy                      =   nodeCoordinates(:,2);
    GDof                    =   2*numberNodes;
    force                   =   zeros(GDof,1);
    displacement            =   zeros(GDof,1);
    force(loadNodesDof,:)   =   loadValues;
    A                       =   pi*0.25*d^2;    % unit:  m^2
    EA                      =   E*A;            % unit:  N;
end

%% Computation of system stiffness matrix
 [stiffness, k1] = detStiffness2Dtruss(GDof, numberElements,...
                   elementNodes,xx, yy, EA, detailedDisplay );

%% Determine Reaction Forces and Nodal Displacements
    detExtForcesAndDisps;

 %% Determine Internal Forces and stresses
     detIntParameters;

 %% Write Output/Results log file
    outputLog

 %% Draw the deformed profile
 if deformedStructure == 0
     detProfile
 end
```

4.3.2 Pre-processor Scripts

As established previously, every FEM solver has a keyword file which drives the simulation. **MATFESE**™ also has a keyword file called: **modelKeywordfile.dat** which specifies the key data that drive the finite element process. A typical model keyword file is given in Algorithm 4.4. The model data (stated

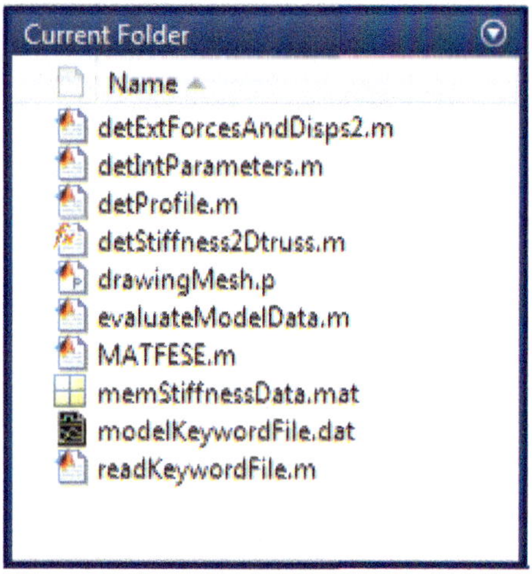

Fig. 4.4 Relevant files drawn from the home drive of **MATFESE**™

in the keyword file) would have to be evaluated and made available using the **readKeywordFile.m** script (please see more discussions on this script later). In the following, we present a discussion of all the different aspects of the **MATFESE**™ keyword file.

4.3.2.1 Material and Geometry Data

This represents the data sets that describe the material that is used in constructing the structure. Since we are dealing with a truss here, the members/bars are assumed to behave in a linear elastic manner and only the Young's Modulus, E, is specified to define the **Hookean** axial behaviour. This section of the keyword file also specifies the cross-sectional geometry (e.g. area, A) of the members that make up the structure. This basic implementation assumes the same E and A values apply for all members that make up the structure. Modifications can be made to this section to accommodate multiple materials and multiple cross-sections as the user may require.

4.3.2.2 Mesh Data

Every keyword file must specify the mesh data, which consists of nodal and element data sets that describe the idealized physical problem. In this current version of the solver, this section of **MATFESE**™ specifies `elementNodes` and `nodeCoordinates`.

Algorithm 4.4 A typical model keyword file for **MATFESE**™

```
%% Keyword/Input File for the Model under investigation
% Author: User
% Date:   16th July, 2016
%%----------------------------------------------------------
%% MATERIAL AND GEOMETRY DATA:  Physical Problem parameters
    E                      =   210e9;  % unit:  N/m^2
    d                      =   2.5e-3; % unit:  m
%%----------------------------------------------------------
%% MESH
DATA: Idealized Problem Parameters
    elementNodes           =   [1 2; 2 3; 3 4; 4 5; 5 6; 6 1];
    nodeCoordinates        =   [0 0; 1 0; 2 0; 2 1; 1 1; 0 1];
%%----------------------------------------------------------
%% BOUNDARY CONDITIONS DATA: Force and Displacement
    loadValues             =   [2000 3000]; %Specify the loads
    loadNodesDof           =   [5 6];       %DoF: Loaded Nodes
    zeroDispNodesDof       =   [1 2 4];     %DoF: Zero disp nodes
%%----------------------------------------------------------
%% MODEL DISPLAY OPTIONS
    scaleMultiplier        =   2; %    scale effect of deformation
    detailedDisplay        =   0; %0 - display results; 1 - do not
        ;
    deformedStructure      =   0; %0 - show deformed structure;
%%----------------------------------------------------------
%% Key to Model Parameters
    %E:                    Modulus of Elasticity
    %d:                    Radius for circle
    %A:                    Area of cross-section of bars
    %L:                    Length of bar
    %GDof:                 Total number of degrees of freedom
    %numberElements:       The Number of Elements
    %numberNodes:          The Numbers of nodes in the structure
    %elementNodes:         List of Nodal points on a given member
    %nodeCoordinates:      List of nodal coordinates
    %loadValues:           Externally applied loads on structure
    %loadNodesDof:         Degree of freedom of loaded nodes
    %zeroDispNodesDof:     Degree of freedom of zero disp nodes
%%----------------------------------------------------------
```

(a) **elementNodes**: This is an $N^e \times 2$ matrix where N^e is number of elements or members, (N^e = numberElements). The was of specifying elementNodes within the **MATFESE**™ keyword file is illustrated in Fig. 4.5. Each member is identified as a sub-matrix of elementNodes thus: $[N_{start} \quad N_{end}]$ where N_{start} and N_{end} are the node numbers for the start and end points of the given member.

(b) **nodeCoordinates**: This is an $N^n \times 2$ matrix where N^n is number of nodes (N^n = numberNodes). The scheme for specifying nodeCoordinates within the **MATFESE**™ keyword file is illustrated in Fig. 4.6. Each node is identified by its nodal coordinates for example, for node N_3 in Fig. 4.6, it is

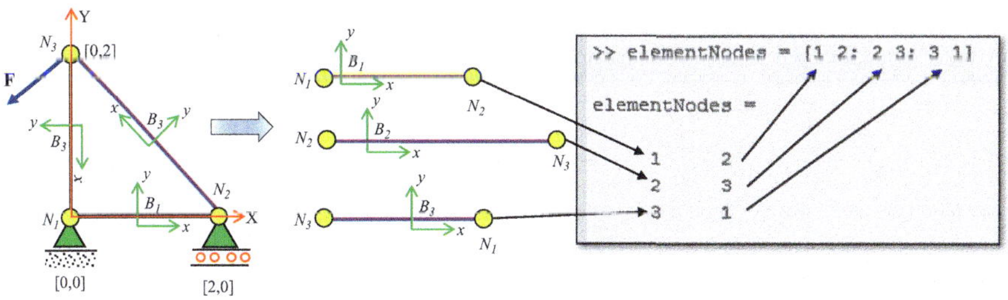

Fig. 4.5 A scheme for specifying the `elementNodes` parameter of **MATFESE**™ keyword file

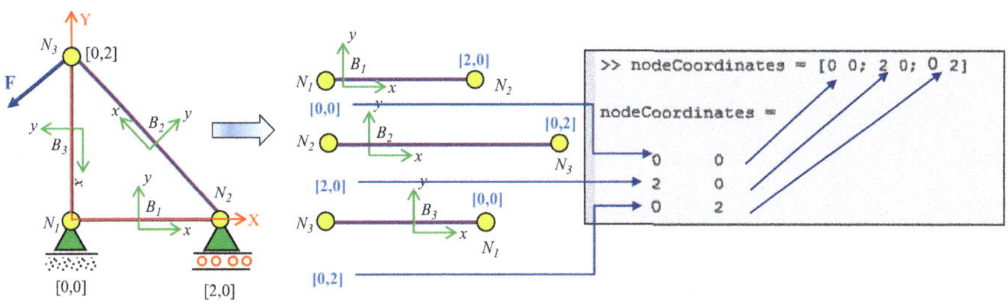

Fig. 4.6 A scheme for specifying the `nodeCoordinates` parameter in the **MATFESE**™ keyword file

represented by: $[x^{N_3}, y^{N_3}]$ where x^{N_3} and y^{N_3} are the x- and y-coordinates of node N_3. A union of all the nodal coordinates sub-sets forms the `nodeCoordinates`.

4.3.2.3 Boundary Conditions Data

This component of the keyword file defines the boundary conditions of the structure. As such, it will specify any existing loads and the degrees of freedom for loaded nodes and support points. The parameters that appear under this are:

(a) **`loadValues`**: This parameter is a column vector that lists all externally imposed loads that appear on the structure. We consider all the loads that exist on roller and fixed supports as well as all externally applied loads. These loads are assembled into a `loadValues` column matrix. For example, the truss shown in Fig. 4.7 consists of three supports and one externally applied load. The figure shows how a `loadValues` column vector can be isolated for the system. Here, we have isolated the second member to show how its externally applied load (as well as supports) results in the `loadValues` matrix.

(b) **`loadNodesDof`**: This represents the degrees of freedom associated with the `loadValues` identified above. The planar truss under consideration is a 2D system hence each nodal point will consist of only $x-$ and $y-$ axes. In the context of extracting the `loadNodesDof`, we continuously number the degrees of freedom from the first node to the last node. Hence for nodes N_1 to

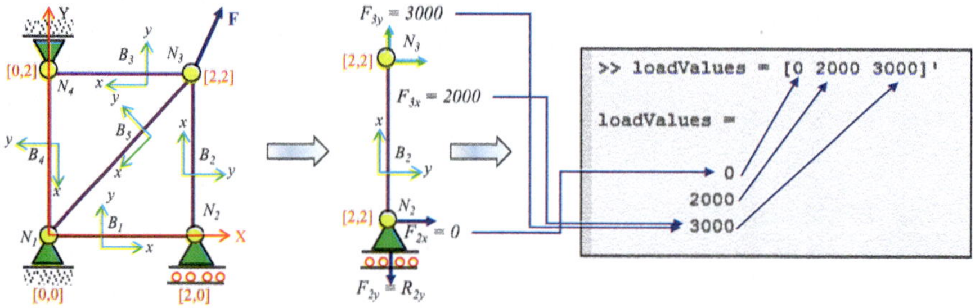

Fig. 4.7 The methodology for specifying `loadValues` in the **MATFESE**™ keyword file

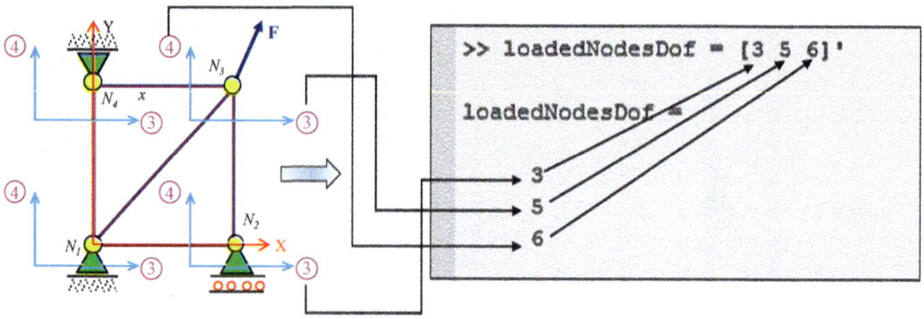

Fig. 4.8 A scheme for specifying the `loadNodesDof` parameter in the **MATFESE**™ keyword file

N_3, the degrees of freedom will be: [1 2 3 4 5 6] where 5 and 6 are the $x-$ and $y-$ axes coordinates for the N_3 node. The `loadNodesDof` matrix is a set of the degrees of freedom for the nodes that have load values imposed on them. The methodology for doing this is illustrated in Fig. 4.8.

(c) **`zeroDispNodesDof`**: This refers to the degree of freedom for those nodes with zero displacements. These displacements are often those that are at fixed supports where they are constrained from moving. We need to inform the simulation engine what these nodes are and what degrees of freedom are affected, so that they can be incorporated during the solution of the force-displacement equation. Recall from Sect. 3.5.3 that these zero-displacement nodes are vital in creating a *reduced stiffness matrix* for subsequent solution of the force-displacement equation. The methodology for labelling them is given in Fig. 4.9.

4.3.2.4 Model Display Options

This section of the keyword file contain optional commands for controlling different aspects of the solver. These optional commands accept two possible answers: `yes` (= 0) and `no` (= 1). The description of the features of each of them are given below:

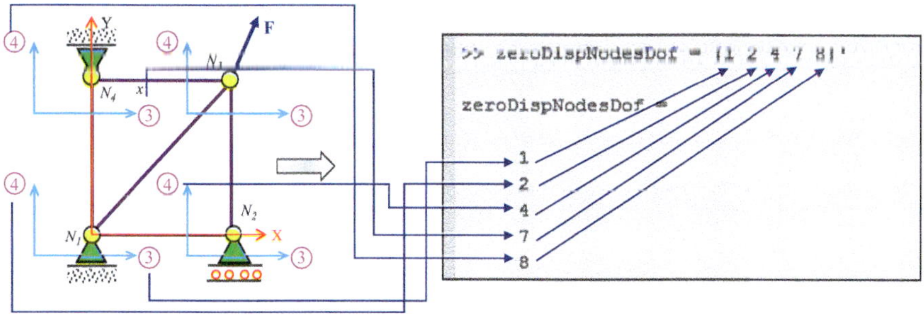

Fig. 4.9 The methodology for assigning degrees of freedom to nodes with zero displacement in the **MATFESE**™ keyword file

(a) **`detailedDisplay`**: This command determines whether the outputs in the MATLAB command window are shown in every possible detail (for all members, nodes, structure) or instead a reduced display of *global stiffness matrix, nodal displacement* and *forces* only.
(b) **`deformedStructure`**: This command determines whether a contour plot of the deformed structure needs be plotted or not.
(c) **`detExtForcesAndDisps`**: This command determines whether or not the FE Simulation Engine should compute the external forces and displacement.
(d) **`detIntParams`**: This command instructs the FE solver to either determine or not the set of internal parameters of the structure under investigation. The parameters include *member forces* and *displacements* as well as *member internal stresses*.

4.3.2.5 Reading of Model Keyword File

The parameters specified in the model keyword input file have to be *read* and made available for **MATFESE**™ at the start of the model execution. This is done through a pre-processor script called `readDATFile.m` that exists within the **MATFESE**™ program tree. The script makes available the model parameters for the *simulation engine* to use for subsequent computational work. It is a M-file that operates on the model keyword file and evaluates the file content within MATLAB™ to make the declared data sets available to the MATLAB™ workspace.

4.3.3 The Simulation Engine

The *simulation engine* of **MATFESE**™ is based on the *direct solution* method. The current version does not incorporate the *iterative solution*. This implies, therefore, that the simulation engine is equipped to determine the global structural stiffness matrix required for multiplication with a displacement column vector to determine the force vector. The approach described in Chap. 3 for determining the global stiffness matrix has been implemented here as part of the simulation engine of **MATFESE**™. The simulation engine is made up of two scripts, namely: `detStiffness2Dtruss.m` and `detExtForcesAndDisps.m`. In the following, these two scripts will be described in more detail.

4.3.3.1 Determination of Stiffness Matrix Script

In **MATFESE**™, the `detStiffness2Dtruss.m` script is the key script of the *simulation engine*. This script is essentially dedicated to determination of the global structural stiffness matrix of the 2D truss under investigation. It is a function script that outputs the *global stiffness matrix*, as well as each *member's stiffness matrix*. The inputs for this function script are seven parameters, which include, among others, information about the idealized/discretized domain (i.e. the mesh data). The script is designed to iterate through the number of elements/members that make up the structure. For each, the script obtains the member's stiffness both in local and global coordinate systems. When all these member-specific stiffness matrices have been determined, the script then *merges/assembles* them to form the *global structural stiffness matrix*. The member-specific and global structural stiffness matrices are both saved to a MATLAB™ data (model output) file called `memStiffnessData.mat`.

4.3.3.2 Determination of External Forces and Displacements Script

A second script called `detExtForcesAndDisps.m` is part of the simulation engine of **MATFESE**™. The script represents the numerical implementation of the theories described in Sect. 3.5. The script takes the stiffness matrices determined by the `detStiffness2Dtruss.m` script and uses the *Gauss-Jordan elimination method* to determine the nodal forces and displacements, i.e. the expected output from a finite element method. The force and displacement outputs are subsequently evaluated within the MATLAB™ command window.

4.3.4 Post-processor Analysis

This is the final analysis stage of the FEM solver. This stage consists of three scripts namely `detIntParameters.m`, `detProfile.m` and `outputLog.m`.

4.3.4.1 Determination of Internal Parameters Script

In **MATFESE**™, a script called `detIntParameters.m` has been develop to undertake the numerical implementation of the principles established in Sect. 3.6 aimed at determining the *internal forces* and *stresses* on the members/bars that make up the structure. The outputs consist of both internal forces and stresses per given member that make up the structure. These outputs are crucial as a design tool for the engineer undertaking the structural analysis.

4.3.4.2 Display of Deformed Profile Script

This script plots the deformed profile of the structure so that visual evidence of the FEM solution is immediately available to the user to use to assess the validity of the FEM solution. It consists of two plots: the initial one consisting of the original undeformed geometry; and, a second plot consisting of the deformed geometry. The deformed geometry is a result of the set of new nodal coordinates formed by adding the initial nodal coordinates to the displacement column vector. In many cases, it was observed that the displacement was not significant with respect to the original geometry, hence a scaling factor, called `scaleMultiplier`, was introduced to scale the plot.

4.3.4.3 Outputs/Results Log File

All finite element tools have an output database file. The is a file or database where all model outputs/results are stored. In `ABAQUS`, this is usually called a `*.odb` file. In such files, all results from the simulation are stored for subsequent post-processor, model validation and parametric studies.

In **`MATFESE`**™, we have also included a script called `outputLog.m` that writes all of the information generated from the given simulation for subsequent post-processor tasks. It is saved as a results file (`*.rez`) and stored in a folder called `Jobi_OutputFolder` created for a specific simulation (where i = job number). The script contains all the material and geometry information as well as stiffness matrices (local and global) for the structure under investigation. Subsequent sections of the output log file contains the external nodal displacements and forces as well as member-specific displacements, forces and stresses.

Example 4.1 The plane triangular truss shown in Fig. 4.10a is subjected to an angular load, $\mathbf{F} = -8000\mathbf{i} + 16{,}000\mathbf{j}$ N. The truss bars are made up of cylindrical steel bars of diameter $d = 4.5$ mm. Take the Young's Modulus of the steel bar, $E_s = 210$ GPa.

Use **`MATFESE`**™ to determine the following parameters:

(a) The member-specific stiffness matrix of all 3 bars (local and global coordinate systems)
(b) The global structural stiffness matrix for the whole structure
(c) The external displacement and force column matrix
(d) The member-specific internal displacement, forces and internal stresses
(e) Show the deformed profile of the truss.

Solution This problem has three nodes (N_1, N_2 and N_3) and three members (B_1, B_2 and B_3). Open an existing model keyword file located in the operating folder of

MATFESE™ and modify relevant sections with details derived from the problem. A snippet of the modified keyword file content adapted for this problem is given in Algorithm 4.5. Also, note the continuous numbering of the nodal degrees of freedom as shown in Fig. 4.10b.

Algorithm 4.5 A snippet of a **MATFESE**™ keyword file for Example 4.1

```
%% Keyword/Input File for the Model under investigation
% Author: User       % Date:   18th July, 2017
%% -----------------------------------------------------------
%% MATERIAL AND GEOMETRY DATA:  Physical Problem parameters
    E                     =   210e9;  % unit:  N/m^2
    d                     =   4.5e-3; % unit:  m
%% -----------------------------------------------------------
%% MESH DATA:  Idealized Problem Parameters
    elementNodes      =   [1 2; 2 3; 3 1];
    nodeCoordinates   =   [0 0;2 2; 0 2]; % unit:  m
%% -----------------------------------------------------------
%% BOUNDARY CONDITIONS DATA: Force and Displacement
    loadValues        =   [0 -8000 16000]; %Specify the loads
    loadNodesDof      =   [4 5 6];          %DoF: Loaded Nodes
    zeroDispNodesDof =   [1 2 4];          %DoF: Zero disp nodes
%% -----------------------------------------------------------
%% -----------------------------------------------------------
%% MODEL DISPLAY AND DEBUGGING OPTIONS
    scaleMultiplier  =   30;    % The scale factor of disp
    detailedDisplay         =   1; %0 - display results; 1 - no
        ;
    deformedStructure       =   0; %0 - sketch deformed
        structure
%% -----------------------------------------------------------
```

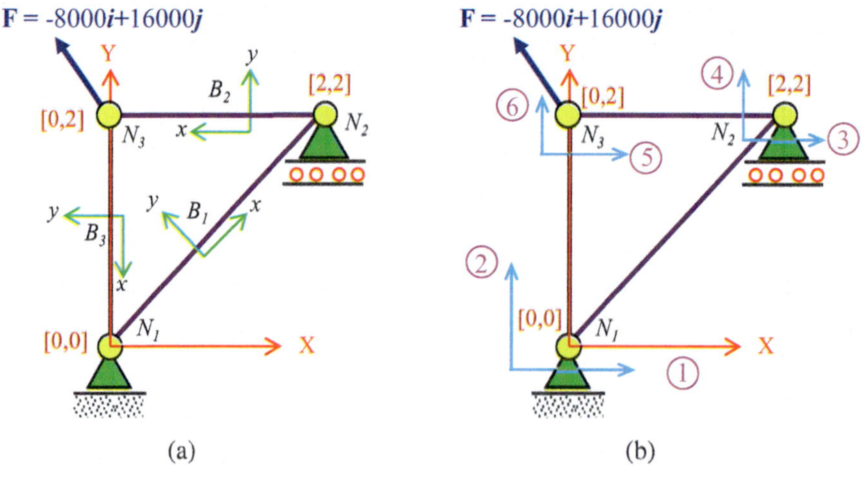

Fig. 4.10 (**a**) A triangular planar truss subjected to angular load, and (**b**) illustration of the continuous numbering system of the degrees of freedom of the structure.

- With the correct model keyword file, run the **MATFESE**™ script for the whole solution to be determined.
- After running the script, typical results for the whole structure, extracted from the `Job1_Outputs.rez` stored in the `Job1_OutputFolder` are given below.

Member-specific stiffness matrices in local coordinates, $\mathbf{k}_i^e$

```
#For Member >> 1
#---- >>  Local Member stiffness matrix, k1 = 1180834.9809*
[
    0.5       0.5      -0.5      -0.5
    0.5       0.5      -0.5      -0.5
   -0.5      -0.5       0.5       0.5
   -0.5      -0.5       0.5       0.5
]
#For Member >> 2
#---- >>  Local Member stiffness matrix, k2 = 1669952.8449*
[
      1        -0        -1         0
     -0         0         0        -0
     -1         0         1        -0
      0        -0        -0         0
]
#For Member >> 3
#---- >>  Local Member stiffness matrix, k3 = 1669952.8449*
[
      0        -0        -0         0
     -0         1         0        -1
     -0         0         0        -0
      0        -1        -0         1
]
```

Member-specific stiffness matrices in global coordinates, $\mathbf{K}_i^s$

```
#Global Member Stiffness matrix, K1 = 1180834.9809*
[
    0.5       0.5      -0.5      -0.5         0         0
    0.5       0.5      -0.5      -0.5         0         0
   -0.5      -0.5       0.5       0.5         0         0
   -0.5      -0.5       0.5       0.5         0         0
      0         0         0         0         0         0
      0         0         0         0         0         0
]


#Global Member Stiffness matrix, K2 = 1669952.8449*
[
      0         0         0         0         0         0
      0         0         0         0         0         0
      0         0         1         0        -1         0
      0         0         0         0         0         0
      0         0        -1         0         1         0
      0         0         0         0         0         0
]
```

```
#Global Member Stiffness matrix, K3 = 1669952.8449*
[
        0       0       0       0       0       0
        0       1       0       0       0      -1
        0       0       0       0       0       0
        0       0       0       0       0       0
        0       0       0       0       0       0
        0      -1       0       0       0       1
]
```

Global structural stiffness matrices, $\mathbf{K}^s$

```
Structure Stiffness matrix, Ks = 10^6*
[
  0.59042    0.59042   -0.59042    -0.59042        0        0
  0.59042    2.2604    -0.59042    -0.59042        0    -1.67
 -0.59042   -0.59042    2.2604      0.59042    -1.67        0
 -0.59042   -0.59042    0.59042     0.59042        0        0
        0          0      -1.67           0     1.67        0
        0      -1.67          0           0        0     1.67
]
```

External nodal forces, $\mathbf{F}$*, and displacement,* $\mathbf{U}$*, matrices*

```
The nodal displacements are   >>>  U =
[
        0
        0
-0.01355
        0
-0.01834
0.0095811
]

The nodal forces are          >>> F =
[
     8000
    -8000
        0
    -8000
    -8000
    16000
]
```

Internal member displacement, $\mathbf{u}$*, forces,* $\mathbf{f}$*, and stresses,* $\boldsymbol{\sigma}$

```
#For Member >> 1 << node [1-2]
Member Displacements, u1 =
[
      0
      0
-0.0095811
0.0095811
]
```

```
Member Internal Force, F1 = -11313.7085 N
Member Internal Stresses, Sigma1 = -711.361 MPa

#For Member >> 2 << node [2-3]
Member Displacements, u2 =
[
0.01355
      0
0.01834
-0.0095811
]
Member Internal Force, F2 = 8000 N
Member Internal Stresses, Sigma2 = 503.0082 MPa

#For Member >> 3 << node [3-1]
Member Displacements, u3 =
[
-0.0095811
-0.01834
      0
      0
]
Member Internal Force, F3 = 16000 N
Member Internal Stresses, Sigma3 = 1006.0164 MPa
```

Sketch of deformed profile

The displacement of the structure under the given load values are quite small. Therefore, below we show graphical representations (from **MATFESE**™) of the displacement without any scale effect on the displacement (see Fig. 4.11a).

Figure 4.11b shows a deformed structure with 30× scaling effect on the deformation. In the later case, the scale multiplier was set to 30 during the **MATFESE**™ simulation (i.e. `scaleMultiplier = 30`).

Example 4.2 The idealized representation of a planar roof truss structure is given in Fig. 4.12. The roof is made from cylindrical steel bars of diameter, $d = 6\,\text{mm}$. The roof is supported by three pillars located at nodes, N_1, N_3 and N_5. The roof supports two structural decks of weights, $\mathbf{F}_1 = \mathbf{F}_2 = -4000\mathbf{j}$ N. The top ends of the roof is anchored by two cables that subject the truss to force, $|\mathbf{F}_3| = |\mathbf{F}_4|$. Take the Young's modulus of steel, $E_s = 210\,\text{GPa}$.

Use **MATFESE**™ to run the 2D truss analyses of the structure:

(a) Show the deformed profile, assuming an acceptable deformation scaling factor.
(b) What is the maximum external nodal displacement in the truss and which node(s) experience this?
(c) In order to strengthen the roof structure, two more trusses were introduced to connect nodes N_7 to N_2 and N_7 to N_4. Determine the new maximum external nodal displacement. Does this re-designing achieve the expected improvement of the roof structure?

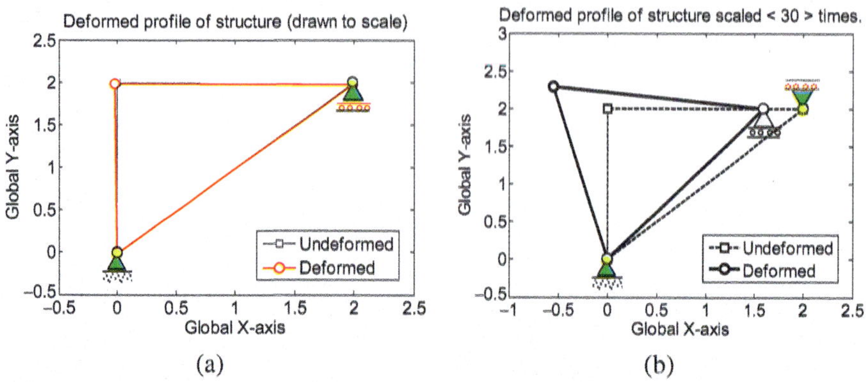

Fig. 4.11 Output from **MATFESE**™ showing deformed and undeformed profile of structure for where (**a**) there is no scaling of deformed structure and (**b**) deformed structure has been scaled by 30× (with supports system included)

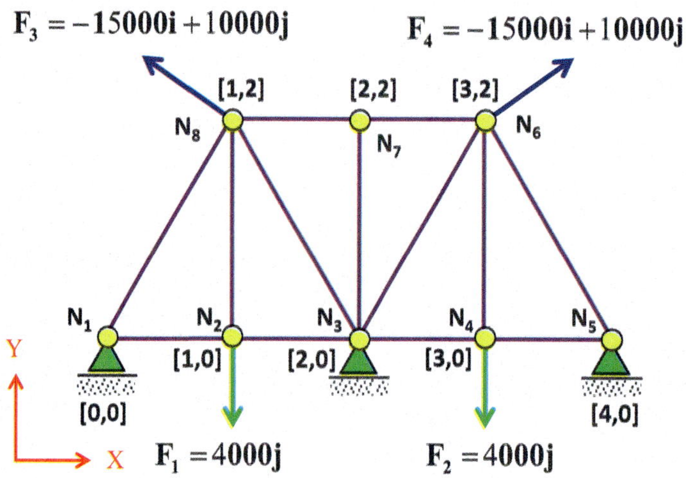

Fig. 4.12 An idealized representation of a planar roof truss structure

Solution

- To begin the analysis, we will label all the nodes with the required 2D degrees of freedom, but in a continuous manner starting from node, N_1 to node, N_8.
- For the 8 nodes, there will be 16 degrees of freedom.
- Counting all the bars in the roof structure, the total number of members in the structure is 13.
- All these information will be used to update the **MATFESE**™ model keyword file and a snippet of it is shown in Algorithm 4.6.

Algorithm 4.6 A snippet of an **MATFESE**™ keyword file for Example 4.2

```
%% Keyword/Input File for the Model under investigation
% Author: User        % Date:   20th July, 2017
%% -----------------------------------------------------------
%% MATERIAL AND GEOMETRY DATA:  Physical Problem parameters
  E                = 210e9;   % unit:  N/m^2
  d                = 6.0e-3;  % unit:  m
%% -----------------------------------------------------------
%% MESH DATA:  Idealized Problem Parameters
  elementNodes    = [1 2; 2 3; 3 4; 4 5; 5 6; 6 7; 7 8;
                          8 1; 8 2; 8 3; 7 3; 6 3; 6 4];
  nodeCoordinates = [0 0; 1 0; 2 0; 3 0; 4 0; 3 2; 2 2; 1 2];
%% -----------------------------------------------------------
%% BOUNDARY CONDITIONS DATA: Force and Displacement
  loadValues       = 1000*[-4 -4 15 10 -15 10]; %Loads
  loadNodesDof     = [4 8 11 12 15 16]; %DoF: Loaded Nodes
  zeroDispNodesDof= [1 2 5 6 9 10]; %DoF: Zero disp nodes
%% -----------------------------------------------------------
%% MODEL DISPLAY AND DEBUGGING OPTIONS
  scaleMultiplier =    100;    % The scale factor of disp
  detailedDisplay =    1; %0 - display results; 1 - no;
%% -----------------------------------------------------------
```

- With these in place, we can now run **MATFESE**™ and the following are typical outputs (extracted from the `Job2_Outputs.rez` results file) for this problem.

Maximum external nodal displacement, U_{max}

By inspecting the displacement matrix for this structure by typing `displacement` into the MATLAB command window, one can find out that the maximum external nodal displacements are:

$$U_{x,max}^{N_6} = 0.021\text{m} \quad \text{and} \quad U_{x,max}^{N_8} = -0.021\text{m}$$

where the superscript (N_6 and N_8) are the nodal positions where the maxima exist. These maxima are in the x-axis. The absolute maximum therefore is: $|U_{x,\max}| = 0.021\text{m} = 21$ mm, and these occur at nodes, N_6 and N_8 (Fig. 4.13).

Sketch of deformed profile

Maximum external nodal displacement for re-designed roof structure

To re-design the roof structure, extra truss bars are introduced to connect nodes N_7 to N_2 and N_7 to N_4. This effectively implies adding extra terms to the `elementNodes` value of the model keyword file, as illustrated below:

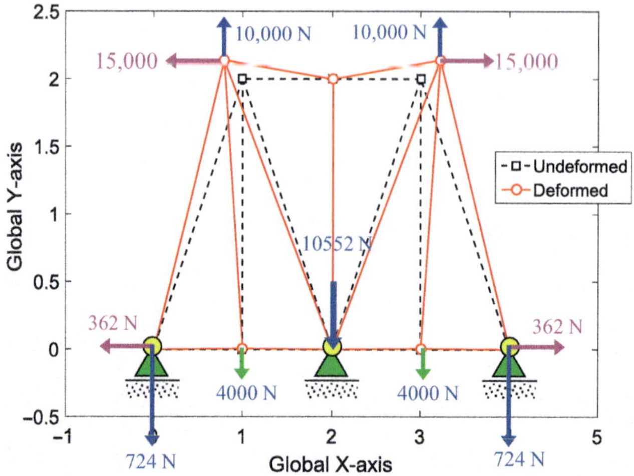

Fig. 4.13 **MATFESE**™ output of the deformation profile of a truss roof structure, with deformation scaled 100×. The required boundary conditions have been inserted to demonstrate the equilibrium of the structure

```
%% -------------------------------------------------------------
%% MESH DATA:  Idealized Problem Parameters
  elementNodes     = [1 2; 2 3; 3 4; 4 5; 5 6; 6 7; 7 8;
                         8 1; 8 2; 8 3; 7 3; 6 3; 6 4;
                         7 2; 7 4];
  nodeCoordinates = [0 0; 1 0; 2 0; 3 0; 4 0; 3 2; 2 2; 1 2];
%% -------------------------------------------------------------
```

After making these modifications to the original keyword file, then re-run **MATFESE**™ such that the resulting sketchy of the deformation profile of the re-designed roof structure is given in Fig. 4.14:

By inspecting the displacement matrix, the maximum external nodal displacements are:

$$U_{x,max}^{N_6} = 0.021\text{m} \quad \text{and} \quad U_{x,max}^{N_8} = -0.021\text{m}$$

This is the same as the previous case so the presence of the extra truss bars has not contributed in limiting excessive displacements at nodes: N_6 and N_8.

4.4 Limitations of MATFESE™ As a Finite Element Modelling Tool

In the above, we have illustrated how a simple 2D truss FEM solver developed by a few lines of code within MATLAB™ can be remarkably powerful as a tool for investigating the external and internal parameters of a truss structure. We have also used it to solve a 16 degree of freedom problem. It has been shown to be quite

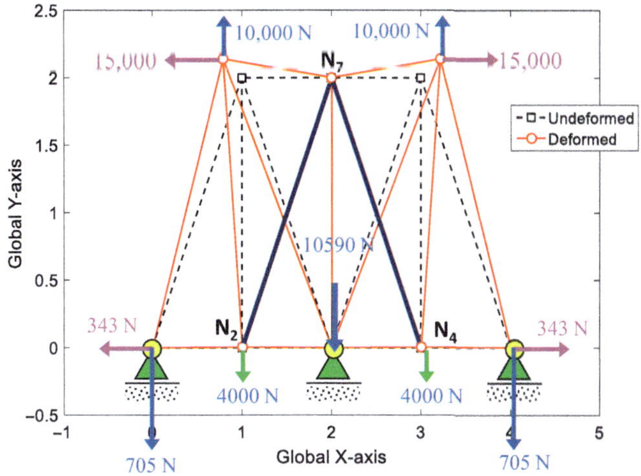

Fig. 4.14 **MATFESE**™ output of the deformation profile of a **re-designed** truss roof structure, with deformation scaled 100×. The extra truss bars are highlighted as bold lines and connect node N_7 to nodes, N_2 and N_4

robust and easy to use. In the following problem section, more challenging practical scenarios can be solved using this basic FE solver.

However, there are major limitations to this FEM Solver which readers need to be aware of and these should serve as motivation for the next chapter where commercial FEM solvers are discussed and features of professional FEM solvers are presented thus providing deeper understanding of the FEM solver design.

The limitations of **MATFESE**™ include the following:

- **Two-dimensional virtual domain:** The FE solver is limited to 2D structures, but real structures are 3D systems. Therefore, **MATFESE**™ need to be re-designed so that 3D domains can be modelled. This can be done by extending the current model framework from 2D to 3D coordinate system representation. This falls under the area of creating representative *virtual domains* for a given problem.
- **Truss-only element library:** The FE solver deals with trusses, so **MATFESE**™ neglects structural effects resulting from bending, torsion. All these must be considered comprehensively in order to design responsibly. The *element library* of **MATFESE**™ need to be enlarged to include quadrilateral and triangular elements.
- **Linear elastic material library:** The FE solver deals with only linear elastic behaviour of the structure. Real structure have much more complex material behaviour. There is need to enlarge the *material library* of **MATFESE**™.
- **Tedious keyword file generation:** The current model keyword file is cumbersome to use and susceptible to user errors. As structures become more and more complicated, creating a keyword file becomes extremely difficult. It is therefore

necessary that as an immediate improvement, a *graphic user interface* need to be created for **MATFESE**™ to help with inputting of the model data.
- **Limited meshing module:** The *meshing module* for **MATFESE**™ is very limited especially since only truss elements have been used in this model. As the *element library* is extended, a more robust meshing algorithm need to be incorporated within **MATFESE**™ to cater for adequate meshing of the virtual domain before the *simulation engine* stage of the FE study.
- **Basic discretization:** The discretization finite elements used in **MATFESE**™ are simply one-dimensional truss elements. Each truss-segment is discretized by one-element only, spanning the length of the member. There is need to extend this discretization exercise to include more mesh refinements per member. Also, the discretization can include complex elements comprising other *element formulations* other than the truss formulation.
- **Limited post-processor:** The post-processor exercises are quite basic with only nodal displacements, internal forces and stresses as well as member displacements recorded as raw numbers. There is need to store the model outputs in a modular relative database style which will make for easy querying of the model outputs and subsequent parametric and validation studies.
- **No contour plots:** There are no *contour plots* for the stress variation in the members. The implementation has only constant single-stress state within the structural members. This is not representative of real structures, which can experience stress concentrations, plastic deformations and fractures. These are crucial for design and therefore the spectrum of stresses in the elements/bars/members need to be widened to account for localized stress changes.

The above limitations have identified such terms as: *virtual domains, element library, material library, element formulation, meshing module, contour plots*, etc. These are the key pillars for any finite element modelling tool and a lot of model development time has been spent in developing each of these with varying levels of success and robustness.

The next chapter will discuss each of the pillars of the FEM process. This chapter has simply demonstrated how a simple FEM tool can be created to demonstrate the philosophy of development of FEM solvers, albeit for only 2D truss problems.

4.5 Conclusions

This chapter has explored the rudiments of finite element solver development based on the principles of the *direct stiffness method*. We have illustrated this using a simple 2D truss solver which has helped demonstrate the essential features of an FEM solver and how these can be implemented in a really simple MATLAB™ FEM tool.

The chapter has also illustrated the solution process using some examples. Limitations of the current FEM solver were identified and these would serve as motivation for the next set of chapters as we explore the features of professionally implemented FEM solvers and how results can be assessed to ensure validity and reliability of FE model predictions.

At the end of this chapter, the reader should be able to:

(a) Understand how the direct stiffness method can be used in development of FEM solver.
(b) Use a MATLAB™ based 2D truss solver developed by the authors for 2D structural analyses.
(c) Assess the validity of results generated from **MATFESE**™ with a view to undertaking model verification and validation studies of real structures, following an FE study.
(d) Use the principles presented in this chapter to attempt the creation of their version of simple finite element modelling tool.

4.6 Problems: Design of Simple Finite Element Modelling Solver

Problem 4.1 A simple 2D truss is shown in Fig. 4.15. Each of the bars is made from steel of Young's Modulus, $E_s = 210\,\text{GPa}$. The bars have a diameter, $d = 3.5\,\text{mm}$. The truss is subjected to a horizontal load, $\mathbf{F} = 5000\mathbf{i}\text{N}$.

Use **MATFESE**™ to determine the following:

(a) The member-specific stiffness matrix of all three bars in both local and global coordinates.

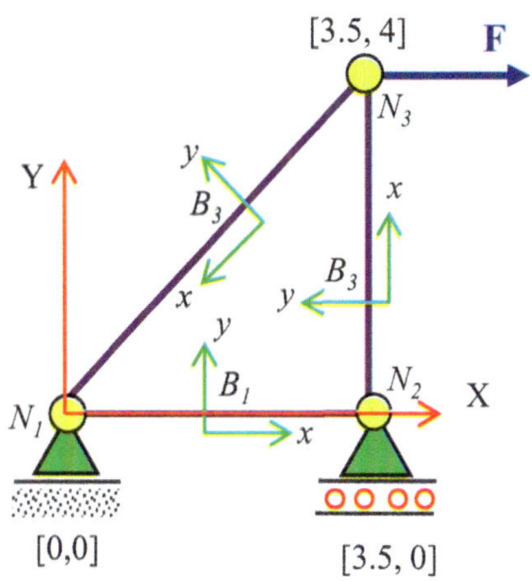

Fig. 4.15 A triangular planar truss subjected to a horizontal load

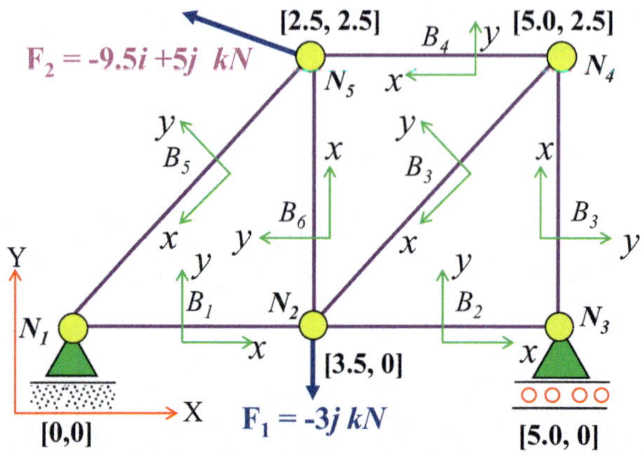

Fig. 4.16 A planar truss subjected to a two external loads

(b) The Global Stiffness Matrix of the structure.
(c) The external nodal displacement and forces on the structure.
(d) The internal displacement, forces and stresses on all three bars of the structure.
(e) The deformed profile of the structure

Problem 4.2 A 2D steel truss structure is made of cylindrical bars of diameter, $d = 2.5$ mm and Young's Modulus, $E_s = 210$ GPa. The structure experiences two external loads, $\mathbf{F}_1$ and $\mathbf{F}_2$, as shown in Fig. 4.16.

By running **MATFESE**™ determine answers to the following:

(a) Nodal $x-$ displacement of node N_5.
(b) Internal stress of bar B_7.
(c) If a bar is added to connect nodes N_5 and N_3, what is the percentage change in $y-$ displacement of N_5?

Problem 4.3 The lifting arm of an industrial crane can be approximately represented as a 2D truss, as shown in Fig. 4.17, consisting of cylindrical 4 mm diameter steel bars with Young's modulus, $E_s = 210$ GPa. The load imposed on the crane can be approximated to a distributed load of 3 kN/m.

Use **MATFESE**™ to determine the following:

(a) The maximum $x-$ and $y-$ displacements experienced by the crane. Identify the nodes where these maximum displacements exist.
(b) Identify which bars, if any, experience zero internal forces and are hence redundant to the structure.
(c) Show the deformed profile of the truss.
(d) Let us support with a $y-$ axis roller support, the node that experiences the largest nodal $y-$ axis displacement. Determine the value and node of the maximum $y-$ displacement for the new arrangement.

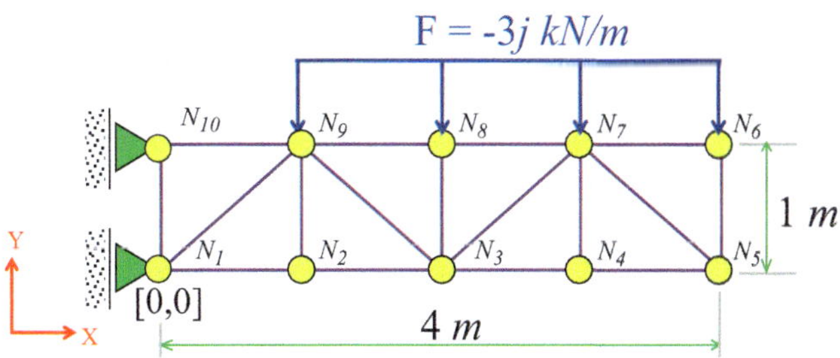

Fig. 4.17 An idealized representation of a lifting arm of a crane

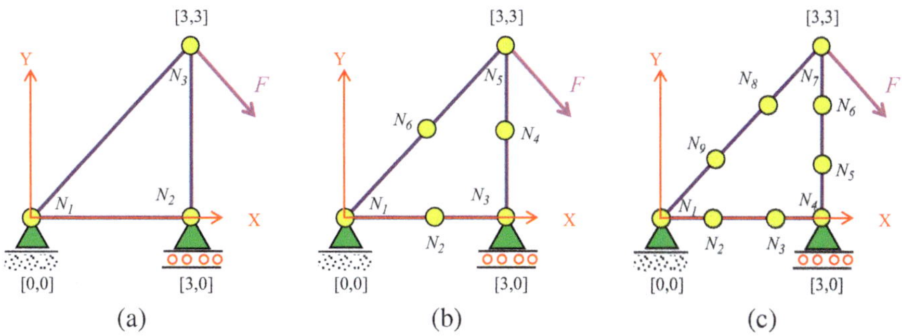

Fig. 4.18 Three instances of meshed models of a triangular truss showing discretization from node-to-node of: (**a**) one unit, (**b**) half unit, and (**c**) one-third unit lengths

(e) A re-design of the crane involves introducing extra bars the original truss. For example, connect nodes N_{10} to N_2, N_8 to N_2, N_8 to N_4 and N_6 to N_4. Using the redesigned structure, find its new maximum $y-$ displacement.

Problem 4.4 A planar triangular truss is made of 5 mm diameter steel bars and subjected to a load, $\mathbf{F} = 10\mathbf{i} - 15\mathbf{j}$ kN for Young's modulus, $E_s = 210$ GPa. To investigate the effect of discretization of the idealized problem (i.e. mesh sensitivity), set up model keyword files for the three cases shown in Fig. 4.18.

Run **`MATFESE`**™ and determine:

(a) The deformed profile from the three cases.
(b) Compare $x-$ and $y-$ displacements for node at $[3, 3]$ m position.
(c) Comment on limitations, if any, of this meshing approach.

Problem 4.5 An L-shaped concert tent support structure is made of cylindrical aluminium bars of diameter, d. The tent is designed to support a distributed load of 3000 kN at the top-side of the truss-structure, as shown in Fig. 4.19. Take the Young's modulus of aluminium, $E_{al} = 70$ GPa. As a design specification, none of the nodes in the structure should sustain a maximum displacement, $\Delta u_{max} \leq$ 10 mm.

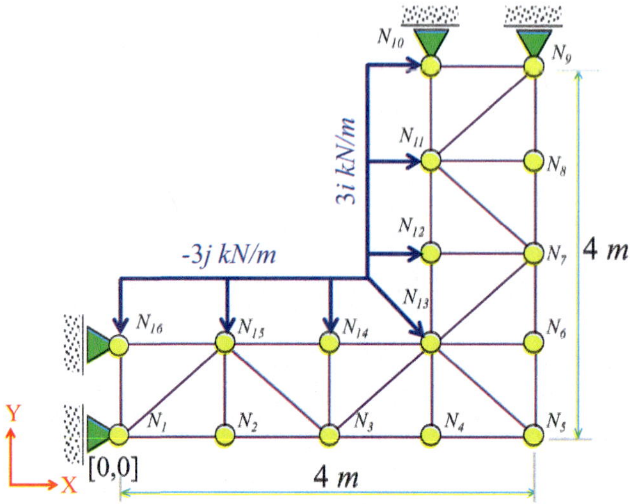

Fig. 4.19 An idealized representation of a lifting arm of a crane

(a) What is the diameter, d of the aluminium bars that meet nodal displacement design specification above? *Hint: Run* **MATFESE**™ *repeatedly as you iterate through different probable diameters until you obtain the correct value that meets the design specification.*
(b) Identify the nodes that are very prone to large displacements.
(c) What are the support reactions that satisfy the design specification?
(d) Show the deformed profile based on your chosen scaling factor.
(e) To reduce the localized deformation at some of the nodes, it was recommended that roller supports are imposed in $y-$ direction on node, N_3 and $x-$ direction on node N_7. What is the new maximum displacement for this re-designed structure?
(f) If extra bars are added in the free-diagonal directions of all the square trusses (e.g. N_{16} to N_2, N_{14} to N_2, N_{14} to N_4, etc.), what is the maximum displacement of the model? Could this be a more acceptable solution than (e) above?

Problem 4.6 The England Wembley stadium arch is a composite of trusses arranged to form a semi-circle. To undertake a structural analysis, the arch is discretised into a collection of different trusses, as shown in Fig. 4.20. The trusses are cylindrical in shape and of diameter, $d = 5.0\,\text{mm}$. Each truss is made of steel material of Young's modulus, $E_s = 210\,\text{GPa}$. The arch is to be analyzed for a design scenario where radial distributed forces, $\mathbf{F} = -5000\mathbf{e_r}$ N act on the structure. $\mathbf{e_r}$ is the unit vector in the radial direction. The arch is fixed securely at the base at nodes N_1, N_2, N_{13} and N_{14}. The inside and outside radii of the arches are, $R_i = 80$ m and $R_o = 90$ m respectively.

(a) Run **MATFESE**™ assuming an appropriate scaling factor and show the resultant deformed profile of the arch.

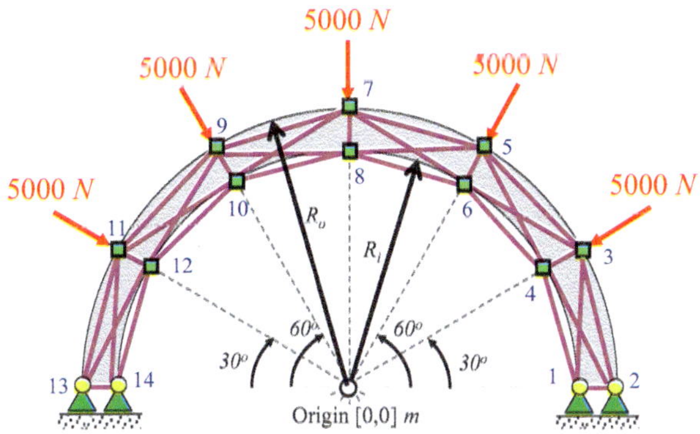

Fig. 4.20 A discretized representation of the Wembley arch subjected to distributed forces

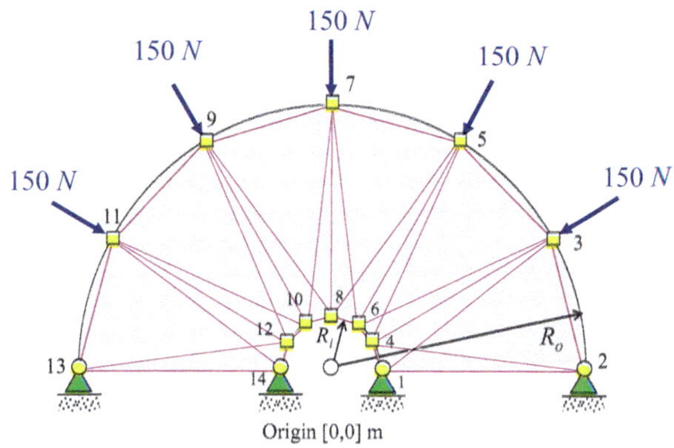

Fig. 4.21 A truss-based discretization of a steel disc

(b) Determine the absolute maximum x- and y-displacements of the structure. Which node(s) relate to these absolute maximum displacements?

(c) Identify all bars, if any, that have zero internal forces and are hence redundant to the structure. Re-run the simulations by removing the redundant bars to prove that these bars are truly redundant. Comment on your findings.

(d) A re-design of the arch was to be implemented with a smaller internal arch radii, $R_i = 60$ m. What is the absolute maximum $y-$ displacement for this re-designed arch?

(e) Demonstrate how you can modify the above arch problem to undertake the FEM study of the disc-like structure shown in Fig. 4.21 which has external radius, $R_o = 1.5$ m and an internal radius of $R_i = 1.0$ m.

Let us assume that the disc-like structure is subjected to similar loading arrangement as the arch, but of lesser load value, $\mathbf{F}_{disc} = -150\mathbf{e_r}$ N. Also, assume that the discretising trusses are of similar material and dimensions as the arch. Determine the absolute maximum displacement in the disc-like structure.

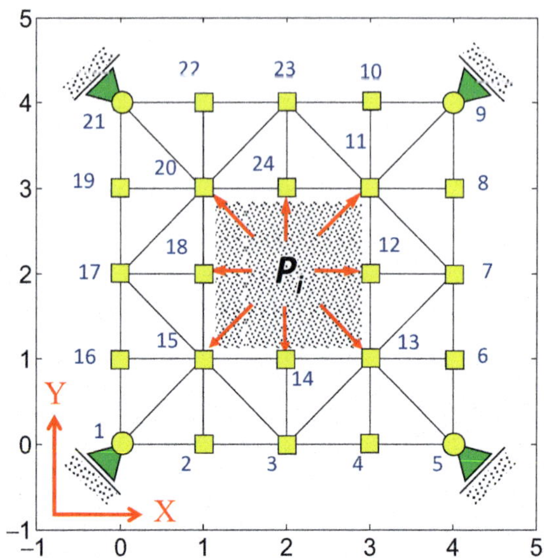

Fig. 4.22 A pressurized container held within a steel structural containment

Problem 4.7 A pressurized container of dimensions $2 \times 2 \times 1\ \text{m}^3$ is rigidly enclosed within a truss-based containment structure, as shown in Fig. 4.22. The pressure inside the container is: $\mathbf{P}_i = 6\,\text{kPa}$. The trusses are made up of steel cylindrical bars of diameter, $d = 2\,\text{mm}$. The containment structure is rigidly fixed at the corners nodes. Assume the steel material has a Young's modulus, $E_s = 210\,\text{GPa}$ and Poisson Ratio, $\nu = 0.33$.

(a) Run **MATFESE**™ to undertake the numerical investigation to determine the maximum displacement experienced within the structure for the given loading case. Show your deformed profile, assuming a scaling factor of your choosing.
(b) This initial result suggests an over-design of the containment structure with a number of redundant trusses (i.e. members with zero internal stresses). Identify those bars and remove them from the structure. Re-run the simulation to assess if they are truly redundant.

Problem 4.8 As part of verification testing of the original steel structure of Problem 4.7, and to explore the response of the structural containment to different aggressive loads, variable distributed loads were applied on all four sides of the structure. These loads are identified as F_R, F_L, F_T and F_B to signify external forces, F on right, left, top and bottom sides of the structure. The expressions for the loads are given as:

- Linear distributed load: $F_R = 1000y$
- Exponential distributed load: $F_L = 72e^y$
- Parabolic distributed load: $F_T = 1000(4x - x^2)$
- Sinusoidal distributed load: $F_B = 4000\sin(0.5\pi x)$

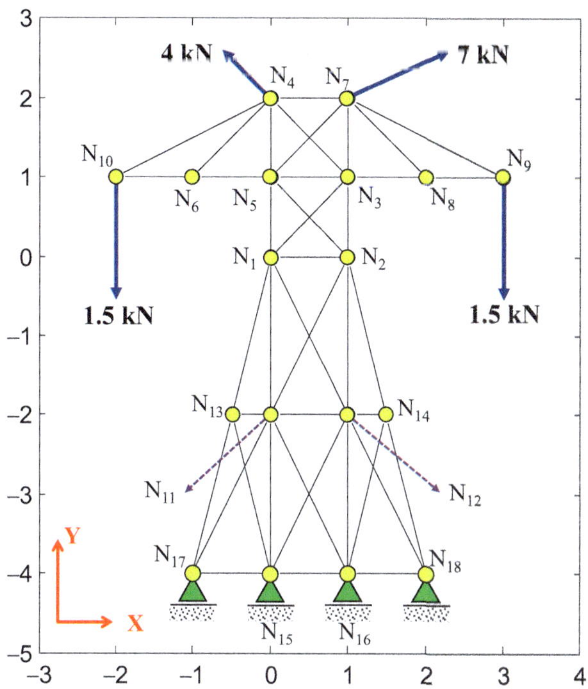

Fig. 4.23 An electricity pylon subjected to different tension cable loads

where x and y are vertical and horizontal positions measured from the origin of the container.

Run **MATFESE**™ and determine the following:

(a) The deformed profile under the new load conditions.
(b) The maximum displacement in the structure under these new load conditions.

Problem 4.9 The truss structure of Fig. 4.23 is a concept design for a medium-height electricity pylon. All distances are in metres. The bars in the truss structure are all cylindrical 5 mm diameter steel bars of Young's modulus, $E_s = 210\,\text{GPa}$. Two 1.5 kN dead weights are attached at nodes N_9 and N_{10}. Tension cables attached at nodes N_4 and N_7 and impose tensile loads of 4 kN and 7 kN respectively.

Run **MATFESE**™ and based on your simulation:

(a) Show the deformed profile of the mast assuming a scaling factor of your choosing.
(b) Determine the absolute maximum x- and y-displacements of the structure. Which node(s) of the mast experience these displacements?

(c) Identify all bars, if any, that have zero internal forces and hence are redundant to the structure.
(d) Re-run the simulations by removing the redundant bars to show if these bars are truly redundant.

Problem 4.10 A re-design of the electricity pylon of Problem 4.9 was to be implemented for three different scenarios namely:

- Change of diameter of the trusses.
- Change of material type.
- Introduce more bars to further reinforce the structure.

A design specification demands that the absolute maximum displacement of the pylon must not exceed a given value i.e. $|\Delta u_{max}| \leq 10\,\text{mm}$.

(a) For each of the three scenarios above, run iteratively simulations to obtain the following that meet the design specifications:

    (i) Optimal diameter of the bars.
    (ii) Optimal Young'S Modulus of another material.
    (iii) Optimal number of extra bars need be added to the structure.

(b) Which of these three scenarios is a most probable solution?
(c) In order to assess the effect of gale force winds on the structure, the variation of wind force, F with vertical position, y of the structure is given as: $F = 100y$ kN. Calculate the distributed wind loads and apply these to the left-hand-side boundary nodes of the structure. Determine the absolute maximum displacement of the structure?

    Hint: Let $y = 0$ at node N_{17} and $y = 6$ at node N_4.

Part II
Finite Element Modelling Principles

This part of the *Finite Element Applications* textbook focuses on principles of the finite element method. It presents a more in-depth look at the key components of a finite element problem. A typical finite element model structure should have elements as shown in Fig. II.1. The chapters under this second part of the book, will explore each of these component parts in much detail. The discussion starts from Chap. 5, generation of virtual domain, and concludes with Chap. 10, where the principles of material models are introduced. We have chosen not to include the *Contact Mechanics pillar* in this edition of the textbook. The reader should expect to understand what is 'under-the-hood' in a typical finite element solver and how solutions are obtained for each of these key FEM principles. There will be no focus on any particular FEM solver, although in most of the presentation, examples are drawn from authors' work using ABAQUS.

Access to Finite Element Software

It is a pre-requisite for this part that in order to attempt the problems in this part, the user should spend some time getting familiar with the workings of any of the finite element software such as ABAQUS, ANSYS, LS-DYNA, COSMOL, etc. It is not the objective of this textbook to teach readers how these software packages are

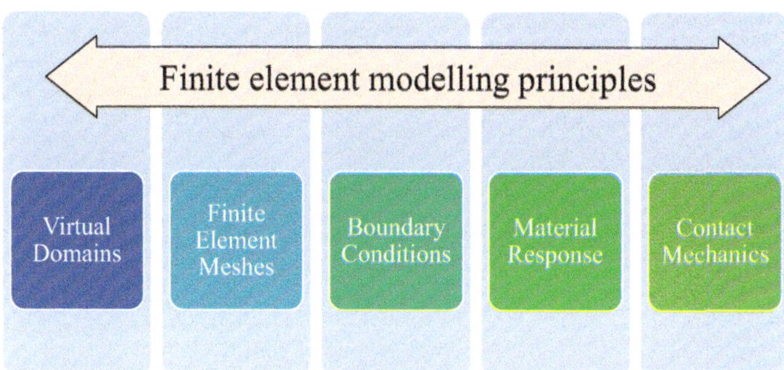

Fig. II.1 The constituent parts of a typical FEM architecture

used, but it pre-supposes that readers should have some basic knowledge of such software.

These FEM software have virtual domain development environments which will be used in tackling the problems in this chapter. To attempt these problems, it is recommended that the reader should get access to these software and follow their help manuals to understand how to develop virtual domains.

Most of these software packages have *student* or *trial licenses* that the user can get access to, which will help them get started. If the reader is a university student, most universities have access to these software packages as part of a typical FEM teaching suite and so the student is encouraged to start practising with them.

Chapter 5
Design of Virtual Domains

Abstract A *virtual domain* is the geometric representation of the boundaries that enclose a given physical system under consideration. All FEM implementations must include a virtual domain, which is the basis upon which the finite element solution to the physical problem is determined. This chapter presents the considerations that an FE user/developer has to make in generating simple and/or complex *virtual domains*. The chapter presents the strategies for converting a *physical domain* into a virtual/computer-based domain that can be solved using finite element methods. We present the generation of virtual domains based on simple sketches or use of computer-aided-design softwares. Also, the authors have provided insights into use of computed tomography images for domain creation. The objective of the domain generation FEM pillar is to generate as near accurate as possible the virtual domains to a physical problem space so that computational errors arising from this idealization process can be minimized.

Keywords Virtual domain • Representative volume element • Computer-aided design (CAD) • Computed tomography • Geometric modelling • Domain generation

5.1 Introduction

Prior to this chapter, cursory references have been made to the **FEM** modelling principles. It is important that readers are introduced to the pillars of the FEM process and how these pillars/principles lead to the FEM solution of complex physical problems. From this chapter onwards, the components of the FEM process will be outlined and discussed. Such insights will help FEM users create realistic FEM models and also query the results obtain from FEM simulations. Figure II.1 shows that a typical FEM model (as available in commercial FE solvers) includes different features: *geometry, meshes, boundary conditions, material models* and *contact mechanics*. This chapter is dedicated to the first principle, namely *geometry*, herein described as an *FEM virtual domain*.

M. Okereke, S. Keates, *Finite Element Applications*, Springer Tracts in Mechanical Engineering, https://doi.org/10.1007/978-3-319-67125-3_5

5.2 Chapter Objectives

At the end of this chapter, readers should be able to:

(a) Understand the importance of a virtual domain to the FEM process.
(b) Understand the different geometry inputs required for creating an idealized representation of the physical problem.
(c) Understand the different approaches that can be used in developing virtual domains for a physical problem.
(d) Assess the effect of length scale consideration while developing virtual domain for multi-scale physical problems.
(e) Understand the different approaches for automating the creation of virtual domains especially for tasks where the FEM user requires repetitive creation/modification of a virtual domain for example during parametric studies.

5.3 FEM Virtual Domain

In this section, we will give a definition of a *virtual domain* and the *representative volume element.* The influence on length scales to domain creation will be considered and the difference between sections and parts discussed.

5.3.1 Defining a Virtual Domain of a Physical Problem

Virtual domains are idealized constructs that house the physical problem so that numerical computations can be undertaken on them. The physical geometry of the problem is described as the *physical domain* whilst the computer-version of the domain is the virtual domain. The finite element process is initiated from the definition of the virtual domain. A virtual domain must therefore be a close representation of the physical problem in dimension and orientation.

Modern FEM solvers have in-built virtual domain generating interfaces, similar to those used in CAD programmes. A lot of care has to be given when developing virtual domains, as a poorly represented domain will adversely affect the validity of any FE solutions significantly, irrespective of the robustness of the whole FEM process. The defining of virtual domains therefore becomes a major task for the **FEA** user.

Consider an example where a design engineer is to undertake a numerical study on the effect of end-cap sizes and material type of three soda cans shown in Fig. 5.1a.

(a) (b)

Fig. 5.1 An example of a virtual domain showing (**a**) the physical domain (soda cans) and (**b**) the virtual domain (the equivalent representation of the soda cans)

The mechanics of pressure vessels equations for a thin-walled cylinder can be used in determining these effects. However, the analytical equations rely on fully cylindrical vessels all through the length of the soda can. As the figure shows, although majority of the cans are quite cylindrical in shape, the end-segments (at the top and the bottom) deviate from a cylindrical to a conical shape at the top and a complex non-linear shape at the bottom. Reliable predictions cannot therefore be made based on analytical formulations in undergraduate mechanics of materials textbooks.

To understand what end-cap material and sizes will be sufficient to retain the internal pressure of the soda can, then the user might want to undertake a finite element analysis of the 'awkward' soda can shape. The first step in such analysis will be to replicate accurately as possible the exact soda geometry i.e. create the FEM virtual domain for the problem. Figure 5.1b shows a typical FEM virtual domain of the soda cans. It is not an exact representation, but captures most of the features of the physical domain. The geometric difference between the physical and virtual domains constitutes a geometric error associated with the finite element solution. It is always the objective for reliable FEM studies to minimize all errors especially the geometric errors. Some of the strategies for recreating accurate virtual domains are presented in Sect. 5.4.

5.3.2 Representative Volume Element: A Subset of a Virtual Domain

Physical domains can be quite complex and onerous to recreate exactly. Consequently, a common approach used by FEM users is to create what is commonly referred to as a *Representative Volume Element* (RVE). This is a subset of a virtual domain that captures the entirety of the physical problem, such that the FEM solution can still be obtained without loss of generality or accuracy. In most cases,

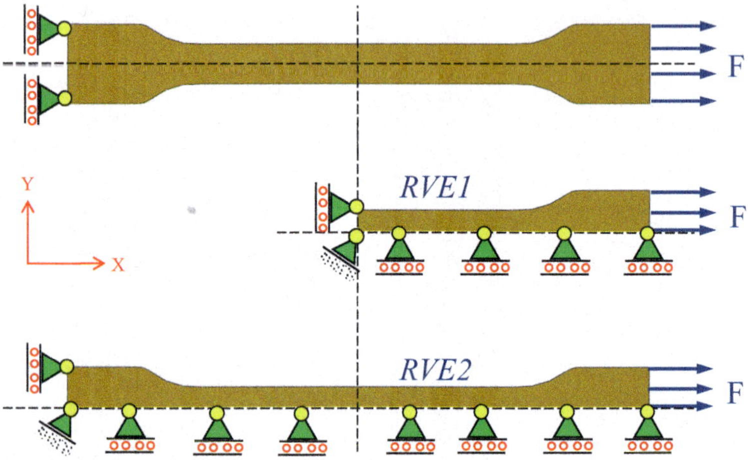

Fig. 5.2 A virtual domain creation for a tensile test specimen with full model and two RVE variants with appropriate boundary conditions

the RVE is never exactly the original geometry, because such comprehensive virtual domains may be expensive on computational resources. Therefore, it is important to identify an equivalent volume that represents the original structure sufficiently closely. The subset may consist a quarter or half of the original physical domain, for instance. To accommodate the *unrepresented* parts of the virtual domain, it is customary to introduce appropriate boundary conditions that should lead to convergent numerical solutions.

For example, if we consider the tensile test specimen shown in Fig. 5.2 which is a dogbone shape specimen. To undertake an FEM study of the test, one can carry out the FE process on the whole structure, but this will lead to many finite elements and hence require more computational resources.

To minimize the computational efforts, let us consider the symmetry of the physical domain and identify a subset that can be analyzed and yet still yield reliable results. This, we can do by considering, say, a quarter of the original model (along lines of symmetry shown). This quarter model creates a typical RVE of the test specimen. Appropriate boundary conditions have been chosen and imposed on the model to allow for the expected tensile deformation of the tensile specimen, as shown in Fig. 5.2.

Some other users may decide to limit the study solely to a half-model, as shown in Fig. 5.2 too. Each of these, *RVE1* and *RVE2*, are RVEs of the dogbone tensile test specimen. However, the *RVE1* variant will give results using fewer finite elements and with a quicker convergence to a solution.

Similarly, consider a impact resistance study of a bespoke sandwich structure, with the core made from cylindrical shaped cells, as shown in Fig. 5.3a. The physical problem has been idealized into a virtual domain shown in Fig. 5.3b. This virtual domain will require a lot of computational resources to solve, hence an RVE is

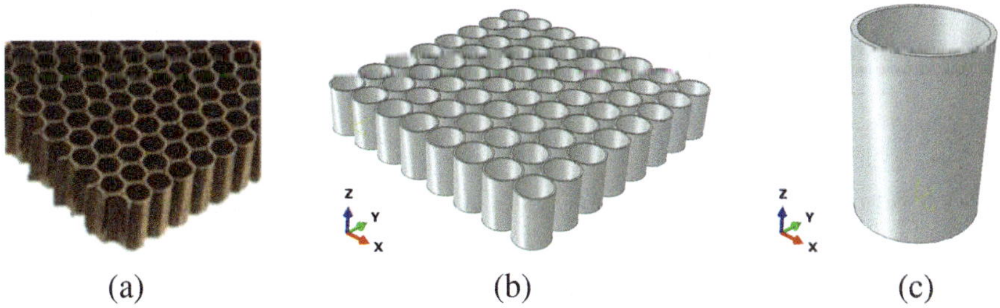

Fig. 5.3 RVE creation for a sandwich core structure showing (**a**) the physical domain, (**b**) the virtual domain and (**c**) a typical unit cell RVE of the virtual domain

isolated as shown in Fig. 5.3c. Appropriate boundary conditions need to be imposed on the top and bottom faces of the cylindrical cell so that the expected deformation profile, comparable with the full model, will be achieved on the RVE.

5.3.3 Length Scale Considerations When Creating Virtual Domains

In creating virtual domains of a problem, length scale considerations become important. The virtual domain can be defined at atomic, molecular, nanoscale, microscale, macroscale or structural length scales as previously illustrated in Fig. 1.4. It is left for the user to choose a relevant length scale for the problem in creating the virtual domain.

The choice of length scale for a given problem depends on the level of analysis detail the user is interested in obtaining. In some cases, the solution may require structural-scale results, for example, the displacement of a building under seismic loading. It will therefore make sense to define the virtual domain at this structural scale. In another study, it might have been noticed that a particular failure initiated in a welded section of a structure. To understand the origin and propagation of the failure, the user might then decide to undertake a more detailed analysis within FEM. In such a case, it is recommended that the user sets up the model at a lower length scale (say microscale).

Let us consider a **unidirectional composite** (**UD**) and the applicable length scales. At structural scale, the virtual domain may consist of laminates. This is the macroscale representation. Each laminate consists of individual laminae. This is referred to as a mesoscale representation. Each lamina is made up of fibres and matrix represented at the microscale length scales.

The macroscale and microscale representation of a typical UD composite is shown in Fig. 5.4. Each of these virtual domains can be used for finite element modelling of the composite, however the level of details of model outputs obtained from each domain will be different. For the higher length scale domains (laminates),

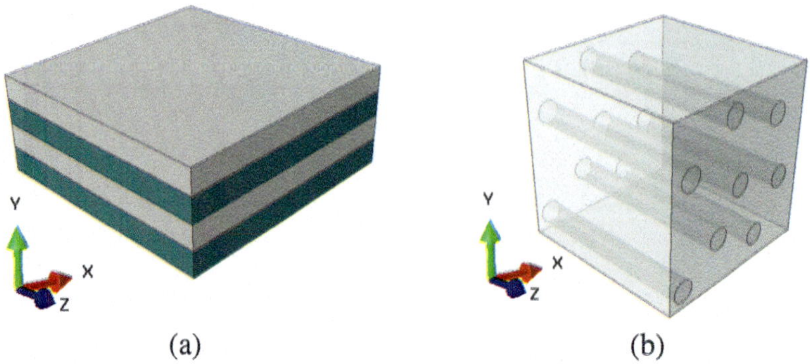

Fig. 5.4 A representation of a unidirectional composite at (**a**) macroscale (laminates) and (**b**) microscale (fibre and matrix) length scales

averaged stresses and strains can be determined whilst for the lower length scale, onset of damage and other nonlinear responses of the material can be obtained. The choice of length scale and the virtual domain to use always depends on the level of detail the user is interested in generating following the FEM study.

Finally, it should be noted that the more refined the virtual domain is (e.g. the microscale), generally, the more accurate the predictions from the model (e.g. stress distribution). However, lower length scale models require extensive computational resources. Also, it is much harder to validate lower length scale models with experimental data derived from macroscale or even mesoscale length scale experiments. This is the paradox facing multiscale modelling [3, 6].

Although predictions from the FEM process is significantly improved at the microscale, the challenges of model validation, computational cost and technical know-how required to undertake simulations at this scale limits its widespread application. In the end, the FEM user will make a compromise on what scale the model has to be defined at and the type of outputs that can be generated. These will depend on the application, level of model output details required from the model, technical know-how of the user and availability of computational resources for running the FEM model. If it is a specialist application, then it might be important to invest the significant resources required into a lower length scale analysis for more accurate prediction. For less expensive applications, a structural-level analysis may be acceptable.

5.3.4 Parts and Sections

Geometric models are usually represented as ***parts*** or ***sections*** in the finite element process. In an FEM solver keyword file, this is often introduced by the commands `*PART` or `*SECTION`. Often for complex structures, the parts are fragments (or

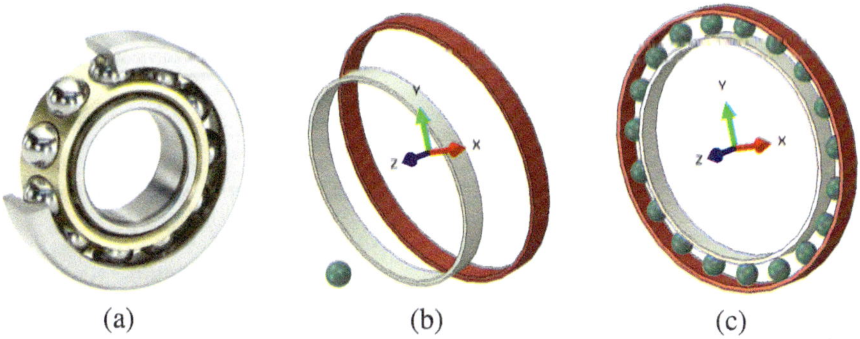

Fig. 5.5 A lubrication ball bearing: (**a**) physical domain (Image source: Bearing Boys Ltd), (**b**) virtual domain: three-component parts. (**c**) virtual domain: assembled component

sub-domains) of the main structure. To be able to analyse the structure, the different parts would have to be assembled together with appropriate compatibility criteria enforced on such models.

A good habit in developing virtual domains is to split a complex structure into parts that represent simple units that can be assembled subsequently. We illustrate this principle using a lubrication ball bearing, which consists of two discs and 18 ball bearings as shown in Fig. 5.5. The balls were created only once as a part and duplicated 17 times to create all the 18 balls. It was then subsequently assembled together with the inner and outer rings to create the ball-bearing assembly.

Once the virtual domain has been created, it is ready for use in a finite element solution. However, the geometric model alone is not sufficient to begin the FE solution. That virtual domain must be associated with a material for the expected physical response to be studied. For other kinds of FE analysis, such as fluid mechanics or thermal response, this material association may not be necessary. However thermodynamic or fluid mechanics properties need to be associated with the virtual domain before using it for an FEM study.

In ABAQUS, the ***Sections*** and ***Section Assignment*** options are used to link the virtual domain to some material response. This is a vital step as it establishes the **constitutive behaviours** of the virtual domain. The same virtual domain can be used to simulate metallic, ceramic, plastic and fluid response, with only the *Sections* and *Section Assignments* defined to capture the needed material behaviour.

5.4 Virtual Domain Creation Strategies

Having established the rudiments of a virtual domain, in this section we will begin to explore the different ways of transporting physical models into virtual domains. The strategies include both standard and some specialist approaches. Some of these approaches lead to automatic virtual domain creation, especially for repetitive tasks while others do not. The choice of which strategy to use depends on the material resources available to the FEM user and their technical abilities.

5.4.1 *Virtual Domain Creation Using Model Development Environment in FE Software*

This is the most common strategy for the creation of FEM geometric models. It involves using the in-built **CAD** environment of the FEM solver to create the virtual domain. The CAD packages usually have highly specialized model creation capabilities, so, in many instances, there is no need for companies to invest in CAD software package as part of their product design and FEM suite.

Figure 5.6 shows a typical ***ABAQUS CAE*** which contains quite a large variety of specialized model creation tools. Most of the models created in this textbook have been created using this interface. Similar schemes exist for ANSYS, LS-DYNA, COSMOL.

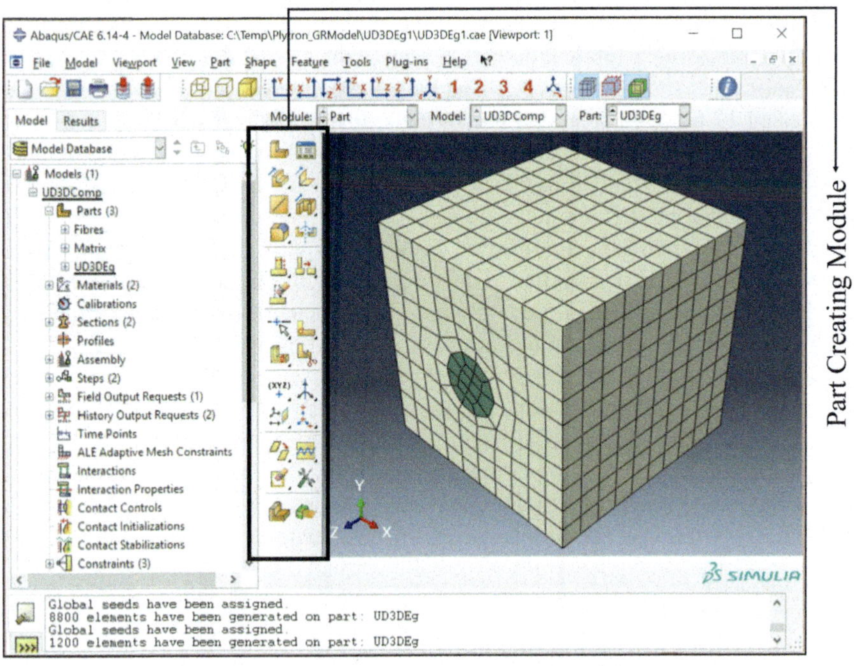

Fig. 5.6 An ABAQUS Complete Analysis Environment showing Part/Model creating module

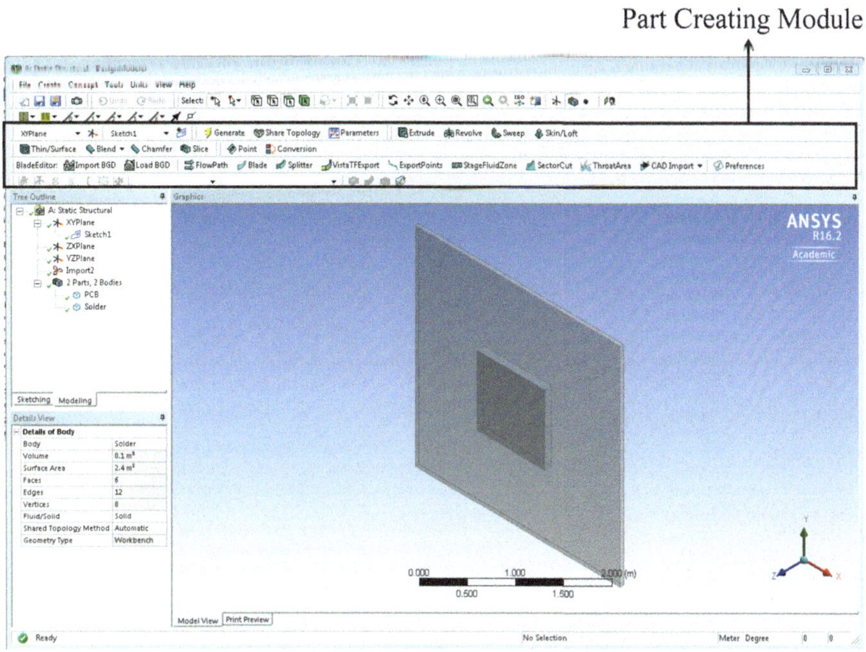

Fig. 5.7 An ANSYS DesignModeler environment showing Part creating module

The ***DesignModeler***, shown in Fig. 5.7, is a CAD-style development environment for ANSYS, whilst ***Ls-Prepost*** shown in Fig. 5.8 provides similar model creation interface for LS-DYNA. Gone are the days when FEM solvers did not have a CAD environment to support the pre-processor aspect of the FEM process. Most interactions with FEM framework are at this model development stage and it is commonplace for FEM users of a given software package to spend time learning how to use the development environment for creating a virtual domain to a given problem.

5.4.2 Automated Virtual Domain Creation Using User-Defined Scripts

The model creation strategy of Sect. 5.4.1, is often targeted to simpler geometries and first-stage creation of the virtual domain. However, in industrial and product design/validation processes, geometries might need to be created quickly and according to already laid down principles. Also, during parametric and model

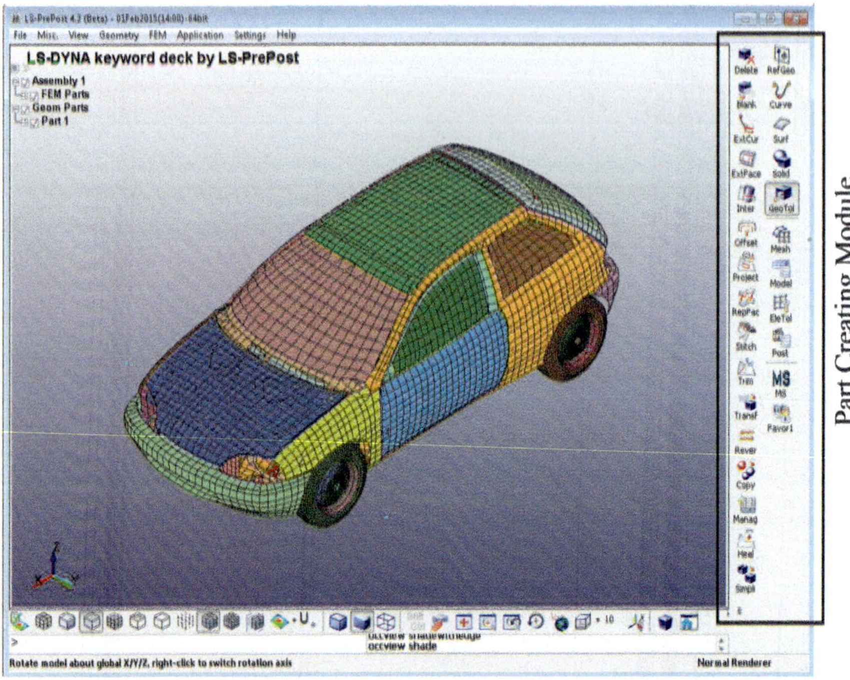

Fig. 5.8 An LS-Prepost model creation environment showing Part/Model creating module

development stages of the FEM process, automation of virtual domain creation is usually required. One of the ways of doing this is to use user-defined scripts, embedded within the FE in-built model development environment. The use of such user-defined scripts is quite commonplace in many of the flagship FEM software packages in the market today. In ABAQUS, this is usually a Python script, whilst in ANSYS, it is a JavaScript. These user-defined scripts contain the commands that drive the model creation.

The user-defined model creation script is usually for repetitive virtual domain creation. For example, let us consider a quality assurance procedure within an organization in which as part of the product design, different loading conditions are imposed on the same geometry to assess its load-bearing capacity. For this instance, the virtual domain (product shape) is the same except for the imposed boundary conditions. Therefore, the domain creation (per batch) can be integrated with the relevant FEM process through a user-defined script. In fact, when new products are designed and accepted for production, the processes involved in the product development within a CAD environment are often documented and saved by the product development team. These are usually in form of a user-defined script/object file, which employees can use for further studies.

A typical top-end of a Python script used for model creation in ABAQUS is shown in Algorithm 5.1. The scripting language used here (i.e. Python) is an object-oriented programming language. At the top-end of the Python script, several objects files that control the ABAQUS CAE are called using the `*import` command. The user can create this script from scratch and use same for automatic virtual domain creation. However, the developers of ABAQUS CAE have included an interface that creates a similar user-defined file during every ABAQUS model creation exercise. This way, the user can simply create the model once in ABAQUS and the resulting user-defined script (called `*.jnl` journal file) can be subsequently used to re-create the geometry if needed. A similar procedure applies for ANSYS and the top-end of a typical ANSYS JavaScript is shown in Algorithm 5.2.

Although these user-defined scripts are just snippets, when the complete script for ABAQUS is executed, it will result in the image shown below in Fig. 5.9a. This represents a **2-D** unidirectional composite with the fibres (represented as circles) randomly distributed within a polymer matrix (represented as the bounding box) [3]. Also, the execution of the full ANSYS JavaScript will result in the output shown in Fig. 5.9b, i.e. randomly distributed circular voids within an intermetallic layer, formed in solder joints during electronic components assembly [4].

Algorithm 5.1 A snippet of ABAQUS user-defined Python script.

```
"""
#*****************************************************
RVE2DFibre Creation Script: UDComposite2D
Author: Dr. MI Okereke
Date:  02-Aug-2017
Purpose: Creates Abaqus model (RVE2DFibre)
using inputs generated from MCRVEGen algorithm
#*****************************************************
"""
#Import Abaqus-related (Python) Object files
from abaqus import *
from abaqusConstants import *
import __main__
import section
import regionToolset
import displayGroupMdbToolset as dgm
import part
import material
import assembly
import step
import interaction
import load
import mesh
import job
import sketch
import visualization
import xyPlot
import displayGroupOdbToolset as dgo
import connectorBehavior
```

```
#*****************************************************
# CREATE MATRIX AND FIBRE MATERIALS/SECTIONS HERE
#*****************************************************
mdb.models['Model-1'].Material(name='matrix')
mdb.models['Model-1'].Material(name='fibre')
mdb.models['Model-1'].HomogeneousSolidSection(name=
              'matrixSection', material='matrix', thickness=None)
mdb.models['Model-1'].HomogeneousSolidSection(name='fibreSection',
                            material='fibre', thickness=None)

#Create Viewport
session.viewports['Viewport: 1'].partDisplay.setValues(mesh=OFF)
session.viewports['Viewport: 1'].partDisplay.meshOptions.
                                     setValues(meshTechnique=OFF)

s = mdb.models['Model-1'].ConstrainedSketch(name='__profile__',
                                              sheetSize=266.573)
g, v, d, c = s.geometry, s.vertices, s.dimensions, s.constraints
s.setPrimaryObject(option=STANDALONE)
session.viewports['Viewport: 1'].view.setValues(width=30,
                                                   height=15)

#*****************************************************
```

Algorithm 5.2 A snippet of ANSYS user-defined Javascript.

```
//*************************************************************
// RVE2DVoids  Creation Script: ANSYS_Model_Snippet.js
// Authors:    Dr. MI Okereke & Dr. KC Otiaba
// Date:       3-Aug-2017
// Purpose:    Creates 2D voids using inputs generated
//                          from MCRVEGen algorithm
//*************************************************************
function planeSketchesOnly (p)
{
// Create a 2D Plane; setting both the X- and Y-Axis
p.Plane     =   agb.GetActivePlane();
p.Origin    =   p.Plane.GetOrigin();
p.XAxis     =   p.Plane.GetXAxis();
p.YAxis     =   p.Plane.GetYAxis();

// Create a Sketch upon which voids and solders are plotted
p.Sk1       =   p.Plane.NewSketch();
p.Sk1.Name  =   "RVE2DVoidedSolder";

// Create the edges that represent VOIDS and SOLDER Systems
with (p.Sk1)
{
// -------------------------------------------------------------
// Create VOIDS for given coordinates and radius, R

  p.Cr1 = Circle(2.151190082890, 8.4495014731352, 2);
  p.Cr2 = Circle(3.4403049559025, 15.979986759024, 2);
  p.Cr3 = Circle(3.7927144781531, 2.6975059333303, 2);
```

```
  p.Cr4 = Circle(3.908776437659, 22.133913488772, 2);

// ----------------------------------------------------------
//Create Boundary Lines
  p.Ln62   = Line(0, 0, 0, 26.89871173662);
  p.Ln63   = Line(0, 29.22126753434, 0, 30.98194398506);
  p.Ln5001 = Line(50, 0, 50, 26.89871173662);
  p.Ln5002 = Line(50, 29.22126753434, 50, 30.98194398506);
  p.Ln6001 = Line(0, 0, 16.46797781616, 0);
  p.Ln6002 = Line(20.277604238346324, 0, 24.92661792641, 0);
  p.Ln7001 = Line(0, 50, 16.46797781616, 50);
  p.Ln7002 = Line(20.27760423834, 50, 24.92661792641, 50);

// ----------------------------------------------------------
//Create Construction Points as centres of all surface voids
  p.Pt9001 = ConstructionPoint(-1.35127226180, 32.4495746356);
  p.Pt9002 = ConstructionPoint(0.0261187678472, 43.271929550);
}
//Finish
agb.Regen(); //To insure model validity
//End DesignModeller JScript for creating numerical voids
//**********************************************************
```

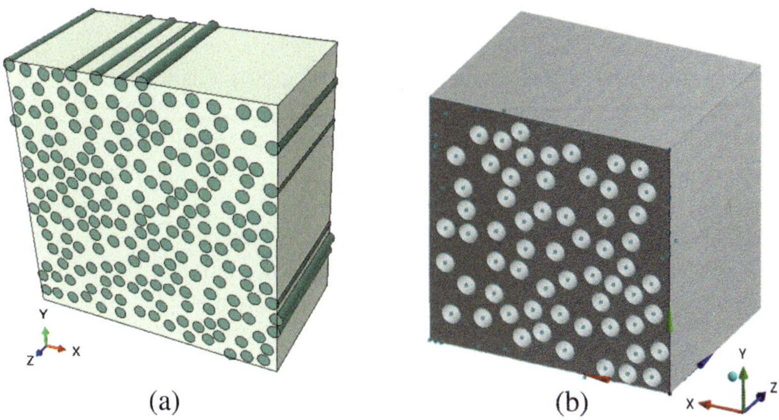

Fig. 5.9 Example virtual domains: (**a**) unidirectional composite, generated from ABAQUS Python script (i.e. Algorithm 5.1); (**b**) voided inter-metallic layer, recreated from ANSYS JavaScript (i.e. Algorithm 5.2)

5.4.3 Import Virtual Domain from Third Party CAD Software

This approach of creating a virtual domain is similar to creating the virtual domain using the FEM software package's in-built CAD interface. The only difference here is that the FEM geometry model would have been created in a third party CAD software package such as SolidWorks, AutoCAD, Google (Trimble) SketchUp, CATIA V5, Pro/Engineer, etc. Complex structures are best modelled in a CAD-dedicated package with comparative ease and much flexibility. Such tools have extra virtual domain-creating functionalities that make it straightforward to create complex models. Usually, there is an interface module (see Sect. 4.2.4) in most FEM software packages to import these third-party CAD-generated geometric models. Once generated, they are then able to be used for running simulations provided appropriate boundary conditions, material model definitions and contact issues are implemented.

5.4.4 Import Virtual Domain from Micro-computed Tomography Images

To undertake multi-scale analysis of the mechanics of a typical engineering material, it is important that any *Representative Volume Element* (**RVE**) of the material must reflect the exact microscale structure of the test material. To this end, any modality that will provide a photographic or volumetric information of the microstructure of the test material becomes absolutely essential to the FEM virtual domain creation exercise. Typically, different imaging modalities are used to capture the microstructure of test materials.

The *micro-computed tomography* (μCT) or similar modalities, have continued to gain importance as viable methodologies for obtaining volumetric data on a material's microstructure. The μCT initially acquires the images of the test material by planar scans through the whole material. Following that, the planar images are assembled/reconstructed into a **3-D** image system. This image is then imported into an FEM solver as the 'exact' virtual domain for the problem.

This domain-generating approach has become quite commonplace in biomedical engineering research. Such research involves materials that have complex shapes and microstructures and these must be used exactly during the FEM process. The *Materialise Mimics Innovation Suite* adopts this approach during the creation of virtual domain required for FEM studies of the human anatomy. Figure 5.10 shows examples of FEM outputs derived from μCT images of woven ceramic composites [5], bone [2] and foam [1] structures. All these structures are subjected to compression.

5.5 Conclusions

In this chapter, we have discussed the first of the FEM principles and this relates to creation of virtual domains. The virtual domain is defined as an idealized construct that houses the physical problem under investigation. We have discussed the concept

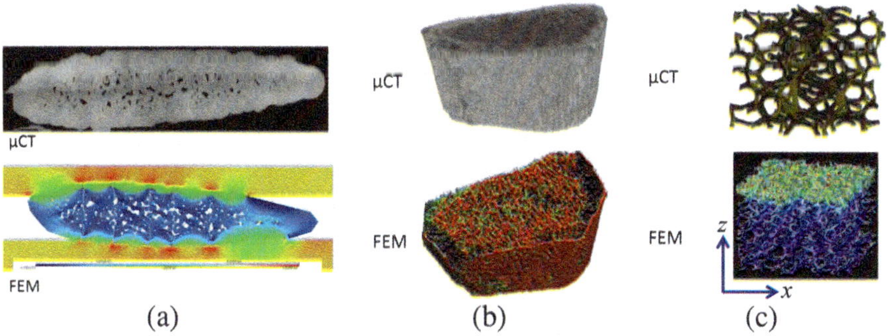

Fig. 5.10 Examples of virtual domains created using micro-CT images: (**a**) woven ceramic composite [5], (**b**) bone microstructure [2] and (**c**) segment of a foam structure [1]. Note: Each of these shows a comparison of the μCT *images and the corresponding FEM output. All test materials were subjected to compression*

of representative volume elements as a subset of a virtual domain. Also, the effect of length scale in isolating an RVE for a given problem has also been presented. The chapter concludes with a discussion of the different strategies for creating virtual domain. In the next chapter, we will explore the next FEM principle i.e. FEM meshes.

At the end of this chapter, the reader should be able to:

(a) Define virtual domains and explain its place in the FEM process.
(b) Explain the importance of RVEs and length scales in describing a virtual domain.
(c) Identify and deploy different automatic and manual strategies for creating virtual domains.

5.6 Problems: Design of Virtual Domains

Problem 5.1 As part of the product design process for the manufacture of new Cocraft Ratchet spanner designs, the engineer is required to undertake finite element analysis of candidate designs to ensure they are structurally sound, as well as easy to use. Develop a virtual domain representation of the mid-range spanner shown in Fig. 5.11 below. Assume a **2-D** representation of the problem

Problem 5.2 During the compression moulding of polypropylene (PP) plates, voids were found to form within the bulk of the material. The voids were randomly distributed and of different shapes. To investigate the effect of these voids on the

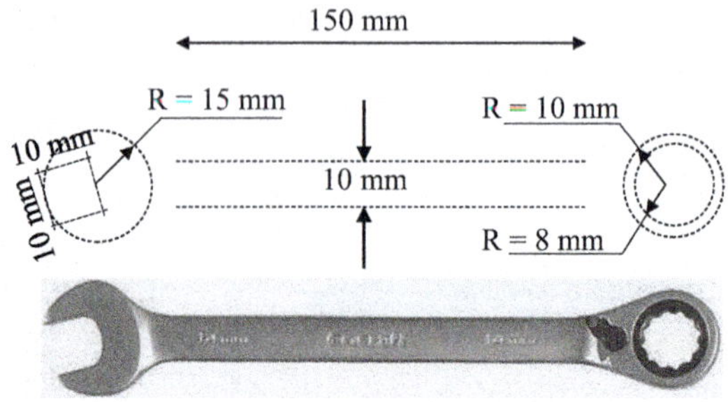

Fig. 5.11 A Cocraft Ratchet spanner with its associated dimensions

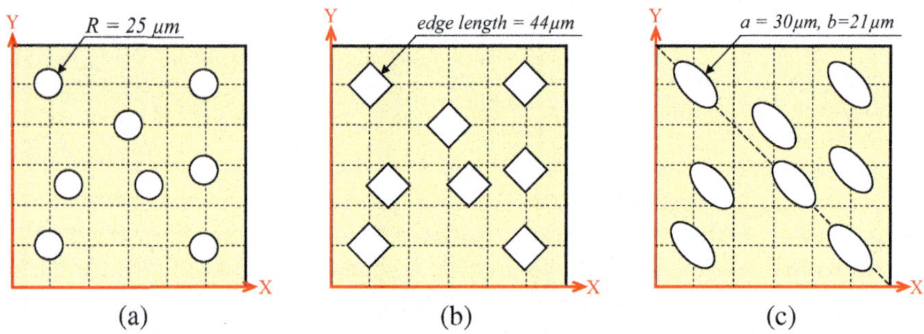

Fig. 5.12 Representative volume elements of voided PP plates for: (**a**) circular, (**b**) rhombic and (**c**) elliptical void shapes. *Each RVE is of size:* 120 μm × 120 μm

mechanical response of the plates, three 2D RVEs were isolated for the study of the effect of void shapes for the void distribution shown in Fig. 5.12. The RVEs are of size: 120μ m × 120μ m. Using any domain creation strategy of your choice, create the virtual domain representation of the three void patterns. *Note: The area fraction, A_f of the voids have been kept constant for all the three RVE types.*

Problem 5.3 The portal frame structure shown in Fig. 5.13 is made of universal I-beam designated as UB203 × 102 × 23. The structure is fixed securely on the ground, and two co-axial cables give further support to the structure. The cables comprise a solid steel core of diameter, $d_s = 20\,\text{mm}$, enclosed by an aluminium sheath of external diameter, $d_a = 22.5\,\text{mm}$. Construct, using your chosen domain-creation strategy, the **3-D** virtual domain for this structure.

Problem 5.4 A short-fibre composite (with the fibres aligned in the longitudinal direction) consists of different lengths of fibre, L_f where $30\,\mu\text{m} \leq L_f \leq 300\,\mu\text{m}$. The fibre is dispersed within a cuboid of dimensions: $L = 800\,\mu\text{m}, W = 240\,\mu\text{m}$, and $H = 180\,\mu\text{m}$, as shown in Fig. 5.14. The fibre diameter, $d_f = 15\,\mu\text{m}$ and the carbon fibres are randomly alligned along the $L-axis$, within the bulk of an epoxy (i.e. plastic) matrix. Construct the 3D RVE of the short-fibre composite using any domain creation strategy of your choosing.

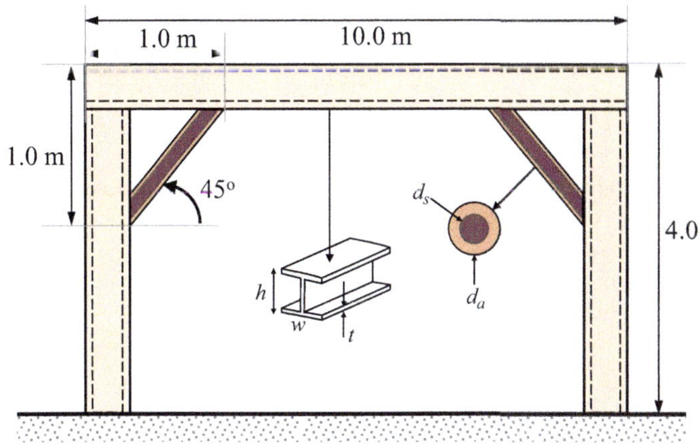

Fig. 5.13 A portal frame structure supported by co-axial cables. *Note: The I-beam has depth,* $h = 203\,\text{mm}$, *width,* $w = 102\,\text{mm}$ *and web thickness,* $t = 5\,\text{mm}$

Fig. 5.14 A 3D RVE of a short-fibre composite material

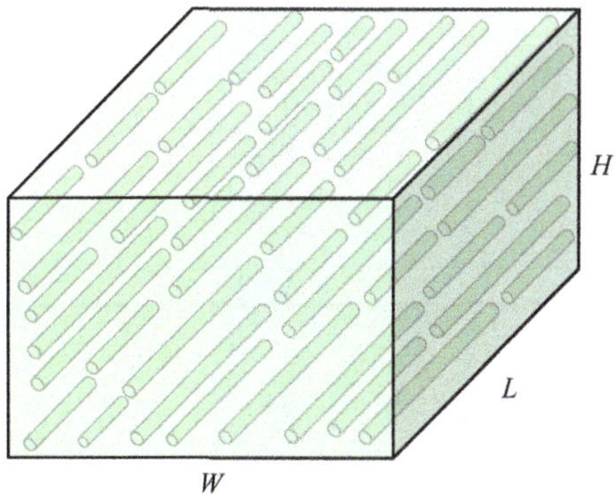

Problem 5.5 A sandwich structure consists of two backing steel plates of dimensions $12 \times 12 \times 0.3\,\text{m}^3$. The plates enclose a honeycomb sandwich core of height, $h = 3\,\text{m}$. The dimensions of each hexagonal cell and the in-plane arrangement is shown in Fig. 5.15, with the radius of the inner and outer circumscribed circles of each cell given as: $R_i = 0.75\,\text{m}, R_o = 0.80\,\text{m}$ respectively.

(a) Construct the virtual domain.
(b) Isolate a representative volume element for this sandwich structure.

Problem 5.6 A complex geometry required for manufacturing machine parts is to be reproduced several times with slight modifications to the positions of some of the components that make up the geometry. The different views one the part and its dimensions are given in Fig. 5.16.

(a) Create the user-defined Python script for the virtual domain creation in ABAQUS/CAE.

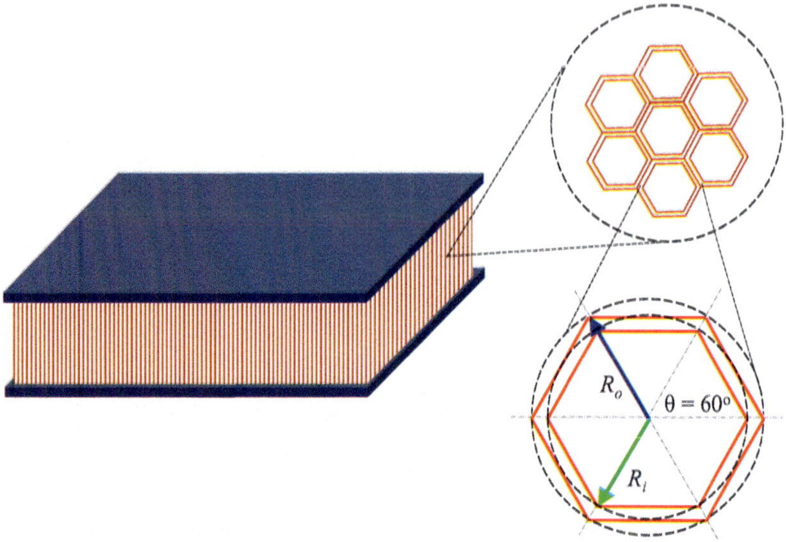

Fig. 5.15 A 3D RVE of a sandwich structure with dimensions of a honeycomb core

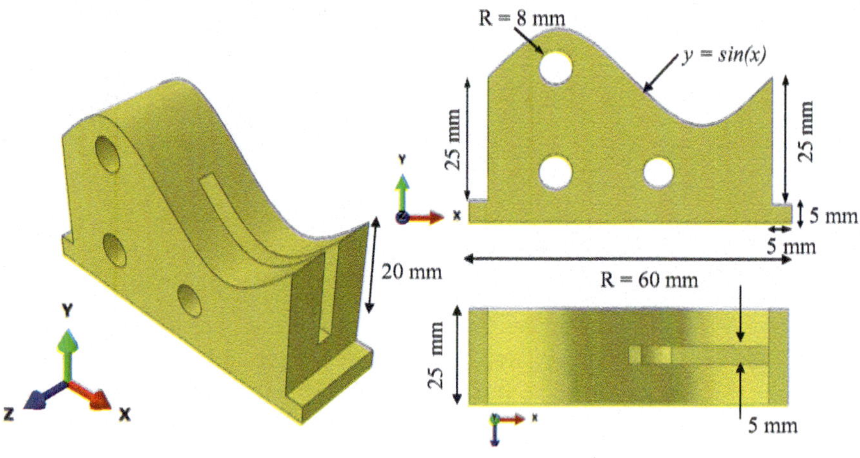

Fig. 5.16 Multiple views of complex geometry of a machine part

(b) Open the saved domain-creation macro (`*.jnl` journal file) from the relevant working directory and make the following modifications:

- Modify the diameter of the circular hole by making it 5% bigger.
- Change the lengths of the webbed feet by making them 20% wider.
- Change the depth of the side extrusion cut making it to cut half-way through.

(c) Save the modified Macro with a new Python script name and re-run it to create a whole new part geometry.

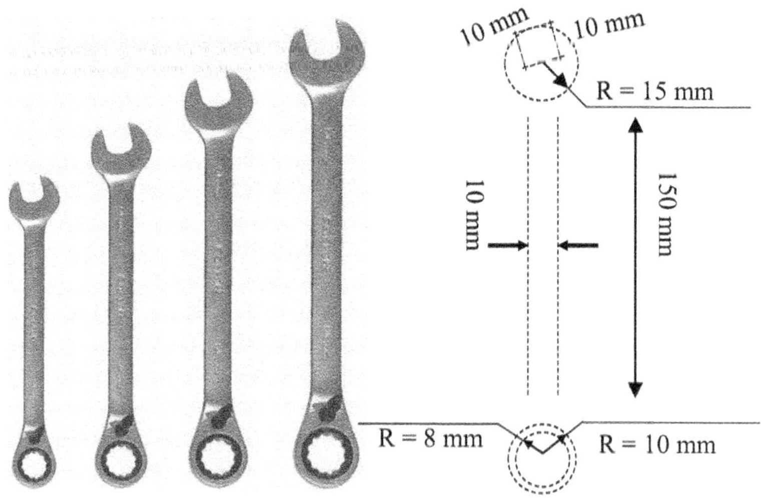

Fig. 5.17 The complete four-set spanner collection. Dimensions given are for the biggest spanner

Hint: During the first creation of the part in ABAQUS, the user should switch on the ***macro*** *function and record all the processes taken to create the part.*

Problem 5.7 Based on the spanner created in Problem 5.1, create a set of four spanners, as shown in Fig. 5.17. The spanners are formed by decreasing the dimensions of the spanner of Problem 5.1 by 10% for all dimensions of the spanner. Use the parametric function in `ABAQUS CAE` such that every dimension in the original sketch is parameterized. Assume a 2D representation of the problem, and create virtual domain of the complete spanner set.

References

1. Brydon, A., Bardenhagen, S., Miller, E., Seidler, G.: Simulation of the densification of real open-celled foam microstructures. J. Mech. Phys. Solids **53**(12), 2638–2660 (2005). doi:http://dx.doi.org/10.1016/j.jmps.2005.07.007. http://www.sciencedirect.com/science/article/pii/S0022509605001298
2. MacNeil, J.A., Boyd, S.K.: Bone strength at the distal radius can be estimated from high-resolution peripheral quantitative computed tomography and the finite element method. Bone **42**(6), 1203–1213 (2008). doi:http://dx.doi.org/10.1016/j.bone.2008.01.017. http://www.sciencedirect.com/science/article/pii/S8756328208000203
3. Okereke, M., Akpoyomare, A.: A virtual framework for prediction of full-field elastic response of unidirectional composites. Comput. Mater. Sci. **70**, 82–99 (2013). doi:http://dx.doi.org/10.1016/j.commatsci.2012.12.036. http://www.sciencedirect.com/science/article/pii/S0927025612007744

4. Otiaba, K.C., Okereke, M., Bhatti, R.: Numerical assessment of the effect of void morphology on thermo-mechanical performance of solder thermal interface material. Appl. Therm. Eng. **64**(1–2), 51–63 (2014). doi:http://dx.doi.org/10.1016/j.applthermaleng.2013.12.006. http://www.sciencedirect.com/science/article/pii/S1359431113008831
5. Pineau, P., Couégnat, G., Lamon, J.: Virtual testing applied to transverse multiple cracking of tows in woven ceramic composites. Mech. Res. Commun. **38**(8), 579–585 (2011). doi:http://dx.doi.org/10.1016/j.mechrescom.2011.08.001. http://www.sciencedirect.com/science/article/pii/S0093641311001716
6. Yu, J.S., Bagheri, N.: Multi-class and multi-scale models of complex biological phenomena. Curr. Opin. Biotechnol. **39**, 167–173 (2016). doi:http://dx.doi.org/10.1016/j.copbio.2016.04.002. http://www.sciencedirect.com/science/article/pii/S0958166916301100. Systems biology Nanobiotechnology

Chapter 6
Finite Element Mesh Generation

Abstract The quality of a mesh is crucially important if **FEM** solutions are to be deemed acceptable. Too coarse a mesh will lead to inaccurate FEM solutions. The finer the mesh, the better the convergence of the numerical solution. However, finer meshes tend to be expensive in terms of computing resources. The experienced user of FEM would have, over time, developed the skills required for creating just the right mesh for a given problem. Becoming proficient users of **FEA**, with the ability to create representative meshes of the idealized physical problem will serve as a motivation for this chapter. This chapter presents fundamentals of finite element meshes by defining nodes and elements, and the different types of elements. The chapter also describes the principle behind meshing algorithms in commercial FEM solvers. This chapter concludes by presenting reflections on quality of meshes and the type of meshes needed for different type of practical problems. It is expected that at the end of this chapter, readers should have developed a holistic understanding of the effects of meshes to the FEM process.

Keywords Finite element meshes • Mesh density • Meshing algorithms • Element types • Convergent solutions

6.1 Introduction

This chapter will continue developing the FEM principles that underpin professional development of FEM solutions to physical problems. In this chapter, we will focus on *finite element meshes* with a specific focus on mesh generation and assessing mesh quality. We document the necessity for meshes and how it can be best generated for a given physical problem. We introduce the concept of nodes and elements, and, in the later sections reflections on meshing algorithms and mesh quality for different design scenarios.

M. Okereke, S. Keates, *Finite Element Applications*, Springer Tracts in Mechanical Engineering, https://doi.org/10.1007/978-3-319-67125-3_6

6.2 Chapter Objectives

At the end of this chapter, the student should be able to:

(a) Justify the place of FE meshes in the FEM process.
(b) Determine the optimal mesh density that should lead to optimal/convergent FE solutions.
(c) Assess the effect of mesh density on the convergence of FE solutions.
(d) Understand different types of elements and what role they play in different FE problems.
(e) Explore the principles behind meshing algorithms.
(f) Justify the effect of element types and mesh density on convergent properties of the FE solutions.

6.3 Introduction to Mesh Elements

Finite element modelling involves the discretization of a geometry into finite elements, simply called *elements*, connected by *nodes*, and used in an FEM solver to obtain *approximate solutions*. The key word here is *approximate* since no FEM solution is ever totally accurate, because FEM solutions are highly dependent on the quality of the mesh created. The *mesh quality* is objectively measured by the *mesh density*.

Mesh density is a modelling parameter that represents the number of elements per unit area of the geometry.

The higher the mesh density, the better the solution and vice versa. This chapter will explore in more detail the role of meshes in the convergence of FEM solutions, the different types of elements that can be modelled and, finally describe specialized strategies for creating *effective meshes*.

6.3.1 The Convergence of FEM Solutions and Mesh Density

The convergence of the numerical solutions is important to the FEM process. Without convergence, the solution is not reliable and so FEM solvers invest a lot of computational powers to ensure that the solver arrives at a converged solution quickly. The FEM user should also develop models where the solution is converged.

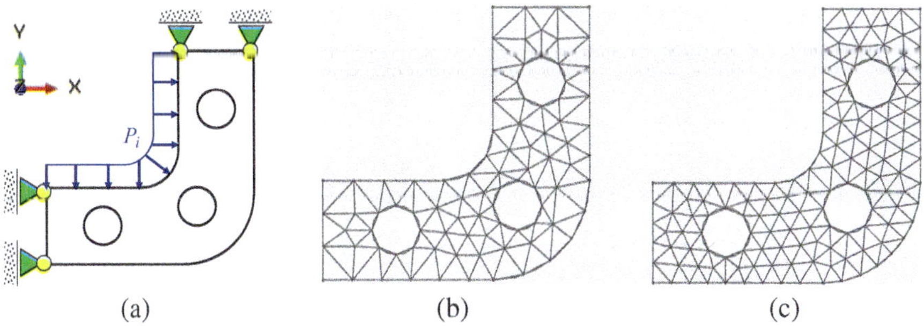

Fig. 6.1 Meshing of an elbow bracket: (**a**) physical problem (unmeshed), (**b**) coarse mesh and (**c**) fine mesh

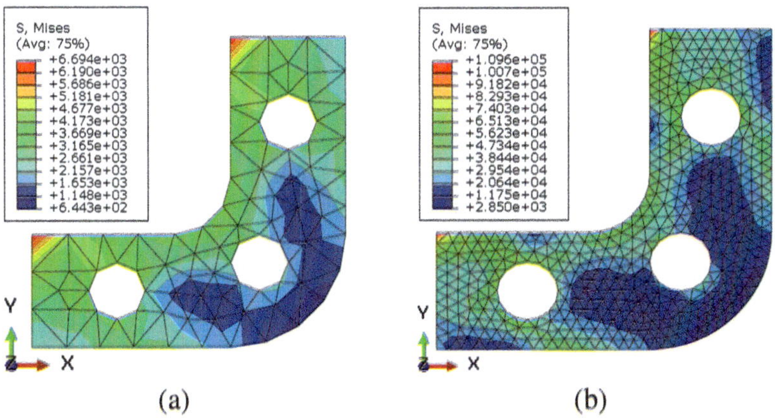

Fig. 6.2 Von Mises stress contours for (**a**) coarse and (**b**) fine meshes

Convergent solutions are as near accurate and approximate to analytical solutions as numerically possible. This is why it is important that in an FEM study, the solution must be *mesh size independent*, which is a feature of convergent solution. Without convergence, the FE solution will vary with changing mesh densities.

Let us consider a steel elbow bracket (with three holes) subjected to pressure of 2 kPa. Two ends of the washer were fixed. Two mesh types were investigated i.e. coarse and fine meshes as shown in Fig. 6.1.

The contour plots of the FEM solution (assuming elastic behaviour of the material) are given in Fig. 6.2. The result shows that the maximum stress in the meshed models changes from 6.7 kPa for the coarse mesh to 110 kPa for the fine mesh. The *intensity* of maximum stress for the fine mesh model is about 16× the maximum stress for the coarse mesh model.

Consequently, the 'spike' in stress is essentially an artefact of the meshing of the model. It is important that a convergent solution has to be derived which is independent of mesh density. As the mesh density increases, the FEM solution will

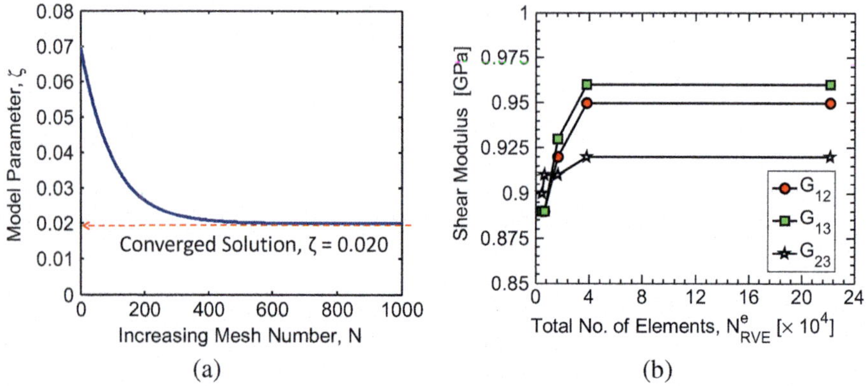

Fig. 6.3 The dependence of FEM solutions on mesh density for (**a**) generic material parameter, ζ and (**b**) Shear Modulus, $G_{ij}(\text{where} i, j = 1 \ldots 3)$ for a UD composite[11]

continue to change until an optimal solution is reached at which point there is no change in the predicted parameter (e.g. stiffness, strength, maximum stress, etc.). This is the *convergent solution* and it is the desired solution.

A graphical illustration of the convergence principle is shown in Fig. 6.3a, where a material parameter, ζ, is predicted following an FEM process for changing mesh number (and consequently density) from a few tens of meshes to hundreds of meshes. The invariant material parameter, $\zeta = 0.020$, represents a convergent solution. Every FEM investigation should explore the convergence of the mesh for different problems. The shape of the convergence is not always exponential as in Fig. 6.3a. It can also be a power law with the convergent solution very much bigger than the initial solution.

Figure 6.3b shows the result of a mesh sensitivity study on all the shear moduli of a unidirectional composite manufactured from polypropylene matrix and E-glass fibre. Mesh density here is defined in terms of the total number of elements, N^e_{RVE} [11]. The same size of **RVE** window was used. The result shows the shear moduli approached a convergent solution at a critical mesh density of $N^e_{RVE} = 4 \times 10^4$ finite elements.

6.3.2 *Understanding Mesh Elements*

The mesh shown in Fig. 6.1 was meshed using *triangular shaped elements*. This is not the only possible shape that can be used. Element shapes can be diverse and an engineer should decide on which element shape to use for the analysis of a given job. Of course, there are always different results and different operating formulations as you switch from one element type to another. This section introduces the different elements that can be used during the meshing operation. It is important for the reader to be aware of these range of elements, and their unique features, in order to choose correctly which element type to use for a given application.

Previously in the discussion under the *direct stiffness method* (see Chap. 3), the discretization of the trusses into members was completed first. Subsequently, each truss/member was analysed as a self-consistent element from which we determined the equations that led to the derivation of the *member-specific stiffness matrix*. This involved converting the member/element formulation to a local coordinate system. As part of the merging/assembly process, the local coordinate system member-specific stiffness matrix was converted to a *global coordinate system (structure) stiffness matrix*. It was at this later stage that different elements/members were summed up to form one representative structural stiffness matrix for the structure under analysis. This same procedure is central to element formulation for each element type chosen for a given FEM problem. However, slightly different steps are taken during a computer implementation of the *direct stiffness method*.

6.3.2.1 Features of Mesh Elements

There are unique features associated with mesh elements. In this section, we will describe these unique features which will inform subsequent derivations of element formulation for different element types.

Elements and Nodes

Every element is made up of nodes, which connect the ends of the elements. A simple illustration of nodes is the end-conditions of a truss member with its two nodes represented by the start and end points of the truss. Similarly, for higher order elements, the nodes could connect the ends of the bars as well as corners of the element. For a triangular shaped element, we can isolate 3 nodes per element. For a cubic or brick element, one can isolate as many as 6 nodes that are used in describing the element formulation. These nodes are also used to describe the total degrees of freedom of the element. For a truss element with two nodes per element, the total degree of freedom (for 2D analysis) is 4 (i.e. $x_1 \rightarrow 1, y_1 \rightarrow 2, x_2 \rightarrow 3$ and $y_2 \rightarrow 4$ for nodes N_1 and N_2). This is illustrated in Fig. 6.4 for truss and quad elements.

Element Dimensions

In defining a finite element, we always have to bear in mind that the FEM process can either be in space or time dimensions. Space dimensions relate to the length scale descriptor of the model while time dimensions relate to the dynamic or static characteristics of the model. As regards the space dimensions, which are required when dealing with spatial discretization of a model, elements can be one, two- or three-dimensions. These dimensions are intrinsic to the element, hence when one wants to define a finite element, a prevailing dimension has to be chosen that will also define the model analysis procedure. However, the reader should be aware

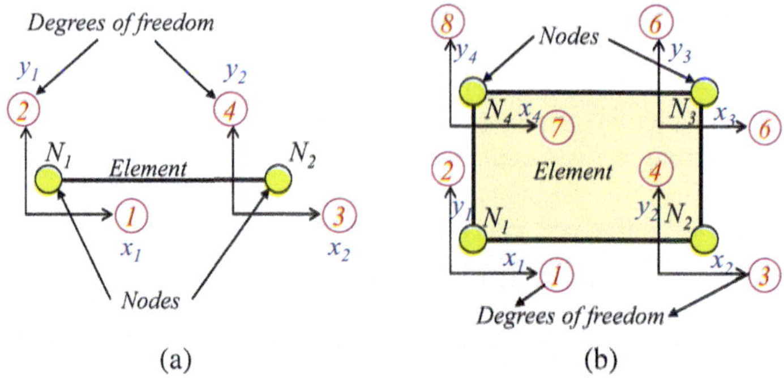

Fig. 6.4 An illustration of elements, nodes and degrees of freedom for: (**a**) 1D (truss) and (**b**) 2D (quad) elements

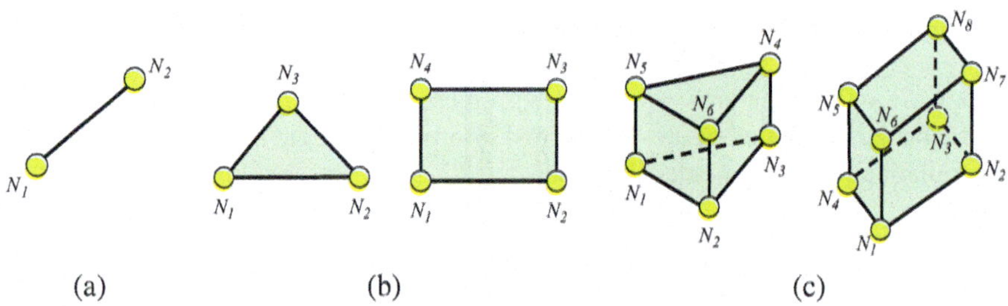

Fig. 6.5 Element dimensions for (**a**) 1D, (**b**) 2D and (**c**) 3D elements

that there are other non-standard element dimensions than those listed above, for example, *zero-dimension elements* such as *lumped springs* and *point masses*. We will not focus on these non-standard elements here.

Element Shapes

As well as the element dimensions, another characteristic of most elements that comes into play when we are describing them is their shape. Their shape generally consists of straight or curved lines/edges and ranges from lines to 2D shapes (triangles and quadrilaterals) and 3D shapes (tetrahedral, hexahedral, bricks, and other polyhedrals) as shown in Fig. 6.6. The shape determines the meshing mechanism that should be applied during the discretization of the model. In later sections on meshing algorithms, we show the role played by an element shape in the discretization of a physical model.

As the FEM user becomes more proficient with the FEM process, the user begins to develop insight into what element shapes are best used in a given problem and which is not to be used. The chosen shape of elements comes with the intrinsic element formulation for that shape and so will influence the FEM solution to some

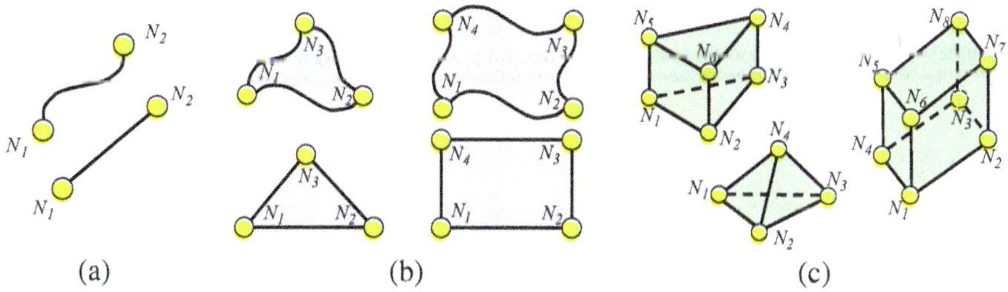

Fig. 6.6 Element shapes used in the FEM process: (**a**) 1D, (**b**) 2D, and, (**c**) 3D finite elements

extent. Consequently, it is important that the user understands and explores the impact of the element shape on the FEM solution, to ensure that the element shape does not adversely influence the FEM convergent solution.

6.3.3 Different Types of Mesh Elements

The finite element used to discretize a virtual domain is crucial for the validity of the problem. For certain domains, e.g. a thin sheet, it might be acceptable to use a 2D element to discretize the domain instead of 3D elements. This is because the stress distribution in such thin structures is dominated by the in-plane, rather than the through-the-thickness, response. Focusing on accurately determining the in-plane stress state is a valid approach here and hence the necessity for using 2D finite elements. However, for structures where a three-dimensional stress state needs to be isolated, for example the fan blade root of a jet engine, it is essential that a 3D finite element is used to discretize such a virtual domain. In this section, we describe some of the common mesh element types that are available for solution of a finite element problem.

6.3.3.1 Structural Elements

These are elements that are used to model structural members and represent fabricated structural members. They represent different sections of a structure formed of articulated members. Examples of structural elements are: *bars/trusses, beams, pipes, frames, spar/web*, and *shear plate/panel*. The previous discussion in Chap. 3 about the *direct stiffness method* treated the members as trusses, which is an instance of a structural element. Figure 6.7 shows the schematic representation of the structural elements.

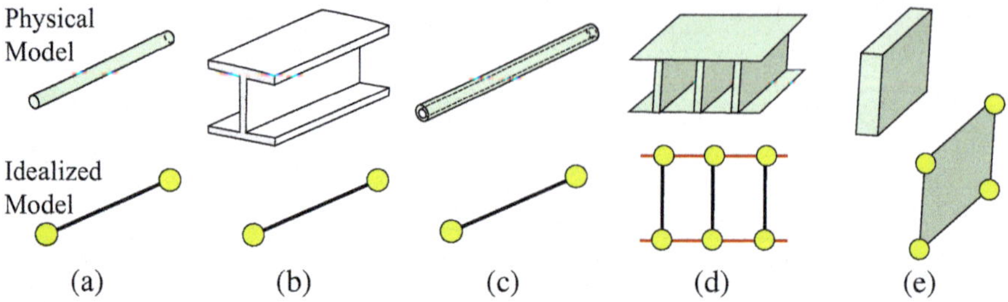

Fig. 6.7 Examples of structural elements showing both physical and idealized models for: (**a**) bars/trusses, (**b**) beams, (**c**) pipes, (**d**) spar/web and (**e**) shear panel elements

6.3.3.2 Continuum Elements

This class of elements aims to model the virtual domain as near a continuum as possible. In other words, the virtual domain is modelled as a continuous mass rather than discrete (or fabricated) parts. It models small blocks of the continuum. It is the commonest element type and can be used in a variety of situations. Continuum elements can be 2D or 3D in shape. For some problems, especially automotive sheet modelling and such like, 2D elements are more applicable for such problems since the deformation is dominated by planar behaviour. On the other hands, for more three-dimensional shaped structures such as engine components, 3D elements are preferred. Typical continuum elements are given in Fig. 6.6b for 2D solid elements (for modelling plates) and Fig. 6.6c for 3D solid elements.

In ABAQUS,[1] there are different types of continuum elements identified, for example,: (a) 1D solid (link), (b) 2D solid, (c) 3D solid, (d) cylindrical solid, and (e) axisymmetric solid elements. The *cylindrical elements* are used for modelling problems where it is best to consider outputs in terms of cylindrical coordinates ($rz\theta$-coordinates).

6.3.3.3 Special Elements

These elements have essentially the features of continuum elements except their specific characteristics comprise the physics of the problem under investigation. For example, a crack element is an example of a special element, but enriched with feature of fracture mechanics, such that crack opening can be predicted using such elements.

In ABAQUS, the following are identified as examples of *special* or *special-purpose elements*, some of which are shown in Fig. 6.8:

[1]Further details can be obtained from the ABAQUS documentation on elements i.e. *Part VI: Elements of ABAQUS Analysis User's Guide*.

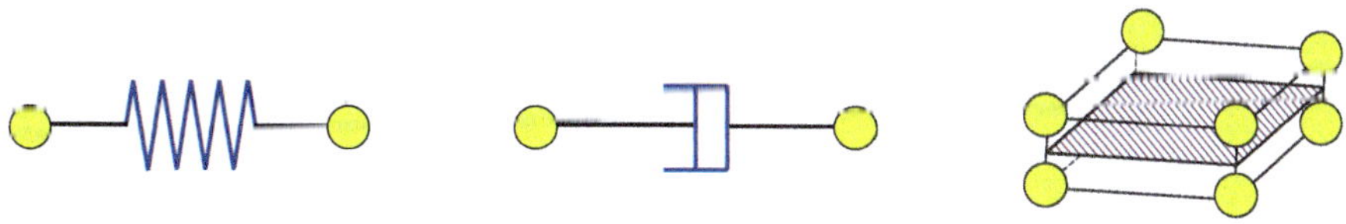

(a) (b) (c)

Fig. 6.8 Examples of special elements: (**a**) spring, (**b**) dashpot, and, (**c**) cohesive elements

(a) *Spring elements* are used to model physical springs and structural dampers.
(b) *Dashpot elements* are used to model physical problems where viscous behaviour through viscous energy dissipation response is required.
(c) *Structural membrane elements* are for modelling fabric-like structures such as tents, canopies, building foundations of stadia. Such applications do not support bending loads. A recent adaptation of membrane elements is for what we describe as *prestressed membrane elements* [7], which capture initial prestress experienced in such fabric-like structures.
(d) *Cohesive elements* are used for modelling the behaviour of bonded interfaces, adhesives, gaskets and rock fractures. This type of element is developed with a unique constitutive behaviour that captures interface/debonding response. The elements are made up of two faces (top and button) separated by an interface thickness.
(e) *Surface elements* are used to introduce thin (no-stiffness) components in solids structures. When used with multi-point constraints, these elements can be used to introduce distributed loading on a surface.
(f) *User-defined elements* are specific element types that the user defines. They are a way of extending the element library of an FEM solver so that the user can formulate a different element **constitutive behaviours** and material definitions required for solving a specific problem.

6.4 Meshing Algorithms

The technique for the discretization of a virtual domain into finite elements is called *meshing*. This is carried out using a set of algorithms which can be quite simple or quite complicated depending on the discretization the FEM user intends to carry out. Complex structures with concavities are usually very difficult to discretize, hence require specialist tools to accomplish the task. The discussion of finite element meshes is not complete without discussing the principle of *meshing algorithms*.

It is important for the reader to be aware that the process of creating meshes is not trivial. It requires significant computational resources and the technical know-how of the programmer who develops the algorithms. A *meshing algorithm* is a set of computer commands/instructions for discretizing a virtual domain.

In most implementation of meshing algorithms, the so-called *Delaunay Triangulation* is central to how meshes are created. The original classic implementation of Delaunay triangulation was strictly for discretization of 2D systems using triangular elements, but the principle has been extended to 3D systems. At the core of the triangulation is the *Rupert's algorithm* [2, 13] or *Delaunay refinement* [8, 15], which defines how rapid meshing can be achieved with minimal Delaunay triangulations [3, 4].

6.4.1 MESH2D Automatic Mesh Generator

Most commercially available FEM solvers are equipped with their patented implementations of the meshing algorithms – most of which can be very complex and can discretize a large range of complex architectures with minimal meshing iterations. Often FE users only interact with the front-end of the FEM software package, where the meshing parameters are specified (for example, mesh size, mesh techniques).

All these mesh parameters are subsequently used by a typical meshing algorithm to create a representative mesh for the FE simulation. The intellectual property protection surrounding the meshing algorithms makes it difficult for users to understand the rudiments of meshing algorithm development using for example the *Delaunay triangulation* There is an increasing list of research work on area of meshing algorithms development.

For our purposes therefore, let us consider a typical meshing algorithm called MESH2D[2] – developed by ***Darren Engwirda*** and made freely available on the Mathworks website. It is an automatic mesh generation routine that requires the MATLAB™ platform to run, and has been developed using the principle of *Delaunay triangulation* for discretizing 2D structures. Readers are encouraged to download a copy of this software and use it to understand the capabilities of a mesh generator.

The MESH2D software gives the user ample opportunities to alter/tweak the mesh result according to any particular design requirement. This gives the user a fair amount of control on how the meshes are generated. Also, a study of the source code shows the implementation of principles of meshing algorithms according to the *Delaunay triangulation* approach. Some of the example results from the MESH2D algorithm are given in Fig. 6.9.

These mesh results shown in Fig. 6.9 are for different scenarios. The developer described the MESH2D meshing algorithm to be specifically adapted for generating 2D meshes for mainly Computational Fluid Dynamics (CFD) applications. Therefore, the examples shown address CFD problems.

[2]MESH2D is MATLAB-based source code, freely available and distributed at: MESH2D – Delaunay-based unstructured mesh-generation.

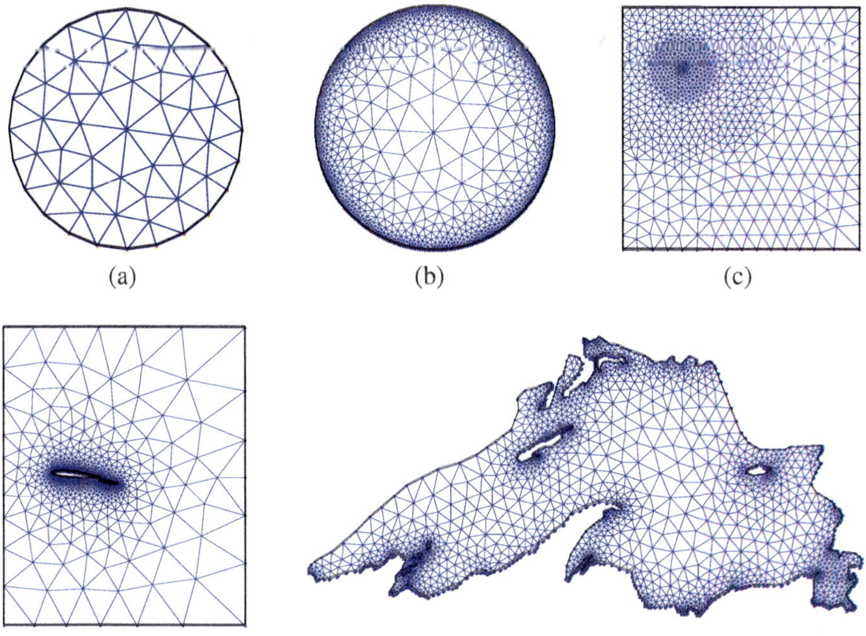

Fig. 6.9 Examples of 2D meshes derived from the MESH2D algorithm, showing: (**a**) uniform and (**b**) non-uniform meshes of a circular structure; (**c**) mesh refinement at a specific region, (**d**) mesh of fluid flow past an airofoil/flap and (**e**) mesh of *Lake Superior, North America*

Once a user has downloaded and installed the MESH2D source code from the given Mathworks repository, the user can run the `tridemo` script bundled with the MESH2D release and then the results in Fig. 6.9 can be obtained. These results include simple geometries such as squares and circles with specific possibilities for local mesh refinement.

Also, MESH2D was used to mesh specific CFD-problems such as *flow around a slender material ('sliver')* and *an airfoil*. A useful test of any 2D mesh generator is the *Lake Superior (North America)* test. This lake has complex boundaries and can be a meshing challenge. MESH2D has also been used successfully to obtain a standard mesh of it, as shown in Fig. 6.9e. In all these cases, specific model parameters were altered to achieve these desired results.

6.5 Creating Effective Meshes

The background behind the necessity of meshing to the FEM process has been established, as well as the different types of elements that a user can choose while undertaking an FEM study. At this point, we will start establishing the considerations one has to make if one is to create *effective meshes*. In this context, effective meshes refer to the adequate mesh type that should lead to quick convergence of the FEM solution for a given geometry.

6.5.1 Specialized Strategies for Creating Effective Meshes

In general, the strategies for meshing can be summarized as follows:

Element choice should be simple. The FEM solution is an approximate process and so the user should always bear in mind that the more complex the element chosen, the higher the assumptions that underpin the solution of the governing equations using approximate functions (and their weighted shape functions). As a consequence, complex elements carry with them inert solution errors and the user should be wary of those. It is important that you only choose the simplest element capable of modelling the solution. Although element choice has to be simple, the user will introduce more error to the solution by using the wrong type of elements. For example, truss structures are best modelled using either truss or beam elements. Using a three-dimensional solid element will introduce some errors and may not necessarily improve the solution obtained using simple truss elements.

The coarsest mesh that leads to convergent solution should always be used. There is no to refine the mesh further when you can obtain a convergent solution from a less refined mesh. The problem is that the user may not always know what mesh density is optimal for the solution. Therefore, always carry out a mesh sensitivity analysis to decide on optimal mesh densities for your solutions. Restrict the solution process to the coarsest mesh possible and save on computational resources. The larger the mesh density, the larger the stiffness matrix for the problem, which will require more computational storage space for operating/inverting the matrix. This makes for an expensive computational cost to the solution and should be avoided where possible.

Higher order or specialized elements should be used carefully. As a result of the complex mechanics that inform the element behaviour for specialized elements (i.e. infinite elements, crack elements, super-elements, macro-elements, etc.), the user should always be careful when working with these elements. Never use them unless you understand what they represent and how best to interpret the FEM output involving such elements. The user should regularly consult the FEM solver's help file or documentations on how the model outputs are best interpreted for the elements used.

The appropriate mesh for complex geometry must be used. Some virtual domains have complex/irregular configurations and a continuous/structured mesh may not always be suitable for the solution. *Continuous meshes* represent meshes where there is minimal variation of the element parameters along the virtual domain axes. For example, the element sizes of a square plate, are kept the same across the whole bulk of the material. The converse to them is an *unstructured mesh* suitable for irregular domains, as in cardiac modelling, for example. The default settings of many meshing algorithms are for continuous meshes, however, it is known that for certain problems, the use of different mesh configurations in the model, can lead to fast convergence of the FEM solution [5, 6, 10].

6.5.2 Dealing with Stress Concentrators

There are areas in a model geometry that require special attention when creating an applicable mesh for the structure. One of such is areas where *stress concentrators* exist. The mechanics of stress concentrators tells us that such areas experience localized increases in stress and are susceptible to initiation of failure of the structures. They need to be given careful attention to ensure that the FEM solution converges accurately without being influenced by the effects of defective/poor meshes.

The approach for dealing with stress-concentrator areas is to refine the mesh in those localized regions. This is most likely going to influence the solution process as illustrated in Fig. 6.10. The V-notch rectangular strip is fixed securely (in $x-$ and $y-$axes) at the base/lower end and pulled by a pressure load of 5 kPa at the upper end of the strip.

For the coarsely meshed stress concentrator (see Fig. 6.10a), the band (darkest region) of interest where failure can happen is quite broad resulting in peak von

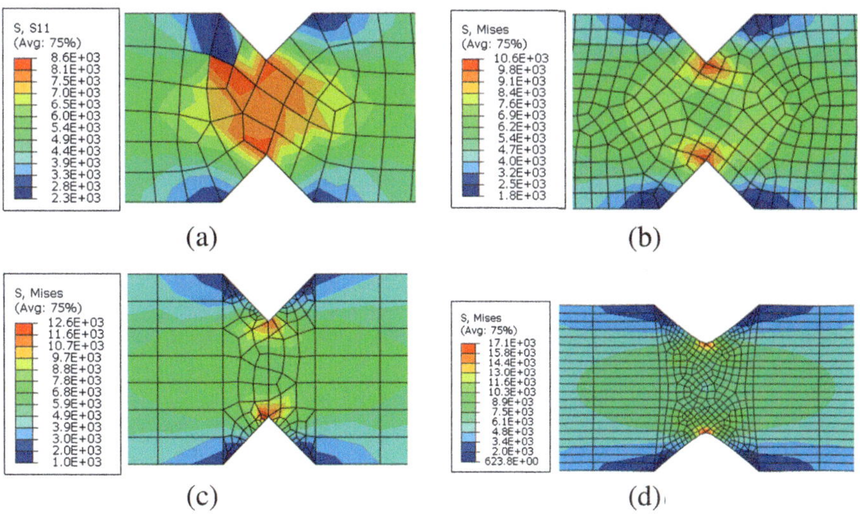

Fig. 6.10 Local mesh refinement effect on tensile loading of metallic strip (**a**) coarse A, (**b**) coarse B, (**c**) coarse C, and (**d**) refined meshes

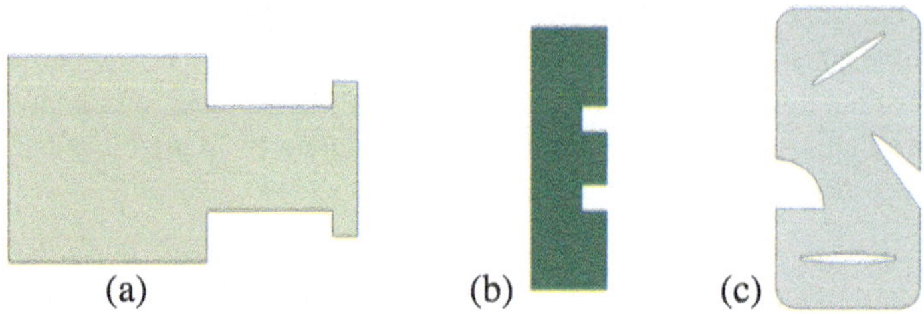

Fig. 6.11 Common examples of stress concentrators: (**a**) sharp non-filleted changes in cross-section (**b**) square notches (**c**) irregular curved notches and grooves

Mises stress of 8.6 kPa. The first refinement can be carried out by meshing the whole structure as shown in Fig. 6.10b, which led to a reduction in the critical band, but the peak stress increased to 10.6 kPa. However, the approach is expensive on computational time. Also, a lot of the mesh was allocated to large portions of the strip with no stress changes, hence making such detailed meshing redundant.

The right strategy is to locally mesh the region of interest as shown in Fig. 6.10c, which led to similar predicted critical failure band as in Fig. 6.10d. This case resulted in a higher peak stress of 12.6 kPa. This study illustrates the importance of local refinement on stress concentrators, especially for the peak stress which, without a mesh refinement, can be under-predicted with disastrous consequences.

Other examples of stress concentrators include: (a) sudden changes in cross-section; (b) square notches; (c) circular notches; and (e) crack shapes as illustrated in Fig. 6.11. In all these cases, the approach should always be locally meshing regions with stress concentrators before undertaking the FEM simulation. Of course, a mesh sensitivity study has to be carried out to ensure that the mesh quality is not influencing significantly the final solution.

6.5.3 Element Validation Tests

It is common practice to always carry out an *element validation* test to assess the suitability of a chosen/designed element for a particular numerical simulation. This has to be carried out as part of the test process during implementation of that element within a computer program. Two tests are commonly used in element validation: *patch* [9] and *single-element* tests [12].

The *element patch test* is a simple test used to check for the quality of the finite element being used. It works on the principle of checking that the finite element solution agrees with the exact solution of the partial differential equation that describes the physical problem. It establishes the necessary condition which all elements have to agree for them to be considered satisfactory for a given solution. It was used to check for any inherent programming errors during the development of the element. Lately, it has become also the standard for checking the ease of convergence of an FE solution as the mesh size changes [14, 16, 17].

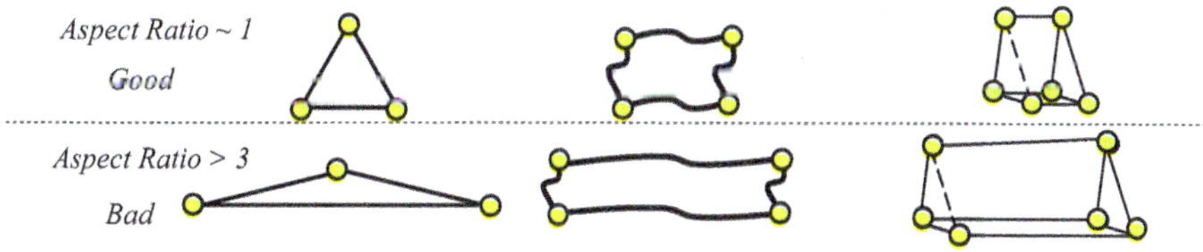

Fig. 6.12 Elements with unbalanced element ratios

The *single-element test* involves the testing a single element alone, with appropriate boundary conditions to assess the constitutive, thermal or acoustic behaviour to be modelled. The results of the tests are usually compared against an available analytical solution. The single element tests usually considers geometrical properties of the element such as convexity, aspect ratio, skewness, taper, and out-of-plane warping [1].

Also, at the core of the single-element test is the assessment of the *element aspect ratio* to ensure it is balanced. The aspect ratio is the ratio of the longest side of the element to its shortest side. Balanced aspect ratios tend to 1, i.e. unity, but unbalanced ones tend to ∞, i.e. infinity. The higher the aspect ratio, the worse the quality of the chosen element.

In practice, most FEM solvers use automatic mesh generation to create nodal points and elements, which ensures that the element aspect ratio is balanced. However, if the user is manually specifying the seeds/nodal points during the discretization of the model, an error can arise from the user specifying an element with unbalanced aspect ratios. Typical examples of the element types the user should be wary of are given in Fig. 6.12 for triangular, quadrilateral and wedge/petrahedral elements. As a rule of thumb, the FE user should choose elements with aspect ratios that are about 3 or less and anything above that becomes worrisome.

However, there are some instances where it is inevitable that elements in such models must have 'slender' geometries. For example, the discretization of 3D structures that are very thin in shape, such as layered composites, pipes/rods and I-section beams, as shown in Fig. 6.13. The structures have inset images showing the unbalanced hexahedral elements.

These *unbalanced* element shapes are acceptable because the mechanics of such structures demands that in-plane properties ($x-$ and $y-$ axes) are more important than through-thickness properties. Therefore, although the element thickness is quite small, the dominant mechanics in the in-plane directions are well captured by balanced (planar) element aspect ratios. Alternatively, instead of using 3D solid elements, engineers would normally use shell elements for discretizing such cases. Shell elements have in-plane dimensions, and through-thickness properties are typically neglected. The right decision has to be made with respect to preference for either the shell elements or unbalanced solid elements. The former option is commonly preferred.

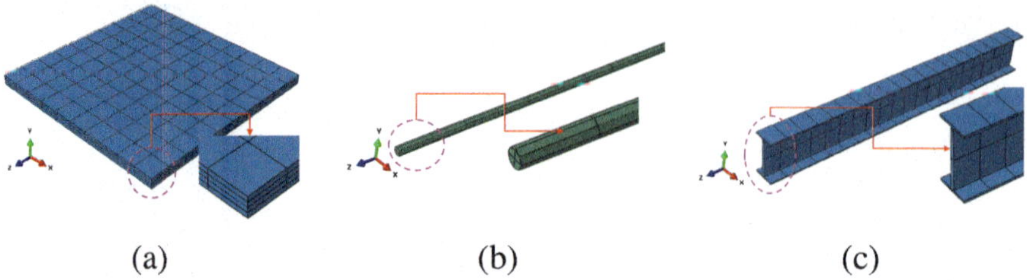

Fig. 6.13 Structural instances where elements with unbalanced aspect ratios are acceptable: (**a**) composite laminate, (**b**) rod-shape trusses and (**c**) I-section beams

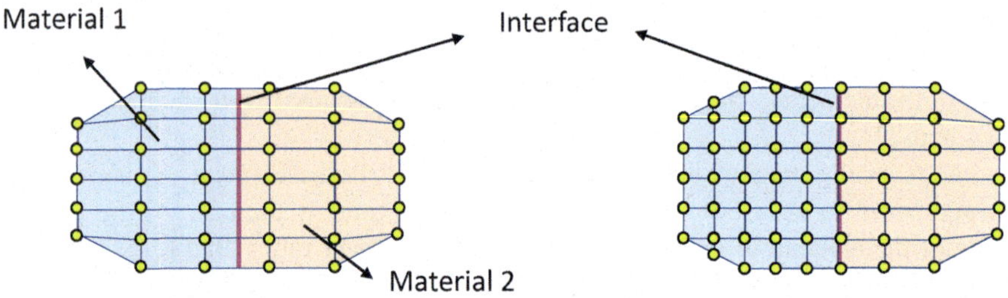

Fig. 6.14 An illustration of the principle of meshing across interfaces: (*left*) poor and (*right*) acceptable meshing

6.5.4 Meshing Across Physical Interfaces

Another important consideration the user makes when deciding on a mesh for a given structure is to consider interfaces. Interfaces can be across weld-lines, where two materials meet. The fact is that the two materials will probably have different material responses, for instance, and so require careful treatment when deciding on what meshing structure to impose on them. As a rule of thumb, one should never mesh across interfaces so that the elements share properties from two different materials. Each element should be self-consistent and must have the same material as shown in Fig. 6.14.

6.5.5 Optimal Meshing Shape

This section takes a philosophical approach by arguing which mesh shape is most suitable for a given problem. The choice of element shape depends on the problems, but in the finite element modelling community, the following two concepts are widely accepted as standard:

(a) Quadrilateral and brick/hexahedral elements are always preferred over triangular or tetrahedral elements for discretization of a virtual domain.
(b) Complex geometrics are difficult to mesh using quadrilateral or brick elements, instead triangles or tetrahedral elements are best for such geometries.

The choice you make when deciding on the mesh structure for any given problem will influence the type of solution you will obtain in the end. Any element that you choose will have a different element formulation from any other and any error in the element formulation, for a given problem, will be different from the error for the same problem discretized using a different element shape. Quantifying this associated error is important and must be investigated in your FEM simulation to ensure that the optimal minimal effects of the element shape choice is reached before the convergent solution can be considered acceptable.

Finally, with careful thought, the user can always mesh a geometry with the same element type irrespective of the geometry. As quadrilateral and brick elements are preferred, the reader should aim to discretize the virtual domain using these elements. You may have to divide the job into sub-domains of regular geometries and mesh them independently before merging the sub-domains. The *partition toolbox* in ABAQUS is a useful feature for *partitioning* up the geometry to achieve this purpose.

6.6 Conclusions

In this chapter, we have explored the FEM principle of meshing. This is an important feature of most FEM solutions and require careful attention. The reader has been exposed to meshing techniques and algorithms. We have also reflected about meshing strategies and ideas about which element shape is optimal for a given problem.

At the end of this chapter, the student would have learnt a lot more about meshing in FEM solution and how this can be developed/generated for the FEM process. The next chapter explores FEM meshes in more details by considering the mathematics behind element formulation. It is important to establish these mathematical formulations as it helps the reader understand how displacements, stresses and strains are generated for a given element shape.

At the end of this chapter, the reader should be able to:

(a) Describe the roles of nodes and elements in discretization of a virtual domain.
(b) Explain the impact of mesh density in the convergence of an FEM solution.

(continued)

(c) Understand the importance of meshing algorithms and how they are developed.
(d) Create effective meshes for a wide variety of FEM problems.
(e) Decide on the optimal element shape for a given problem.

6.7 Problems: Finite Element Mesh Generation

Problem 6.1 A plate with holes is to be analysed. The dimensions of the plate and the hole locations are given in Fig. 6.15. The plate is fixed in $x-$ direction at the back and subjected to a distributed load of 500 N/m over the front end. The plate is made from steel of dimensions: $12 \times 12\,\text{m}^2$. Each hole has a radius, $R = 0.75$ m.

Using an appropriate FEM solver of your choice (say, ABAQUS, ANSYS), create 4 mesh densities of the model using these seed lengths on the outer edges of the plate: (i) 2.4 m ; (ii) 1.2 m ; (iii) 0.6 m ; and, (iv) 0.1 m.

Problem 6.2 A steel cylindrical tube with dimensions given in Fig. 6.16, is carrying gas at a pressure of 50 kPa. The cylinder is fixed securely at the base (in all x-, y- and z-directions) and top faces.

Using the ABAQUS meshing interface or similar, undertake an element type study on the maximum stress in the cylinder. Ensure you keep the seed size constant for all three element types. The properties of the steel material are: Young's Modulus, $E = 210$ GPa and Poisson ratio, $\nu = 0.30$, and consider the constitutive behaviour of the steel to be **Hookean**.

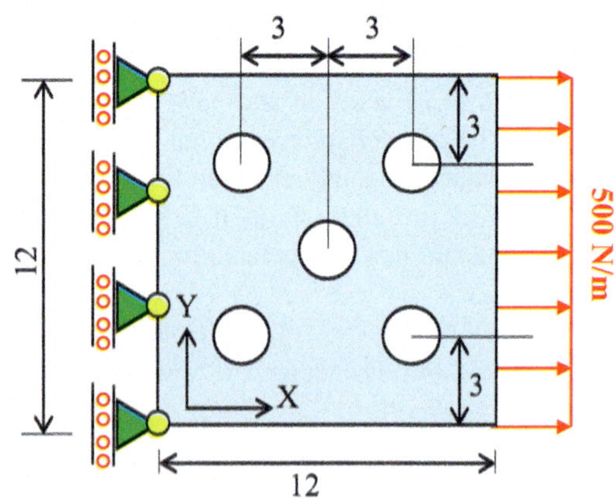

Fig. 6.15 The dimensions, loads and boundary conditions of a plate with holes

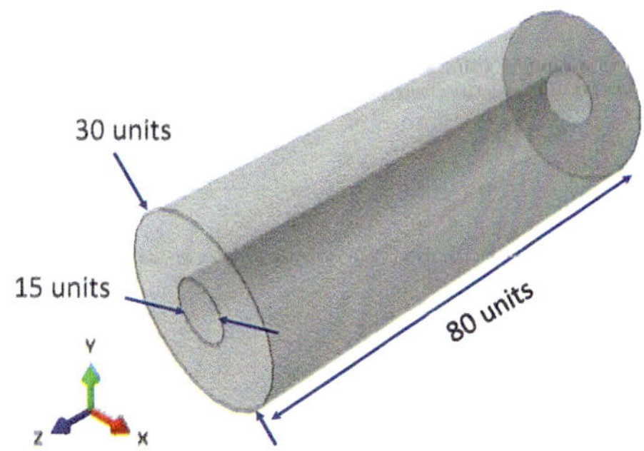

Fig. 6.16 An open-ended thick-walled cylindrical tube

(a) Mesh the structure using ***hexahedral elements*** based on sweep meshing techniques. Investigate the effect of sweeping algorithms of (i) medial axis and (ii) advancing front. Obtain the maximum stress for both cases.
(b) Re-mesh the structure using the same ***hexahedral elements***, but based on a sweep meshing technique of bottom-up meshing. Obtain the maximum stress for this case.
(c) Re-mesh the structure using ***tetrahedral elements*** assuming the free meshing technique (and default meshing algorithm) based on non-standard interior elements with growth rate of: (i) 1.05 and (ii) and 1.52. Obtain the maximum stresses in both cases.
(d) Re-mesh the structure using ***wedge elements*** with a sweep meshing technique. Obtain the maximum stress in this case.
(e) Draw a histogram plot for the comparison of all maximum stresses for the different element types.

Problem 6.3 A screw machine drill bit of US size X has dimensions given in Fig. 6.17. It is to be used in high performance drilling operations. The user intends to subject it to a higher new load, $P_{new} = 20$ kN bit during its operation against the manufacturer's advice, which recommends a maximum limit load, $P_{max} = 15$ kN. Assume the material is made of alloy steel of Young's Modulus, $E = 210$ GPa, Poisson Ratio, $\nu = 0.30$ and a yield strength, $\sigma_{max} = 1000$ MPa.

To assess the response of the drill bit to the proposed new load, the site engineer decides to carry out an FEM analysis of the problem to determine the maximum stress the drill bit can sustain without degradation.

(a) Carry a convergence study of the maximum von Mises stress by considering 5 mesh densities.
(b) Plot a graph of maximum von Mises stress versus $\log_{10}$(total mesh number, N).
(c) Based on your results, comment on whether the suggested 20 kN is an allowable loading for the drill bit.

Note: It might be easiest to create the drill bit in a dedicated CAD-package such as SolidWorks, and import it into an FEM solver of your choosing. The screw geometry is best modelled using a helical spline path with flute length = 50 mm, diameter

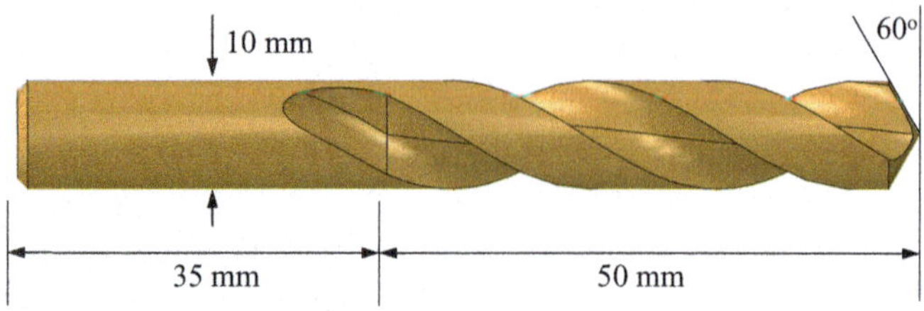

Fig. 6.17 A drill bit with parameters

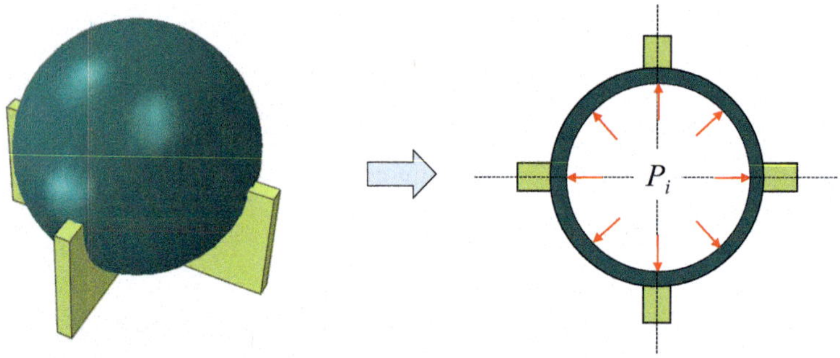

Fig. 6.18 A nuclear reactor, supported on four legs, with the right image showing the section view of the reactor viewed from the top

= 10 mm, start angle = 180°, *and one-revolution of the helix. At the base, the chamfer depth is* 1 *mm and angle,* 45°.

Problem 6.4 Investigate the associated radial stresses developed in a spherical nuclear reactor of thickness, $T = 100$ mm, radius, $R = 1.5$ m made from steel, but subjected to $P_i = 200$ MPa pressure. Design the structure such that it is firmly secured on four concrete bases. Treat all materials as linear elastic with Young's modulus, $E = 210$ GPa and Poisson ratio, $\nu = 0.30$ (Fig. 6.18).

(a) Carry out a convergence study on the radial stress in the walls of the reactor. You may consider using one-quarter of the sphere as an **RVE** for the test.
(b) Undertake a similar convergence study, leading to a convergence plot, of the tangential stress in the structure.

Problem 6.5 Design a cylindrical mould to be used as part of the manufacturing process for a plastics manufacturing company. The mould is made of copper with thermal conductivity, $K = 400$ W/mK, thickness, $T = 10$ mm, and radius, $R = 200$ mm. The mould at the start of the cooling process has wall temperature $T_C = 250$ °C and is water cooled with initial water temperature, $T_W = 5$ °C, with a flux, $Q_1 = 5$W/m^2 (Fig. 6.19).

(a) Using an appropriate FEM solver, undertake a mesh density study for the thermal resistance of the mould, considering the heat transfer process as steady state conditions. Obtain a convergence plot. You can treat the problem using half of the cylinder as an RVE for the test.

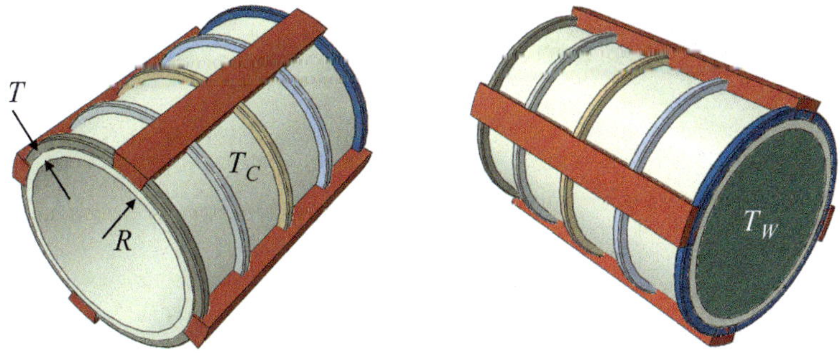

Fig. 6.19 A typical cylindrical mould, with webs and supports attached. The images show the front and back views of the mould

(b) Re-run the problem using a higher flux, $Q_2 = 15\text{W/m}^2$, and determine a convergence plot for this second scenario.
(c) Plot the two graphs (from [a] and [b] above) and comment on the effect of the changing loads on the convergence of the FEM solutions.

References

1. Akin, J.: Finite Element Analysis with Error Estimators: an Introduction to the FEM and Adaptive Error Analysis for Engineering Students. Elsevier Science, Amsterdam/Boston (2005). https://books.google.co.uk/books?id=uUsm99d8WBAC
2. Barth, T., Deconinck, H.: Error Estimation and Adaptive Discretization Methods in Computational Fluid Dynamics. Lecture Notes in Computational Science and Engineering. Springer, Berlin/Heidelberg (2013). https://books.google.co.uk/books?id=hGTsCAAAQBAJ
3. Cheng, S., Dey, T., Shewchuk, J.: Delaunay Mesh Generation. Chapman & Hall/CRC Computer and Information Science Series. CRC Press (2016). https://books.google.co.uk/books?id=oJ3SBQAAQBAJ
4. de Berg, M.: Computational Geometry: Algorithms and Applications. Springer, Berlin/Heidelberg (2008). https://books.google.co.uk/books?id=tkyG8W2163YC
5. Fortunato, M., Persson, P.O.: High-order unstructured curved mesh generation using the winslow equations. J. Comput. Phys. **307**, 1–14 (2016)
6. Gargallo-Peiró, A., Roca, X., Peraire, J., Sarrate, J.: Optimization of a regularized distortion measure to generate curved high-order unstructured tetrahedral meshes. Int. J. Numer. Methods Eng. **103**(5), 342–363 (2015). http://dx.doi.org/10.1002/nme.4888
7. Gil, A.J., Bonet, J.: Finite element analysis of prestressed structural membranes. Finite Elem. Anal. Des. **42**(8), 683–697 (2006)
8. Hjelle, Ø., Dæhlen, M.: Delaunay refinement mesh generation. In: Hjelle, Ø., Dæhlen, M. (eds.) Triangulations and Applications, pp. 131–155. Springer, Berlin (2006)
9. Irons, B., Loikkanen, M.: An engineers' defence of the patch test. Int. J. Numer. Methods Eng. **19**(9), 1391–1401 (1983)
10. Ngo, L.C., Choi, H.G.: A multi-level adaptive mesh refinement method for level set simulations of multiphase flow on unstructured meshes. Int. J. Numer. Methods Eng. pp. n/a–n/a (2016). http://dx.doi.org/10.1002/nme.5442. NME-Jul-16-0477.R2

11. Okereke, M., Akpoyomare, A.: A virtual framework for prediction of full-field elastic response of unidirectional composites. Comput. Mater. Sci. **70**, 82–99 (2013). http://dx.doi.org/10.1016/j.commatsci.2012.12.036. http://www.sciencedirect.com/science/article/pii/S0927025612007744
12. Robinson, J.: A single element test. Comput. Methods Appl. Mech. Eng. **7**(2), 191–200 (1976)
13. Ruppert, J.: A delaunay refinement algorithm for quality 2-dimensional mesh generation. J. Algorithms **18**(3), 548–585 (1995)
14. Sacco, R., Gatti, E., Gotusso, L.: The patch test as a validation of a new finite element for the solution of convection-diffusion equations. Comput. Methods Appl. Mech. Eng. **124**(1), 113–124 (1995). http://dx.doi.org/10.1016/0045-7825(95)00784-X. http://www.sciencedirect.com/science/article/pii/004578259500784X
15. Si, H.: Adaptive tetrahedral mesh generation by constrained delaunay refinement. Int. J. Numer. Methods Eng. **75**(7), 856–880 (2008)
16. Taylor, R., Simo, J., Zienkiewicz, O., Chan, A.: The patch test – a condition for assessing fem convergence. Int. J. Numer. Methods Eng. **22**(1), 39–62 (1986)
17. Zienkiewicz, O., Taylor, R.: The finite element patch test revisited a computer test for convergence, validation and error estimates. Comput. Methods Appl. Mech. Eng. **149**(1), 223–254 (1997). http://dx.doi.org/10.1016/S0045-7825(97)00085-6. http://www.sciencedirect.com/science/article/pii/S0045782597000856

Chapter 7
Mathematics of Element Formulation

Abstract The finite element method seeks to offer solutions of displacements arising when loads are applied on a structure. It can also determine the forces that results when a displacement loading condition is imposed on the structure. Finite element meshes play an important role in the derivation of the displacements for a structure. Element formulation deals with derivations needed to deduce displacements acting on a body when discretized by finite elements. Different formulations apply for different types of elements. Derivation of such displacements is dependent on shape functions. This chapter presents the mathematics that drive the element behaviour. This communication is based on simple element types ranging from one-dimensional linear elements, to high-order three-dimensional elements. This is an important chapter as it helps the reader understand how displacements, nodal forces, element stresses and strains are derived for different element types. Readers will also gain the understanding of the limitations of one element type from another, based on the mathematics of their formulation. The content here can be considered for an advanced finite element course, and if one is new to the FEM process, one can skip this chapter.

Keywords Element formulation • Shape functions • One-dimensional elements • Two-dimensional elements • Three-dimensional elements • Serendipity elements

7.1 Introduction

This chapter will continue developing the FEM principles by considering the mathematics behind simple elements used in the FEM process. *Element formulation* deals with the mathematics and assumptions that drive the development of the displacements for different finite element types. Insights gained from displacements of the elements help determine the strain, stresses and nodal forces for that type of element. Different elements yield different values, and it is this disparity between element types that contribute to the element errors seen in the FEM process. Consequently, any FEM problem, discretized by a given element type, must establish how displacements can be generated for that element type.

For example, if a 2D representation of a football is discretized using triangular elements, the mathematics of element formulation guides the reader through the

M. Okereke, S. Keates, *Finite Element Applications*, Springer Tracts in Mechanical Engineering, https://doi.org/10.1007/978-3-319-67125-3_7

processes of determining the displacements at the three nodes that make up the triangular elements. When this information is obtained, deductions can be made about the overall mechanics of the football by considering the summation of the individual displacements of the discretizing finite elements. At the end of this chapter, the reader should be able to assess the influence of one element type to another, when using the FEM process to solve physical problems.

7.2 Chapter Objectives

At the end of this chapter, the student should be able to:

(a) Develop element formulations for simple element types.
(b) Understand the effect of shape functions on element formulations.
(c) Derive the displacements, strain and stresses that result for different element types.
(d) Explore the higher order elements and appreciate the assumptions made during their development.
(e) Understand the requirements a user has to make when deriving their variants of element behaviour.
(f) Apply the element formulations in solving real life (physical) problems.

7.3 Mathematics of Element Formulation for Simple Elements

In Chap. 3, Sect. 3.4.3, we illustrated the process for formulating the stiffness matrix of a bar/member in global and local coordinates. The equations derived from that process were transposed from local coordinates to global coordinates, and subsequently used in determining the global structural stiffness matrix which was an objective value for describing the behaviour of the structure under any loading condition.

However, for the variety of elements presented above, a similar process has to be undertaken to establish an element formulation for the element type. The formulation for that element type should lead to a stiffness matrix (or similar parameter) that captures the whole behaviour of that element. Depending on the boundary conditions imposed on the structure (discretized by such an element), the expected response will result from interactions with the stiffness matrix of that element. This leads us to consider the mathematics of element formulation for diverse element systems implemented in common FEM solvers. This will only provide an introduction to this wide field of study and it is given here for sake of completeness of the subject matter.

The process of discretizing the original model into discrete equations is not straightforward for the finite element method. This is because in order to establish a holistic behaviour of the structure under investigation, the behaviours of all discrete/finite elements have to be integrated (or merged) to obtain a holistic response. Unfortunately, it is not always easy to carry out the integration on the discrete equations in their original format (called *governing equations*). Since they are required to be integrated, an intermediary form of the original equations is developed for the integration exercise. This intermediary form is called a *weak form* of the governing equation. It is the integral form of the governing equation and used for integration to formulate the finite element method.

Based on the weak forms, approximate solutions to the FEM problem are obtained. These approximate solutions will, of course, depend on the element size, hence as the element size keeps decreasing, the approximate solutions continue to converge to the optimal or convergent solution. In this section of mathematics of element formulation, we will not concern ourselves with the governing equations or the weak forms, but we will rather focus on the approximate functions and how to manipulate them to obtain the stiffness matrix equations of the many elements that make up the element library of the FEM solver. We will consider element formulations for one-, two- and three-dimensional elements.

7.4 Element Formulation for One-Dimensional Elements

In this section, we will detail the process implemented in a finite element solver for deriving and implementing the bar stiffness equations for a two-node one-dimensional linear element. Let us consider for a start a rectangular cross-section prismatic axially loaded bar subjected to a concentrated force, **P**, and fixed at the other end to a wall, as shown in Fig. 7.1. We will be using this prismatic bar/member as an instance of a two-node **1-D** linear element. The member, in terms of FE analysis, can be represented using a two-node 1-D linear element, as shown in Fig. 7.2. These elements are also used to model ropes, cables, chains, and such fabricated structural components.

Consider the deformation of the bar, $u(x)$, of length, L, cross-sectional area, A, Young's Modulus, E, axial rigidity, EA. The prescribed load, **P** over any length in the bar imposes a distributed internal load, $q(x)$, which is position-dependent. Of course, the principles developed in *Mechanics of Materials* tells us that the displacement of any point in the bar can be calculated using the *force-displacement* equation, thus (assuming linear elastic behaviour of the bar):

$$\delta = \frac{PL}{EA} \qquad \Longrightarrow \qquad u(x) = \frac{q(x)dx}{EA} \tag{7.1}$$

However, if we are going to treat the above equation in terms of finite elements, we will have to solve the governing equations (strong and weak forms) for the axial

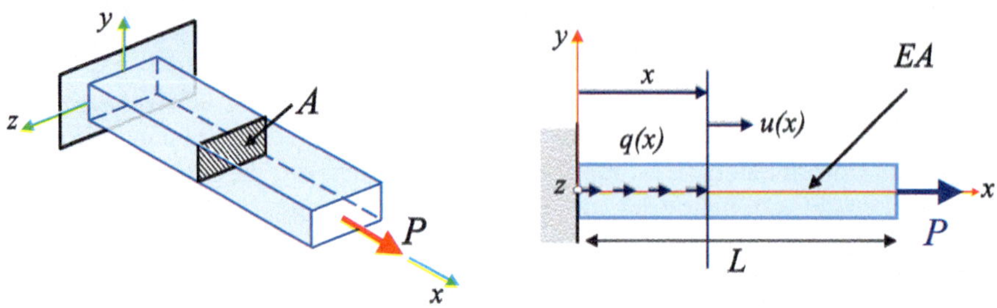

Fig. 7.1 A schematic representation of an axially loaded member showing: (**a**) 3-D view (slice of its cross-sectional area) and (**b**) 2D view with same boundary conditions

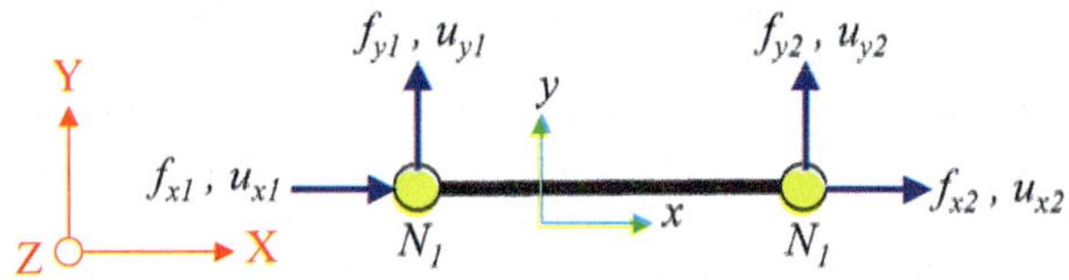

Fig. 7.2 An idealized model of the member of Fig. 7.1, showing its discretization using a two-node one-dimensional linear element. Note the corresponding local coordinates boundary forces and displacements

deformation, $u(x)$. Unfortunately, as already established, the direct integration is not always possible for all physical problems, hence we need to use approximate solutions. The methodology for arriving at these approximate solutions for the two-node one-dimensional linear element under consideration will be presented int he next section.

7.4.1 *Defining the Element Shape Functions for 1-D Elements*

In order to determine the displacement of the member, $u^e(x)$, we assume that this element displacement is a weighted contribution of the displacements at the ends of the bar. According to Fig. 7.2, the displacements at the start and end of the bar are $\mathbf{u}_1$ and $\mathbf{u}_2$ respectively, in vector form. For the bar structural element that experiences only axial deformation, we will neglect the displacement in the $y-$ axis, hence: $\mathbf{u}_1 = u_1^e$ and $\mathbf{u}_2 = u_2^e$.

The new displacement will be linear in terms of both end displacements according to Eq. 7.2:

$$u^e(x) = N_1^e u_1^e + N_2^e u_2^e \implies u^e(x) = \begin{bmatrix} N_1^e & N_2^e \end{bmatrix} \begin{bmatrix} u_1^e \\ u_2^e \end{bmatrix} \implies u^e(x) = \mathbf{N}^e \mathbf{u}^e \tag{7.2}$$

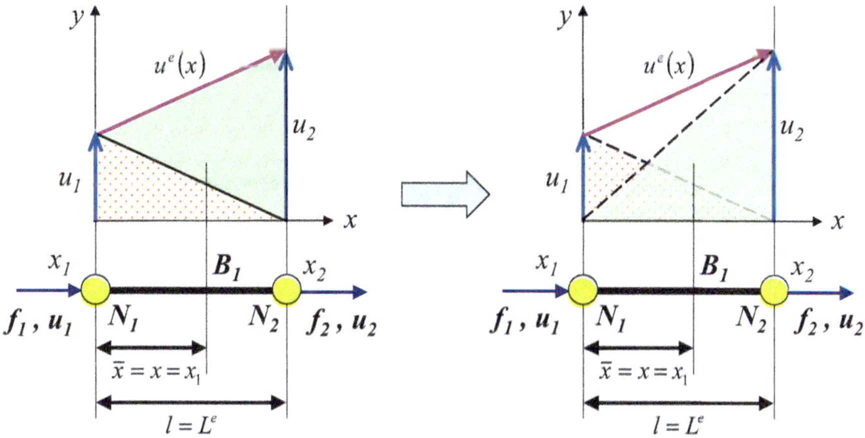

Fig. 7.3 An illustration of the interpolation by shape functions of a one-dimensional linear element nodal displacements (u_1 and u_2)

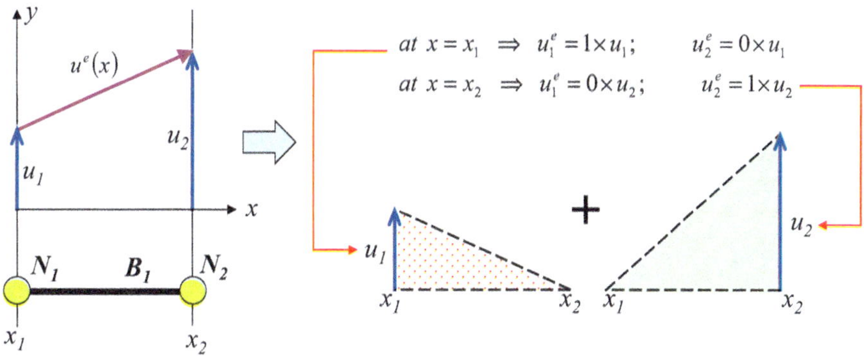

Fig. 7.4 An illustration of the operation of shape functions on the nodal values

In the above, N_1^e and N_2^e are called *shape functions* or *interpolation functions* and when these terms are collated into a matrix form, we have the shape function matrix, $\mathbf{N}^e$. The shape functions are used to interpolate on the nodal displacement such that a final displacement is obtained which represents the displacement of the 1-D element.

To visualize the concept of shape function and its effect on the element displacement, consider Figs. 7.3, 7.4, and 7.5. The displacements in graphical format of nodes N_1 and N_2 are u_1 and u_2 respectively. The contributions of u_1 and u_2 can be interpolated using shape functions, as shown in Fig. 7.5.

Based on Fig. 7.4, we notice that the shape function operates on the nodal values to yield a value or zero depending on the coordinate position. The shape function must therefore be such that it will yield either a 0 or 1 depending on its location.

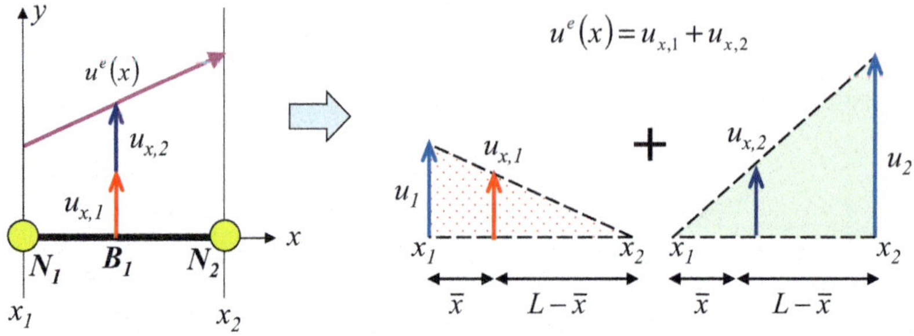

Fig. 7.5 Derivation of the shape functions for the two-node 1-D element

To derive the exact formulations of the shape functions above, we will consider again the two-node 1-D element above, but now isolate an arbitrary position, x for the element of length, L, such that the distance from the left node, N_1, can be given as: $\bar{x} = x - x_1$, as shown in Fig. 7.5. The element displacement at x is a cumulative of displacements from nodes N_1 and N_2.

Based on the right-angled triangles shown in Fig. 7.5, we can use *congruence of similar triangles* to calculate the associated distances from node N_1 displacement ($u_{x,1}$) and node N_2 displacement ($u_{x,2}$). Thus, for both triangles, we write:

$$\begin{aligned} &\text{Node } N_1: \frac{u_{x,1}}{u_1} = \frac{L-\bar{x}}{L} \implies \frac{u_{x,1}}{u_1} = \left[1 - \frac{\bar{x}}{L}\right]u_1 \implies u_{x,1} = N_1^e u_1 \\ &\text{Node } N_2: \frac{u_{x,2}}{u_2} = \frac{\bar{x}}{L} \implies \frac{u_{x,2}}{u_2} = \left[\frac{\bar{x}}{L}\right]u_2 \implies u_{x,2} = N_2^e u_2 \end{aligned} \tag{7.3}$$

Therefore, the expression for shape functions for the two-nodes element becomes:

$$N_1^e = 1 - \frac{\bar{x}}{L} \implies N_1^e = 1 - \zeta \qquad N_2^e = \frac{\bar{x}}{L} \implies N_2^e = \zeta \tag{7.4}$$

In the above expression, we identify a dimensionless quantity:

$$\zeta = \frac{x - x_1}{L} = \frac{\bar{x}}{L} \tag{7.5}$$

where ζ is called the *isoparametric natural coordinate* and whose value varies from 0 to 1 along the length of the 1-D linear element. Note also the following to be true:

$$\zeta = \frac{x - x_1}{L} \implies x = x_1 + L\zeta \tag{7.6}$$

Therefore, we can differentiate Eq. 7.6 such that we obtain:

$$\frac{dx}{d\zeta} = L \implies d\zeta = \frac{dx}{L} \tag{7.7}$$

The above parameters: N_1^e, N_2^e = *shape functions*, ζ = *isoparametric natural coordinate* and $d\zeta$ – *first order derivative of* ζ, are the bases upon which the derivations of the element formulation (in this case the two-node 1-D element). In the following section, we return to derivations of the mechanics of the two-node 1-D linear element, which will be deployed in FEM solvers. These parameters also form the basis of element formulation for beam elements and other higher order elements.

Remark 7.1 The shape function must obey a Kronecker delta function which states:

$$\delta_{ij} = \begin{cases} 0 & \text{if } i \neq j \\ 1 & \text{if } i = j \end{cases} \tag{7.8}$$

where the Kronecker delta δ_{ij}, is a piecewise function of variables i and j. Therefore, a shape function yields a value of 1 at its applicable node and a value of 0 at all other nodes. For example, $N_1 = 1$ at node $x = x_1$ and $N_1 = 0$ at node $x = x_2$. Similarly, $N_2 = 0$ at node of $x = x_1$ and $N_2 = 1$ at node $x = x_2$.

Remark 7.2 Also, a shape function must obey the property of *polynomial completeness* which requires that the following conditions have to be obeyed for any shape function:

$$\sum_i^n N_i = 1 \qquad \sum_i^n N_i\zeta_i = \zeta \qquad \sum_i^n N_i\eta_j = \eta \tag{7.9}$$

where ζ and η are *isoparametric natural coordinates* required for mapping from the local coordinates to the global coordinates; and n = total number of nodes.

The first condition: $\sum_i^n N_i = 1$, demands the element described by the shape functions must obey what are called the *rigid body modes*. The second and third conditions imply that the element described by the shape functions must exhibit a *constant strain state* i.e. any change displacement based on the shape functions should result in the same strain for the element.

7.4.2 Lagrangian Polynomial Description of Shape Functions

Having established the shape functions for 1-D linear elements, we extend the description of shape functions beyond such linear elements to now include any higher order elements. Since shape functions are central to the FEM process, it is vital that the reader is able to derive shape functions for multi-node higher order elements.

The most common way to describe the shape functions is to base them on *Lagrangian polynomial functions* whose terms are derived from *Pascal's triangle*. Elements whose shape functions are defined this way are often called *Lagrangian elements*. Pascal's triangle helps identify the monomials that have to be added to a holistic expression of the finite element trial/approximation expression required for calculating model parameters, say, displacement for example. Equation 7.10 shows a Pascal's triangle and for each level the necessary monomials needed to describe that type of element.

$$\begin{array}{lccccccccccc}
\text{Constant elements:} & \Longrightarrow & & & & & 1 & & & & \\
\text{Linear elements:} & \Longrightarrow & & & & \zeta & & \eta & & & \\
\text{Quadratic elements:} & \Longrightarrow & & & \zeta^2 & & \zeta\eta & & \eta^2 & & \\
\text{Cubic elements:} & \Longrightarrow & & \zeta^3 & & \zeta^2\eta & & \zeta\eta^2 & & \eta^3 & \\
\text{Quartic elements:} & \Longrightarrow & \zeta^4 & & \zeta^3\eta & & \zeta^2\eta^2 & & \zeta\eta^3 & & \eta^4
\end{array} \tag{7.10}$$

To formulate the complete shape functions using the monomials identified in the Pascal's triangle of Eq. 7.10, we will have to multiply, along a given row, the monomials with an unknown coefficient, α_i (where $i = 1, \ldots, n$ with n = order of the element). For example, the polynomial expressions of the shape functions for a linear, quadratic or cubic element are given in Eq. 7.11:

$$\begin{aligned}
N_{linear} &= \alpha_1 + \alpha_2\zeta + \alpha_3\eta \\
N_{quadratic} &= \alpha_1 + \alpha_2\zeta + \alpha_3\eta + \alpha_4\zeta^2 + \alpha_5\zeta\eta + \alpha_6\eta^2 \\
N_{cubic} &= \alpha_1 + \alpha_2\zeta + \alpha_3\eta + \alpha_4\zeta^2 + \quad \ldots \quad + \alpha_{n-1}\zeta\eta^2 + \alpha_n\eta^3
\end{aligned} \tag{7.11}$$

where n = total number of unknowns required to describe the shape function for a given element.

Example 7.1 A truss-based structure is to be analyzed using finite element methods. An idealized representation of one of the trusses that make up the structure is given as a two-node one-dimensional element shown in Fig. 7.6 with coordinates specified along the $x-$ axis.

Determine the shape functions, N_i, for $i = 1, 2$, that will be used for interpolating the displacements of nodes, N_1 and N_2.

Solution

The following are true for the idealized representation of the two-node element:

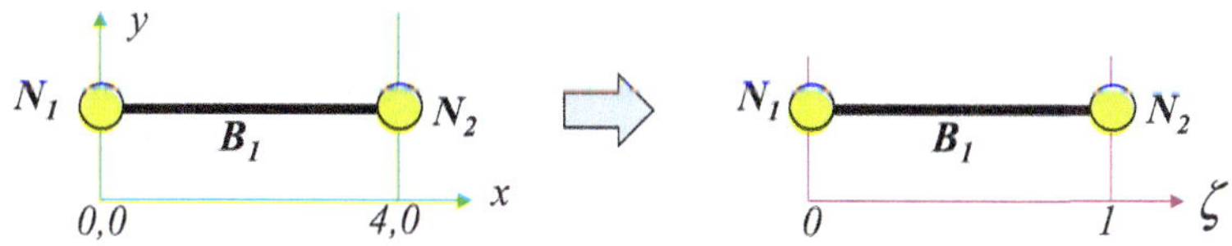

Fig. 7.6 An idealized representation of a two-node 1-D element showing (*left*) *spatial (x,y)* and (*right*) *isoparametric natural, ζ coordinates* representations

(a) For an $x-$ coordinate displacement, $u_{x,i}$, defined along the length of the element, the displacement is a weighted contribution of displacements of nodes N_1 and N_2.
(b) The weighted contribution is driven by the shape function, N_i, where $i = 1, 2$, for the two-node element.
(c) The definition of N_i can be in terms of *spatial (x,y)* or *isoparametric natural*, ζ, *coordinate* representations. Being a one-dimensional element, the displacement will be restricted solely to the one axis in this case the $x-$ axis.
(d) The relationship between spatial and natural coordinates has been defined previously as: $\zeta = \frac{x - x_1}{4}$ and $d\zeta = \frac{dx}{4}$, for element length, $L = 4$ with node N_1 located at $x-$ coordinate, x_1.

Since the element under consideration has two nodes, the expression of the shape functions, in terms of natural coordinates, must be linear functions and can be given as:

$$N_i = \alpha_1^{(i)} + \alpha_2^{(i)}\zeta \tag{7.12}$$

where $\alpha_1^{(i)}$ and $\alpha_2^{(i)}$ are *unknown coefficients* that have to be obtained.

It is imperative that shape functions yields a value of 1 at applicable nodes and 0 at all other nodes. Hence, for the element under consideration:

$$\begin{array}{llll} N_1 = 1 & \text{at } \zeta = 0 & \text{and } N_1 = 0 & \text{at } \zeta = 1 \\ N_2 = 0 & \text{at } \zeta = 0 & \text{and } N_2 = 1 & \text{at } \zeta = 1 \end{array} \tag{7.13}$$

Based on the conditions stated in Eq. 7.13, we substitute these values into Eq. 7.12 thus:

$$\text{At } \zeta = 0 \implies N_1 = \alpha_1^{(1)} + \alpha_2^{(1)}(0) = 1 \implies 1 = \alpha_1^1 \tag{7.14}$$

$$\text{At } \zeta = 1 \implies N_1 = \alpha_1^{(1)} + \alpha_2^{(1)}(1) = 0 \implies 0 = \alpha_1^{(1)} + \alpha_2^{(1)} \tag{7.15}$$

$$\text{At } \zeta = 0 \implies N_2 = \alpha_1^{(2)} + \alpha_2^{(2)}(0) = 0 \implies 0 = \alpha_1^{(2)} \tag{7.16}$$

$$\text{At } \zeta = 4 \implies N_2 = \alpha_1^{(2)} + \alpha_2^{(2)}(1) = 1 \implies 1 = \alpha_1^{(2)} + \alpha_2^{(2)} \tag{7.17}$$

In the above, $\alpha_1^{(1)}$ and $\alpha_2^{(1)}$ are the unknown coefficients for shape function, N_1. Similarly, $\alpha_1^{(2)}$ and $\alpha_2^{(2)}$ are also unknown coefficients for shape function, N_2. Solving Eqs. 7.14, 7.15, 7.16, and 7.17 simultaneously, the resulting unknown coefficients are:

$$\alpha_1^{(1)} = 1 \qquad \alpha_2^{(1)} = -1 \qquad \alpha_1^{(2)} = 0 \qquad \text{and} \qquad \alpha_2^{(2)} = 1$$

Therefore, the resulting shape functions for the two-node 1-D linear element becomes:

$$N_1 = 1 - \zeta \qquad \text{and} \qquad N_2 = \zeta \tag{7.18}$$

Example 7.2 Obtain the shape functions, N_i for $i = 1, \ldots, 3$ required for operating on the nodal displacements for the truss member shown in Fig. 7.6 with coordinates specified along the $x-$ axis.Plot the graph of the derived shape functions.

Solution
The truss-member is a three-node element hence the expression for its shape function should be a quadratic function expressed as:

$$N_i = \alpha_1^{(i)} + \alpha_2^{(i)}\zeta + \alpha_3^{(i)}\zeta^2 \quad \text{for} \quad i = 1, \ldots, 3 \tag{7.19}$$

where $\alpha_1^{(i)}$, $\alpha_2^{(i)}$ and $\alpha_3^{(i)}$ are *unknown coefficients* that have to be obtained. Also, the shape functions have to have values of 1 at applicable nodal coordinates i.e. $N_1 = 1$ at $\zeta = -1$; $N_2 = 1$ at $\zeta = 0$, and, finally, $N_3 = 1$ at $\zeta = 1$. For all other cases, the shape functions take a value of zero.

Shape Function, N_1: Using the applicable end conditions for shape function, $N_1 = \alpha_1^{(1)} + \alpha_2^{(1)}\zeta + \alpha_3^{(1)}\zeta^2$, we obtain the following simultaneous equations.

$$\text{At } \zeta = -1 \implies 1 = \alpha_1^{(1)} - \alpha_2^{(1)} + \alpha_3^{(1)} \tag{7.20}$$

$$\text{At } \zeta = \ \ 0 \implies 0 = \alpha_1^{(1)} \tag{7.21}$$

$$\text{At } \zeta = +1 \implies 0 = \alpha_1^{(1)} + \alpha_2^{(1)} + \alpha_3^{(1)} \tag{7.22}$$

Solving Eqs. 7.20, 7.21, and 7.22 simultaneously, the unknown coefficients are:

$$\alpha_1^{(1)} = 0 \qquad \alpha_2^{(1)} = -\frac{1}{2} \qquad \text{and} \qquad \alpha_3^{(1)} = \frac{1}{2}$$

Therefore, the first shape functions, N_1, for the three-node 1-D element becomes:

$$N_1 = -\frac{1}{2}\zeta + \frac{1}{2}\zeta^2 \implies N_1 = -\frac{1}{2}\zeta(1 - \zeta) \tag{7.23}$$

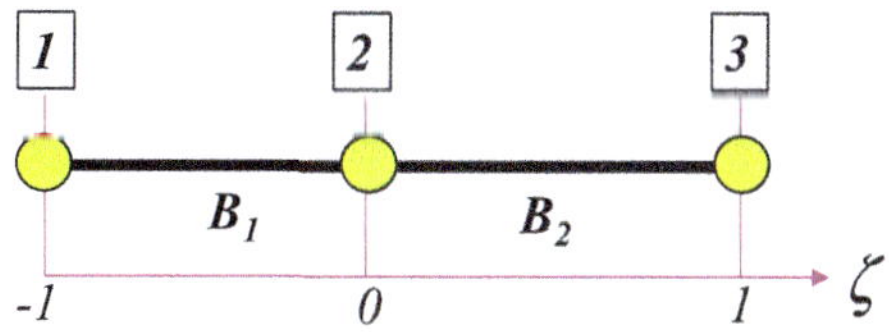

Fig. 7.7 An idealized representation of a three-node 1-D element, with nodal positions defined in terms of *isoparametric natural, coordinate*, i.e. ζ

Shape Function, N_2: Using the applicable end conditions for shape function, $N_2 = \alpha_1^{(2)} + \alpha_2^{(2)}\zeta + \alpha_3^{(2)}\zeta^2$, we should obtain the following simultaneous equations.

$$\text{At } \zeta = -1 \quad \Longrightarrow \quad 0 = \alpha_1^{(2)} - \alpha_2^{(2)} + \alpha_3^{(2)} \tag{7.24}$$

$$\text{At } \zeta = 0 \quad \Longrightarrow \quad 1 = \alpha_1^{(2)} \tag{7.25}$$

$$\text{At } \zeta = +1 \quad \Longrightarrow \quad 0 = \alpha_1^{(2)} + \alpha_2^{(2)} + \alpha_3^{(2)} \tag{7.26}$$

Solving Eqs. 7.24, 7.25, and 7.26 simultaneously, the resulting unknown coefficients are: $\alpha_1^2 = 1, \quad \alpha_2^{(2)} = 0$ and $\alpha_3^{(2)} = -1$

Therefore, the resulting first shape functions, N_2, for the three-node 1-D element becomes (Fig. 7.7):

$$N_2 = 1 - \zeta^2 \tag{7.27}$$

Shape Function, N_3: Following the same procedure as illustrated for N_1 and N_2, the unknown coefficients for the third shape function, N_3 are: $\alpha_1^{(3)} = 0$; $\alpha_2^{(3)} = \frac{1}{2}$ and $\alpha_3^3 = \frac{1}{2}$. Substituting these into the expression of N_3 we obtain:

$$N_3 = \frac{1}{2}\zeta + \frac{1}{2}\zeta^2 \quad \Longrightarrow \quad N_3 = \frac{1}{2}\zeta(1 + \zeta) \tag{7.28}$$

Graph of shape functions: Based on Eqs. 7.23, 7.27, and 7.28, the graphical representations of the three shape functions are given in Fig. 7.8.

Example 7.3 A truss-member, *AB*, is discretized into a five-node one-dimensional linear element shown as in Fig. 7.9. The nodal positions are given at specified ζ−natural coordinate positions.

Consider a case where the displacement: $\mathbf{u} = \left[0.2 \ 0.5 \ 0.8 \ 2.0 \ 3.5\right]$, is imposed on the nodal positions, where $\mathbf{u}$ is a function of the ζ natural coordinates.

(a) Obtain the applicable shape functions, N_i needed for the truss member.
(b) Derive a plot of the interpolated displacements, $\mathbf{u}_i$ for the truss member.

Solution

The shape function for the five noded element will be a polynomial of the format:

$$N_i = \alpha_1^{(i)} + \alpha_2^{(i)}\zeta + \alpha_3^{(i)}\zeta^2 + \alpha_4^{(i)}\zeta^3 + \alpha_5^{(i)}\zeta^4 \tag{7.29}$$

where $i = 1, \ldots, 5$ for all the five nodal coordinate positions and $\alpha_1^{(i)}$ are unknown coefficients of the shape functions.

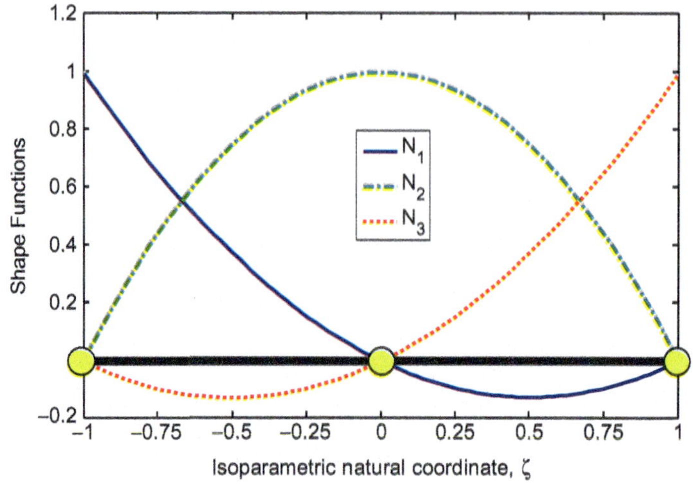

Fig. 7.8 A plot of the three shape functions of a three-node 1-D element

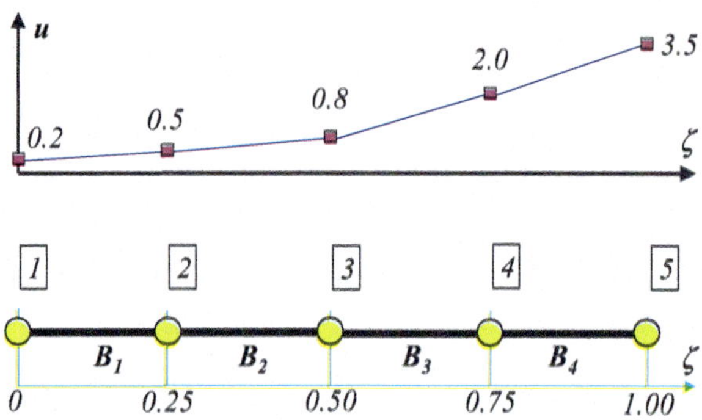

Fig. 7.9 An idealized representation of a five-node 1-D element. The top insert graph shows the displacement profile of the five-node member

Using the same approach as the previous examples, we can determine the unknown constants for all the shape functions. The expected shape functions are therefore given as:

$$
\begin{aligned}
N_1 &= 1.0 - 8.3\zeta + 23.3\zeta^2 - 26.7\zeta^3 + 10.7\zeta^4 \\
N_2 &= + 16.0\zeta - 69.3\zeta^2 + 96.0\zeta^3 - 42.7\zeta^4 \\
N_3 &= - 12.0\zeta + 76.0\zeta^2 - 128.0\zeta^3 - 64.0\zeta^4 \\
N_4 &= + 5.3\zeta - 37.3\zeta^2 + 74.7\zeta^3 - 42.7\zeta^4 \\
N_5 &= - 1.0\zeta + 7.3\zeta^2 - 16.0\zeta^3 + 10.7\zeta^4
\end{aligned}
\tag{7.30}
$$

Recall Eq. 7.2, such that the interpolated displacement for the five-node 1-D linear element becomes:

$$u^e(\zeta) = \mathbf{Nu}' \quad \Longrightarrow \quad u^e(\zeta) = \begin{bmatrix} N_1 & N_2 & N_3 & N_4 & N_5 \end{bmatrix} \begin{bmatrix} 0.2 \\ 0.5 \\ 0.8 \\ 2.0 \\ 3.5 \end{bmatrix}$$

$$\Longrightarrow \quad 0.2N_1 + 0.5N_2 + 0.8N_3 + 2.0N_4 + 3.5N_5$$

The plots of the shape functions and the interpolated displacements are shown in Fig. 7.10.

Remark 7.3 The above examples illustrate how to determine the shape functions, N_i, for a multi-node 1-D elements. The shape functions are dependent on $\zeta-$coordinate position. There is a dependence between the $\zeta-$coordinate and the spatial $x-$ coordinate system. It is essential to obtain the mapping function between the two coordinate systems, even though most element formulations are based on the isoparametric natural coordinate system. It is important during element formulation that the correct shape functions for a given FEM discretization is determined and solved to obtain the reliable nodal displacements for that element type.

Remark 7.4 A MATLAB™ function script called `ShapeFunction1D` has been developed to help generate the shape function equations and plots for the one-dimensional element for multiple nodes. The script is called by typing into the MATLAB™ command window:

```
[alphaValues,xx,funcN]=shapeFunction1D(number
Elements, xStart, xEnd )
```

where the main output, `alphaValues`, is a cell array consisting of all the unknown coefficients, $\alpha_i^{(i)}$, that define the shape functions. `xx` and `funcN` are the $x-$ and $y-$ values needed for plotting the shape functions. The function script requires inputs of: `numberElements` = number of elements or bars/members; (`xStart`) and (`xEnd`) are respectively the start and end nodal coordinate positions for the element under investigation.

For example, to determine the coefficients of the shape functions of Example 7.3, the MATLAB™ script command becomes:

```
[alphaValues,xx,funcN]=shapeFunction1D(4, 0, 1)
```

where `numberElements` = 4 (there are four elements in the five-node discretization of the bar), `xStart` = 0, and `xEnd` = 1 (in terms of isoparametric natural coordinates). To extract the `alphaValues`, type into the command window: `alphaValues{1:5}`, which returns the set of $\alpha_i^{(i)}$ for all shape functions, where $i = 1, \ldots, 5$.

The script can be downloaded from the ***extra resources website*** of this book.

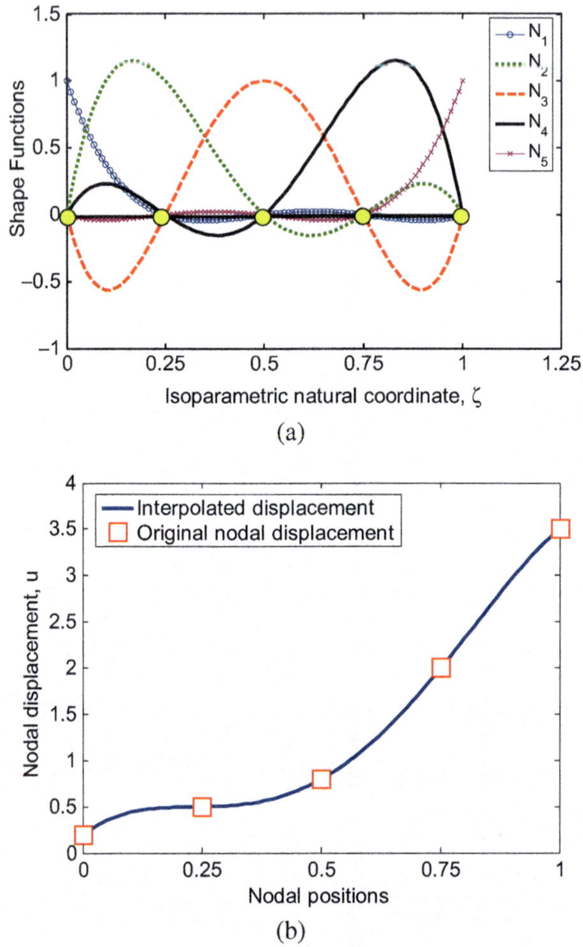

Fig. 7.10 Graphical representations of: (**a**) all shape functions and (**b**) interpolated, $u^e(\zeta)$ and original displacements, $\mathbf{u}$ for Example 7.3

7.4.3 Derivation of Strain-Displacement Matrix for 1-D Elements

Equation 7.2, above, established that the expected displacement of the two-node 1-D element is cumulative displacements at nodes N_1 and N_2 and expressed mathematically via the shape function as:

$$u^e(x) = N_1^e u_1^e + N_2^e u_2^e \quad \Longrightarrow \quad u^e(x) = \begin{bmatrix} N_1^e \ N_2^e \end{bmatrix} \begin{bmatrix} u_1^e \\ u_2^e \end{bmatrix} \quad \Longrightarrow \quad u^e(x) = \mathbf{N}^e \mathbf{u}^e$$

We can describe this element-level displacement, $u^e(x)$, as a *trial displacement* for the element. It is not an accurate value of the true displacement at the nodes since they are computed based on specific shape functions. For a different set of shape functions (for example, higher order functions), a more accurate value of the displacements can be obtained.

Although this *trial displacement* is not accurate, the FEM process allows us to use it and, following subsequent iterations, a *convergent solution* can be obtained. For now, we will follow the FEM process using this trial displacement.

Based on this trial displacement, the next step in the FEM process is to determine the associated axial strain in the element. Thus, axial strain within the element, which is determined based on the first order derivative of Eq. 7.2, becomes:

$$\epsilon^e(x) = \frac{du^e(x)}{dx} \quad \Longrightarrow \quad \epsilon^e(x) = \begin{bmatrix} \dfrac{dN_1^e}{dx} & \dfrac{dN_2^e}{dx} \end{bmatrix} \begin{bmatrix} u_1^e \\ u_2^e \end{bmatrix} \tag{7.31}$$

Recall the expressions of the shape functions and their derivatives, expressed in terms of x as:

$$\begin{aligned} N_1^e &= 1 - \zeta = 1 - \left[\frac{x - x_1}{L}\right] &&\Longrightarrow N_1^e = -\frac{1}{L}x + \frac{1}{L}x_1 + 1 &&\Longrightarrow \frac{dN_1^e}{dx} = -\frac{1}{L} \\ N_2^e &= \zeta = \left[\frac{x - x_1}{L}\right] &&\Longrightarrow N_2^e = +\frac{1}{L}x - \frac{1}{L}x_1 &&\Longrightarrow \frac{dN_2^e}{dx} = +\frac{1}{L} \end{aligned} \tag{7.32}$$

Combining Eqs. 7.31 and 7.32, the expression of the element axial strain becomes:

$$\epsilon^e(x) = \begin{bmatrix} \dfrac{dN_1^e}{dx} & \dfrac{dN_2^e}{dx} \end{bmatrix} \begin{bmatrix} u_1^e \\ u_2^e \end{bmatrix} \quad \Longrightarrow \quad \epsilon^e(x) = \frac{1}{L}\begin{bmatrix} -1 & 1 \end{bmatrix} \begin{bmatrix} u_1^e \\ u_2^e \end{bmatrix} \quad \Longrightarrow \quad \epsilon^e(x) = \mathbf{B}\mathbf{u}^e \tag{7.33}$$

You will have noticed that in Eq. 7.33, we have defined a new parameter called the *strain-displacement matrix*, **B**. Its exact formulation for the two-node 1-D element is given in Eq. 7.34.

$$\mathbf{B} = \begin{bmatrix} \dfrac{dN_1^e}{dx} & \dfrac{dN_2^e}{dx} \end{bmatrix} \quad \Longrightarrow \quad \mathbf{B} = \frac{1}{L}\begin{bmatrix} -1 & 1 \end{bmatrix} \tag{7.34}$$

The strain-displacement matrix is constant over the length of the element. It is used to compute the strains at any point inside the element. For the element under consideration, the axial strain is constant since **B** is not a function of x. Hence, **B** yields a constant value over the element length, L.

In element formulation, the shape functions are typically expressed in terms of isoparametric natural coordinates. Therefore, the strain-displacement matrix need also be described in terms of the isoparametric natural coordinates. We already established in Eq. 7.32 that there exists a relationship between the isoparametric natural coordinates, ζ, and the $x-$ axis coordinate system. We will simply determine strain, ϵ, thus:

$$\epsilon^e(x) = \begin{bmatrix} \frac{dN_1^e}{d\zeta}\frac{d\zeta}{dx} & \frac{dN_2^e}{d\zeta}\frac{d\zeta}{dx} \end{bmatrix} \begin{bmatrix} u_1^e \\ u_2^e \end{bmatrix} \quad \Longrightarrow \quad \epsilon^e(x) = \mathbf{B}\mathbf{u}^e \tag{7.35}$$

The strain-displacement matrix, $\mathbf{B}$, obtained from Eq. 7.35, is expressed as:

$$\mathbf{B} = \begin{bmatrix} \frac{dN_1^e}{d\zeta}\frac{d\zeta}{dx} & \frac{dN_2^e}{d\zeta}\frac{d\zeta}{dx} \end{bmatrix} \quad \Longrightarrow \quad \mathbf{B} = \frac{d\zeta}{dx}\begin{bmatrix} \frac{dN_1^e}{d\zeta} & \frac{dN_2^e}{d\zeta} \end{bmatrix} = \mathbf{J}^{-1}\begin{bmatrix} \frac{dN_1^e}{d\zeta} & \frac{dN_2^e}{d\zeta} \end{bmatrix} \tag{7.36}$$

where $\mathbf{J}$ is *Jacobian matrix* expressed in this case as: $\mathbf{J} = \dfrac{dx}{d\zeta}$ such that $\mathbf{J}^{-1} = \dfrac{d\zeta}{dx}$. The Jacobian matrix maps the spatial $x-$ axis coordinate system to the isoparametric natural $\zeta-$axis coordinate system. It is a single number here since we have only a one-dimensional coordinate system. For $n \times n$ multi-dimensional coordinate system, the $\mathbf{J}$ matrix becomes an $n \times n$ matrix (see Sect. 7.5.2 for Jacobian matrix derivations for a 2-D triangular matrix).

Using the $\mathbf{J}$ matrix in Eq. 7.36 results in a strain-displacement matrix in which the first order derivative of the shape functions are with respect to the isoparametric natural coordinates as against the spatial coordinate system representation of Eq. 7.34. This is the format in which FEM solvers evaluate the strain within the element. It is important that this Jacobian matrix is determined, from first principles, as show here, when one is developing a new element type and its formulation.

Example 7.4 Based on the truss member of Example 7.2:

(a) Determine the strain-displacement matrix.
(b) If the nodal displacement is $\mathbf{u} = \begin{bmatrix} 0.15 & 0.25 & 0.75 \end{bmatrix}'$, determine the plot of the interpolated displacement in terms of natural coordinates, $\mathbf{u}^e(\zeta)$.
(c) Calculate the strains in the three-node truss at $\zeta = -0.5$ and $\zeta = 0.5$.

Solution

Shape functions: Let us recall the expressions of the shape functions for the three-node 1-D linear element thus:

$$\begin{aligned} N_1 &= -\tfrac{1}{2}\zeta(1-\zeta) \\ N_2 &= \quad 1-\zeta^2 \\ N_3 &= \tfrac{1}{2}\zeta(1+\zeta) \end{aligned} \tag{7.37}$$

Strain-displacement matrix: The expression for the strain-displacement matrix, **B** for the three-node element becomes:

$$\mathbf{B} = \begin{bmatrix} \frac{dN_1}{dx} & \frac{dN_2}{dx} & \frac{dN_3}{dx} \end{bmatrix} \implies \mathbf{B} = \mathbf{J}\begin{bmatrix} \frac{dN_1}{d\zeta} & \frac{dN_2}{dx} & \frac{dN_3}{dx} \end{bmatrix} \quad (7.38)$$

where the *Jacobian matrix*, $\mathbf{J} = \frac{d\zeta}{dx}$.

The mapping function that makes Fig. 7.11(*right*) possible can be defined as $\zeta = \frac{2(x - x_2)}{L}$ such that $\mathbf{J} = \frac{d\zeta}{dx} = \frac{2}{L}$. Note that the length of the element, $L = 2$.

Executing the required differentiations for Eq. 7.37 and combining results with Eq. 7.38, the resulting strain-displacement matrix becomes:

$$\mathbf{B} = \frac{2}{L}\begin{bmatrix} (-0.5 + \zeta) & -(2\zeta) & (0.5 + \zeta) \end{bmatrix} \quad (7.39)$$

Interpolated displacement plot: For the truss-member under investigation, there exist three nodal displacements. Using the expression $\mathbf{u}^e(\zeta) = \mathbf{N}\mathbf{u}$, we can determine the expression of the interpolated displacement in terms of natural coordinates as shown in Fig. 7.12.

Strains at specific ζ values: Combining Eq. 7.39 and the specified *nodal displacements*, $\mathbf{u} = \begin{bmatrix} 0.15 & 0.25 & 0.75 \end{bmatrix}'$, we note that the strain function becomes:

$$\epsilon^e(x) = \mathbf{Bu} \implies \epsilon^e(\zeta) = \frac{2}{L}\begin{bmatrix} (-0.5 + \zeta) & -(2\zeta) & (0.5 + \zeta) \end{bmatrix}\begin{bmatrix} 0.15 \\ 0.25 \\ 0.75 \end{bmatrix}$$
$$\implies \epsilon^e(\zeta) = \frac{2}{L}\begin{bmatrix} 0.30 + 0.40\zeta \end{bmatrix} \quad (7.40)$$

Therefore, the strain at specified natural coordinates locations are:

$$\epsilon^e(\zeta = -0.5) = \frac{0.20}{L} \quad \text{and} \quad \epsilon^e(\zeta = +0.5) = \frac{1.00}{L}$$

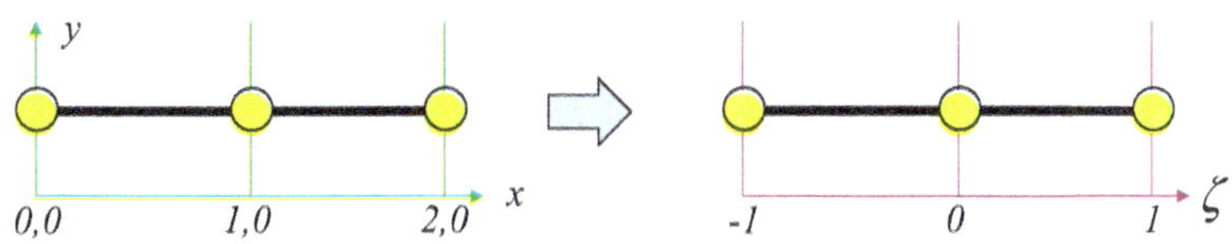

Fig. 7.11 (*left*) Spatial and (*right*) isoparametric natural coordinates system representations of a three-node 1-D linear element.Note $x_1 = 0$ and $x_2 = 2$

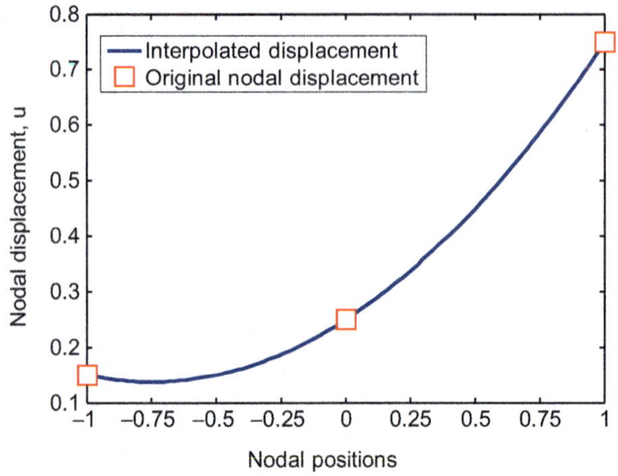

Fig. 7.12 An interpolated displacement profile for the three-node truss member of Example 7.2 where the truss member is subjected to a displacement, $\mathbf{u}$

7.4.4 Derivation of Force-Displacement Equation for 1-D Elements

As well as the strain-displacement equation, we can also attempt to derive the force-displacement equations which are usually referred to as the finite element equations, because they are the first set of equations that we set out to solve using the FEM process. The most common method for determining the force-displacement equation is based on the *variational principles* and it is based on the balance of energies in the system between *total internal energy*, U, and the *external work*, W that is done on the system.

Consider the prismatic bar shown in Fig. 7.1a subjected to an axial force, $\mathbf{P}$, and with rectangular cross-section, A. The rod experiences a deformation, $u(x)$, which creates a strain, $\epsilon(x)$, and stress, $\sigma(x)$, in the bar. In the following, we will use the bar to obtain the equations for the internal and external energies in the 1-D element discretization unit of the problem.

Total internal energy, U

The total internal energy, U can be derived for the 1-D linear element under consideration as follows:

$$\begin{aligned} U = \frac{1}{2}\int_V \sigma(x)\epsilon(x)dV \quad &\Longrightarrow U = \frac{1}{2}\int \sigma(x)\epsilon(x)Adx \\ &\Longrightarrow U = \frac{1}{2}\int_0^L \frac{F(x)}{A}\epsilon(x)Adx \\ &\Longrightarrow U = \frac{1}{2}\int_0^L F(x)\epsilon(x)dx \end{aligned} \tag{7.41}$$

Recall the deformation equation for an axially loaded bar of constant cross-section where the bar material has **Young's Modulus**, E. If we assume the material undergoes a linear elastic deformation, Δu, then:

$$\Delta u = \frac{F(x)dx}{EA} \tag{7.42}$$

Hence, the internal force becomes:

$$F(x) = EA\frac{\Delta u(x)}{dx} = EA\epsilon(x)$$

Substituting Eq. 7.42 into Eq. 7.41, the modified expression of the internal energy with the prismatic bar is:

$$\begin{aligned} U = \frac{1}{2}\int_0^L F(x)\epsilon(x)dx \quad &\Longrightarrow U = \frac{1}{2}\int_0^L \Big[EA\epsilon(x)\Big]\epsilon(x)dx \\ &\Longrightarrow \quad U = \frac{EA}{2}\int_0^L \Big[\epsilon(x)\Big]^2 dx \end{aligned} \tag{7.43}$$

Adapting Eq. 7.44 for the two-node 1-D linear element of Fig. 7.1b, the expression of internal energy becomes:

$$\begin{aligned} U = \frac{1}{2}\int_0^L \Big[\epsilon(x)EA\epsilon(x)\Big]dx \quad &\Longrightarrow U^e = \frac{1}{2}\left[\frac{u^e(x)}{L^e}E^eA^e\frac{u^e(x)}{L^e}\right]L^e \\ &\Longrightarrow U^e = \frac{1}{2}\left[u^e(x)\frac{E^eA^e}{L^e}u^e(x)\right] \end{aligned} \tag{7.44}$$

In the above equation: E^e = element Young's modulus; A^e = cross-sectional area of element; $u^e(x)$ is element displacement, L^e = length of element, and element stiffness; and, $k^e = \dfrac{E^eA^e}{L^e}$.

Equation 7.44 can be re-written in vector format thus:

$$U^e = \frac{1}{2}\left[u^e(x)\frac{E^eA^e}{L^e}u^e(x)\right] \quad \Longrightarrow U^e = \frac{1}{2}\Big[(\mathbf{u}^e)^T\mathbf{k}^e\mathbf{u}^e\Big] \tag{7.45}$$

Total external work done, W

Using the same prismatic bar of Fig. 7.1b, the external work is done by the concentrated force **P** acting on the bar and creating an axial deformation, $u(x)$. For every unit length of the bar, the work done is the product of the distributed load at that point, $q(x)$, and the axial deformation, $u(x)$. To get the total work done, we have to integrate the product: $q(x)u(x)$ over the entire length of the bar. The external work done, W on the bar becomes:

$$W = \int_0^L q(x)u(x)dx \quad \Longrightarrow \quad W = q(x)u(x)L \quad \Longrightarrow \quad W = f(x)u(x) \tag{7.46}$$

Also, the external work done on the 1-D linear element in terms of vector format representation is:

$$W = f(x)u(x) \quad \Longrightarrow \quad W^e = \left(\mathbf{u}^e\right)^T \mathbf{f}^e \tag{7.47}$$

In Eqs. 7.45 and 7.47, the *transpose* of element displacement, $\left(\mathbf{u}^e\right)^T$, is a necessary matrix manipulation so that the resulting product will be a scalar quantity. Recall that energy and work are both scalar quantities, and the product of a row matrix multiplied by the transpose of another row matrix (i.e. a column matrix) will result in a scalar quantity.

Total potential energy of the bar, T

The elastic potential energy stored in the material will consist of the difference between the external work done and the internal energy of the system. If the total number of bar elements for a given structure is N^e, the total potential energy of the bar becomes:

$$\sum_{i=1}^{N^e} T_i^e = \sum_{i=1}^{N^e} U_i^e - \sum_{i=1}^{N^e} W_i^e \quad \Longrightarrow \quad T^e = U^e - W^e \tag{7.48}$$

Re-writing Eq. 7.48 in vector format results in:

$$\begin{aligned} T^e = U^e - W^e \quad &\Longrightarrow T^e = \frac{1}{2}\left[\left(\mathbf{u}^e\right)^T \mathbf{k}^e \mathbf{u}^e\right] - \left(\mathbf{u}^e\right)^T \mathbf{f}^e \\ &\Longrightarrow T^e = \left(\mathbf{u}^e\right)^T \left[\frac{1}{2}\mathbf{k}^e \mathbf{u}^e - \mathbf{f}^e\right] \end{aligned} \tag{7.49}$$

From conservation of energy, the derivative of the total potential energy should tend to zero. Applying the partial derivative of the total potential energy to Eq. 7.49:

$$\frac{\partial T^e}{\left(\mathbf{u}^e\right)^T} = \mathbf{0} \quad \Longrightarrow \quad \left[\frac{1}{2}\mathbf{k}^e \mathbf{u}^e - \mathbf{f}^e\right] = \mathbf{0} \tag{7.50}$$

The above derivative leads to $\mathbf{k}^e \mathbf{u}^e - \mathbf{f}^e = \mathbf{0}$, which establishes the element-level force-displacement equation given in Eq. 7.51 and the basis for obtaining solutions in the FEM process. The equation connects the resultant nodal forces on a body and the nodal displacement as well as the stiffness matrix (material property) of the system.

$$\mathbf{f}^e = \mathbf{k}^e \mathbf{u}^e \tag{7.51}$$

The reader should note that the ½ factor in Eq. 7.50 has disappeared following the operation of the partial differentiation of the total internal energy, U^e at the element level. This is because U^e is quadratic in the node displacements, u^e.

Consider Eq. 7.44 and the strain in the element, $\epsilon = \frac{u}{L}$. For a bar of constant cross-sectional area, A, and length, L, the volume is V. Thus, the internal energy from Eq. 7.44 becomes: $U = \frac{1}{2}EVu^2$. Hence, taking the partial derivative of U with respect to u will cause the ½ factor to disappear.

7.4.5 Derivation of Stiffness Matrix Equation for 1-D Elements

From the previous two sections, we showed the derivation of the *strain-displacement matrix* and the *force-displacement* equation needed for the FEM process. In this section, we will now begin to build the *stiffness matrix* for the 1-D linear element. As previously established, determining the stiffness matrix is at the core of the FEM implementation of commercial solvers.

Recall the *total internal energy,* U^e, expression of an element given in Eq. 7.44 and let us re-write it in terms of the element/truss member shown in Fig. 7.2:

$$U^e = \frac{1}{2}\int_{x_1}^{x_2} \sigma^e \epsilon^e A dx \quad \Longrightarrow U^e = \frac{1}{2}\int_{x_1}^{x_2} \left[\epsilon^e EA\epsilon^e\right] dx \tag{7.52}$$

We will now introduce into Eq. 7.52 the natural coordinates, ζ for 1-D element, where $dx = L\,d\zeta$. Note ζ takes values from 0 to 1 and so the element-level internal energy equation in terms of ζ becomes:

$$U^e = \frac{1}{2}\int_{x_1}^{x_2} \left[\epsilon^e EA\epsilon^e\right] dx \quad \Longrightarrow U^e = \frac{1}{2}\int_{0}^{1} \left[\left(\epsilon^e\right)^T EA\epsilon^e\right] L\,d\zeta \tag{7.53}$$

From Eq. 7.54, with strain measures: $\epsilon^e = \mathbf{B}\mathbf{u}^e$ and $\left(\epsilon^e\right)^T = \left(\mathbf{u}^e\right)\mathbf{B}^T$, we obtain the following:

$$\begin{aligned} U^e = \frac{1}{2}\int_{0}^{1} \left[\left(\epsilon^e\right)^T EA\epsilon^e\right] Ld\zeta \quad &\Longrightarrow U^e = \frac{1}{2}\int_{0}^{1} \left(\mathbf{u}^e\right)^T \mathbf{B}^T EA\mathbf{B}\mathbf{u}^e Ld\zeta \\ &\Longrightarrow U^e = \frac{1}{2}\left(\mathbf{u}^e\right)^T\left[\int_{0}^{1} EA\mathbf{B}^T\mathbf{B}Ld\zeta\right]\mathbf{u}^e \\ &\Longrightarrow U^e = \frac{1}{2}\left(\mathbf{u}^e\right)^T\mathbf{K}^e\mathbf{u}^e \end{aligned} \tag{7.54}$$

In the above, we have derived the stiffness matrix, $\mathbf{K}^e$, and this is defined as:

$$\text{Stiffness matrix, } \mathbf{K}^e = \int_0^1 EA\mathbf{B}^T\mathbf{B}L\,d\zeta \tag{7.55}$$

For the homogeneous, prismatic 1-D linear element of Fig. 7.2, the following are true:

$$\mathbf{u}^e = \begin{bmatrix} u_1^e \\ u_2^e \end{bmatrix} \quad \text{and} \quad \mathbf{B} = \frac{1}{L}\begin{bmatrix} -1 & 1 \end{bmatrix}$$

Substituting these parameters into the stiffness matrix equation (refer to Eq. 7.55) for the 1-D linear element, the result becomes:

$$\begin{aligned}
\mathbf{K}^e = \int_0^1 EA\mathbf{B}^T\mathbf{B}Ld\zeta \quad &\Longrightarrow \mathbf{K}^e = \int_0^1 EA\left\{\frac{1}{L}\begin{bmatrix} -1 \\ 1 \end{bmatrix}\right\}\left\{\frac{1}{L}\begin{bmatrix} -1 & 1 \end{bmatrix}\right\}L\,d\zeta \\
&\Longrightarrow \mathbf{K}^e = \frac{EA}{L^2}\begin{bmatrix} 1 & -1 \\ -1 & 1 \end{bmatrix}\int_0^1 L\,d\zeta \\
&\Longrightarrow \mathbf{K}^e = \frac{EA}{L}\begin{bmatrix} 1 & -1 \\ -1 & 1 \end{bmatrix}
\end{aligned} \tag{7.56}$$

Example 7.5 A three-node idealization of a truss structure is a one-dimensional linear element shown in Fig. 7.13 where the natural coordinate is bounded by $-1 \leq \zeta \leq 1$. The truss structure is 1.2 m long and is made of cylindrical steel bar of diameter, $d = 4\,\text{mm}$ and of Young's Modulus, $E = 210\,\text{GPa}$. At a stage during the deformation, the truss is subjected to an external nodal displacement, $\mathbf{u} = \begin{bmatrix} 0 & 1.2 & 1.5 \end{bmatrix}$ mm, as shown also in Fig. 7.13.

(a) Determine the applicable shape functions for the three-node element.
(b) Obtain a plot of the interpolated displacement, $\mathbf{u}^e$.
(c) Calculate the strain-displacement matrix, $\mathbf{B}$ for the structure.
(d) Calculate the strain within the bar at $\zeta = 0.5$.
(e) Calculate the stiffness matrix, $\mathbf{K}^e$ for the structure.

Solution

- *Shape functions, N_i*: The shape function for a three-node member will have to be a quadratic function of the form:

$$N_i = \alpha_1^{(i)} + \alpha_2^{(i)}\zeta + \alpha_3^{(i)}\zeta^2$$

where $i = 1, \ldots, 3$. We have to determine all $\alpha_k^{(i)}$ for $k = 1, \ldots, 3$.

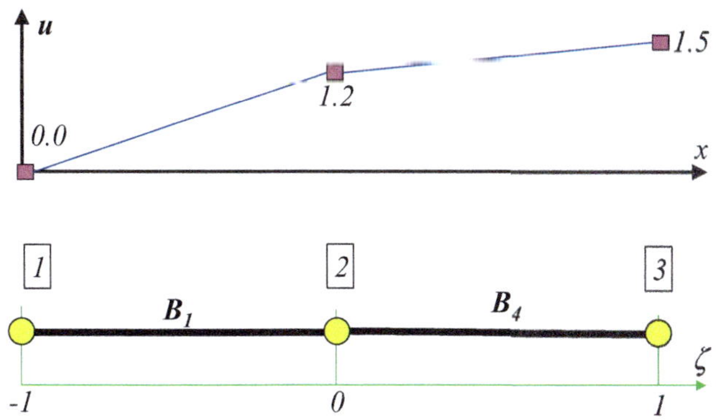

Fig. 7.13 A truss structure idealized with a three-node one-dimensional linear element, with the inserted plot showing the imposed nodal displacement, **u** on the structure

Using the approach already explained in Example 7.2 above or by running the `ShapeFunction1D` MATLAB™ script,[1] we determine that the applicable shape functions are:

$$\begin{aligned} N_1 &= \quad -0.5\zeta(1-\zeta) \\ N_2 &= \quad \zeta(\zeta^{-1}-\zeta) \\ N_3 &= \quad 0.5\zeta(1+\zeta) \end{aligned} \tag{7.57}$$

- *Interpolated displacements,* $\mathbf{u}^e$ *plot*: The interpolated displacement, in terms of natural coordinates, ζ is calculated based on the equation: $\mathbf{u}^e(\zeta) = \mathbf{Nu}$. This can also be plotted following previous approach or using the MATLAB function script called `interpolateDisplacements`.[2] The MATLAB™ command-window evaluation instruction for this particular problem is:

```
[uNI,xx,funcN]=interpolateDisplacements([0 1.2 1.5],
0,1)
```

where the function output parameters are: `uNI` $= u^e(\zeta)$ is interpolated displacement; `xx` $= \zeta$ is isoparametric natural coordinates; and, `funcN` $= [N_1 \ N_2 \ N_3]$ is a cell-array that contains the evaluated shape functions.

The input parameters are: nodal displacement: $u = [0 \ 1.2 \ 1.5]$, and start, (i.e. `xStart` $= 0$), and end, (i.e. `xEnd` $= 1.0$) isoparametric nodal positions.

Following the execution of the `interpolateDisplacements` script, the resulting plot is given in Fig. 7.14.

[1]This script is available on the ***extra resources website*** of the textbook.

[2]The interpolation script is also available on the ***extra resources website*** of the textbook. You will need it to quickly evaluate the interpolated displacements.

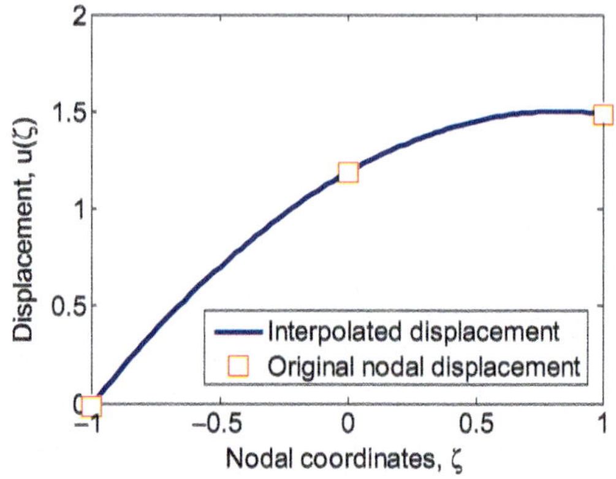

Fig. 7.14 The interpolated displacement profile of a truss member subjected to a nodal displacement, **u**

- *Strain-displacement matrix,* **B**: We will need to determine the *Jacobian matrix* for this element, so that we can map the spatial $x-$ axis coordinate system to the isoparametric $\zeta-$axis natural coordinate system. The mapping function is:

$$\zeta = \frac{2x - x_2}{L} \quad \Longrightarrow \quad \mathbf{J} = \frac{d\zeta}{dx} = \frac{2}{L} \tag{7.58}$$

Using the above mapping functions, the resulting strain-displacement matrix, **B**, becomes:

$$\begin{aligned} \mathbf{B} = \begin{bmatrix} \frac{dN_1}{dx} & \frac{dN_2}{dx} & \frac{dN_3}{dx} \end{bmatrix} \quad &\Longrightarrow \mathbf{B} = \frac{d\zeta}{dx}\begin{bmatrix} \frac{dN_1}{d\zeta} & \frac{dN_2}{dx} & \frac{dN_3}{dx} \end{bmatrix} \\ &\Longrightarrow \mathbf{B} = \frac{2}{L}\begin{bmatrix} (-0.5+\zeta) & -2\zeta & (0.5+\zeta) \end{bmatrix} \end{aligned} \tag{7.59}$$

- *Strain, ϵ at specific location*: The strain is calculated using: $\epsilon = \mathbf{B}\mathbf{u}$ and is calculated thus:

$$\begin{aligned} \epsilon^e = \mathbf{B}u^e \quad &\Longrightarrow \epsilon^e = \frac{2}{L}\begin{bmatrix} (-0.5+\zeta) & -2\zeta & (0.5+\zeta) \end{bmatrix}\begin{bmatrix} 0.0 \\ 1.2 \\ 1.5 \end{bmatrix} \\ &\Longrightarrow \epsilon^e = \frac{1.5 - 1.8\zeta}{L} \\ &\Longrightarrow \epsilon^e(\zeta = 0.5) = \frac{1.5 - 1.8 * 0.5}{1200} = 5.0 \times 10^{-4} \end{aligned} \tag{7.60}$$

- *Stiffness matrix,* $\mathbf{K}^e$: The expression for the stiffness matrix is:

$$\mathbf{K}^e = \int_0^1 EA\mathbf{B}^T\mathbf{B}L\, d\zeta$$

We have determined the value for **B** and as a consequence $\mathbf{B}^T$. The results of the substitutions become:

$$\mathbf{K}^e = EAL \int_0^1 \mathbf{B}' \times \mathbf{B}\, d\zeta$$

$$= EAL \times \frac{4}{L^2} \int_0^1 \begin{bmatrix} \zeta^2 - \zeta + 0.25 & -2\zeta^2 + \zeta & \zeta^2 - 0.25 \\ -2\zeta^2 + \zeta & 4\zeta^2 & -2\zeta^2 - \zeta \\ \zeta^2 - 0.25 & -2\zeta^2 - \zeta & \zeta^2 + \zeta + 0.25 \end{bmatrix} d\zeta$$

$$= \frac{4EA}{L} \left\| \begin{bmatrix} \frac{1}{3}\zeta^3 - \frac{1}{2}\zeta^2 + 0.25\zeta & -\frac{2}{3}\zeta^3 + \frac{1}{2}\zeta^2 & \frac{1}{3}\zeta^3 - 0.25\zeta \\ -\frac{2}{3}\zeta^3 + \frac{1}{2}\zeta^2 & \frac{4}{3}\zeta^3 & -\frac{2}{3}\zeta^3 - \frac{1}{2}\zeta^2 \\ \frac{1}{3}\zeta^3 - 0.25\zeta & -\frac{2}{3}\zeta^3 - \frac{1}{2}\zeta^2 & \frac{1}{3}\zeta^3 + \frac{1}{2}\zeta^2 + 0.25\zeta \end{bmatrix} \right\|_0^1 \tag{7.61}$$

Evaluating the above integral gives:

$$\mathbf{K}^e = 8.7965 \times 10^6 \begin{bmatrix} 0.0833 & -0.1667 & 0.0833 \\ -0.1667 & 1.3333 & -1.1667 \\ 0.0833 & -1.1667 & 1.0833 \end{bmatrix} \mathrm{Nm}^{-1} \tag{7.62}$$

7.5 Element Formulation for Two-Dimensional Triangular Elements

Physical domains are easily discretized using triangular elements using the *Delaunay triangulation* as discussed earlier. Understanding the element formulation for such elements is essential in this introductory stage of the FEM process. This section will follow similar procedure as used for 1-D elements to determine the element formulation for **2-D** triangular elements.

Let us consider a typical 2-D triangular element as shown in Fig. 7.15 consisting of three nodes, $(1, 2, 3)$. The element is described in both global $(XY-)$ and isoparametric natural $(\zeta\eta-)$ coordinate systems. Each of the nodes has two degrees of freedom such that X-axis displacement is u_i for $i = 1, 2, 3$ while the $Y-$axis displacement is v_i for $i = 1, 2, 3$.

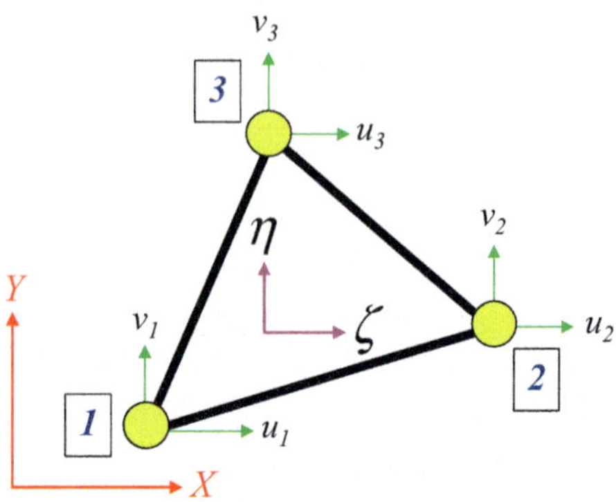

Fig. 7.15 A 2-D triangular element showing nodal displacements in two directions (u_i and v_i), nodal numbers (square boxes), global ($XY-$axis) and isoparametric natural coordinates ($\zeta\eta$-axis)

Since the element is a 2-D element, the two isoparametric natural coordinates, namely ζ and η, have been introduced to map local element level $x-$ axis and $y-$ axis coordinates respectively to their global coordinate counterparts. In the following section, we will present the derivations of all the corresponding element parameters needed for the FEM process.

7.5.1 Derivation of Shape Functions for 2-D Triangular Elements

The expression of the shape functions for 2-D triangular elements follows the same process as the 1-D linear elements except we consider the two axes applicable here. For the 2-D triangular elements, there exist two *trial* or *interpolated displacements*, namely u_i and v_i for $i = 1, 2, 3$. The expressions for these displacements are:

$$\begin{aligned} u(\zeta,\eta) &= N_1u_1 + N_2u_2 + N_3u_3 \quad &\Longrightarrow u(\zeta,\eta) = \mathbf{Nu} \\ v(\zeta,\eta) &= N_1v_1 + N_2v_2 + N_3v_3 \quad &\Longrightarrow v(\zeta,\eta) = \mathbf{Nv} \end{aligned} \tag{7.63}$$

where N_1, N_2 and N_3 are the yet-to-be determined shape functions for the 2-D triangular element. The above expressions can be written in *matrix format* as follows:

$$\mathbf{u} = \begin{bmatrix} u(\zeta,\eta) \\ \\ v(\zeta,\eta) \end{bmatrix} \quad \Longrightarrow \mathbf{u} = \begin{bmatrix} N_1 & 0 & N_2 & 0 & N_3 & 0 \\ \\ 0 & N_1 & 0 & N_2 & 0 & N_3 \end{bmatrix} \begin{bmatrix} u_1 \\ v_2 \\ u_2 \\ v_2 \\ u_3 \\ v_3 \end{bmatrix}. \tag{7.64}$$

The task is to determine the unknown shape functions. Similar to the approach used in Sect. 7.4.1 for a 1-D element, let us define a linear relationship between the shape functions and the *isoparametric natural coordinates* thus:

$$N_i = \alpha_1^{(i)} + \alpha_2^{(i)}\zeta + \alpha_3^{(i)}\eta \quad \text{where} \quad i = 1, 2, 3 \tag{7.65}$$

Equation 7.65 is a linear element formulation of the shape function needed for the interpolating nodal displacements of the triangular elements. It is linear in the applicable isoparametric natural coordinates, ζ and η. Consequently, the linear element formulation leads to a constant strain triangular element where the strain is invariant within the element. For other types of analysis in which more resolution of strain map in the element is desired, it might be acceptable to use either a quadratic or cubic dependence of shape functions with isoparametric natural coordinates.

For the triangular element of Fig. 7.15, we will use three pairs of (ζ, η) isoparametric natural coordinates to identify the exact formulation of the shape functions. Doing so, we will obtain three simultaneous equation which will be solved to determine the unknown coefficients, $\alpha_i^{(i)}$, for $i = 1, 2, 3$. In the following example, we illustrates how the three shape functions that describe the given triangular element, can be determined using the Lagrangian (polynomial interpolation) approach.

Example 7.6 A washer plate is discretized using 2-D triangular elements, as shown in Fig. 7.16. In order to use this element type in an FE analysis, derive all the shape functions for the 2-D triangular elements. Draw a graph of the shape functions.

Solution
The shape function for first node is: $N_1 = \alpha_1^{(1)} + \alpha_2^{(1)}\zeta + \alpha_3^{(1)}\eta$. Substituting the three pairs of isoparametric nodal coordinates into the equation, we obtain the following simultaneous equations.

$$\text{At } \zeta = 0.1, \eta = 0 \quad \Longrightarrow \quad 1 = \alpha_1^{(1)} + 0.1\alpha_2^{(1)} \tag{7.66}$$

$$\text{At } \zeta = 0.4, \eta = 0.3 \quad \Longrightarrow \quad 0 = \alpha_1^{(1)} + 0.4\alpha_2^{(1)} + 0.3\alpha_3^{(1)} \tag{7.67}$$

$$\text{At } \zeta = 0, \eta = 0.5 \quad \Longrightarrow \quad 0 = \alpha_1^{(1)} + 0.5\alpha_3^{(1)} \tag{7.68}$$

Solving Eqs. 7.66, 7.67, and 7.68 simultaneously, the resulting unknown coefficients are: $\alpha_1^{(1)} = 1.111$, $\alpha_2^{(1)} = -1.111$ and $\alpha_3^{(1)} = -2.222$. The same process will also be used to determine the remaining shape functions and all shape functions include:

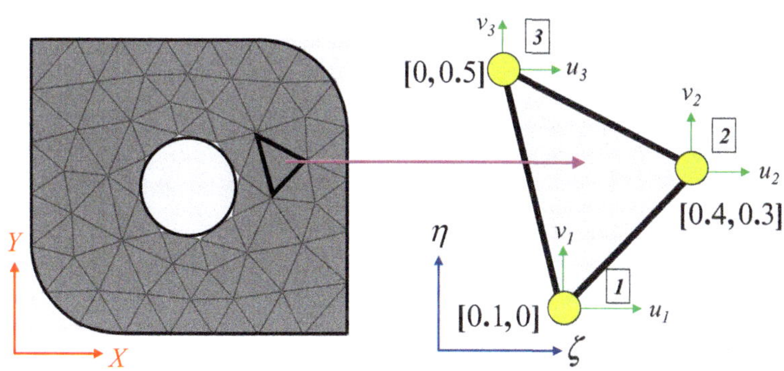

Fig. 7.16 A washer plate discretized with a triangular element where the element is described in isoparametric natural coordinates

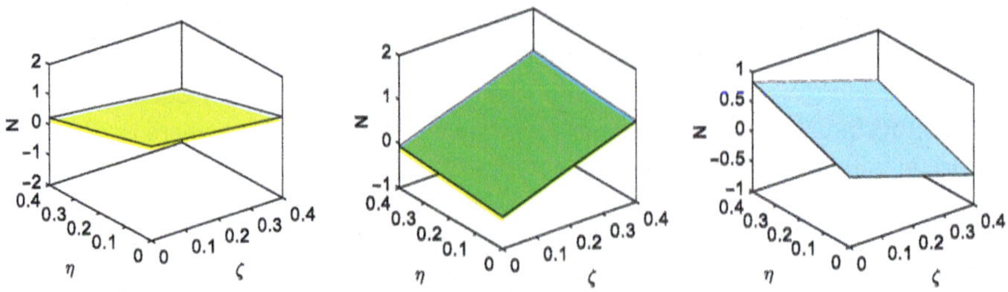

(a) (b) (c)

Fig. 7.17 Shape functions of the 2-D triangular element of Example 7.6 showing profiles for: (**a**) N_1; (**b**) N_2; and, (**c**) N_3

$$
\begin{aligned}
N_1 &= 1.111(\ \ 1 - \zeta - 2\eta) \\
N_2 &= 0.278(-1 + 10\zeta + 2\eta) \\
N_3 &= 0.167(\ \ 1 - 10\zeta + 10\eta)
\end{aligned}
\tag{7.69}
$$

A graph of the three shape functions can be generated as a 3-D plot using the `surf` function in MATLAB and the resulting plots are given in Fig. 7.17.

Remark 7.5 **Analytical formulation for deriving shape functions**: Alternatively, one may wish to use analytical formulations to obtain the shape functions for the 2-D triangular elements. This particularly important if one is looking to incorporate these into an existing in-house FEM solver, such as **MATFESE** discussed previously.

Given the triangular element shown in Fig. 7.15, whose natural coordinates are defined as: (ζ_1, η_1), (ζ_2, η_2) and (ζ_3, η_3), the expressions for the three shape functions can be analytically derived thus:

$$N_1 = \frac{a_1 + b_1\zeta + c_1\eta}{2A} \tag{7.70}$$

$$N_2 = \frac{a_2 + b_2\zeta + c_2\eta}{2A} \tag{7.71}$$

$$N_3 = \frac{a_3 + b_3\zeta + c_3\eta}{2A} \tag{7.72}$$

where the constants, a_i, b_i and c_i for $i = 1, 2, 3$ are defined as follows:

$$
\begin{array}{lll}
a_1 = \zeta_2\eta_3 - \zeta_3\eta_2 & b_1 = \eta_2 - \eta_3 & c_1 = \zeta_3 - \zeta_2 \\
a_2 = \zeta_3\eta_1 - \zeta_1\eta_3 & b_2 = \eta_3 - \eta_1 & c_2 = \zeta_1 - \zeta_3 \\
a_3 = \zeta_1\eta_2 - \zeta_2\eta_1 & b_3 = \eta_1 - \eta_2 & c_3 = \zeta_2 - \zeta_1
\end{array}
\tag{7.73}
$$

and the area of the triangular element, A, is obtained thus:

$$\text{Area of Triangular element, } A = \frac{1}{2}\det\left(\begin{bmatrix} 1 & \zeta_1 & \eta_1 \\ 1 & \zeta_2 & \eta_2 \\ 1 & \zeta_3 & \eta_3 \end{bmatrix}\right) \tag{7.74}$$

where det is the **determinant** of the matrix.

Remark 7.6 **MATLAB™ script for deriving shape functions**: A MATLAB™ script has been developed to automate the process of determining the shape functions for triangular elements. It is a function script called `shapeFunction2DTri` and is available in the ***extra resources website*** of this textbook. It is called by typing in the MATLAB™ command window the command:

```
alphaValues = shapeFunction2DTri(N_1, N_2, N_3)
```

where, `N_1`, `N_2` and `N_3` are the natural coordinates of the three nodes of the element. For example `N_2 = [0.4 0.5]` for the element of Fig. 7.16. The script command for evaluating Fig. 7.16 is:

```
alphaValues = shapeFunction2DTri([0.1 0],[0.4
0.3],[0 0.5])
```

The function script output (`alphaValues`) is a cell array consisting of all the unknown coefficients, $\alpha_i^{(i)}$ that describe the shape functions. Once the script is evaluated, the unknown coefficients, for example shape function N_1, is evaluated using the command: `alphaValues{1}`.

7.5.2 *Jacobian Matrix for 2-D Triangular Elements*

The Jacobian matrix of the 2-D triangular elements is essential in the derivation of element formulation for this type of element. Let us consider the 2-D triangular element of Fig. 7.15, such that the $x-$ and $y-$ axes displacements are u and v respectively. These are in Cartesian coordinate system, but element formulation is usually in isoparametric natural coordinates. The corresponding coordinates to u and v are ζ and η as shown in Fig. 7.15. We will now determine the Jacobian matrix that converts the (u, v) coordinates to the (ζ, η) coordinates and vice versa.

Consider the relationship between the two coordinate systems to be related according to Eq. 7.75. We determine the partial differentials of (u, v) with respect to (ζ, η) as follows:

$$\begin{aligned} u = f(\zeta, \eta) \quad &\longrightarrow \quad du = \frac{\partial u}{\partial \zeta} d\zeta + \frac{\partial u}{\partial \eta} d\eta \\ v = f(\zeta, \eta) \quad &\longrightarrow \quad dv = \frac{\partial v}{\partial \zeta} d\zeta + \frac{\partial v}{\partial \eta} d\eta \end{aligned} \tag{7.75}$$

The matrix format of Eq. 7.75 is:

$$\begin{bmatrix} du \\ dv \end{bmatrix} = \begin{bmatrix} \frac{\partial u}{\partial \zeta} & \frac{\partial u}{\partial \eta} \\ \frac{\partial v}{\partial \zeta} & \frac{\partial v}{\partial \eta} \end{bmatrix} \begin{bmatrix} d\zeta \\ d\eta \end{bmatrix} \implies \begin{bmatrix} du \\ dv \end{bmatrix} = \mathbf{J}' \begin{bmatrix} d\zeta \\ d\eta \end{bmatrix} \tag{7.76}$$

where $\mathbf{J}'$ is the transpose of the Jacobian matrix of (u, v) that converts to (ζ, η) and it is expressed as:

$$\mathbf{J} = \frac{\partial(u, v)}{\partial(\zeta, \eta)} = \begin{bmatrix} \frac{\partial u}{\partial \zeta} & \frac{\partial v}{\partial \eta} \\ \frac{\partial u}{\partial \zeta} & \frac{\partial v}{\partial \eta} \end{bmatrix} \implies \begin{bmatrix} J_{11} & J_{12} \\ J_{21} & J_{22} \end{bmatrix} \tag{7.77}$$

where J_{ij} for $i, j = 1, 2$ are the individual terms of the Jacobian matrix.

It might also be essential as part of the element formulation process to convert isoparametric natural coordinates to spatial Cartesian coordinates, the Jacobian matrix is also used for this. Although here, the *inverse Jacobian*, $\mathbf{J}^{-1}$, is used. The inverse Jacobian is expressed as:

$$\mathbf{J}^{-1} = \frac{\partial(\zeta, \eta)}{\partial(u, v)} = \begin{bmatrix} \frac{\partial \zeta}{\partial u} & \frac{\partial \zeta}{\partial v} \\ \frac{\partial \eta}{\partial u} & \frac{\partial \eta}{\partial v} \end{bmatrix} \implies \frac{1}{J} \begin{bmatrix} J_{22} & -J_{12} \\ -J_{21} & J_{11} \end{bmatrix} \tag{7.78}$$

where $J = \det(\mathbf{J})$ is the determinant of the Jacobian matrix. In earlier one-dimensional element formulation, the Jacobian matrix was a scalar multiplier that makes the conversion from one coordinate system to another. This scalar quantity is equal to the determinant, J of the Jacobian matrix, $\mathbf{J}$. However, in the two-dimensional and the multi-dimensional element formulation cases, the Jacobian matrix is a complete matrix and need to be carefully determined.

Finally, given a set of shape functions, N_i for $i = 1, 2, 3$, for a triangular element, the individual terms of the Jacobian matrix can be obtained as follows:

$$\begin{aligned} J_{11} &= \frac{\partial u}{\partial \zeta} = \sum_{i=1}^{3} u_i \frac{\partial N_i}{\partial \zeta} \qquad & J_{12} &= \frac{\partial v}{\partial \zeta} = \sum_{i=1}^{3} v_i \frac{\partial N_i}{\partial \zeta} \\ J_{21} &= \frac{\partial u}{\partial \eta} = \sum_{i=1}^{3} u_i \frac{\partial N_i}{\partial \eta} \qquad & J_{22} &= \frac{\partial v}{\partial \eta} = \sum_{i=1}^{3} v_i \frac{\partial N_i}{\partial \eta} \end{aligned} \tag{7.79}$$

where u_i and v_i are respectively the $x-$ and $y-$ nodal displacements of the element.

7.5.3 *Strain-Displacement Matrix for 2-D Triangular Elements*

Following similar derivations of the strain-displacement matrix, $\mathbf{B}^e$, shown in Eq. 7.31 for a 1-D element, we will attempt to derive the $\mathbf{B}^e$ matrix for a 2-D triangular element too. The trial displacement for such matrix is a function of two isoparametric natural coordinates, namely ζ and η, hence the first order partial derivative of the displacements will be with respect to those two coordinates.

Given an interpolated displacement of the element, $\mathbf{u}^e = N_1^e \mathbf{u}_1^e + N_2^e \mathbf{u}_2^e + N_3^e \mathbf{u}_3^e$, where, for example, $\mathbf{u}_1^e = u_1 \mathbf{i} + v_1 \mathbf{j}$ where $\mathbf{i}$ and $\mathbf{j}$ are unit vectors in Cartesian reference frame. Let us take the derivative of all three shape functions for the 2-D triangular element according to Eq. 7.65 such that we obtain:

$$\epsilon^e = \begin{bmatrix} \dfrac{\partial \mathbf{u}^e}{\partial \zeta} \\ \\ \dfrac{\partial \mathbf{u}^e}{\partial \eta} \end{bmatrix} \Longrightarrow \epsilon^e = \begin{bmatrix} \dfrac{\partial N_1^e}{\partial \zeta}\mathbf{u}_1^e + \dfrac{\partial N_2^e}{\partial \zeta}\mathbf{u}_2^e + \dfrac{\partial N_3^e}{\partial \zeta}\mathbf{u}_3^e \\ \\ \dfrac{\partial N_1^e}{\partial \eta}\mathbf{u}_1^e + \dfrac{\partial N_1^e}{\partial \eta}\mathbf{u}_2^e + \dfrac{\partial N_1^e}{\partial \eta}\mathbf{u}_3^e \end{bmatrix} \tag{7.80}$$

$$\epsilon^e = \begin{bmatrix} \dfrac{\partial \mathbf{u}^e}{\partial \zeta} \\ \\ \dfrac{\partial \mathbf{u}^e}{\partial \eta} \end{bmatrix} \Longrightarrow \epsilon^e = \underbrace{\begin{bmatrix} \dfrac{\partial N_1^e}{\partial \zeta} & \dfrac{\partial N_2^e}{\partial \zeta} & \dfrac{\partial N_3^e}{\partial \zeta} \\ \\ \dfrac{\partial N_1^e}{\partial \eta} & \dfrac{\partial N_2^e}{\partial \eta} & \dfrac{\partial N_3^e}{\partial \eta} \end{bmatrix}}_{\mathbf{B}^e} \underbrace{\begin{bmatrix} \mathbf{u}_1^e \\ \mathbf{u}_2^e \\ \mathbf{u}_3^e \end{bmatrix}}_{\mathbf{u}^e} \tag{7.81}$$

$$\Longrightarrow \epsilon^e = \mathbf{B}^e \mathbf{u}^e$$

The $\mathbf{B}^e$ matrix is constant in each element since it was derived based on linear relationships between the natural coordinates. Its values solely depend on the coordinates of the nodes of the element rather than ζ and η coordinates. The 2-D triangular element can be considered to be an element type equivalent to the 1-D linear element presented previously, with both showing invariant strain across the element. This can be changed by considering higher order triangular elements, which will be discussed in subsequent sections.

Remark 7.7 **Analytical formulation for calculating $\mathbf{B}^e$ matrix**: The evaluation of the strain-displacement matrix of a 2-D triangular element analytically can be carried out using the following equation:

$$\mathbf{B}^e = \frac{1}{2A^e} \begin{bmatrix} (\zeta_2 - \zeta_3) & (\zeta_3 - \zeta_1) & (\zeta_1 - \zeta_2) \\ \\ (\eta_3 - \eta_2) & (\eta_1 - \eta_3) & (\eta_2 - \eta_1) \end{bmatrix} \tag{7.82}$$

where A^e = Area of triangular element defined previously in Eq. 7.74.

7.5.4 Stiffness Matrix for 2-D Triangular Elements

The stiffness matrix for a 2-D triangular element can be derived by similar approach as Eq. 7.55. However, in this instance the integration is over the surface area of the element (A^e), as against the linear segment (L) used for the 1-D linear element. The derivation is as follows:

$$\mathbf{K}^e\Big|_{tri} = \int_V \mathbf{B}^T E \mathbf{B} \partial V \quad \Longrightarrow \mathbf{K}^e\Big|_{tri} = \mathbf{B}^T E \mathbf{B} A^e \tag{7.83}$$

where E = Young's modulus of the element, A^e = area of the triangular element. The above stiffness matrix is calculated by assuming unit thickness of the element such that: $\partial V = 1 \times \partial A^e$. Note also that just like $\mathbf{B}^e$, $\mathbf{K}^e$ does not depend on coordinates position but solely on the isoparametric (mapping) natural coordinates. As a result, the stresses and strains within the triangular element will be constant.

To derive the stiffness matrix of the constant strain triangular element of Eq. 7.83 above, we made assumptions about the thickness of the element to evaluate the volume integral. It is not always this simple and consequently numerical methods are used to evaluate volume integrals for such cases where the formulation is dependent on multiple natural coordinates. For higher order elements, determining the volume integral is not always straightforward, a more robust approach is required. This leads to the principle of *Gauss Quadrature*, which is commonly used in commercial FEM solvers [9].

7.5.5 Higher Order Triangular Elements

The 2-D constant strain triangular elements described in Sect. 7.5 are suitable for FEM studies where model geometries can be discretized into straight edge segments. Also, such elements are applicable for problems where the strain map within the element is constant and does not change with coordinate positions. However, real physical systems often have curved edges and areas that are dominated by stress concentration, plastic deformation, and, the onset and evolution of damage.

Consequently, there is a need to adapt the element formulation such that these specific design scenarios can be modelled reliably. This leads to the introduction of ***higher order elements***. Examples of higher order triangular elements can include the (a) *six-node quadratic* and (b) *ten-node cubic* triangular elements shown in Fig. 7.18.

The polynomial approach for describing shape functions can be utilized here in defining the shape functions for a higher order triangular element. According to Pascal's triangle, the shape function for the six-node quadratic element shown in Fig. 7.18a, is defined in terms of six unknown parameters, α_i^i for $i = 1, \ldots, 6$.

$$N_i^{6T}(\zeta, \eta) = \alpha_1^{(i)} + \alpha_2^{(i)}\zeta + \alpha_3^{(i)}\eta + \alpha_4^{(i)}\zeta^2 + \alpha_5^{(i)}\zeta\eta + \alpha_6^{(i)}\eta^2 \tag{7.84}$$

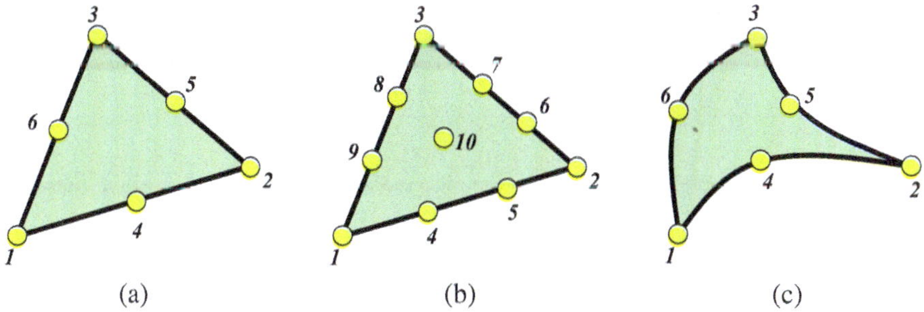

Fig. 7.18 Examples of higher order triangular elements showing: (**a**) six-node quadratic; (**b**) ten-node cubic; and (**c**) *curved-edge* quadratic elements. *Note: The node labelling is in an anti-clockwise manner with the corner nodes labelled first before the edge nodes*

where $N_i^{6T}(\zeta, \eta)$ is the shape function for the i-th node of a six-node triangular element. The evaluation of the unknown coefficients, α_i^i is done using the approach already explained in Example 7.3. According to the Kronecker delta property, the value of N_i^{6T} is 1, for example, at nodes at which the shape function is defined. Using all *isoparametric natural coordinates* for the six nodes of the 2-D triangular element, we obtain six equations, which when solved yield the unknown coefficients.

Similarly, according to Pascal's triangle of Eq. 7.10, the polynomial expression for the ten-node cubic triangular element will consist of ten unknown coefficients, $\alpha_i^{(i)}$, for $i = 1, \ldots, 10$. The difference between the *quadratic* and *cubic* elements is that the total number of elements required to fully describe the cubic elements exceed the number of *corner* and *edge* nodes associated with the element. The extra unknown (linked to the 10th node) is placed somewhere inside the cubic element as shown in Fig. 7.18b. The expression for the ten-node triangular element shape function, N_i^{10T} is given as:

$$\begin{aligned} N_i^{10T}(\zeta, \eta) =& \alpha_1^{(i)} + \alpha_2^{(i)}\zeta + \alpha_3^{(i)}\eta + \alpha_4^{(i)}\zeta^2 + \alpha_5^{(i)}\zeta\eta \\ &+ \alpha_6^{(i)}\eta^2 + \alpha_7^{(i)}\zeta^3 + \alpha_8^{(i)}\zeta^2\eta + \alpha_9^{(i)}\zeta\eta^2 + \alpha_{10}^{(i)}\eta^3 \end{aligned} \tag{7.85}$$

Example 7.7 A virtual blood vessel segment is discretized using higher order elements of the six-node quadratic triangular element type, as shown in Fig. 7.19. Derive all the six shape functions for the higher order 2-D triangular element. Draw all the graphs of the shape functions.

Solution

The polynomial expression for this six-node triangular element is given in Eq. 7.84. To evaluate the unknown coefficients, $\alpha_i^{(1)}$, for the first shape function, N_1^{6T}, let us substitute all the six pairs of isoparametric natural coordinates.

At $\zeta = 0.0, \eta = 0.0 \Longrightarrow 1 = \alpha_1^{(1)}$

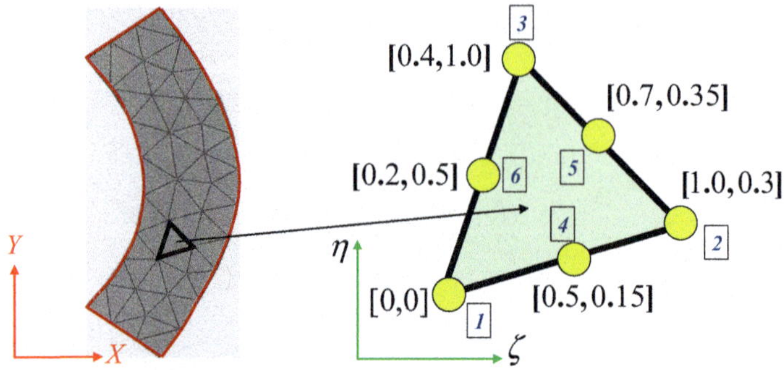

Fig. 7.19 A virtual blood vessel segment discretized with higher order elements

$$\text{At } \zeta = 1.0, \eta = 0.3 \Longrightarrow 0 = \alpha_1^{(1)} + \alpha_2^{(1)} + 0.3\alpha_3^{(1)} + \alpha_4^{(1)} + 0.3\alpha_5^{(1)} + 0.09\alpha_6^{(1)}$$

$$\text{At } \zeta = 0.4, \eta = 1.0 \Longrightarrow 0 = \alpha_1^{(1)} + 0.4\alpha_2^{(1)} + \alpha_3^{(1)} + 0.16\alpha_4^{(1)} + 0.4\alpha_5^{(1)} + \alpha_6^{(1)}$$

$$\text{At } \zeta = 0.5, \eta = 0.15 \Longrightarrow 0 = \alpha_1^{(1)} + 0.5\alpha_2^{(1)} + 0.15\alpha_3^{(1)} + 0.25\alpha_4^{(1)} + 0.075\alpha_5^{(1)} + 0.0225\alpha_6^{(1)}$$

$$\text{At } \zeta = 0.7, \eta = 0.35 \Longrightarrow 0 = \alpha_1^{(1)} + 0.7\alpha_2^{(1)} + 0.35\alpha_3^{(1)} + 0.49\alpha_4^{(1)} + 0.245\alpha_5^{(1)} + 0.1225\alpha_6^{(1)}$$

$$\text{At } \zeta = 0.2, \eta = 0.5 \Longrightarrow 0 = \alpha_1^{(1)} + 0.2\alpha_2^{(1)} + 0.5\alpha_3^{(1)} + 0.04\alpha_4^{(1)} + 0.1\alpha_5^{(1)} + 0.25\alpha_6^{(1)}$$

We can then solve for the unknown coefficients using the approach for solving a system of linear equations described in Sect. 2.4. The unknown coefficients are:

$$\alpha_1^{(1)} = +1.0000 \quad \alpha_2^{(1)} = -2.3864 \quad \alpha_3^{(1)} = -2.0455$$

$$\alpha_4^{(1)} = +0.8030 \quad \alpha_5^{(1)} = +3.8961 \quad \alpha_6^{(1)} = +0.3131$$

We can then repeat the above process to determine all the unknown coefficients for the remaining five shape functions. This is challenging and so a MATLAB™ script, called `shapeFunction2DTriHigherOrder` has been developed to help with this process.

The resulting shape functions are given thus:

$$\begin{aligned}
N_1 &= 1.00 - 2.39\zeta - 2.05\eta + 0.80\zeta^2 + 3.90\zeta\eta + 0.31\eta^2 \\
N_2 &= \quad\;\; - 1.14\zeta + 0.45\eta + 3.25\zeta^2 - 4.55\zeta\eta + 1.30\eta^2 \\
N_3 &= \quad\;\; + 0.34\zeta - 1.14\eta - 0.18\zeta^2 + 0.00\zeta\eta + 2.02\eta^2 \\
N_4 &= \quad\;\; + 4.55\zeta - 1.82\eta - 1.62\zeta^2 - 9.10\zeta\eta + 3.90\eta^2 \\
N_5 &= \quad\;\; - 0.00\zeta + 0.00\eta - 3.83\zeta^2 + 14.29\zeta\eta - 5.10\eta^2 \\
N_6 &= \quad\;\; - 1.36\zeta + 4.55\eta + 1.58\zeta^2 - 4.55\zeta\eta - 2.44\eta^2
\end{aligned} \tag{7.86}$$

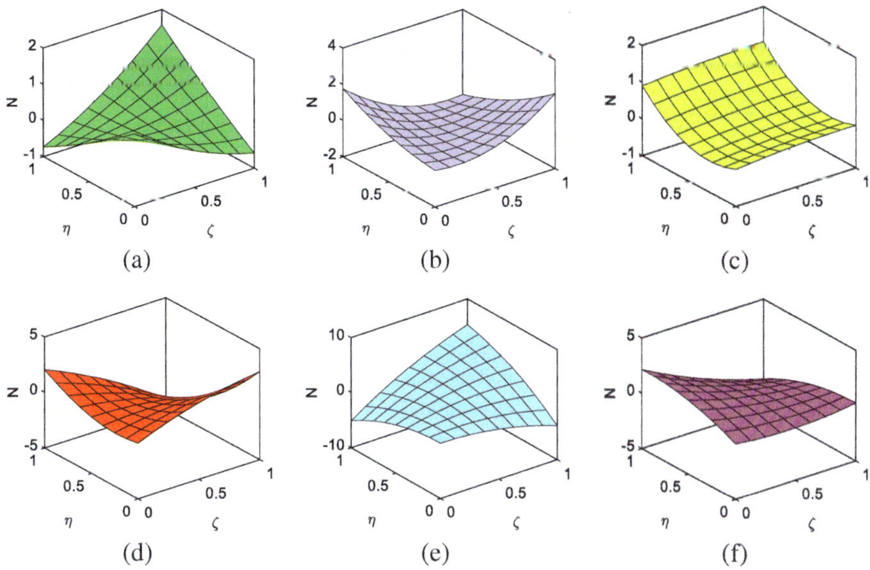

Fig. 7.20 Shape functions for: (**a**) N_1; (**b**) N_2; (**c**) N_3; (**d**) N_4; (**e**) N_5; and, (**f**) N_6

The plots of all the six shape functions are given in Fig. 7.20.

Remark 7.8 **MATLAB Script for Higher order triangular elements**: A MATLAB™ script called: `shapeFunction2DTriHigherOrder` has been developed to help with determining the shape function expressions and plots for the 2-D quadratic higher order triangular elements. The script requires inputs of all the isoparametric natural coordinates for the triangular element of interest. The coordinates are specified as, say, `N_6 = [0.5 2.0]` for the sixth node in the triangular element of Example 7.7. The command that was used for the Example 7.7 is:

```
alphaValues=shapeFunction2DTriHigherOrder([0 0],
[1 0.3],[0.4 1.0],[0.5 0.15],[0.7 0.35],[0.2 0.5])
```

The script generates an output called `alphaValues`, which is a cell array of all the unknown coefficients for all the shape functions. The set of unknown coefficients for each shape function, say, N_1 is obtained by typing in the MATLAB™ command window the instruction: `alphaValues{1}`, for the first shape function. Also, the six plots, as shown in Fig. 7.20, and a combined plot of all the shape functions are also generated from the script.

7.6 Element Formulation for Quadrilateral Elements

Another element type that is widely used in FEM studies is quadrilateral elements, *quad* for short. These are four-sided elements with the most elementary type having four nodes, located at the corners of the element. Higher order quadrilateral elements consist of more nodes located at the edges and internal perimeter of the elements. A degenerate form of the quadrilateral elements is the rectangular elements. However, the presentation here will adopt a generalized approach based on *quad* elements of any edge lengths, as shown in Fig. 7.21.

Figure 7.21 is a sketch of a typical 2-D four-node quadrilateral element described in two isoparametric natural coordinates, namely ζ and η. Let us consider that each of the four nodes consist of u_i and v_i displacements (for $i = 1, \ldots, 4$). Let us assume that the *approximate* or *trial displacement* for this element type can be described thus:

$$
\begin{aligned}
u(\zeta,\eta) &= N_1u_1 + N_2u_2 + N_3u_3 + N_4u_4 \quad \Longrightarrow u(\zeta,\eta) = \mathbf{Nu} \\
v(\zeta,\eta) &= N_1v_1 + N_2v_2 + N_3v_3 + N_4u_4 \quad \Longrightarrow v(\zeta,\eta) = \mathbf{Nv}
\end{aligned}
\tag{7.87}
$$

where N_1, N_2, N_3 and N_4 are the yet-to-be determined shape functions for the 2-D quadrilateral element. The above expressions can be written in *matrix format* as:

$$
\mathbf{u} = \begin{bmatrix} u(\zeta,\eta) \\ \\ v(\zeta,\eta) \end{bmatrix} \Longrightarrow \mathbf{u} = \begin{bmatrix} N_1 & 0 & N_2 & 0 & N_3 & 0 & N_4 & 0 \\ \\ 0 & N_1 & 0 & N_2 & 0 & 0 & 0 & N_4 \end{bmatrix} \begin{bmatrix} u_1 \\ v_1 \\ u_2 \\ v_2 \\ u_3 \\ v_3 \\ u_4 \\ v_4 \end{bmatrix}
\tag{7.88}
$$

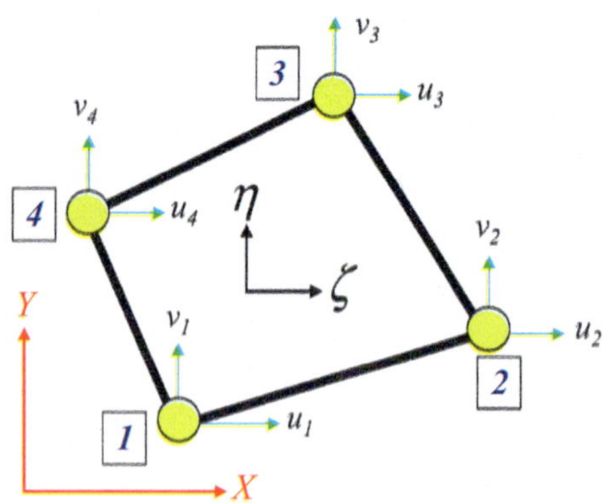

Fig. 7.21 A quadrilateral element showing nodal displacements in two directions (u_i and v_i), nodal numbers (square-boxed numbers), global ($XY-$axis) and isoparametric natural coordinates ($\zeta\eta-$axis)

7.6.1 Shape Functions for Quadrilateral Elements

In order to define explicitly the shape functions given in Eq. 7.88 in terms of isoparametric natural coordinates (ζ, η), we will select the relevant monomials in the Pascal triangle (see Eq. 7.10) for a four-node element. The resulting polynomial expression of the shape function, N_i^{4Q}, of the four-node quadrilateral element becomes:

$$N_i^{4Q} = \alpha_1^{(i)} + \alpha_2^{(i)}\zeta + \alpha_3^{(i)}\eta + \alpha_4^{(i)}\zeta\eta \tag{7.89}$$

where $i = 1, \ldots, 4$ and $\alpha_i^{(i)}$ is the unknown coefficients for the shape functions.

Here, we have assumed a linear dependence between shape functions and isoparametric natural coordinates. This is because there is need for linearity to be maintained in approximating the displacement (for example) along the edges of the quadrilateral element. This element type consists of only two nodes per edge which demands that only a straight-line profile, in our case linear polynomial expansion, can be isolated for describing the displacement profile per quad element edge.

The above ideas were first published by Zienkiewicz's team at Swansea University in 1968 [6]. For further studies about this, the reader should refer to Fish and Belytschko [7] and Davies [5]. We can follow the process described previously to determine the values of the unknown coefficients, $\alpha_i^{(i)}$, bearing in mind that the Kronecker delta property of shape functions has to be met for all four shape functions.

7.6.2 Strain-Displacement Matrix for Quadrilateral Elements

The interpolated, or trial, displacement of the four-node quadrilateral element is given as: $\mathbf{u}^e = N_1\mathbf{u}_1^e + N_2\mathbf{u}_2^e + N_3\mathbf{u}_3^e + N_4\mathbf{u}_4^e$. Using a similar approach to that used in deriving the $\mathbf{B}^e$ matrix of Eq. 7.81, we see that the strain-displacement matrix in terms of isoparametric natural coordinates becomes:

$$\mathbf{B}^e = \begin{bmatrix} \dfrac{\partial N_1^{4Q}}{\partial \zeta} & \dfrac{\partial N_2^{4Q}}{\partial \zeta} & \dfrac{\partial N_3^{4Q}}{\partial \zeta} & \dfrac{\partial N_4^{4Q}}{\partial \zeta} \\ \\ \dfrac{\partial N_1^{4Q}}{\partial \eta} & \dfrac{\partial N_2^{4Q}}{\partial \eta} & \dfrac{\partial N_3^{4Q}}{\partial \eta} & \dfrac{\partial N_4^{4Q}}{\partial \eta} \end{bmatrix} \tag{7.90}$$

The above can be expressed in terms of a global coordinate system by using the chain rule to determine the partial derivatives of the shape function. The resulting derivatives in terms of the Cartesian coordinate system are:

$$\begin{aligned} \frac{\partial N_i^{4Q}}{\partial \zeta} &= \frac{\partial N_i^{4Q}}{\partial x}\frac{\partial x}{\partial \zeta} + \frac{\partial N_i^{4Q}}{\partial y}\frac{\partial y}{\partial \zeta} \\ \frac{\partial N_i^{4Q}}{\partial \eta} &= \frac{\partial N_i^{4Q}}{\partial x}\frac{\partial x}{\partial \eta} + \frac{\partial N_i^{4Q}}{\partial y}\frac{\partial y}{\partial \eta} \end{aligned} \tag{7.91}$$

We re-write Eq. 7.91 in matrix-format to give:

$$\begin{bmatrix} \dfrac{\partial N_i^{4Q}}{\partial \zeta} \\ \\ \dfrac{\partial N_i^{4Q}}{\partial \eta} \end{bmatrix} = \underbrace{\begin{bmatrix} \dfrac{\partial x}{\partial \zeta} & \dfrac{\partial y}{\partial \zeta} \\ \\ \dfrac{\partial x}{\partial \eta} & \dfrac{\partial y}{\partial \eta} \end{bmatrix}}_{\mathbf{J}} \begin{bmatrix} \dfrac{\partial N_i^{4Q}}{\partial x} \\ \\ \dfrac{\partial N_i^{4Q}}{\partial y} \end{bmatrix} \tag{7.92}$$

We have introduced a new matrix called the *Jacobian matrix*, **J**, which is a matrix that relates the derivatives of the global Cartesian coordinates to the isoparametric natural coordinates. It has to be evaluated for every element type as part of that element formulation.

7.6.3 Higher Order Quadrilateral Elements

Similar to the 2-D triangular elements, we can also isolate higher order quadrilateral elements. This will allow for element formulation to be described for discretization of curved boundaries, such as regions around holes. Such elements show the widest range of features useful in discretizing actual curved edges, hence are very appealing to FEM users. Figure 7.22 shows three examples of higher order quadrilateral elements. For all these element types, we can follow the approach described previously for triangular elements to derive their applicable shape functions.

For instance, consider the quadrilateral element shown in Fig. 7.22a. Each edge consists of a three-node arrangement hence, using Pascal's triangle, one can describe the applicable shape function using a quadratic function. Also, the twelve-node *quad* element of Fig. 7.22b consists of five nodes on two edges and three-nodes on two edges, as well as two internal nodes.

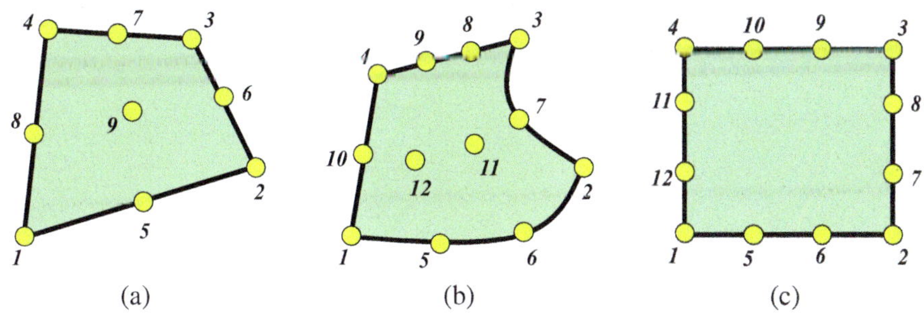

Fig. 7.22 Examples of higher order quadrilateral elements showing: (**a**) nine-node quadratic; (**b**) twelve-node cubic; and, (**c**) twelve-node ***serendipity*** quadrilateral elements

The nodes numbering convention for these higher order elements are as follows:

(a) The corner nodes are numbered first in an counterclockwise manner beginning from 1.
(b) The mid-side nodes are numbered next with the first of the mid-side nodes located in between corner nodes 1 and 2.
(c) The internal nodes are numbered last.

The principles hold for all higher order elements, whether they are 2-D triangles, quadrilaterals or 3-D elements.

For the nine-node quadrilateral of Fig. 7.22a, we can use the Pascal triangle of Eq. 7.10 to formulate the applicable shape function. In doing this, we have to isolate nine distinct monomials (in terms of ζ and η) that are needed to account for the nine nodes that make up this element. It should be noted that in isolating the nine terms, the chosen monomials have to be quadratic in terms of ζ and η, hence the formulation precludes a cubic term. This therefore means that the expected shape function will be *biquadratic* in terms of ζ and η and the form of the highest monomial will be $\zeta^2\eta^2$. Therefore, the probable element shape function is:

$$\begin{aligned} N_i^{9Q}(\zeta,\eta) =& \alpha_1^{(i)} + \alpha_2^{(i)}\zeta + \alpha_3^{(i)}\eta + \alpha_4^{(i)}\zeta^2 + \alpha_5^{(i)}\zeta\eta \\ & + \alpha_6^{(i)}\eta^2 + \alpha_7^{(i)}\zeta^2\eta + \alpha_8^{(i)}\zeta\eta^2 + \alpha_9^{(i)}\zeta^2\eta^2 \end{aligned} \tag{7.93}$$

where $i = 1, 2, \ldots, 9$, and $\alpha_i^{(i)}$ are the unknown coefficients for the shape functions. Also, N_i^{9Q} is the i-th node shape function for a nine-node quadrilateral element. The reader may refer to Chapter 7 of Fish and Belytschko [7] for a detailed discussion on the use of the *tensor product method* to determine shape functions for this type of element.

Using the shape functions of Eq. 7.93, we can map the local/cartesian coordinate system to the isoparametric natural coordinates system thus:

$$x(\zeta,\eta) = \mathbf{N}^{9Q}(\zeta,\eta)\mathbf{x}^e \tag{7.94}$$

$$y(\zeta,\eta) = \mathbf{N}^{9Q}(\zeta,\eta)\mathbf{y}^e \tag{7.95}$$

$$u^e(\zeta,\eta) = \mathbf{N}^{9Q}(\zeta,\eta)\mathbf{u}^e \tag{7.96}$$

where x, y are mapped isoparametric natural coordinates; $\mathbf{x}^e, \mathbf{y}^e$ are vectors of all local coordinates nodal positions; u^e is the interpolated/trial displacement for the element in isoparametric natural coordinates; $\mathbf{u}^e$ is set of nodal displacement in local coordinates system and finally, $\mathbf{N}^{9Q}$ is a vector of all nine shape functions for the element under investigation.

Once the shape functions are determined, the processes shown in Eqs. 7.90 and 7.92 can be used in conjunction with Eq. 7.93 to determine the strain-displacement matrix, $\mathbf{B}^e$, as well as stiffness matrix expression, $\mathbf{K}^e$. A similar approach can be used to develop element formulations for other higher order quadrilateral elements, such as the twelve-node element shown in Fig. 7.22b.

7.6.4 Serendipity Elements

In Fig. 7.22c, we illustrate a special type of quadrilateral elements called ***serendipity elements***.[3] They are quadrilateral elements where majority of the nodes are as much as possible located on the edges and corners of the element away from the interior of the element. They are commonly used in commercial final element codes as they provide the functionality of a quadrilateral element.

However, the absence of the internal nodes implies that they do not contribute to the element connectivity. As a result of presence of the full-set of boundary nodes, inter-element connectivities between an element and its neighbours are preserved, without significant compromises to the validity of the FEM solution. As a result, the element matrix for serendipity elements is reduced hence improving the speed of convergence of results during the FEM solution.

We can have quadratic, cubic and quartic serendipity element and these are illustrated in Fig. 7.23. The full set of shape functions for these element types can also be determined. The simplest form of serendipity elements are the *linear serendipity elements*, which consists of four nodes at the corners of a quadrilateral. This is exactly the same as the element type shown in Fig. 7.21. The interpolating shape function for such linear serendipity elements is exactly the same given in Eq. 7.89.

[3]The word *serendipity* means "fortunate happenstance" or "accidental discovery." This type of element was discovered by Ergatoudis et al. [6] accidentally as they observed the element formulation of a fully-described quadrilateral element.

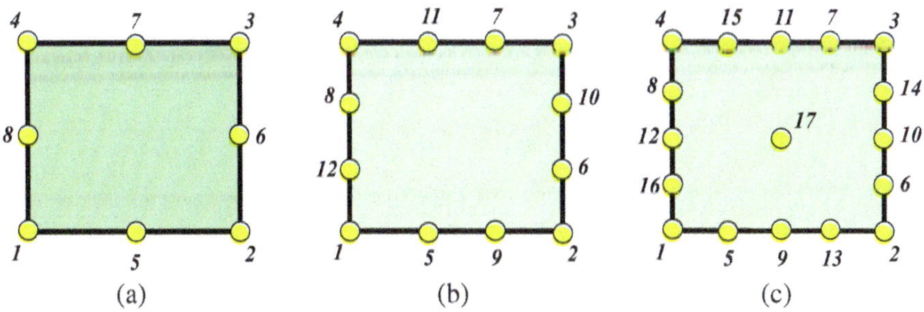

Fig. 7.23 Examples of serendipity elements showing: (**a**) eight-node quadratic; (**b**) twelve-node cubic; and, (**c**) seventeen-node quartic elements

Quadratic serendipity element: For the quadratic serendipity element shown in Fig. 7.23a, there exists eight nodes all located on the boundaries of the element. A typical shape function for this type of element will consist all the terms of the linear formulation of Eq. 7.89 as well as four extra terms to account for the four extra nodes. We will now consult the Pascal triangle, to identify which extra terms those would be. Two conditions have to be obeyed when assembling the applicable terms for the element namely:

(a) The shape function must be in the order of the element. For example, if it is a cubic function, the power of the shape function has to be of order $n = 3$.
(b) For the shape functions of Fig. 7.23, the nodes were positioned in a symmetrical manner so it is essential that spatial symmetry is also maintained in the isoparametric natural coordinates. This condition is called a *spatial isotropy* condition.

Based on the above, the extra four terms that will comprise the shape functions of the quadratic serendipity elements are: $\zeta^2, \eta^2, \zeta^2\eta$ and $\zeta\eta^2$. Thus, the shape function of the quadratic serendipity element becomes:

$$\begin{aligned} N_i^{8S} = \alpha_1^{(i)} + \alpha_2^{(i)}\zeta \; + \alpha_3^{(i)}\eta + \alpha_4^{(i)}\zeta\eta \\ + \alpha_5^{(i)}\zeta^2 + \alpha_6^{(i)}\eta^2 + \alpha_7^{(i)}\zeta^2\eta + \alpha_8^{(i)}\zeta\eta^2 \end{aligned} \tag{7.97}$$

where $i = 1, \ldots, 8$ and $\alpha_i^{(i)}$ are the unknown coefficients for the shape functions, and N_i^{8S} is the shape-function for an 8-node serendipity element.

Cubic serendipity element: Using a similar approach as for the quadratic serendipity elements, we notice that the cubic quadratic elements of Fig. 7.23b consist of twelve nodes. To fully describe their shape functions, four extra terms have to be isolated from the Pascal triangle whilst preserving the cubic-power of the interpolating function and also spatial isotropy. These extra terms will consist of: $\zeta^3, \eta^3, \zeta^3\eta$ and $\zeta\eta^3$. Thus, the full shape function for the cubic serendipity element becomes:

$$
\begin{aligned}
N_i^{12S} = \alpha_1^{(i)} + \alpha_2^{(i)}\zeta \ + \alpha_3^{(i)}\eta + \alpha_4^{(i)}\zeta\eta \\
+ \alpha_5^{(i)}\zeta^2 + \alpha_6^{(i)}\eta^2 + \alpha_7^{(i)}\zeta^2\eta + \alpha_8^{(i)}\zeta\eta^2 \\
+ \alpha_9^{(i)}\zeta^3 + \alpha_{10}^{(i)}\eta^3 + \alpha_{11}^{(i)}\zeta^3\eta + \alpha_{12}^{(i)}\zeta\eta^3
\end{aligned} \tag{7.98}
$$

Quartic serendipity element: As established previously for the quadratic and cubic serendipity elements, the quartic element consists of at least four-more nodes, as shown in Fig. 7.23c. If we go ahead and define the terms for the expected shape functions, we will expect at least sixteen-terms to appear in the shape function expression. We will also expect that the highest order of such a Lagrangian polynomial shape function is $n = 4$.

However, the condition of spatial isotropy will fail for such a sixteen-node serendipity element. Consider the Pascal's triangle for a quartic element shown in Eq. 7.10, we need as well as the $\zeta^4, \eta^4, \zeta^3\eta, \zeta\eta^4$ terms, we will also need to include the $\zeta^2\eta^2$ term. This forces us to include a seventeenth node for the quartic serendipity element. This will have to be an interior node, as shown in Fig. 7.23c. The comprehensive shape function for the seventeen-node quartic serendipity element becomes:

$$
\begin{aligned}
N_i^{17S} = \alpha_1^{(i)} + \alpha_2^{(i)}\zeta \ + \alpha_3^{(i)}\eta + \alpha_4^{(i)}\zeta\eta \\
+ \alpha_5^{(i)}\zeta^2 + \alpha_6^{(i)}\eta^2 + \alpha_7^{(i)}\zeta^2\eta + \alpha_8^{(i)}\zeta\eta^2 \\
+ \alpha_9^{(i)}\zeta^3 + \alpha_{10}^{(i)}\eta^3 + \alpha_{11}^{(i)}\zeta^3\eta + \alpha_{12}^{(i)}\zeta\eta^3 \\
+ \alpha_{13}^{(i)}\zeta^4 + \alpha_{14}^{(i)}\eta^4 + \alpha_{15}^{(i)}\zeta^4\eta + \alpha_{16}^{(i)}\zeta\eta^4 + \alpha_{17}^{(i)}\zeta^2\eta^2
\end{aligned} \tag{7.99}
$$

The unknown coefficients, $\alpha_i^{(i)}$, can be derived as we have shown previously by solving the set of linear equations resulting from substituting known shape function values at specific natural coordinates positions. For more details about this, for not only 2-D but also 3-D serendipity elements, the interested reader should consult the Kaliakin textbook [8]. More details about this can be gleaned from recent re-formulations of the serendipity elements by the following authors [1, 4, 10, 11].

7.7 Element Formulation for Three-Dimensional Elements

The next type of element formulation that needs to be considered is the *three-dimensional elements*. There are two categories to this type of element namely: *hexahedral or brick*; and, *tetrahedral elements*. The hexahedral elements are cubic or cuboid in shape and are generalizations of quadrilateral elements. This implies that they comprise of essentially the fundamental assumptions of quadrilateral elements except that they have been extended from a **2-D** to **3-D** coordinate system. Figure 7.24 shows the schematics of three types of 3-D elements.

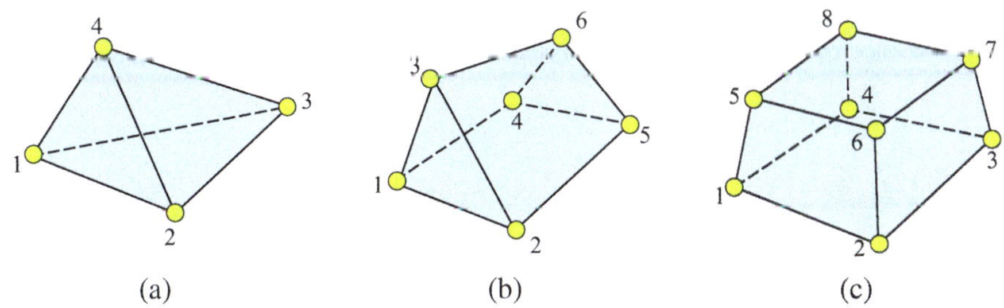

Fig. 7.24 Examples of three-dimensional solid elements showing: (**a**) tetrahedrons ('tet'); (**b**) pentahedrons ('wedges'); and, (**c**) hexahedrons ('brick') elements

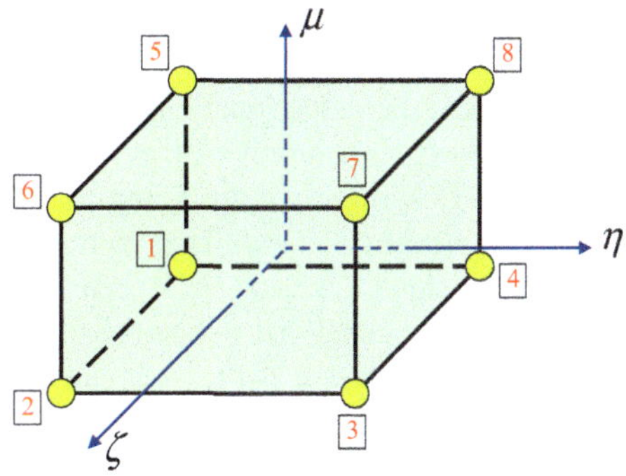

Fig. 7.25 An illustration of the mapped isoparametric nodal coordinates for a hexahedral element. Note: The isoparametric natural coordinates ζ, η, μ *take values from* -1 to $+1$

Similarly, the tetrahedral elements are generalizations of triangular elements. We can identify a degenerate form of hexahedral elements called *pentahedral* or *wedge* elements. The wedge-shaped elements are created by collapsing nodes of the a hexahedral element such that the three sides of the wedge-elements are a quadrilateral, but the other sides are triangular elements.

7.7.1 *Shape Functions for Eight-Node (Trilinear) Hexahedral Element*

We will start by deriving the shape function of an eight-node hexahedral or 'brick' element. We will allow for the element to experience three degrees of freedom displacement, as shown in Fig. 7.25. Each node experiences a displacement in three directions namely, u_i, v_i, w_i (where i = node number), corresponding to $x-$, $y-$, and $z-$ axes displacements.

For mapping the local coordinates to isoparametric natural coordinates, we have to identify mapping expressions for the three Cartesian coordinates such that equivalent isoparametric coordinate system becomes ζ, η, and μ natural coordinates as shown in Fig. 7.25.

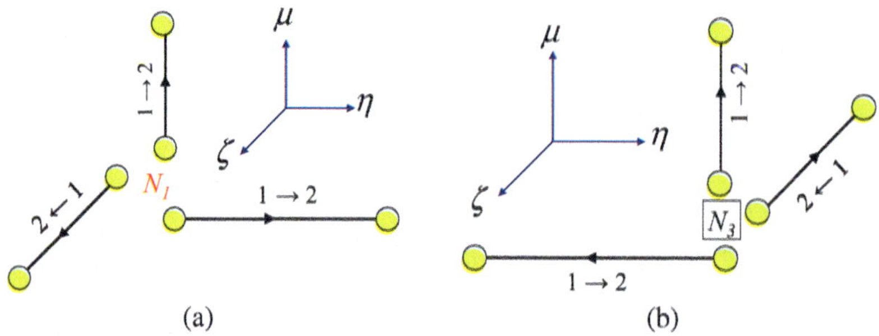

Fig. 7.26 An illustration of the three constituent 1-D two-node linear elements for nodes: (**a**) N_1; and, (**b**) N_3, of the hexahedron shown in Fig. 7.25

Previously, the shape functions for 2-D elements were derived based on monomials extracted from a Pascal's triangle. That approach was suitable for 2-D systems, but for a 3-D system, we are going to use a different approach. This will involve the *tensor product method*. This method states that, for an eight-node hexahedral element, its shape function, N^{8Q}, can be expressed as a product of the three one-dimensional elements that define the three isoparametric natural coordinates of the hexahedral. This can be expressed mathematically as:

$$N^{8H}(\zeta, \eta, \mu) \quad = \quad N^{2L}(\zeta)N^{2L}(\eta)N^{2L}(\mu) \tag{7.100}$$

where $N^{2L}(\zeta)$ is the shape function for a two-node one-dimensional linear element along the ζ natural coordinates. Similar shape functions apply for the η and μ natural coordinates. To evaluate the terms that make up Eq. 7.103, we must evaluate the shape functions for the one-dimensional two-node linear elements.

The eight-node hexahedral element will consist of eight shape functions. Therefore, the contributory two-node linear shape functions that define the expression of Eq. 7.103 will change from node to node. To understand this, consider Fig. 7.26 for three representative 1-D two-node elements that are needed when defining the shape functions of a 3-D hexahedral element.

In Fig. 7.26, node N_1 of the eight-node hexahedron, for example, consists of three 1-D linear elements that align with the ζ, η and μ natural coordinate axes. To define the shape function, N_1^{8Q}, at this node will require defining the shape functions for these three 1-D two-node linear elements. Also, observe that node N_1 is located at $\zeta = -1$, $\eta = -1$ and $\mu = -1$. Similarly, for node N_3, the natural coordinates location is: $\zeta = 1$, $\eta = 1$ and $\mu = -1$.

As shown also in Fig. 7.26, the label of $1 \rightarrow 2$ is attached to each of the 1-D elements. This indicates the direction of action from the start to end nodes for that 1-D element. If the direction coincides with the positive direction of the natural coordinates, it is $1 \rightarrow 2$ whilst the converse is labelled $2 \rightarrow 1$.

Using the attached numberings from Fig. 7.26, the equivalent 1-D linear element node numbering is $I = 1, J = 1$ and $K = 1$, for node N_1. Similarly, the equivalent 1-D linear element node numbering for node N_3 is: $I = 2, J = 2$, and $K = 1$

Table 7.1 Element parameters for an eight-node hexahedron

Node numbering	Natural coordinates			Equivalent 1-D element numbering		
N_i	ζ	η	μ	I	J	K
1	−1	−1	−1	1	1	1
2	+1	−1	−1	2	1	1
3	+1	+1	−1	2	2	1
4	−1	+1	−1	1	2	1
5	−1	−1	+1	1	1	2
6	+1	−1	+1	2	1	2
7	+1	+1	−1	2	2	1
8	−1	+1	+1	1	2	2

where I, J, and K are the Cartesian coordinates equivalent to ζ, η and μ in natural coordinates system. Table 7.1 shows the list of the natural coordinates locations for all eight nodes as well as the equivalent 1-D linear element node number.

Based on the one-dimensional two-node linear element along the ζ-axis of Fig. 7.26a, with $-1 \leq \zeta \leq +1$, the expression of the shape function is:

$$N_i^{2L} = \alpha_1^{(i)} + \alpha_2^{(i)}\zeta \tag{7.101}$$

Using a similar approach to that used in Example 7.1, it can be shown that the expressions for the applicable shape functions for the two-node 1-D linear element are:

$$N_1^{2L}(\zeta) = \frac{1}{2}(1-\zeta) \qquad \text{and} \qquad N_2^{2L}(\zeta) = \frac{1}{2}(1+\zeta) \tag{7.102}$$

The same is true for the other η- and μ- natural coordinates directions. For the eight-node hexahedron, we can now apply the tensor product expression of Eq. 7.103 to determine the shape function for node, N_1 thus:

$$N_1^{8H}(\zeta, \eta, \mu) \quad = \quad \frac{1}{8}(1-\zeta)(1-\eta)(1-\mu) \tag{7.103}$$

Combining Eq. 7.103 with Table 7.1, the complete set of shape functions for all eight node hexahedron is:

$$N_1^{8H} = \frac{1}{8}(1-\zeta)(1-\eta)(1-\mu) \qquad N_2^{8H} = \frac{1}{8}(1+\zeta)(1-\eta)(1-\mu) \tag{7.104}$$

$$N_3^{8H} = \frac{1}{8}(1+\zeta)(1+\eta)(1-\mu) \qquad N_4^{8H} = \frac{1}{8}(1-\zeta)(1+\eta)(1-\mu) \tag{7.105}$$

$$N_5^{8H} = \frac{1}{8}(1-\zeta)(1-\eta)(1+\mu) \qquad N_6^{8H} = \frac{1}{8}(1+\zeta)(1-\eta)(1+\mu) \tag{7.106}$$

$$N_7^{8H} = \frac{1}{8}(1+\zeta)(1+\eta)(1+\mu) \qquad N_8^{8H} = \frac{1}{8}(1-\zeta)(1+\eta)(1+\mu) \tag{7.107}$$

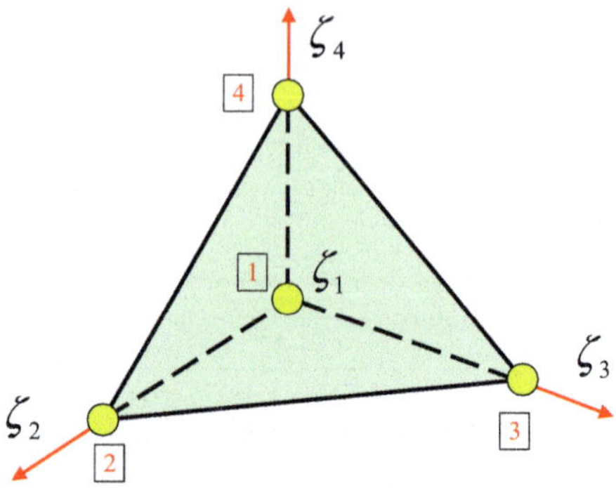

Fig. 7.27 An illustration of the isoparametric natural coordinates for a tetrahedral element

We can condense the above set of equations into a single set of equations defined in terms of i-th node thus:

$$N_i^{8H} = \frac{1}{8}(1 + \zeta\zeta_i)(1 + \eta\eta_i)(1 + \mu\mu_i) \tag{7.108}$$

where ζ_i, η_i and μ_i are the coordinates of the i-th node.

Consider, for example, the sixth node of the hexahedron of Fig. 7.25. We observe from Table 7.1 that: $\zeta_i = +1, \eta_i = -1$, and $\mu_i = +1$. Substituting these into Eq. 7.108, the shape function for this sixth node becomes: $N_6^{8H} = \frac{1}{8}(1 + \zeta)(1 - \eta)(1 + \mu)$, which is the same as obtained previously. In a similar way, we can also obtain all the shape functions for all higher order hexahedral elements.

7.7.2 *Shape Functions for Four-Node Tetrahedral Element*

Similar to the hexahedron, here we derive the shape functions for the four-node tetrahedron. The isoparametric natural coordinates for this type of element is a four-dimension coordinate system expressed in terms of $\zeta_1, \zeta_2, \zeta_3$ and ζ_4 as shown in Fig. 7.27.

Tetrahedral and triangular elements are generally regarded as *unstructured elements*. The tensor product method used for hexahedral elements are not always easy to deploy for such unstructured elements, neither can the Pascal triangle approach be used for a 3-D system as a tetrahedron. Therefore, we will use a different approach based on a *barycentric coordinate system*.

The ***barycentric coordinate system*** is a coordinate system in which the location of a position in the system, say point, P for a particular geometry (e.g.

(continued)

triangles, tetrahedrons), is given as the centre of mass (called *barycentre*) of unequal masses placed at the vertices of that shape.

For the triangle of Fig. 7.28a with a total area, A^{tri}, the triangle is sub-divided into three sub-triangles of area, A_i^{subtri}, where $i = 1, 2, 3$ such that $A^{tri} = \sum_{i=1}^{3} A_i^{subtri}$. We can define the barycentric (area) coordinates for a position, P, as ζ_i such that $0 \leq \zeta_i \leq 1$ with $(i = 1, 2, 3)$. The area coordinates are defined as follows:

$$\sum_{i=1}^{3} \frac{A_i^{subtri}}{A^{tri}} = 1 \quad \longrightarrow \quad \zeta_1 + \zeta_2 + \zeta_3 = 1 \tag{7.109}$$

where

$$\zeta_i = \frac{A_i^{subtri}}{A^{tri}}, \quad \text{where} \quad i = 1, 2, 3 \tag{7.110}$$

Similarly, we can also deduce the barycentric coordinate system (volume coordinates) for the tetrahedron shown in Fig. 7.28b, but here we illustrate the procedure for position, P. The full barycentric coordinate system of position, P is fully defined by four coordinates namely: $\zeta_1, \zeta_2, \zeta_3$ and ζ_4. According to Eq. 7.109:

$$\sum_{i=1}^{4} \frac{V_i^{subtet}}{V^{tet}} = \frac{V_1^{subtet}}{V^{tet}} + \cdots + \frac{V_4^{subtet}}{V^{tet}} = 1 \longrightarrow \zeta_1 + \zeta_2 + \zeta_3 + \zeta_4 = 1 \tag{7.111}$$

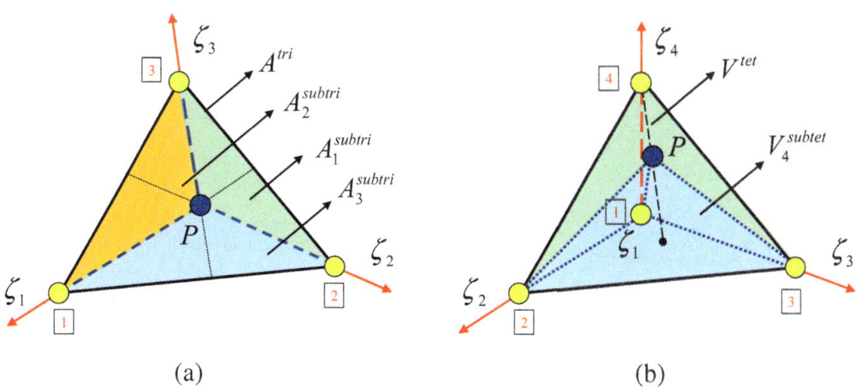

Fig. 7.28 An illustration of barycentric coordinate system of a point, P, for: (**a**) triangles (area coordinates); and, (**b**) tetrahedrons (volume coordinates)

where V^{tet} is the volume of the tetrahedral element, and V_i^{subtet} is the i-th sub-tetrahedron formed by using position, P and the triangular plane directly opposite point, P. For example, as shown in Fig. 7.28b, V_4^{subtri} = Volume $\overline{P123}$ which is the volume of the shaded tetrahedron bounded by nodal points, P, 1, 2, and 3. Therefore, for the tetrahedron, the barycentric coordinate, ζ_i, is defined as:

$$\zeta_i = \frac{V_i^{subtet}}{V^{tet}} \quad \text{where} \quad i = 1, 2, 3, 4 \tag{7.112}$$

The shape functions that define the four-node tetrahedron must obey the Kronecker Delta condition, which demands that the shape functions have a value of 1 at the applicable nodes and zeros at every other nodes. For example, $N_1^{4Tet} = 1$ at node 1 and zero at nodes 2 to 4. Similarly, $N_2^{4Tet} = 1$ at node 2 and zeros at nodes 1, 3 and 4. The barycentric natural coordinates representation in terms of *lines* for triangle and *planes* for the tetrahedron is shown in Fig. 7.29.

A similar linear or planar representation also holds for ζ_2, ζ_3 and ζ_4 (for tetrahedrons) natural coordinates. In all cases, it can be observed that:

$$\zeta_i = constant \qquad \text{where, } i = 1, 2, \ldots, 4, \text{ for a tetrahedron} \tag{7.113}$$

We see that Eq. 7.113 represents the set of planes parallel to the face opposite the $\underline{i\text{-th}}$ corner. For example in Fig. 7.29b, three planes are shown parallel to the face $\overline{234}$ which is opposite the 1–nodal corner.

In order to formulate the shape function for the tetrahedron, in terms of barycentric natural coordinates $\zeta_1, \zeta_2, \zeta_3$ and ζ_4, we will consider a coordinate position, P, located within the tetrahedron as shown in Fig. 7.30a. This coordinate position can be interpolated according to all four contributory planes of the tetrahedron that are shown in Fig. 7.30b.

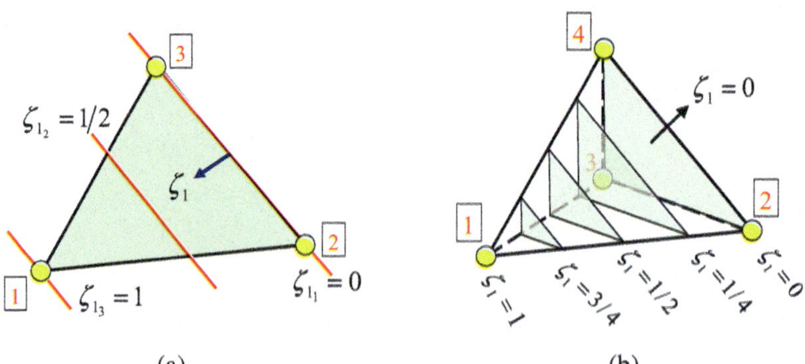

Fig. 7.29 The barycentric natural coordinate representation of: (**a**) triangular element (showing *lines* of constant ζ_1–values); and, (**b**) tetrahedral elements (showing *planes* of constant ζ_1–values). Square-boxed numbers are nodal numbers

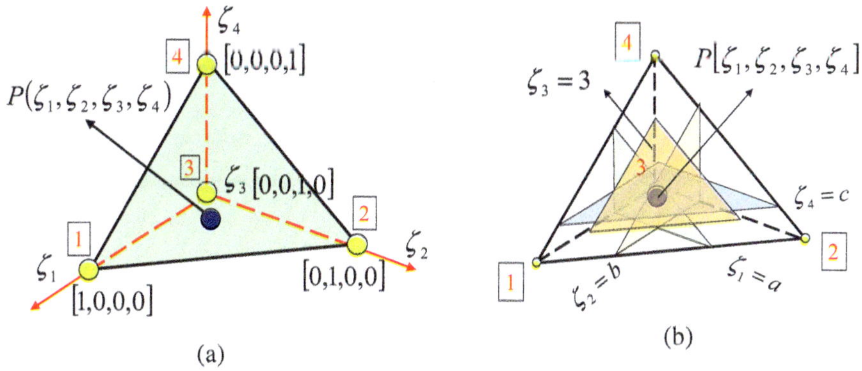

Fig. 7.30 The barycentric natural coordinate representation for: (**a**) coordinate position, P; and, (**b**) illustration of four distinct planes that contribute to the coordinate position, P. Square-boxed numbers are nodal numbers

Each of the four planes are located at definite locations, a, b, c and d, and are parallel to the i-th nodal coordinate that is directly opposite the plane. Each of these planes can be represented by the expression: $\alpha_i^{(i)}\zeta_i$ where $\alpha_i^{(i)}$ is an unknown coefficient that defines the plane and ζ_i is the applicable barycentric natural coordinates for the i-th node that is directly opposite the plane of interest. It can be shown that the shape function for the tetrahedral element in terms of the contributing four planes becomes:

$$N^{4Tet} = \alpha_1^{(i)}\zeta_1 + \alpha_2^{(i)}\zeta_2 + \alpha_3^{(i)}\zeta_3 + \alpha_4^{(i)}\zeta_4 \tag{7.114}$$

The next objective is to determine the unknown coefficients, α_i, for $i = 1, 2, \ldots, 4$. We will adopt the approach that was shown previously for Lagrangian polynomial representation of shape functions by substituting into Eq. 7.114 the known coordinates values for nodes $1, 2, 3$ and 4. This will help us establish the values of the unknown coefficients, α_i. The reader should prove that for the tetrahedron shown in Fig. 7.30a, the unknown coefficients for node 1 are: $\alpha_1^{(1)} = 1, \alpha_2^{(1)} = 0, \alpha_3^{(1)} = 0$, and $\alpha_4^{(1)} = 0$. Therefore, the expression of the shape functions, N_i^{4Tet} (for $i = 1, 2, \ldots, 4$) for the four-node tetrahedron becomes:

$$\begin{aligned} N_1^{4Tet} &= \zeta_1 \\ N_2^{4Tet} &= \zeta_2 \\ N_3^{4Tet} &= \zeta_3 \\ N_4^{4Tet} &= \zeta_4 = 1 - \zeta_1 - \zeta_2 - \zeta_3 \end{aligned} \tag{7.115}$$

A similar approach can be used to determine the shape functions for higher order tetrahedral elements. Interested readers should refer to Bittencourt [2, 3] who showed extensively how a tensor product method can be used to determine shape functions for triangles and tetrahedrons based on the barycentric natural coordinates representation shown here.

Also, for pentahedral elements, a similar implementation can be used to determine their shape functions. Such elements will also consist of $\zeta_1, \zeta_2, \zeta_3$ and ζ_4 barycentric coordinate systems. We will not concern ourselves with demonstrating these, but interested readers should consult some of the textbooks in the references in this chapter for a more in-depth look at the shape function derivation for pentahedral elements.

7.7.3 *Mapping in Three-Dimensions*

Having derived the applicable shape functions of the hexahedral, tetrahedral or pentahedral shape elements, the next challenge is to map the coordinate positions from global to element reference frames. Similar to Eq. 7.94, let us define the global coordinate system for a 3-D virtual domain in terms of X-, Y-, and Z-axes, and the element displacement as: u^e. The mapping expression for a four-node tetrahedral element is:

$$x(\zeta_1, \zeta_2, \zeta_3, \zeta_4) = \mathbf{N}^{4Tet}(\zeta_1, \zeta_2, \zeta_3, \zeta_4)\mathbf{x}^e \tag{7.116}$$

$$y(\zeta_1, \zeta_2, \zeta_3, \zeta_4) = \mathbf{N}^{4Tet}(\zeta_1, \zeta_2, \zeta_3, \zeta_4)\mathbf{y}^e \tag{7.117}$$

$$z(\zeta_1, \zeta_2, \zeta_3, \zeta_4) = \mathbf{N}^{4Tet}(\zeta_1, \zeta_2, \zeta_3, \zeta_4)\mathbf{z}^e \tag{7.118}$$

$$u^e(\zeta_1, \zeta_2, \zeta_3, \zeta_4) = \mathbf{N}^{4Tet}(\zeta_1, \zeta_2, \zeta_3, \zeta_4)\mathbf{u}^e \tag{7.119}$$

where x, y, and z are mapped barycentric coordinates; $\mathbf{x}^e, \mathbf{y}^e$, and $\mathbf{z}^e$ are vectors of all local coordinates nodal positions; u^e is the interpolated/trial displacement for the element in barycentric coordinates; $\mathbf{u}^e$ is set of nodal displacement in the local coordinate system and, finally, $\mathbf{N}^{4Tet}$ is a vector of all four shape functions for the tetrahedral element under consideration. The above will then lead to derivations of the Jacobian matrix, surface tractions, volume forces and volume integrals required for the description of the holistic element behaviour needed for the FEM process.

7.8 Conclusions

In this chapter, we have explored the FEM principle of meshing with a specific focus on the mathematics of element formulation. The mathematical expressions needed to describe one type of element must be developed and implemented within the element library of an FEM solver. This chapter has demonstrated the steps required in deriving the applicable mathematical element formulation for 1-D, 2-D and 3-D elements. We have also shown the role shape functions play in deducing correct element formulations for different element types.

At the end of this chapter, the student should have learnt extended their understanding of the mathematics of element formulation within the FEM process.

We feel that the student is now able to design 'new elements',[4] derive their relevant shape functions and hopefully consider deploying the 'new' elements as part of a be-spoke FEM solver's element library. The next chapter introduces the challenge of incorporating boundary conditions with the meshes.

At the end of this chapter, the reader should be able to:

(a) Undertake the derivation of element formulations for simple 1-D, 2-D and 3-D element types.
(b) Appreciate the importance shape functions on element formulations.
(c) Determine the displacements, strain and stresses that result for different element types, based solely on the shape functions for such elements.
(d) Distinguish between first order and higher order elements.
(e) Begin developing user-defined element formulations for 'new' element types.
(f) Use the knowledge gained in solving real life (physical) problems.

7.9 Problems: Mathematics of Element Formulation

Problem 7.1 A truss-based roof structure, shown in Fig. 7.31, consists of three truss arrangements namely *horizontal, vertical* and *inclined* arrangements. In order to undertake an FEM study of the roof structure, one-dimensional linear truss elements are to be used to discretize the structure.

To determine the appropriate element formulations for this element type:

(a) Determine the *mapping expression*, ζ, that maps the local $x-$ and $y-$ coordinates to the *isoparametric nodal coordinate system*, for the three element types.
(b) Determine the first order derivative of the mapping expression.
(c) Determine the shape function of the three types of truss elements.

Problem 7.2 A concept design of a stent structure is shown in Fig. 7.32. Displacement sensors attached via annular discs inserted at specific locations along the main axis of the stent. In one test, a researcher obtained readings of axial displacements of the stent as given in the inserted table. You are required to undertake a finite element study of this stent by idealizing its response using one-dimensional linear elements.

[4]The term 'new elements' is used here to represent adaptations of existing standard FEM solver element types. For example, the reader can create a 'new element' by modifying some of the element parameters of standard 1-D, 2-D, or 3-D element types discussed in this chapter; thereby prescribing a new/modified element behaviour to such elements.

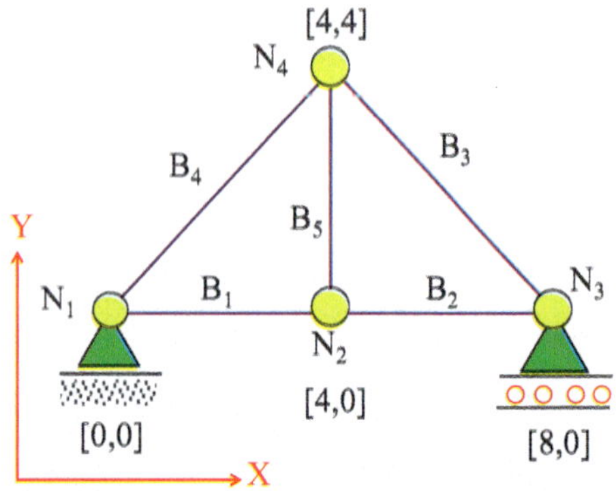

Fig. 7.31 A truss-based roof structure showing three truss element types

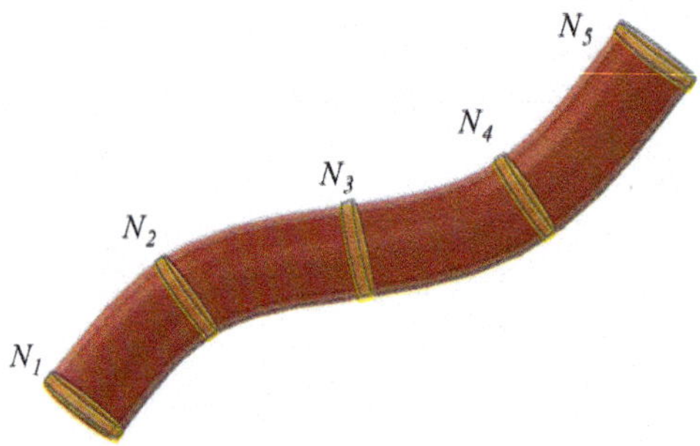

N_i	ζ	u
1	0	0.000
2	0.25	0.003
3	0.50	0.004
4	0.75	0.006
5	1.00	0.008

Fig. 7.32 A stent with displacement sensors attached at nodal positions, N_i for $i = 1, 2, \ldots, 5$. The table shows the nodal axial displacements

Discretize the stent using:

(a) A two-node element and find the interpolated displacement, u^e_{N2}.
(b) A three-node element and find the interpolated displacement, u^e_{N3}.
(c) A four-node element and find the interpolated displacement, u^e_{N5}.
(d) The researcher attached a sensor at $\zeta = 0.56$ and derived a displacement of $u = 0.43$. Determine the discretization error associated with all three displacement predictions at this ζ-value.

Problem 7.3 A 2-D representation of a rugby ball is discretized using triangular elements. The ball is made from a material of Young's modulus, $E = 300$ MPa. We are interested in developing the element formulation for the three-node triangular element isolated from the meshed rugby ball as shown in Fig. 7.33.

Determine, for the triangular element, the following:

(a) The relevant element shape functions, N_i where $i = 1, 2, 3$.
(b) The plot of the shape functions.
(c) The strain-displacement matrix, $\mathbf{B}$.
(d) The stiffness matrix, $\mathbf{K}^e$.

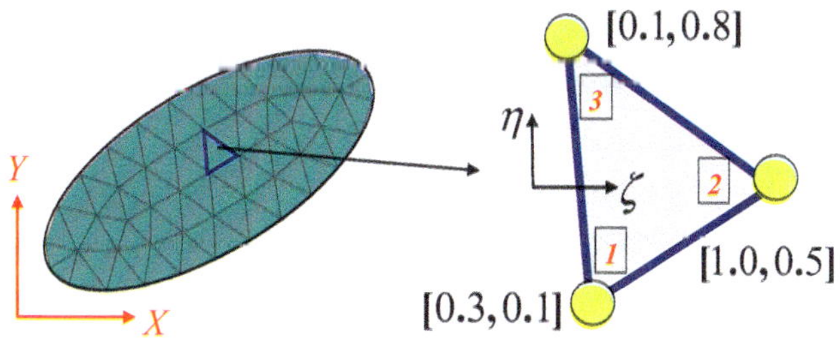

Fig. 7.33 A rugby ball discretized with triangular elements, with natural coordinates values for isolated triangular element given

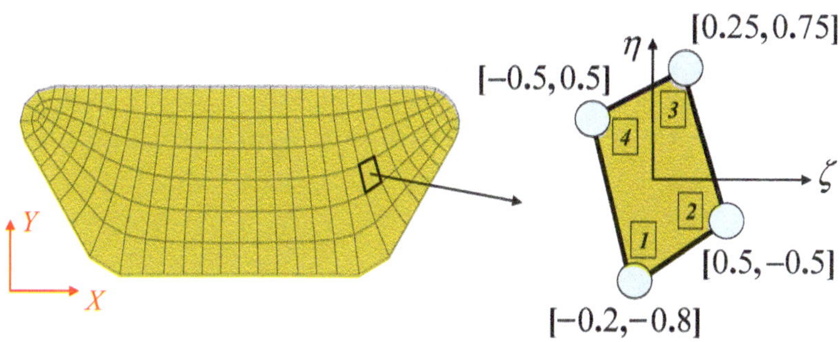

Fig. 7.34 A bath model discretized with quadrilateral elements. The isolated elements is defined in terms of values of its isoparametric natural coordinates

Problem 7.4 The meshed model of a bath is shown in Fig. 7.34. The bath is made of steel of Young's Modulus, $E = 210\,\text{GPa}$. The mesh is based on 2-D quadrilateral elements. To develop the element formulation for this element type, let us isolate one of the elements in isoparametric natural coordinates, as shown in Fig. 7.34.

Based on the isolated quadrilateral elements, determine:

(a) The relevant element shape functions, N_i where $i = 1, 2, \ldots, 4$.
(b) The plot of the shape functions.

Problem 7.5 Based on the quadrilateral elements of Fig. 7.34, let us determine:

(a) The Jacobian matrix of the element.
(b) The strain-displacement matrix, **B**.
(c) The stiffness matrix, $\mathbf{K}^e$.

Problem 7.6 A contact lens design is meshed using four-node quadrilateral elements, as shown in Fig. 7.34. Initial observations of the mesh show that the curved edges of the lens were not appropriately captured during the meshing due to the limitations of the linear-edge four-node quadrilaterals. To improve the meshing and FEM solution, you designed a higher order 2-D quadrilateral element with isoparametric nodal coordinates, shown as the isolated element of Fig. 7.34.

Using this element, determine:

(a) The relevant element shape functions, N_i where $i = 1, 2, \ldots, 9$.
(b) The plot of the shape functions.
(c) The strain-displacement matrix, $\mathbf{B}^e$ for the element.
(d) Quantify and comment on any improvements to the FEM simulation.

Problem 7.7 A twenty-node (serendipity) higher order 3-D hexahedron is shown in Fig. 7.36. Each of the three natural coordinate directions take values from -1 to $+1$ and the mid-edge nodes take a value of 0.

(a) Determine the three shape functions for the equivalent 1-D three-node linear element for this hexahedron.
(b) Determine the shape functions for nodes 1, 7, 15 and 20.
(c) Show that the condensed form of the shape functions for all corner nodes is:

$$N_i^{20H} = \frac{1}{8}(1 + \zeta\zeta_i)(1 + \eta\eta_i)(1 + \mu\mu_i)(\zeta\zeta_i + \eta\eta_i + \mu\mu_i - 2)$$

for nodes, $i = 1, 2, \ldots, 8$.

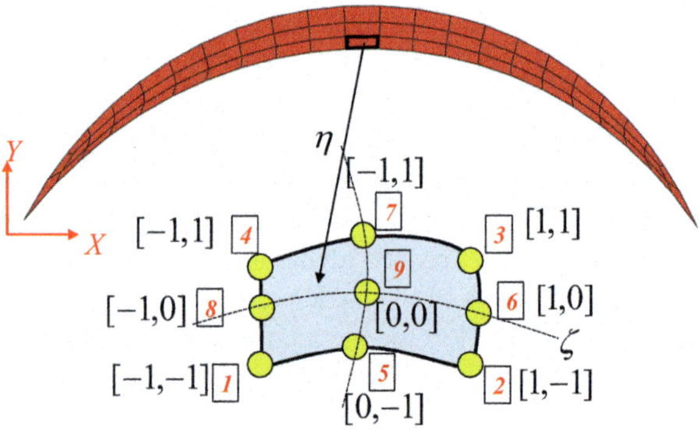

Fig. 7.35 A model of a contact lens meshed with quadrilateral elements. The selected element is ten-noded higher order quadrilateral element

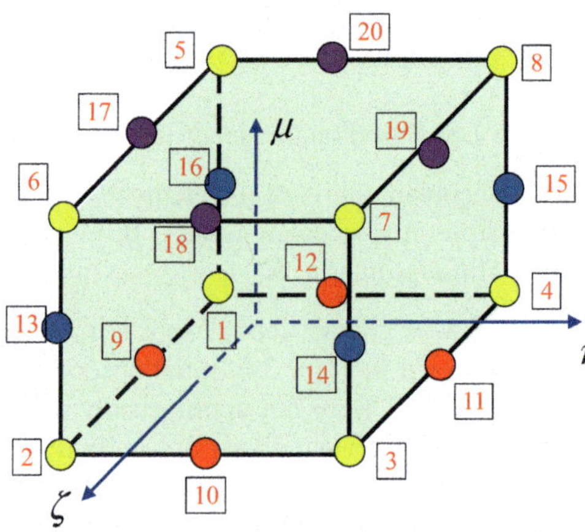

Fig. 7.36 A twenty-node (serendipity) hexahedron with nodes labelled from 1 to 20 in a counter clockwise manner

References

1. Arnold, D.N., Awanou, G.: The serendipity family of finite elements. Found. Comput. Math. **11**(3), 337–344 (2011)
2. Bittencourt, M.: Computational Solid Mechanics: Variational Formulation and High Order Approximation. Taylor & Francis, Boca Raton (2014). https://books.google.co.uk/books?id=643aBAAAQBAJ
3. Bittencourt, M.L., Vazquez, M.G., Vazquez, T.G.: Construction of shape functions for the h- and p-versions of the FEM using tensorial product. Int. J. Numer. Methods Eng. **71**(5), 529–563 (2007). http://dx.doi.org/10.1002/nme.1955
4. Da Veiga, L.B., Brezzi, F., Marini, L., Russo, A.: Serendipity nodal VEM spaces. Comput. Fluids **141**, 2–12 (2016)
5. Davies, A.: The Finite Element Method: An Introduction with Partial Differential Equations. Oxford University Press, Oxford (2011). https://books.google.co.uk/books?id=Z1dI2TiVvSgC
6. Ergatoudis, I., Irons, B., Zienkiewicz, O.: Curved, isoparametric, "quadrilateral" elements for finite element analysis. Int. J. Solids Struct. **4**(1), 31–42 (1968)
7. Fish, J., Belytschko, T.: A First Course in Finite Elements. Wiley, Chichester (2007). https://books.google.co.uk/books?id=kqwMmAEACAAJ
8. Kaliakin, V.: Introduction to Approximate Solution Techniques, Numerical Modeling, and Finite Element Methods. Civil and Environmental Engineering. Taylor & Francis (2001). https://books.google.co.uk/books?id=joMDL60B0KAC
9. Kovvali, N.: Theory and Applications of Gaussian Quadrature Methods. Synthesis Digital Library of Engineering and Computer Science. Morgan & Claypool, San Rafael (2011). https://books.google.co.uk/books?id=uTCdW8L6i8sC
10. Rand, A., Gillette, A., Bajaj, C.: Quadratic serendipity finite elements on polygons using generalized barycentric coordinates. Math. Comput. **83**(290), 2691–2716 (2014)
11. Utku, M.: An improved transformation for universal serendipity elements. Comput. Struct. **73**(1), 199–206 (1999)

Chapter 8
Boundary Conditions

Abstract No FEM solution is possible without defining the behaviour/conditions of the nodes at the boundaries of the virtual domain, the so-called *boundary conditions* of the FEM model. Therefore, the solution of any FEM problem is highly dependent on the boundary conditions implemented. This chapter explores this extremely important feature of an FEM model. Although *loads* and *boundary conditions* are distinguished in certain FEM solvers, such as ABAQUS, the presentation here considers both of them under the broad description of boundary conditions. This chapter introduces the different types of boundary conditions commonly in use in many FEM solvers. Special attention is given to *periodic boundary conditions* since it has been shown in literature that they can quickly arrive at the convergent solution of an FEM problem. A methodology for imposing this type of boundary condition for a 2D representative volume element (RVE) is shown too. The ability to define correctly an appropriate boundary condition for a given problem is a vital skill for anyone who wants to use the FEM process to determine reliable solutions of any physical problem.

Keywords Boundary condition • Periodic boundary condition • Dirichlet boundary condition • Neumann boundary condition • Effective properties • **PBC2DGEN**

8.1 Introduction

In continuation of the discussion on the FEM principles, this chapter is dedicated to discussions about *boundary conditions*. The nature of the applied boundary conditions influences the outcome of the solution significantly, hence this chapter. Extreme care has to be taken when defining the boundary conditions of a problem as they can make or break ones solution. It is important that FEM users, who are new to the FEM process, understand the different types of boundary conditions that can be imposed on a virtual domain and their impact on the validity of FEM solutions.

M. Okereke, S. Keates, *Finite Element Applications*, Springer Tracts
in Mechanical Engineering, https://doi.org/10.1007/978-3-319-67125-3_8

In the context of finite element modelling, a *boundary condition* (**BC**) is a set of constraints imposed on nodal coordinates located at the boundaries of a virtual domain. It is also the condition that must be satisfied by a solution to a differential equation associated with the physical problem under investigation.

The set of constraints can be in $x-$, $y-$, and $z-$ axes as well as different rotational reference frames. Certain FEM solvers distinguish between *loads* and *boundary conditions*. Loads are externally applied constraints on the virtual domain, which cause a change in the original configuration of the body. On the other hand, boundary conditions here do not cause a change in configuration of the body, but provide support reactions to the body. Some FEM solvers (e.g. ANSYS) do not distinguish between loads and classic support-originating boundary conditions. Within ABAQUS, however, loads and boundary conditions are clearly distinguished. Figure 8.1 shows typical sub-divisions of the loads and boundary conditions that are commonly used under the boundary condition module of many FEM solvers.

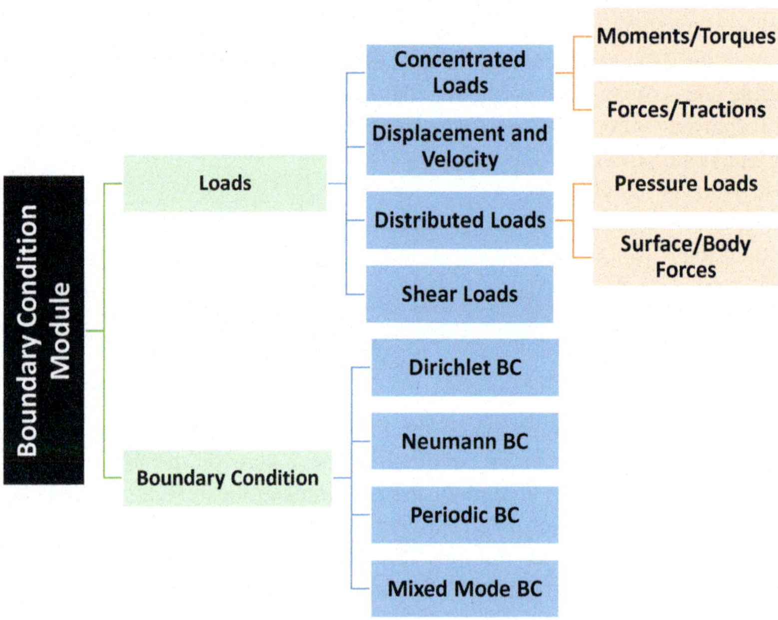

Fig. 8.1 A schematic representation of the sub-divisions of loads and boundary conditions (**BC**) in a typical boundary condition module of an FEM solver

8.2 Chapter Objectives

At the end of this chapter, the reader should be able to:

(a) Appreciate the difference between a load and a classic boundary condition.
(b) Understand the different types of loads that can be applied in a virtual domain, e.g. force, displacement.
(c) Understand how a physical problem can be idealized to reflect the appropriate boundary conditions and load states for the problem.
(d) Understand the different types of boundary conditions and the effect each type of boundary condition has on the convergent FEM solutions.
(e) Demonstrate a pragmatic way of imposing periodic boundary conditions for a 2D RVE.
(f) Assess the impact of periodic boundary conditions on the effective properties of a material system.
(g) Interpret the FE solutions by comparing numerical predictions with expected analytical solutions.

8.3 Types of Boundary Conditions

We established earlier that numerical simulations become necessary where the limitations of analytical predictive tools exist. Generally, in numerical simulations, a region of interest is chosen for study (typically an **RVE**) and this region has defined boundaries that are associated with it (i.e. the points of intersection with the parent material's surroundings). In a numerical analysis, the physical processes around the boundaries of an RVE have to be considered judiciously by applying appropriate *boundary conditions* (**BCs**). This leads us to consider what is a boundary condition represents in reality.

Quite often, many engineering problems are governed by ordinary and partial differential equations. These equations describe the rate at which variables, for example stresses and strains, change within the domain of interest. In addition, it might be a requirement that certain conditions are satisfied at the boundaries of this region of interest (say, no extra external body forces). This specific requirement is a *boundary condition*. The requirement is fitted into the governing differential equation of the domain and a unique solution to the *boundary value problem* (**BVP**) is obtained, provided that the stipulated boundary condition is well defined and satisfied.

Hence, different boundary conditions give rise to different solutions for the BVP. Generally, this is true, but there are pieces of evidence in literature by He [6] and Huet [8] that postulate that if the RVE is indeed representative, the solution to the BVP should be invariant of the boundary conditions imposed. These authors did acknowledge, though, that for the boundary condition effect to be insignificant, the size of the RVE has to be extremely large and thus infeasible in many applications given the enormous computational and human resources needed and the impracticalities associated with obtaining a convergent solution.

Consequently, it is not only essential to isolate the correct RVE for the test material, but it is equally, if not more, important to define the proper boundary conditions that govern the physical processes of the material's deformation. Boundary conditions essentially portray the **constitutive behaviour** of a structure under a given loading condition and they represent the type and magnitude of constraints that control the physical mechanism to be studied. Different variants of boundary conditions are used in the literature and these are closely examined in the following section.

8.3.1 Dirichlet Boundary Conditions

This type of boundary condition is implemented by imposing a uniform displacement over the external boundaries of the RVE in the absence of external body forces. It is often called a *first-type boundary condition*. Other books call it the *kinematic uniform boundary conditions* (KUBCs) or *homogeneous displacement boundary conditions*. It is named after a German mathematician, Johann Peter Gustav Lejeune Dirichlet (1805–1859), who invented it.

Mathematically, over a domain $\Omega \subset \Re^n$, the Dirichlet BC for a model variable, ς on an infinitesimal boundary domain, $\partial\Omega$, is defined as:

$$\varsigma = c \quad \text{where } c = \text{ constant} \tag{8.1}$$

For a given governing equation of an FEM problem, defined in terms of ordinary or partial differential equations, the Dirichlet boundary condition establishes the value of the governing equation along the boundary of the virtual domain. It is a known condition that the equation takes at those known locations.

Consider a given structural solid mechanics problem, the *Dirichlet boundary condition* will prescribe a priori the values of displacement at the boundaries even before the FEM simulation starts. This prescribed value is not randomly chosen, but is instead informed by either experimental data or design specifications to ensure that static equilibrium and compatibility constraints of the structure are maintained.

An illustration of this is given in Fig. 8.2, showing a typical 2D (meshed) virtual domain subjected to a uniform $x-$ axis displacement, δ_x on all right edge nodes while the back and bottom edge nodes were constrained in $x-$ and $y-$ axes respectively. This feature is typical of Dirichlet BCs with nodes prescribed with a fixed displacement value.

Since the imposed displacement (and, as a consequence strain) is uniform over the chosen edge, the deformed profile of the RVE's boundaries will remain straight

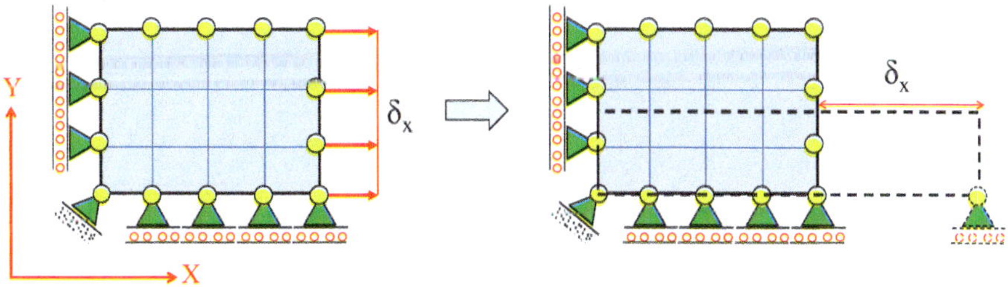

Fig. 8.2 An illustration of a typical Dirichlet boundary condition and the deformation profile

(or planar for a deformed surface). This is one of the most common forms of boundary conditions applied to RVEs in micromechanical analysis. Several authors [2, 3, 12] have used these boundary conditions in their studies to determine the effective properties of composite materials based on the chosen microscale RVE of the test material.

In its simplest forms, this type of boundary condition can be regarded as strain-controlled. It is well documented in the literature that this type of boundary condition overestimates the predicted effective properties of composite materials [6, 8, 10, 18]. Also, Xia et al. [21] showed that the application of this type of boundary condition on RVEs violates traction and stress continuity across the boundaries of an RVE, resulting in a violation of conditions for static equilibrium.

From a structural analysis assessment, this type of displacement boundary condition may be essential if a structural component is to experience a certain displacement as part of its in-service design specification. During the FEM modelling phase, a Dirichlet BC is imposed on that particular component to enforce the design-specified displacement behaviour.

Practical examples of Dirichlet boundary conditions are given in Fig. 8.3 showing:

- Figure 8.3a shows a tensile test specimen, with front and back end edge of the specimen, prescribed with Dirichlet boundary conditions. The front edge is prescribed with a constant displacement, δ_x whilst the back edge is prescribed with zero displacement.
- Figure 8.3b shows a fluid flow between two walls. A Dirichlet BC was imposed such that one wall has zero velocity (i.e. is fixed) and another wall is prescribed with say, a velocity, $V_x = 3$ m/s.
- Figure 8.3c shows the heat conduction profile along a plate with the nodes on the bottom and top edges prescribed with temperature-based Dirichlet BC.
- Figure 8.3d shows a ball-bearing assembly around a rotating shaft. The inner walls of the bearing have a transverse velocity of say, ω rad/s while the external walls are rigid: both specified using Dirichlet BCs.

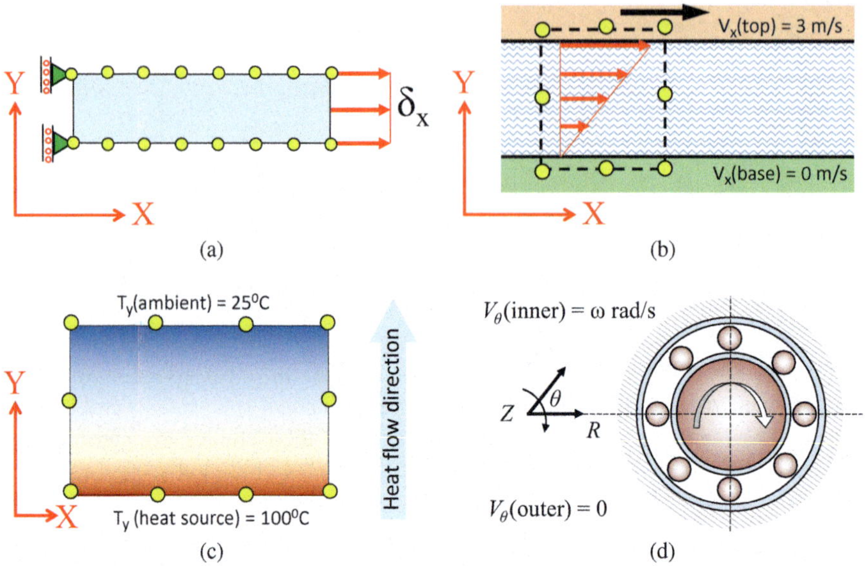

Fig. 8.3 Practical scenarios where Dirichlet BCs are required: (**a**) tensile test; (**b**) fluid flow in-between two walls; (**c**) heat conduction; and, (**d**) ball-bearing assembly with inner rotating shaft

8.3.2 Neumann Boundary Conditions

In solving the ordinary or partial differential equations that result from the governing equations of a particular physical problem, previously we enforced that the boundary nodes take up a constant value of, say, displacement, velocity, temperature. *Neumann boundary conditions* are named after the inventor, a German mathematician, Carl Gottfried Neumann (1832–1925). The Neumann BCs are also called *second-type BCs* or in some circles, *static uniform boundary conditions* (SUBCs). For the Neumann BCs, the boundary nodes are enforced with a condition that establishes that the degrees of freedom of those nodes are equal to the derivative/gradient of the solution.

Consider a boundary domain, $\Omega \subset \Re^n$, which encloses a physical problem. We can define mathematically the Neumann BC over an infinitesimal subset of that domain, $\partial\Omega$ which has an outward normal vector, $\mathbf{n}$, for a domain variable, ς, thus:

$$\frac{\partial \varsigma}{\partial \mathbf{n}} \equiv \alpha \quad \text{where } \alpha = \text{ constant} \tag{8.2}$$

The α constant is usually a scalar quantity. For most load cases, the normal derivative of a variable is equivalent to the flux of that variable.

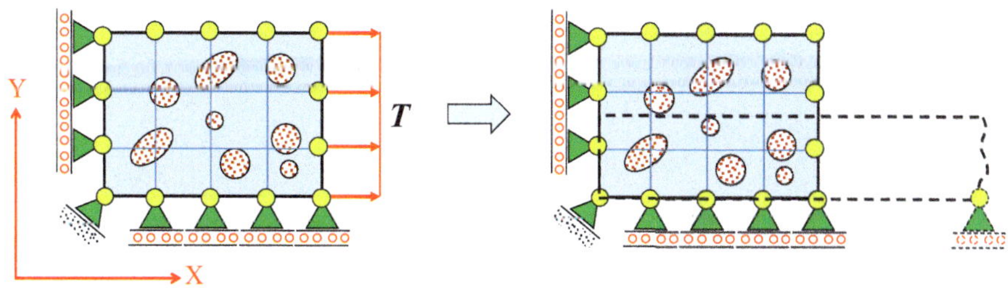

Fig. 8.4 A typical Neumann boundary condition and the associated deformation profile

For example, the normal derivative of a force is equivalent to the *traction* i.e. $\mathbf{T} = \dfrac{\partial \mathbf{F}}{\partial \mathbf{n}}$. The converse to a Neumann BC is called a *no-flux BC* as there is no flux associated with it. In the context of structural solid mechanics, this type of boundary condition is prescribed by imposing a uniform traction vector, **T**, over the external boundaries of the RVE. An illustration of this type of boundary condition is shown in Fig. 8.4.

The reader will notice that the deformed edge consists of an irregular profile, for the heterogeneous material of Fig. 8.4. Since the imposed traction is uniform over a given surface, the deformed profile of that surface bounding the RVE may not remain planar due to the differences in stiffness between the different components that make up the heterogeneous system. However, if the microstructure under investigation is homogeneous, and hence constant stiffness across the material volume, then the deformed profile will be regular – similar to a Dirichlet BC deformed profile.

Neumann BCs are another commonly used form of boundary conditions applied to RVEs in micromechanical analysis. It is known that Neumann BCs typically underestimate the predicted effective properties of composite materials [20]. Examples of Neumann BCs include: *heat flux, diffusion flux, concentration gradient, normal and shear traction, osmotic pressure gradient, magnetic flux, luminous flux, volumetric flux.*

8.3.3 Robin or Mixed Boundary Conditions

This is a third type boundary condition and developed by French mathematician Victor Gustave Robin (1855–1897). It is also called a *mixed BC*, because it is a weighted linear contribution of the two previous boundary conditions. In the context of a finite element scheme where solutions of ordinary or partial differential equations are required, the *Robin BC* is a linear combination of Dirichlet and Neumann BCs.

Given a virtual domain, $\Omega \subset \Re^n$, we can define mathematically a generic form of the *Robin BC* for a model variable, ς, in the vicinity of an infinitesimal boundary domain, $\partial\Omega$ defined as:

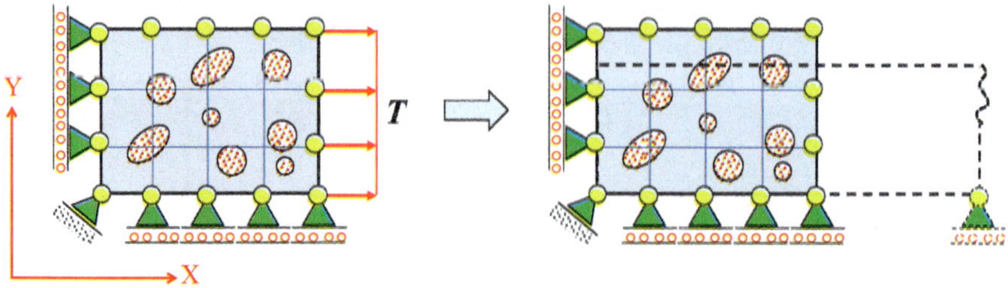

Fig. 8.5 A typical Robin boundary condition and the associated deformation profile

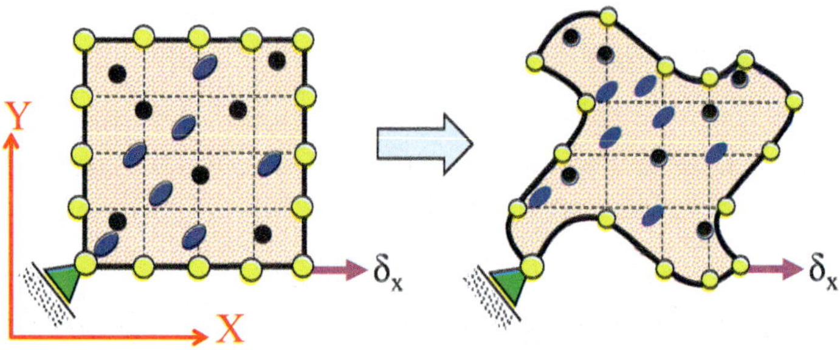

Fig. 8.6 A typical Periodic boundary condition and the deformation profile

$$\varepsilon \frac{\partial \varsigma}{\partial \mathbf{n}} + \phi \varsigma \equiv \beta \tag{8.3}$$

where ε and ϕ are non-zero weighting constants that operate on *Dirichlet* and *Neumann* BCs respectively. The right hand side, β, is a mathematical expression that is defined on the boundary domain, Ω, and can take the value of zero for certain problems.

An illustration of the *mixed BC* for an RVE of a heterogeneous system is shown in Fig. 8.5. The structure is subjected to a uniaxial tensile deformation using a combination of displacement-based and traction-based load cases. One will notice that the deformed edge consists of straight and irregular profile, with the former corresponding to the Dirichlet BC and the latter to the Neumann BC.

8.3.4 Periodic Boundary Conditions

The *periodic boundary condition* (**PBC**) stipulates that opposite pairs of edges or surfaces on the boundary of an RVE should deform identically under a given loading history [4]. The PBC essentially precludes the constraint of enforcing edges to remain planar after deformation, as is the case for KUBC. Figure 8.6 shows a typical 2D RVE of a heterogeneous material subjected to uniaxial tensile deformation and imposed with periodic BCs.

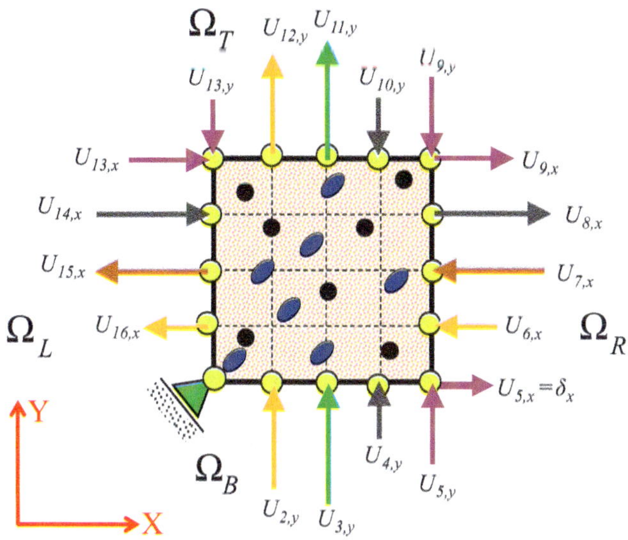

Fig. 8.7 A virtual domain, showing periodic boundary conditions defined by applicable imposed displacements

In order to illustrate how the PBC can be implemented, let us consider the 2D RVE of Fig. 8.7. The domain encloses a heterogeneous material with nodes N_1 to N_{16}. The 2DRVE virtual domain is bounded by the boundary $\Omega \subset \Re^n$ where:

$$\Omega = \Omega_L \cup \Omega_R \cup \Omega_T \cup \Omega_B \tag{8.4}$$

where $\Omega_L, \Omega_R, \Omega_T$ and Ω_B are the boundaries that enclose the *left, right, top* and *bottom* edges of the 2D RVE respectively. Periodic boundary conditions demand that the left boundary, Ω_L, deformation profile must be similar in shape and value to the right boundary, Ω_R. Similar deformation profiles are also expected for the top boundary, Ω_T, and the bottom boundary, Ω_B. Figure 8.6 (left) shows the expected deformed profile for the system. Note that this deformation profile applies if the material is not homogeneous – otherwise the edges will be straight – in which case the stiffness of the material will be constant throughout the virtual domain.

Using the virtual domain, Ω, shown in Fig. 8.7, we can define mathematically the periodic boundary condition of a model variable, ς:

$$\varsigma^{N_a}_{(x,y)} = \varsigma^{N_b}_{(x,y)} \quad \text{on boundary } \Omega_k \tag{8.5}$$

where N_a and N_b are nodes that are kinematically linked on opposite edges and also where $k = L, R, T$ and B. With respect to the virtual domain shown in Fig. 8.7, the model variable is a nodal displacement, i.e. $\varsigma = u_{(x,y)}$.

8.4 Periodic Boundary Conditions and Mesh Types

The implication of Eq. 8.5 is that all the nodes on one edge must match with the corresponding nodes on the opposite edge. With such a plan, it is possible to kinematically tie the two corresponding nodes. Great effort has to be invested in ensuring that nodes on one edge or surface (for 3-D problems) match with same nodes on a corresponding opposite edge (or surface). This is a major limitation in the periodic BC approach, since a lot of effort is invested during the meshing stage.

The discretization of the virtual domain has to happen such that each edge or surface is seeded or split up into same number of divisions for two opposite edges/surfaces. Inside `ABAQUS`, the meshing algorithm is equipped to do this by using the `Seeding by Number` option. Such meshes that have matching number of nodes on opposite edges or surfaces are described as ***periodic meshes***. They are created easily for simple geometries as squares, rectangles, etc. If the geometry is complex, it is not always easy to create periodic meshes, although this can still be done by using advanced meshing techniques. Examples of different types of meshes for the same virtual domain of a voided polymeric plate are given in Fig. 8.8.

The converse to periodic meshes are ***non-periodic meshes*** and these are the most common mesh types. They require *unstructured*[1] meshing of the geometry and the resulting meshed virtual domain would have non-matching number of nodes on corresponding opposite edges. Unfortunately, the classic definition of the periodic boundary condition as Eq. 8.5 cannot be implemented on such meshes without other modifications.

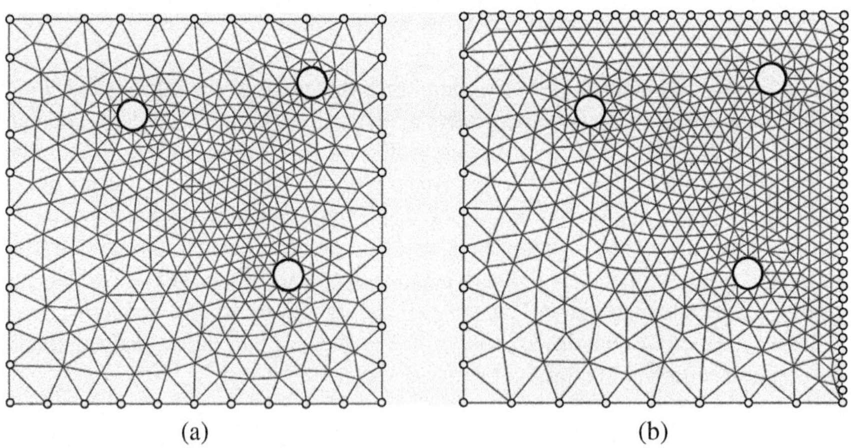

Fig. 8.8 (**a**) Periodic and (**b**) non-periodic mesh types for a voided polymeric plate

[1]Unstructured or free meshing refers to a random discretization of a virtual domain such that the resulting mesh is made in an irregular pattern. Its commonly used when the domain has a complex and irregular shape.

In Fig. 8.8a, the four edges have 11 nodes each while in Fig. 8.8b, each of the four edges have different number of nodes with *top* and *right* edges finely meshed and *bottom* and *left* edge coarsely meshed. The latter mesh is a *non-periodic mesh* and cannot be used for the classic formulation the periodic BC. In Sect. 8.7, we will introduce the technique for imposing periodic BCs on non-periodic meshes. The next section demonstrates a mechanism for imposing periodic boundary conditions on periodic mesh types.

8.5 Implementation of Periodic Boundary Conditions on Periodic Meshes

There are many ways of imposing the periodic BC constraints on periodic meshes. These ways are all aligned according to Eq. 8.5. Some of the key contributors to this include: for 2D RVE – Sluis et al. [20], Kouznetsova et al. [11], Gitman et al. [4] and for 3-D RVE – Okereke and Akpoyomare [14]. These implementations do not simply use Eq. 8.5 but extend it slightly by linking the inherent kinematical coupling to known (Dirichlet-style) boundary conditions at what is called *retained nodes* at the corners of the RVE. As a result, the comparative deformation of the retained nodes will be homogeneously equivalent to the deformation of the nodes of interest. This is repeated for all relevant axes of the RVE.

This section starts with a brief introduction of *multi-freedom constraints*: an essential feature of the nodal displacement equations needed for imposing periodic boundary condition on virtual domains. Then the mathematics of periodic boundary conditions will be derived before the implementation of the PBC within a numerical scheme is demonstrated. The focus here is solely on periodic meshes.

8.5.1 *Multi-freedom Constraint (MFC)*

If a nodal displacement is set such that the a component of the nodal displacement (say $x-$ axis) is set to a defined value according to a Dirichlet-style BC formulation, we describe this type of constraint as a *single-freedom constraint* (SFC). Typical examples of a single-freedom constraint for nodes N_A, N_{B2} and N_{100} in $x-$, $y-$ and $z-$ axes respectively are:

$$u_x^{N_A} = 0\ \text{m} \qquad u_y^{N_{B2}} = 200\ \text{mm} \qquad \text{and} \qquad u_z^{N_{100}} = 35\ \mu\text{m} \tag{8.6}$$

where u is the nodal displacement variable. The first SFC example above is described as *homogeneous* while the other two are *non-homogeneous*. Homogeneous constraints result when the terms of the right hand-side of the constraint equation is zero. If this condition is not obeyed, the constraint is non-homogeneous.

In practice, such nodal dependence as in Eq. 8.6 is not very common. Rather, the displacement of a node (particularly for a 3-D RVE) is dependent on the deformation of other nodes that make up the RVE. We define a *multi-freedom constraint (**MFC**)* as a type of constraint where two or more displacement components are combined to equate to a prescribed value. We can also define homogeneous and non-homogeneous variants of MFCs.

When an MFC is defined such that all the interacting nodal displacement components can be assembled to the right side of the equation and the left-hand side is either a zero or a non-zero prescribed value, then such an MFC is described as the *canonical form* of the constraint. Unlike the SFC representation, the MFC may require that the equation involves interaction of nodal displacements at multiple nodes, in which case, the MFC is said to be *multi-point* or *multi-node*. When the interacting nodal displacements combine in a linear manner, the MFC is described as *linear* and when this condition is violated, then the MFC is *nonlinear*.

Examples of typical MFCs are given in Eqs. 8.7, 8.8, 8.9 for nodes N_A, N_{B2} and N_{100} in $x-$, $y-$ and $z-$ axes respectively.

$$\frac{1}{2}u_x^{N_A} + 2 \times 10^{-2} u_y^{N_A} = 0 \tag{8.7}$$

$$0.5u_z^{N_A} - 8au_x^{N_{B2}} + 3u_y^{N_{100}} = 2.83 \quad \text{and} \tag{8.8}$$

$$e^{a^2bc}\left(\frac{u_x^{N_{B2}}}{ab} + \frac{c}{d}u_y^{N_A}\right)^2 = u_y^{N_{100}}\left(u_z^{N_A}\right)^{-2d} \tag{8.9}$$

where $a,\ b,\ c$ and d are constants. Equation 8.7 is a *linear, homogeneous, canonical form* MFC while Eq. 8.8 is also *linear*, but a *non-homogeneous, multi-point* MFC. Equation 8.9 represents a complex MFC which is *nonlinear, homogeneous* and *multi-point*. It might be necessary to introduce model parameters such as $a,\ b,\ c$ and d into the model, and then using the MFC approach shown above is a viable way to introduce these nodal displacement constraints to the model.

8.5.2 Numerical Implementation of Multi-freedom Constraints

The main challenge in using the **MFCs** within a numerical framework like an FE solver is the onerous task of developing the MFC equations for all nodes in the boundaries of the chosen virtual domain. Once these MFC equations are derived, they have to be passed back into the numerical framework for use as standard BCs during the analysis. In order words, for an FEM solver, the FEM equation: $\mathbf{F} = \mathbf{KU}$, will have to be solved where the right hand side of the equation contains the nodal column vector, $\mathbf{U}$, of all nodes in the model. The FEM solver needs to know which of these nodes belong to the boundary and which subset needs to be imposed with a defined MFC. The large number of such equations presents a real challengewithout recourse to automatic implementation. Once the MFC equation is passed into the numerical framework, the solution of the FEM equation progresses naturally and a convergent solution should be obtained.

Some of the ways of implementing the MFC equations in many numerical frameworks (i.e. FEM solvers, in particular) include:

(a) *Master-slave elimination*: Nodes that are masters have prescribed values to their degrees of freedom. The *slave nodes* derive their degrees of freedom based on the values assigned to the *master nodes*. These slave nodes are often eliminated in the MFC equation hence simplifying the nodal displacement vector, **U**.
(b) *Penalty function method:* At times, it may be possible to introduce penalty node(s) that connect(s) a specific node to the penalty node(s). The structural response for this spurious connection is usually introduced as an elastic spring and the element that connects the two nodes does not play a part during the FEM solution. This is why it is called a *penalty element*. However, the presence of the reference nodes helps establish the degrees of freedom for the nodes of interest. The introduction of the penalty elements and nodes help simplify further the unknown nodal displacements at the boundaries of **U**, hence leading to a solution.
(c) *Lagrangian multiplier method*: This method involves the multiplication of the FEM equation by a *Lagrangian multiplier*. The FE equation ($\mathbf{F} = \mathbf{KU}$) is altered to ensure that the FEM solution is achieved. The modification is achieved by adding unknowns to the master stiffness equation, **K**. This is a complex process and beyond the scope of this text.

In the following, we establish the mathematics of periodic boundary conditions. This will result in a multi-freedom constraint equation.

8.5.3 Mathematical Formulation for Periodic Boundary Conditions

Consider the 2D RVE of the microstructure of, say, a particulate composite represented by domain, $\Omega \subset \Re^n$, as shown in Fig. 8.9. The domain is divided into four sub-domains, namely Ω_L, Ω_R, Ω_T and Ω_B, which corresponds to the *left, right, top* and *bottom* edges of the RVE respectively. The domain also has of four *corner* (or *retained*) *nodes* , N_1, N_2, N_3 and N_4, and four *internal/reference nodes* N_A, N_B, N_C and N_D. It is rigidly fixed at node N_1 to prevent rigid body motion.

We will now establish the mathematical formulation of the periodic BC, where the deformation of the reference nodes are *kinematically tied* (or linearly coupled) to the homogeneous deformation about the *retained nodes*.[2]

[2]Note that node N_3 is excluded from the list of retained nodes as its deformation is influenced by the deformations of nodes N_2 and N_4. Node N_3 is commonly regarded as a slave/dummy node

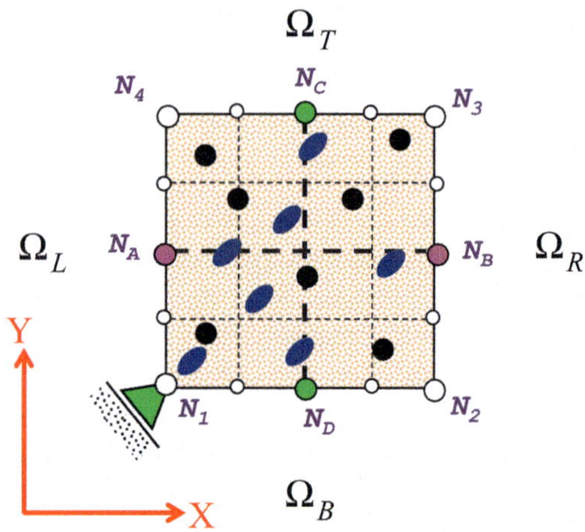

Fig. 8.9 A 2D RVE with *reference* and *retained nodes* for creating the periodic BC

The periodic BC implementation for displacement of the reference nodes can be written thus:

$$\mathbf{U}^{N_B}_{(x,y)} = \mathbf{U}^{N_A}_{(x,y)} \quad \Longrightarrow \quad \mathbf{U}^{N_B}_{(x,y)} - \mathbf{U}^{N_A}_{(x,y)} = \mathbf{0} \tag{8.10}$$

$$\mathbf{U}^{N_C}_{(x,y)} = \mathbf{U}^{N_D}_{(x,y)} \quad \Longrightarrow \quad \mathbf{U}^{N_C}_{(x,y)} - \mathbf{U}^{N_D}_{(x,y)} = \mathbf{0} \tag{8.11}$$

where $\mathbf{U}^{N_A}_{(x,y)}$ is the nodal displacement vector of node N_A in $x-$ and $y-$ axes. Also for the retained nodes, Eqs. 8.10 and 8.11 can likewise be written thus:

$$\mathbf{U}^{N_2}_{(x,y)} = \mathbf{U}^{N_1}_{(x,y)} \quad \Longrightarrow \quad \mathbf{U}^{N_2}_{(x,y)} - \mathbf{U}^{N_1}_{(x,y)} = \mathbf{0} \tag{8.12}$$

$$\mathbf{U}^{N_4}_{(x,y)} = \mathbf{U}^{N_1}_{(x,y)} \quad \Longrightarrow \quad \mathbf{U}^{N_4}_{(x,y)} - \mathbf{U}^{N_1}_{(x,y)} = \mathbf{0} \tag{8.13}$$

It is clear that both Eqs. 8.12 and 8.13 are equivalent and the difference in their nodal displacements are equal for the *left* and *right* boundaries. Similarly, Eqs. 8.12 and 8.13 are also equal and the difference in nodal displacement for the edge nodes on the *top* and *bottom* edges are the same.

The above implementation enforces the mathematical formulation of periodic boundary conditions established in Eq. 8.5. Consequently, the 2D RVE will deform in a periodic manner as shown in Fig. 8.6. We define here that the domain that is implemented with periodic boundary conditions (PBCs) is $\Omega_{PBC} \subset \Re^n$, as again the original domain, Ω, without PBC imposed on it.

---

since its deformation degrees of freedom are constrained and kinematically linked to the N_2 and N_4 retained nodes.

If we are interested in, say, the axial deformation simulation of the Ω_{PBC}, we will introduce known Dirichlet-style boundary condition, say, $\delta_x = 0.005$ m, on any of the retained nodes. As a result, the deformation of any of the retained nodes will follow the mathematical formulations of Eqs. 8.10, 8.11, 8.12, and 8.13. Here, we equate Eqs. 8.10–8.12 and 8.11–8.13, which results in the following:

$$\mathbf{U}^{N_B}_{(x,y)} - \mathbf{U}^{N_A}_{(x,y)} - \mathbf{U}^{N_2}_{(x,y)} + \mathbf{U}^{N_1}_{(x,y)} = \mathbf{0} \tag{8.14}$$

$$\mathbf{U}^{N_C}_{(x,y)} - \mathbf{U}^{N_D}_{(x,y)} - \mathbf{U}^{N_4}_{(x,y)} + \mathbf{U}^{N_1}_{(x,y)} = \mathbf{0} \tag{8.15}$$

The above equations are the basis upon which the periodic boundary condition is imposed on Ω_{PBC}. According to the discussions of Sect. 8.5.1, Eqs. 8.14 and 8.15 are *linear, homogeneous, multi-point, canonical form* multi-freedom constraint equations.

If we re-arrange Eq. 8.14 to isolate the nodal displacements of node, N_B, we obtain:

$$\mathbf{U}^{N_B}_{(x,y)} = \mathbf{U}^{N_A}_{(x,y)} + \mathbf{U}^{N_2}_{(x,y)} - \mathbf{U}^{N_1}_{(x,y)} \tag{8.16}$$

The above equation establishes that the displacement of node N_B is based on displacement degrees of freedom of nodes N_1, N_2 and N_A. According to Fig. 8.9, the degrees of freedom for node N_1 is known, since it is prescribed to prevent rigid body rotation as: $\mathbf{U}^{N_1}_{(x,y)} = \mathbf{0}$.

The displacements of nodes N_2 and N_A are unknown and are solved during the FEM solution. For a discretized virtual domain consisting of multiple nodes, it is important that a matrix of equations such as Eqs. 8.14 and 8.15 are developed for all nodes in the boundaries of a chosen 2D RVE.

MFC equations for all nodes have to be developed and these will establish a matrix of equations for their interactions. These interactions become known during the FEM solution and will change based on the prescribed boundary conditions imposed on some of the nodes. This last statement presents an opportunity for introducing loads to a system subjected to periodic BCs. A known load introduced, say, at node N_2 via a Dirichlet-style BC, will set off the analysis procedure which will ultimately yield the required displacement values for all nodes in the FEM model. The next section shows how practically this can be accomplished in a given numerical scheme, using say, `ABAQUS` FEM Solver.

8.5.4 Implementation of Periodic BCs in FEM Solvers

Several FEM solvers have their different ways of introducing the multi-freedom constraint equations into the FE solver. The one that we will be emphasizing here is the implementation used by ABAQUS but any FEM solver can be used.[3] Details of how ABAQUS handles MFC equations can be obtained from *Section 35.2* of the *ABAQUS/CAE Users Analysis Manual.*

In ABAQUS, the implementation of the linear constraint Eqs. 8.14 and 8.15 was achieved by using the ⋆EQUATION command. This is a *linear, homogeneous, multi-point constraint* command, which enforces that the linear combination of variables at the nodes of interest must be equal to zero. In other words, if we consider a typical mathematical expression representative of Eqs. 8.14 and 8.15 as:

$$A_1 U_1^P + A_2 U_2^Q + \ldots + A_N U_k^R = 0 \tag{8.17}$$

where U_i^P is a variable (in this case nodal displacement) at node P and degree of freedom, i. The constants, $A_1, A_2, \ldots, A_N$ are the coefficients of the constraint equation, and they define the relative displacement magnitude of each of the nodes. To introduce Eq. 8.17 within ABAQUS, we specify the following:

(a) The total number of terms in the equation.
(b) The node, P and its degree of freedom, i for the nodal variable, U_i^P.
(c) The coefficients, A_N.

The ABAQUS-style linear constraint expression based on the ⋆EQUATION command for Eq. 8.17 becomes:

```
*EQUATION
N
P,   i,   A1
Q,   j,   A2
.
.
.
R,   k,   AN
```

The above linear constraint will be fed back into the ABAQUS keyword file as an update such that this new set of boundary conditions is incorporated within the original model. In this way, the model will have information about the periodic BCs.

Also, implementing the ⋆EQUATION format for Eqs. 8.14 and 8.15 reveals that the number of coefficients required is four (4) and their magnitudes are $[+1, -1, -1, +1]$. If this is required for a given solution, the user would have to convert the original MFC equations into ABAQUS-style linear constraint equation for use in the analysis. In the following sections, we will illustrate this procedure with a few examples.

Example 8.1 The multi-freedom constraint equations for nodal displacements of a 3-D RVE is specified as:

[3]Different FEM solvers have different implementations and keyword commands for enforcing the multi-freedom constraints equations. Interested readers should consult documentations of the said FEM solver to identify the format of the solver's MFC commands.

(a) $u_3^{N_8} - u_3^{N_7} + u_3^{N_6} = 0$
(b) $u_2^{N_{50}} = u_1^{N_{10}} - 8u_2^{N_{18}} + 2u_3^{N_5} \quad 6u_1^{N_{49}}$

Solution

(a) **Case (a)**: $u_3^{N_8} - u_3^{N_7} + u_3^{N_6} = 0$. The equation is a *linear, homogeneous, canonical form* constraint equation in which, if it were to become equivalent to Eq. 8.17, the following conclusions are true:

- Number of terms in the equation, $N = 3$.
- Coefficients of the three terms: $[A_1, A_2, A_3] = [+1, -1, +1]$.
- Node labels in the equation are: N_8, N_{10} and N_6.
- The nodal displacements are along the 3−axes material reference frame. This corresponds to the $z-$ axes of a Cartesian coordinate system.

It is always good practice to put each of the terms in the MFC equation on a single line. Combining the above information with `*EQUATION`, the resulting command for incorporating the MFC equation into the ABAQUS keyword file becomes:

```
*EQUATION
3
N8,      3,      +1
N10,     3,      -1
N6,      3,      +1
```

(b) **Case (b)**: $u_2^{N_{50}} = u_1^{N_{10}} - 8u_2^{N_{18}} + 2u_3^{N_5} - 6u_1^{N_{49}}$. To develop an ABAQUS-style MFC representation of the above equation, we have to re-write the expression into its canonical form with all nodal displacements components on the left-hand side of the equality sign, and the right hand side prescribed value equals zero.

The resulting equation is:

$$u_1^{N_{10}} + 8u_2^{N_{18}} - 2u_3^{N_5} + 6u_1^{N_{49}} = 0$$

Using similar approach as *Case (a)* above, we identify that:

- Number of terms, $N = 5$.
- Coefficients of the five terms are: $[A_1, A_2, A_3, A_4, A_5]$ $= [+1, -1, +8, -2, +6]$.
- Node labels in the equation are: $N_{50}, N_{10}, N_{18}, N_5$ and N_{49}.
- The nodal displacement axes are: $[2, 1, 2, 3, 1] \equiv [y, x, y, z, x]-$ axes.

The resulting ABAQUS-style MFC equation is:

```
*EQUATION
5
N50,     2,      +1
N10,     1,      -1
N18,     2,      +8
N5,      3,      -2
N49,     1,      +6
```

Example 8.2 A 60 × 60 units^2 2-D RVE of a unidirectional composite material consists of fibres (shown as circles) dispersed randomly within a polymer matrix represented by a square window. The coarsely meshed virtual domain of the composite is as shown in Fig. 8.10. You are required to determine the Young's modulus in the x—-direction. Periodic BCs were to be imposed on the system and load applied on node N_2 with a prescribed x—-axis displacement, $\delta_x = 3$ units.

(a) Write the Multi-freedom constraint (MFC) equations for all model boundary nodes.
(b) Convert the MFC equations of Nodes C, D, I and J to ABAQUS-style linear constraint equations.

Solution

(a) **Case (a): MFC Equations** – The 2D RVE consists of four edges: Ω_L, Ω_R, Ω_T and Ω_B corresponding to *left, right, top* and *bottom* of the boundary domain, $\Omega \subset \Re^n$, respectively. To change the original domain, Ω to the *PBC-imposed* domain, Ω_{PBC}, we will create the following MFC equations according to Eqs. 8.14 and 8.15 above. Along the Ω_L and Ω_R domains, we collate the following MFC equations using nodes N_1 and N_2 as retained nodes:

$$\mathbf{U}^{N_B}_{(x,y)} - \mathbf{U}^{N_A}_{(x,y)} - \mathbf{U}^{N_2}_{(x,y)} + \mathbf{U}^{N_1}_{(x,y)} = \mathbf{0} \tag{8.18}$$

$$\mathbf{U}^{N_D}_{(x,y)} - \mathbf{U}^{N_C}_{(x,y)} - \mathbf{U}^{N_2}_{(x,y)} + \mathbf{U}^{N_1}_{(x,y)} = \mathbf{0} \tag{8.19}$$

$$\mathbf{U}^{N_F}_{(x,y)} - \mathbf{U}^{N_E}_{(x,y)} - \mathbf{U}^{N_2}_{(x,y)} + \mathbf{U}^{N_1}_{(x,y)} = \mathbf{0} \tag{8.20}$$

$$\mathbf{U}^{N_3}_{(x,y)} - \mathbf{U}^{N_4}_{(x,y)} - \mathbf{U}^{N_2}_{(x,y)} + \mathbf{U}^{N_1}_{(x,y)} = \mathbf{0} \tag{8.21}$$

Similarly, we collate MFC equations for the Ω_T and Ω_B domains using nodes N_4 and N_1 as retained nodes.

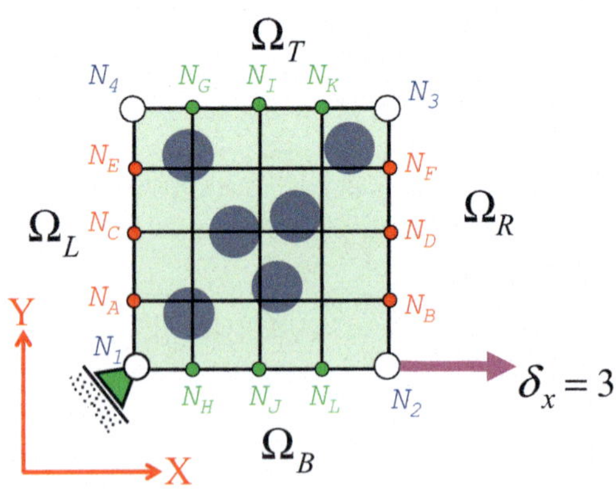

Fig. 8.10 A 2D RVE window in transverse-to-fibre direction for a unidirectional (UD) composite material

$$\mathbf{U}^{N_G}_{(x,y)} - \mathbf{U}^{N_H}_{(x,y)} - \mathbf{U}^{N_4}_{(x,y)} + \mathbf{U}^{N_1}_{(x,y)} = \mathbf{0} \tag{8.22}$$

$$\mathbf{U}^{N_I}_{(x,y)} - \mathbf{U}^{N_{I'}}_{(x,y)} - \mathbf{U}^{N_4}_{(x,y)} + \mathbf{U}^{N_1}_{(x,y)} = \mathbf{0} \tag{8.23}$$

$$\mathbf{U}^{N_K}_{(x,y)} - \mathbf{U}^{N_L}_{(x,y)} - \mathbf{U}^{N_4}_{(x,y)} + \mathbf{U}^{N_1}_{(x,y)} = \mathbf{0} \tag{8.24}$$

Note that in this formulation of the MFC equations, we omit establishing kinematical tyings and constraints for node N_3, because its kinematics have been established from Eq. 8.21, in which case it is a *slave* node.

(b) **Case (b): ABAQUS-style linear constraint equations** – With the implementation of periodic BCs on the original domain, we obtain the Ω_{PBC} domain. To translate this information into the ABAQUS keyword file, we have to modify the nodal displacements of all boundary nodes according to the MFC Eqs. 8.18, 8.19, 8.20, 8.21, 8.22, 8.23, and 8.24. These consist of 12 nodes.

For the sake of brevity, we consider only 4 of those nodes, namely nodes *C*, *D*, *I* and *J*. Each node has two degrees of freedom and this has to be accounted for in *EQUATION for all the nodes.

For example, for nodes *C* and *D*, we will have the following MFC equations (in the two nodal displacement directions):

$$U_1^{N_D} - U_1^{N_C} - U_1^{N_2} + U_1^{N_1} = 0 \qquad \text{and} \qquad U_2^{N_D} - U_2^{N_C} - U_2^{N_2} + U_2^{N_1} = 0$$

Similar expressions apply for nodes *I* and *J* with retained nodes 4 and 1.

- The ABAQUS-style constraint equations for nodes *C* and *D* are:

```
*EQUATION                   *EQUATION
4                           4
ND,    1,   +1              ND,    2,        +1
NC,    1,   -1              NC,    2,        -1
N2,    1,   -1              N2,    2,        -1
N1,    1,   +1              N1,    2,        +1
```

- The ABAQUS-style constraint equations for nodes *I* and *J* are:

```
*EQUATION                   *EQUATION
4                           4
NI,    1,   +1              NI,    2,        +1
NJ,    1,   -1              NJ,    2,        -1
N4,    1,   -1              N4,    2,        -1
N1,    1,   +1              N1,    2,        +1
```

8.6 PBC2DGEN: Periodic Boundary Conditions for 2-D RVE Generator

On a finite element modelling platform, the number of surface and edge nodes on a typical RVE can be enormous, which leads to many degrees of freedom and many MFC equations. In fact, in Example 8.2 above, with 7 nodal pairs that can be kinematically tied, there will be 14 `*EQUATION` linear constraints equations that will be used in the `ABAQUS` solution for the 2D RVE of that unidirectional composite material. Consequently, implementing the linear constraint equations can be very challenging.

To bypass the above difficulty, Okereke and Akpoyomare [14] developed an algorithm called the ***P**eriodic **B**oundary **C**onditions **GEN**enerator for **2-D** RVEs* (**PBC2DGEN**). The software was developed for automating the creation and assembly of all the `*EQUATION` linear constraints for any 2-D RVE, hence generating periodic BCs to such 2D virtual domains. **PBC2DGEN** is implemented as a MATLAB™ script and is freely available on the ***extra resources website*** of this book. This section details the development principles behind the code and demonstrates how **PBC2DGEN** can be used to generate elastic properties of heterogeneous materials.

PBC2DGEN is designed for *periodic meshes* and for square RVEs, specifically for the ABAQUS FEM solver. The same principle can be extended to other FEM solvers, but the current release of **PBC2DGEN** is strictly for `ABAQUS`. Initially, **PBC2DGEN** requires `ABAQUS` model keyword file, which has a file extension identified by`*.INP`. This file contains all the important model information of the typical 2-D RVE model to be analysed, which includes, but is not limited to element number, the coordinate positions of the nodes within the model, material properties, contact arguments as well as boundary conditions. The **PBC2DGEN** will seek to update only the boundary conditions of the 2-D RVE to enforce the requirement of periodic boundary conditions on the virtual domain modelled.

Using a user-supplied `*.INP` `ABAQUS` model file, the **PBC2DGEN** algorithm scans through the whole model and sorts the edge nodes according to their nodal coordinates. **PBC2DGEN** is designed to prescribe known load cases to the 2-D RVE as part of its implementation. The current version of the code is set to impose *uniaxial deformation* in $x-$ and $y-$-axes as well as *shear deformation* in the $xy-$plane. Based on the load case being considered, **PBC2DGEN** identifies and implements the appropriate linear constraint equations on the FEM model. With this information, the **PBC2DGEN** algorithm finally updates the previously supplied `*.INP` file with the constraint equations and boundary conditions for later use in the `ABAQUS` FEM model. A flow chart of how the algorithm works is shown in Fig. 8.11.

8.6.1 Structure of PBC2DGEN

PBC2DGEN has been developed in modular format in MATLAB™. In its current implementation, **PBC2DGEN** requires MATLAB™ to run, although it can also be re-compiled into a stand-alone executable.[4] Figure 8.12 shows the layout of **PBC2DGEN** in the MATLAB™ current folder/directory.

[4]If the user does not have a MATLAB™ license, visit the *extra resources website* to download a compiled executable of **PBC2DGEN** and use this instead.

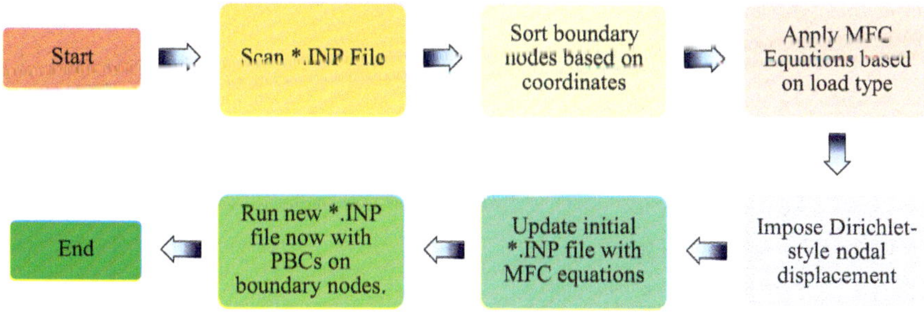

Fig. 8.11 A flow chart of the **PBC2DGEN** algorithm

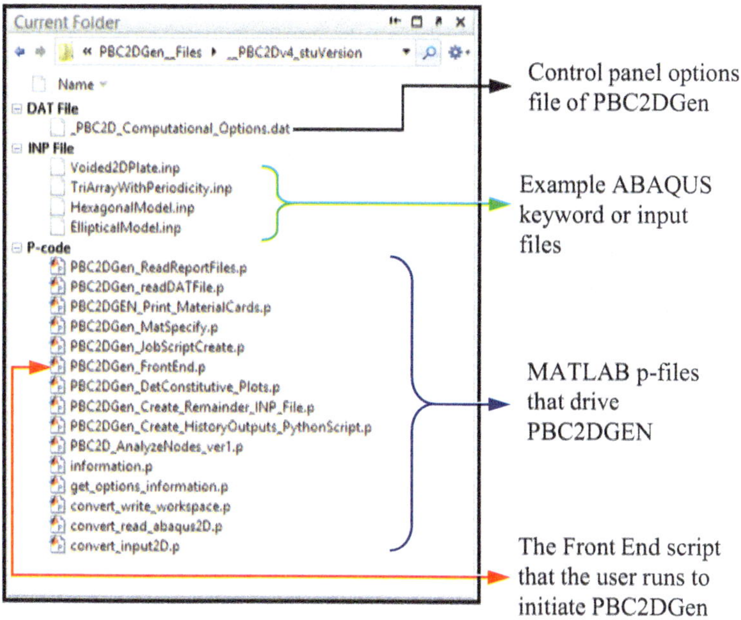

Fig. 8.12 The layout of **PBC2DGEN** in MATLAB™'s current working directory

As shown in Fig. 8.12, **PBC2DGEN** consists of the following of file types:

(a) **MATLAB p-files**[5]**:** These are the same as scripts or function files, but the contents of the scripts have been obfuscated in order to protect the source code. Using p-files ensures that the integrity of the source code is maintained while also maintaining secure distribution of code to end users. **PBC2DGEN** has 15

[5]The `convert*.p`, `information.p` and `get_option_information.p` scripts of Table 8.1 were developed in 2009 by Dr. Ben Elliot of the Impact Engineering Team, Department of Engineering Science, University of Oxford. His contribution here is gratefully acknowledged.

Table 8.1 Description of the component `*.P` files of **PBC2DGEN**

PBC2DGEN * P file	Description of the purpose of the individual PBC2DGEN scripts
PBC2DGen_FrontEnd	The main script that initiates the **PBC2DGEN** model implementation
PBC2DGen_ReadReportFiles	Reads reports created by ABAQUS based on Output Database files
PBC2DGen_readDATFile	Reads and evaluates contents of the computational options (*.DAT) file
PBC2DGen_Print_MaterialCards	Writes relevant material commands in modified *.INP file
PBC2DGen_MatSpecify	Specifies properties of selected materials
PBC2DGen_JobScriptCreate	Creates shell-scripting based automated job submission scripts
PBC2DGen_DetConstitutive_Plots	Determines the constitutive plots of material (for **UD** composites)
PBC2DGen_Create_Remainder_INP	Updates current *.INP file with *.EQUATION constraints
PBC2DGen_Create_HistoryOut	Creates a Python Script to be associated with a given simulation
PBC2D_AnalyzeNodes_ver1	Sorts out nodes, their coordinates, and element connectivities
Information	Displays output information in MATLAB™ command window
get_options_information	Extracts information from **PBC2DGEN** for the execution process
Convert_write_workspace	Writes node and element data to the workspace
Convert_read_abaqus2D	Reads node and element data from ABAQUS input file
Convert_input2D	A function to convert between different FEM input file formats

*.p files with descriptive names that indicate their purpose. The most important *.p file is PBC2DGen_FrontEnd.p, which is the main script that connects all the other *.p files. A brief description of each of the *.p files shown in Fig. 8.12 is given in Table 8.1.

(b) **Computational options file:** This is an important pre-filled *.DAT file that must accompany the operation of **PBC2DGEN**. It details the user's specifications for a given simulation. Such specifications include: *material type, load conditions, level of strain, display outputs, time-step instructions*, etc. In fact, it represents a control panel for **PBC2DGEN**. It should always be opened and its contents modified and saved before running **PBC2DGEN**.

(c) **ABAQUS Keyword File:** This is a keyword file with file extension *.INP, obtained from ABAQUS, which contains geometric and mesh information about the model. It is obtained when the user instructs ABAQUS to write the keyword/input file to the ABAQUS current working directory. This will be passed onto **PBC2DGEN** before execution. You can have as many `*.INP` files in the current MATLAB **PBC2DGEN** directory as you want. The original release of **PBC2DGEN** (downloaded from the *extra resources website*) already comes pre-bundled with three example keyword files for the reader to try out **PBC2DGEN**.

8.6.2 Understanding the PBC2DGEN Computational Options File

The *computational options file* is the main control panel for **PBC2DGEN**. It is important that the parameters identified within it are understood. This section will explain some of the parameters and advise on how best to assign values to them. Algorithm 8.1 shows the key top end of the computational options file. It is typically saved with a file extension of `*.DAT`.

Algorithm 8.1 A snippet of **PBC2DGEN** computational options file

```
%% PBC2DGEN Computational Options File
%-------------------------------------------------------------
%  Author: Michael I. Okereke  Date:      28/07/2017
%-------------------------------------------------------------
%BOUNDARY CONDITIONS OPTIONS
PBC2D.LCType4Unial  = [0 1 2];  %0-X,1-Y; 2-ShearXY;
PBC2D.Strain        = 0.3;      %Maximum strain
PBC2D.linearLim     = 0.005;    %Limit for Modulus
%-------------------------------------------------------------
%MATERIAL OPTIONS
PBC2D.PrintMatType  = 1;   %0 - None,1-Elastic PP+Fibre,
PBC2D.MatIDNum      = 0;   %0 - GFRP(Plytron);
                           %1 - Boron-Aluminium;10-Steel
                                (Elastic)
%-------------------------------------------------------------
%GENERAL OPTIONS
PBC2D.pickINP       = 1;   %0 - automatic, 1-user selects *.inp
PBC2D.plotRVE       = 0;   %0 - Plot graphs of RVE; 1-dont
PBC2D.dispRVE2DInfo = 1;   %0 - display Nodes info, 1-dont
PBC2D.PrintRemINP   = 0;   %0 - Print remainder of INP file
PBC2D.EleOutputHist = 1;   %0 - Print output history, 1-do not
PBC2D.ChoseBCMode   = 0;   %0 - disp, 1 -  velocity
PBC2D.ChoseTStep    = 0;   %0 - Automatic time step size ,
                           %1 - user-controlled time-step size
PBC2D.UMATPath     = '../';%  - Set path before the UMAT file
%-------------------------------------------------------------
%JOBSCRIPT CREATING OPTIONS
PBC2D.JobScript     = 1;   %0 - Make *.tcsh Jobscript, 1 - dont
PBC2D.HistScript    = 0;   %0 - Make *.py script, 1 - don't
%-------------------------------------------------------------
```

The computational options file is divided into four parts: ***(a)*** *Boundary conditions Options;* ***(b)*** *Material Options;* ***(c)*** *General Options; and,* ***(d)*** *Job script Creation* *options*. The first two sections need to be regularly changed as necessary to undertake a given simulation type.

The following describes the meaning of the file parameters and what possible values the user can specify. The 2-D RVE used for the simulation is square and oriented in a global *XY* reference frame, as shown in Fig. 8.7.

- **`PBC2D.LCType4Uniax:`** This option is for specifying which load type one would like to prescribe on the RVE. Probable values are: 0 = *Uniaxial deformation along the x-axes*; 1 = *Uniaxial deformation along the y-axes*; and, 2 = *shear deformation on the xy-plane*.
- **`PBC2D.Strain:`** This option specifies the maximum nominal strain that the user wants to impose on the 2-D RVE. This strain is used to define the Dirichlet-style boundary conditions that has to be prescribed on one of the retained nodes. For example, if strain, $\varepsilon = 0.025$, and the RVE is of edge length, L, then according to Fig. 8.10, the $x-$ axis uniaxial deformation prescribed displacement, $\delta_x = 0.025L$.
- **`PBC2D.LinearLim:`** This parameter specifies the linear limit of strain at which the algorithm automatically calculates the **Young's Modulus** of the test material. For a strain limit value of 0.005, this is equivalent to 0.5% strain.
- **`PBC2D.PrintMatType:`** This option instructs **`PBC2DGEN`** to write into the `ABAQUS` original keyword file the material model for a specific material type. Implemented material types are:

  * 1 – Unidirectional composite (E-glass fibre and polypropylene matrix)
  * 2 – User-defined material model for Polypropylene called UMAT_PP
  * 21 – UD Composite: UMAT_PP for PP matrix and E-glass for fibers.
  * 3 – UD lamina format for shell elements.
  * 4 – Engineering constant format for a homogenized composite
  * 5 – Linear elastic (isotropic) steel material
  * 6 – Aluminum-SiC particle reinforced aluminum composite (PRAC)

- **`PBC2D.MatIDNum:`** This option supplies specific material property values to the keyword file. Here, the values would have been originally hard-coded into **`PBC2DGEN`** as part of the release bundle. For the current version, the available material data are for the following:

  * 0 – Glass fibre reinforce plastic [14]
  * 1 – Boron-Aluminium composite [17]
  * 10 – Homogeneous isotropic steel [7]
  * 3 – Aluminum-SiC PRAC with SiC volume fraction of 15%

- **`PBC2D.pickINP:`** This option allows PBC2DGen to automatically select any of the `*.INP` file for the execution with **`PBC2DGEN`** (i.e. option = 0) or allow for user-selection of input file (i.e. option = 1).
- **`PBC2D.plotRVE:`** This option is used to determine if the nodes (inside and on the boundary) of the RVE need to be plotted (option = 0) or not (option = 1).

- **`PBC2D.ChoseBCMode:`** This option allows users to use either displacement-based (option = 0) load cases to drive the simulation or alternatively use velocity-based ones (option = 1).
- **`PBC2D.UMATPath:`** This option is used to supply the full operating system path to the location of a user-defined material (**UMAT**) required for the simulation process. This is only essential if the user is running a model based on a user-defined material model.
- **`PBC2D.Jobscript:`** This option instructs **PBC2DGEN** to write a shell-script with (`*.tcsh`) extension. This is used to automatically submit, in batches, jobs created based on the **PBC2DGEN** script.
- **`PBC2D.HistScript:`** If option = 0, this means that **PBC2DGEN** should automatically generates two Python script: (**a**) job submission script and (**b**) post-processor analyses script for the resulting ABAQUS output database (`*.ODB`) files.

We will now run a few example cases to illustrate the robustness of **PBC2DGEN**. The examples considered here are drawn from micromechanical analysis of heterogeneous materials, since mechanics of RVEs of such materials are best analysed using periodic boundary conditions.

Example 8.3 An isotropic homogeneous steel material is to be used to verify the suitability of **PBC2DGEN** as a design tool for prescribing periodic BCs on 2-D square RVEs of structures. Three different load cases were prescribed on the 100×100 units^2 RVE which include: (**i**) uniaxial $x-$ axis strain, $\varepsilon_x = 0.300$; (**ii**) uniaxial $y-$ axis strain, $\varepsilon_y = 0.100$; and, (**iii**) shear deformation on the $xy-$plane, $\gamma_{xy} = 0.250$.

(a) Using an appropriate FEM solver, obtain the strain contour plots for all three load cases.
(b) Determine the displacement contour plot for the prescribed strains.

Solution
Here, we will use ABAQUS as the FEM solver of choice. The virtual domain is the 2-D RVE shown in Fig. 8.13a – meshed using quadrilateral elements. Figure 8.13b shows **PBC2DGEN**-generated plot of all nodal coordinates of the domain.

Let us define a boundary domain, Ω, which represents all the nodal points bounded by the 2-D RVE. The **PBC2DGEN** code divides the cloud of nodal points into X_{pos} and X_{neg}, which represent nodal points along the *positive* and *negative* edges perpendicular to the $X-$axis as well as Y_{pos} and Y_{neg}, for *positive* and *negative* edges perpendicular to the $Y-$axis. Hence, the domain Ω must obey the argument:

$$\Omega = \Omega_{X_{pos}} \cup \Omega_{X_{neg}} \cup \Omega_{Y_{pos}} \cup \Omega_{Y_{neg}} \tag{8.25}$$

where Ω_k is the domain that encloses the $k-$th edge for $k = \left[X_{pos},\ X_{neg},\ Y_{pos},\ Y_{neg}\right]$.

PBC2DGEN is used to enforce the periodic BC, the load case type, create the material cards, generate job-submission and analysis scripts, and also subsequent

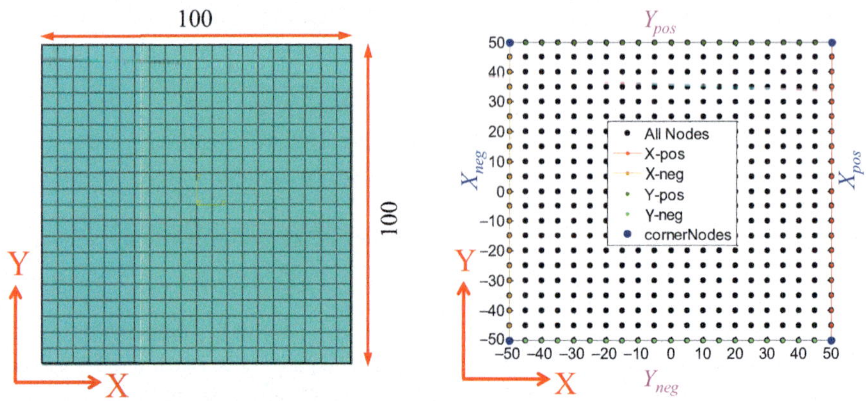

(a) (b)

Fig. 8.13 A 2-D RVE window for steel showing: (**a**) model in ABAQUS discretized using quadrilateral elements; and, (**b**) the corresponding nodal positions from the **PBC2DGEN** plotting option

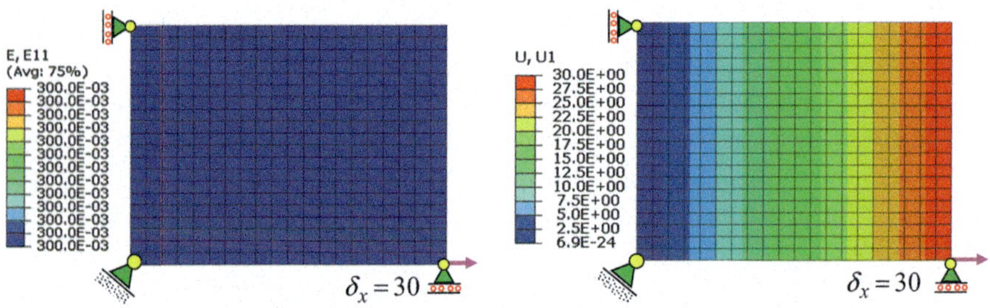

(a) (b)

Fig. 8.14 Uniaxial deformation along the $X-$axis showing: (**a**) the homogeneous strain, ε_{11} contour plot, where $\varepsilon_{11} = 300 \times 10^{-3}$; and, (**b**) the displacement contour plot, where $\delta_x = U_{1,max} = 30$ units

post-processing of the results. The following presents the results generated from using **PBC2DGEN**.

(a) **Case** (a_1, b_1) **Uniaxial** $X-$**axis deformation:** Using **PBC2DGEN**, the results obtained for the strain[6] and displacement contour plots are shown in Fig. 8.14. Since the test material is homogeneous, the expected strain plot is homogeneous and this is shown in Fig. 8.14a.

[6]Note: In ABAQUS, the *XYZ* reference frame is equivalent to the 123−material reference frame. Hence, $\varepsilon_{11} \equiv \varepsilon_{XX}$

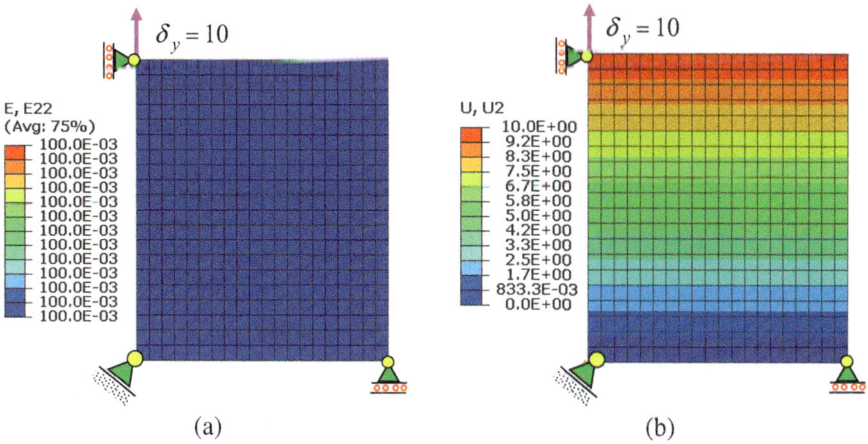

Fig. 8.15 Uniaxial deformation along the $Y-$axis showing: (**a**) homogeneous strain, ε_{22} contour plot, where $\varepsilon_{22} = 100 \times 10^{-3}$; and, (**b**) displacement contour plot, where $\delta_y = U_{2,max} = 10$ units

(b) **Case** (a_2, b_2) **Uniaxial** $Y-$**axis deformation:** **PBC2DGEN** is again used here to impose uniaxial deformation along the $y-$ axis and the results of strain and displacement obtained are shown in Fig. 8.15a, b respectively.

(c) **Case** (a_3, b_3) **Shear deformation on the** $XY-$**plane:** **PBC2DGEN** is again used to impose shear deformation on the $xy-$plane and the results of strain and displacement contour plots obtained are shown in Fig. 8.15a, b respectively.

Remark 8.1 **Comments on model verification results**: Visual verification showed that for all three load cases, the steel material showed homogeneous deformation as expected for a homogeneous material subjected to periodic boundary conditions.

Also, for the $x-$ axis uniaxial deformation load case, the input strain of $\varepsilon = 0.300$ was returned after the test as $\varepsilon_{11} = 300 \times 10^{-3}$, as shown in Fig. 8.14a.

For the original edge length, $L_o = 100$ units, the RVE when subjected to a nominal strain of $\varepsilon_{11} = 0.300$ yields the displacement, $\Delta L = \varepsilon_{11} L = 30$ units. This is the same maximum displacement, $\delta_x = U_{1,max} = 30$ units, shown in Fig. 8.14b.

Similar conclusions can be drawn for the other two load cases (Figs. 8.15 and 8.16). This demonstrates that the current implementation of **PBC2DGEN** returns reliable results. This, in term, gives the reader confidence to use **PBC2DGEN** to explore the micromechanics of other heterogeneous systems.

8.6.3 Using PBC2DGEN to Obtain Mechanical Response of Materials

In Solid Mechanics, a key objective most of the time is to determine the stress-strain response of a chosen test material. The current implementation of **PBC2DGEN** contains scripts for obtaining the mechanical response of some test materials. The

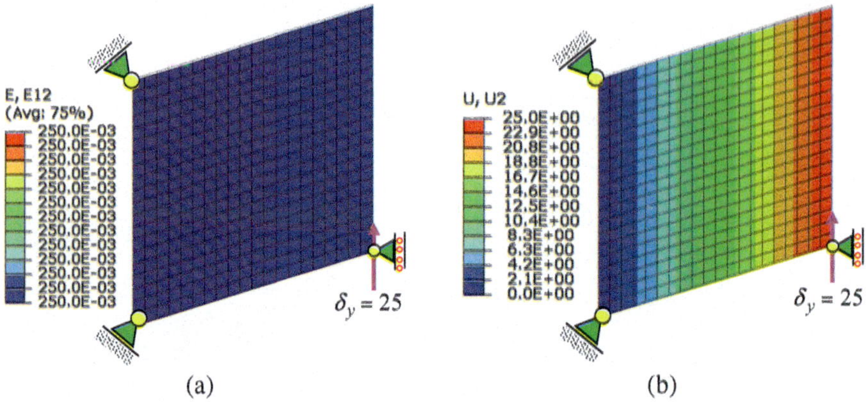

Fig. 8.16 Shear deformation about the XY–plane showing: (**a**) the homogeneous shear strain, γ_{12} contour plot, where $\varepsilon_{12} = 250 \times 10^{-3}$; and, (**b**) displacement contour plot where $\delta_y = U_{2,max} = 25$ units

implementation here is based on **Micromechanics** – an aspect of Engineering Mechanics, that explores the behaviour of engineering materials at a microscale length scale. The information generated at microscale needs to be related to macroscale or structural scale experiments.

Computational homogenization[7] is used to bridge the gap between microscale predictions and macroscale experimental data. The **PBC2DGEN** implementation for obtaining the mechanical response, is based on the principle of computational homogenization. In the following, we present two approaches that can be used to generate the mechanical response of a heterogeneous system.

8.6.3.1 Mechanical Response Through Computational Homogenization

After an FEM simulation is completed, the output database (`*.ODB` in ABAQUS) file contains information about stresses, strains, displacements, etc. of the test. The next step in determining the stress-strain plot is to undertake the *computational homogenization*. Within an FEM scheme such as the ones under consideration here, we can achieve this by summing up, for a given model variable, all the values of specific model variable across all elements in the model, and dividing by the total volume of all number of elements in the model.

Consider a model variable, ϱ, of a finite element simulation, we can determine the homogenized value of this variable, $\langle \varrho \rangle$, according to Eq. 8.26:

[7]This term should not be confused with *asymptotic homogenization* – an analytical homogenization method – which requires mathematical stress-strain derivations of simplistic RVEs of the microstructure under investigation. For a UD composite, such representation can include simple ordered arrangement of the fibre within the matrix medium.

$$\langle \varrho \rangle = \frac{\sum_{i=1}^{N} \varrho_i}{\sum_{i=1}^{N} V_i} \implies \langle \varrho \rangle = \frac{1}{Vol} \int_V \varrho \partial V \tag{8.26}$$

where Vol = total volume of all elements in the model which is equivalent to the sum of each element volume, V_i; i = element number, and, N = total number of elements in the model. The homogenized model variable, $\langle \varrho \rangle$ is also called the *volume-averaged model variable*. It can be stress, strain, temperature, etc., as may be required for any simulation. For our purposes here, $\langle \varrho \rangle$ represents the stress and strain values of the model.

Consider a 2-D RVE of heterogeneous media subjected to a uniaxial deformation along the $x-$ axis, we may be interested in obtaining the effective elastic properties for the material (e.g. Young's Modulus). To do so, we will use the homogenized stress and strain forms of Eq. 8.26:

$$E_{11} = \frac{\langle \sigma_{11} \rangle}{\langle \varepsilon_{11} \rangle} \implies E_{11} = \frac{\frac{1}{Vol} \int_V \sigma_{11} \partial V}{\frac{1}{Vol} \int_V \varepsilon 11 \partial V} \implies E_{11} = \frac{\int_V \sigma_{11} \partial V}{\int_V \varepsilon 11} \partial V \tag{8.27}$$

The above expression essentially implies the division of the volume-average values of the stresses and strains to obtain the effective properties. Equation 8.27 can be directly used within ABAQUS to determine the averaged value of stresses and strains, and subsequently determine the stress-strain diagrams and the relevant effective properties.

8.6.3.2 Mechanical Response Through the Virtual Work Principle

Another way of determining the mechanical properties of heterogeneous systems is based on the *virtual work principle*. This approach uses only the model variables (e.g. nodal forces and displacements) at the retained (or corner nodes) to compute the work done within the virtual domain. This will be used to subsequently determine the stresses, strains, and other effective properties of the system. In the following, we provide a brief description of this approach although detailed description can be obtained from [14].

Consider a 2-D RVE window enclosing a heterogeneous material by boundary edges identified by Ω_k where k = $[L,\ T,\ R,\ B]$, as shown in Fig. 8.17. Nodal reaction forces and displacements for 3 retained nodes are given as $\mathbf{f}_i$ and $\mathbf{x}_i$ where $i = 1, 2, 3$ on retained nodes N_1, N_2 and N_3.

The mathematical formulation for deriving the homogenized (volume-averaged) stresses using the principle of virtual work (as in [14]) is given as:

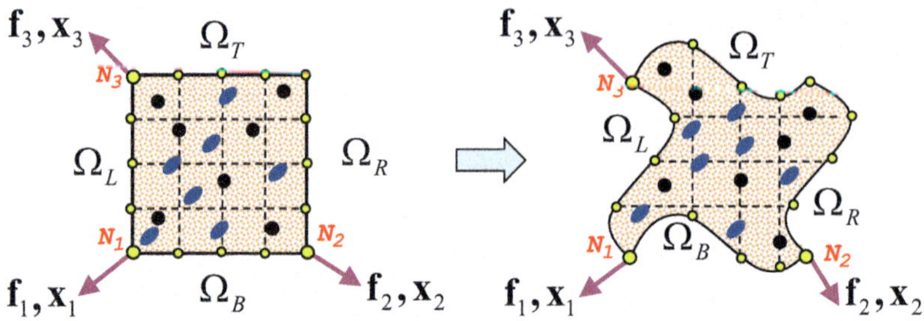

Fig. 8.17 A 2-D RVE window of a heterogeneous material, showing nodal forces and reactions in [*left*] undeformed [*right*] periodically deformed profile

$$\langle\sigma_{11}\rangle \quad = \quad \frac{1}{Vol}\left[\mathbf{x}_1 \otimes \mathbf{f}_{N1} \;+\; \mathbf{x}_2 \otimes \mathbf{f}_{N2} \;+\; \mathbf{x}_3 \otimes \mathbf{f}_{N3}\right] \tag{8.28}$$

Within **PBC2DGEN**, the mechanical response of the material being tested is generated according to Eq. 8.28. Only nodal displacement and force values for 3 retained nodes will be needed to generate the **constitutive behaviour** (i.e. the stress-strain response) of the tested material.

Within **PBC2DGEN**, Eq. 8.28 is implemented using two MATLAB™ p-files. These files are 'called' after running the two Python scripts: when the simulation has completed and the output database file has been generated. The two `*.P` MATLAB™ files are then used to determine the stresses and strains as well as effective properties of the test material.

These `*.P` MATLAB™ files are:

- **`PBC2DGen_ReadReportFiles:`** The script reads through `ABAQUS` generated `*.RPT` files, which were created when the `Analyze*.PY` script was executed. This `PBC2DGEN_ReadReportFiles` MATLAB™ p-file operates on `ABAQUS *.RPT` file which prints MATLAB™ plots of the reaction forces and nodal displacements of all the corner nodes. Several graphs will result following the operation on the `PBC2DGEN_ReadReportFiles` p-file. It produces a `rxnForce_RefNodes*.mat` data file as well as another file called `coordPos_RefNodes*.mat` which is also a data file. The first file contains nodal reaction information while the later contains corner (retained) nodal displacement information.
- **`PBC2DGen_DetConstitutive_Plots:`** This script is a MATLAB™ p-file that operates on outputs from the `PBC2D_ReadReportFiles` based on the principle of virtual work. It is the script that generates the stress-strain diagrams of the simulation and the effective properties associated with the simulation.

8.6.4 *Micromechanics Exemplars Involving* **PBC2DGEN**

In this section, we will present some exemplars of micromechanics of heterogeneous systems. The exemplars will investigate the constitutive behaviour of example heterogeneous materials using **PBC2DGEN** as a tool for incorporating the periodic boundary conditions. This section illustrates the extended application of **PBC2DGEN** as a design and research tool.

Example 8.4 A probable RVE for investigating the effective properties of an E-glass fibre, polypropylene matrix unidirectional composite is chosen to be a body-centred arrangement with a fibre located centrally within a square RVE window. The volume fraction of the fibre reinforcement within the composite is 35%. The material properties of the composite constituents are given in Table 8.2. Assume that the mechanical behaviour of the constituents is elastic.

(a) Using a square RVE window size of 80 μm, determine the diameter of the fibre that should lead to a fibre volume fraction of 35%. Assume unit thickness for the RVE.
(b) Develop the virtual domain of the composite.
(c) Obtain both a coarse and fine mesh of the test composite for subsequent simulation.
(d) Using **PBC2DGEN**, impose periodic boundary conditions on the two meshes for uniaxial $X-$ and $Y-$ deformation, as well as the shear load case on the $XY-$plane.
(e) Based on simulation results from the three load cases, determine the Young's Moduli, E_{11}, and E_{22} as well as shear modulus, G_{12} of the test composite.

Solution

(a) **Diameter of fibre:** Volume fraction of the fibre is defined as the volume occupied in space by the fibres divided by the overall volume of the test composite.

Assuming unit thicknesses, the volume fraction is equivalent to the area fraction. If the RVE has dimensions of length, L, and width, W, and the fibre has a diameter, D_f, and the chosen RVE window has N[8] strands of fibre, the fibre area fraction is:

Table 8.2 Elastic properties for an E-glass fibre polypropylene matrix composite

Material	Young's Modulus, E (GPa)	Poisson ratio
E-glass fibre	75	0.25
Isotactic polypropylene	1.4	0.40

[8]For this example, $N = 1$, since the chosen RVE window encloses only one single fibre as stated in the question.

$$\text{Area fraction}, A_f = \frac{\text{Total area of fibres}}{\text{Total area of RVE window}} = \frac{0.25N\pi D_f^2}{LW}$$

Re-writing the expression, making D_f the subject of the formula:

$$\text{Fibre diameter}, D_f = \sqrt{\frac{LWA_f}{N\pi}} = \sqrt{\frac{4 \times 80 \times 80 \times 0.35}{1 * \pi}} = 53.4\,\mu\text{m}$$

(b) **Virtual domain of the test composite:** There are different ways of creating the virtual domain, for example, **(a)** create it in a dedicated **CAD** software and import into an FEM solver such as `ABAQUS`; or, **(b)** use the `ABAQUS/CAE` design environment to create the parts. Here, we will use the latter option. This will involve these processes:

- Create the part geometry of a fibre with diameter, $D_f = 53.4$ units.
- Create the part geometry of the matrix with dimensions 80×80 units^2.
- Create instances of both parts and merge them in the `Assembly` module of `ABAQUS`. This merging operation (retaining the parts boundaries) creates the composite part geometry that will be the basis of the simulation. This is illustrated in Fig. 8.18.

(c) **Meshing of the test composite:** Two mesh types were obtained: *fine* and *coarse* meshes. The discretization of the model was based on triangular elements, but similar arguments can be made for quadrilateral elements. Figure 8.19 shows the two meshes, such that:

- Coarse Mesh: 27 nodes, 36 elements $\longrightarrow$ Fibre is poorly meshed
- Fine Mesh: 491 nodes, 900 elements $\longrightarrow$ Fibre is finely meshed

(d) **Periodic BCs, load cases and `PBC2DGEN`:** The **`PBC2DGEN`** code was used to impose periodic BCs for the three load cases for the two mesh types. The contour plots are given in Figs. 8.20a and 8.21b.

Comment on contour plots: All the uniaxial deformation simulations show straight edges due to the symmetry of the virtual domain (microstructure). This is not the same for shear deformation, which show curved periodic edges. The

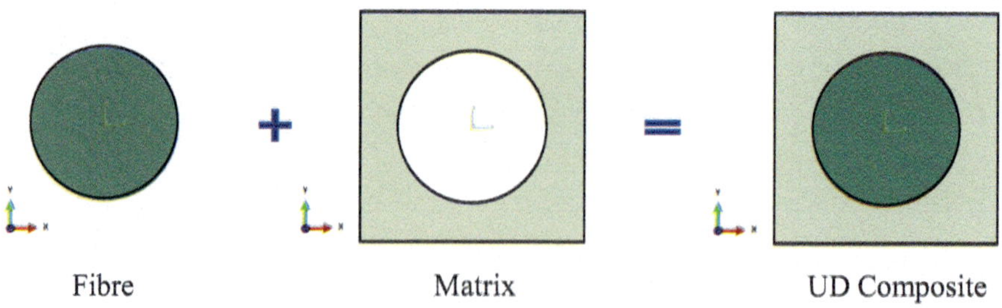

Fig. 8.18 An assembly process for creating a 2-D RVE of a **UDD** composite in `ABAQUS`

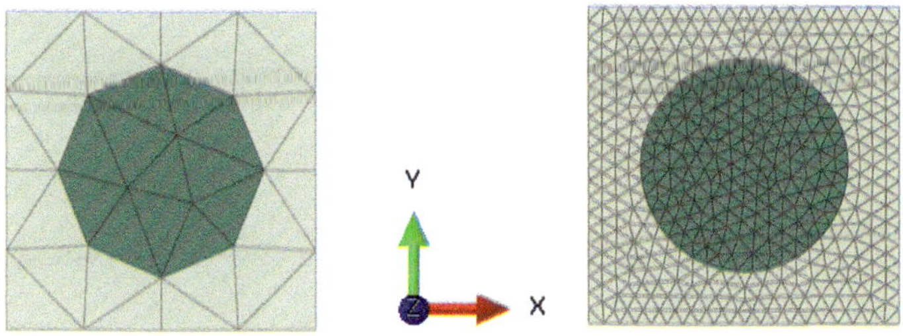

Fig. 8.19 Coarse and fine meshes of the 2-D RVE of test composite

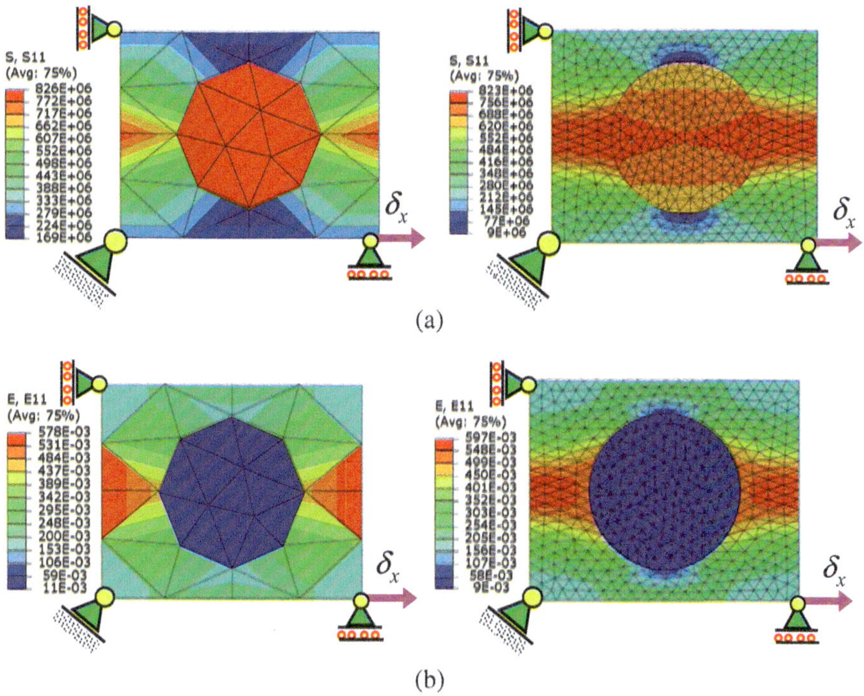

Fig. 8.20 Contour plots for *uniaxial deformation along the x-axis* for coarse and fine meshes. The plots are for: (**a**) stress, $\sigma_{11} \equiv \sigma_{XX}$ and (**b**) strain, $\varepsilon_{11} \equiv \varepsilon_{XX}$

maximum and minimum values of stresses and strains in uniaxial cases for both mesh types did not show much difference, but this is not so for shear loads.

Also, the shape of the critical failure band (darkest regions[9]) is significantly different with changing mesh types. This suggests a different failure pattern being created for changing mesh types thus justifying the necessity of mesh

[9]The e-book version of these contour plots will show these regions as bright red shear bands.

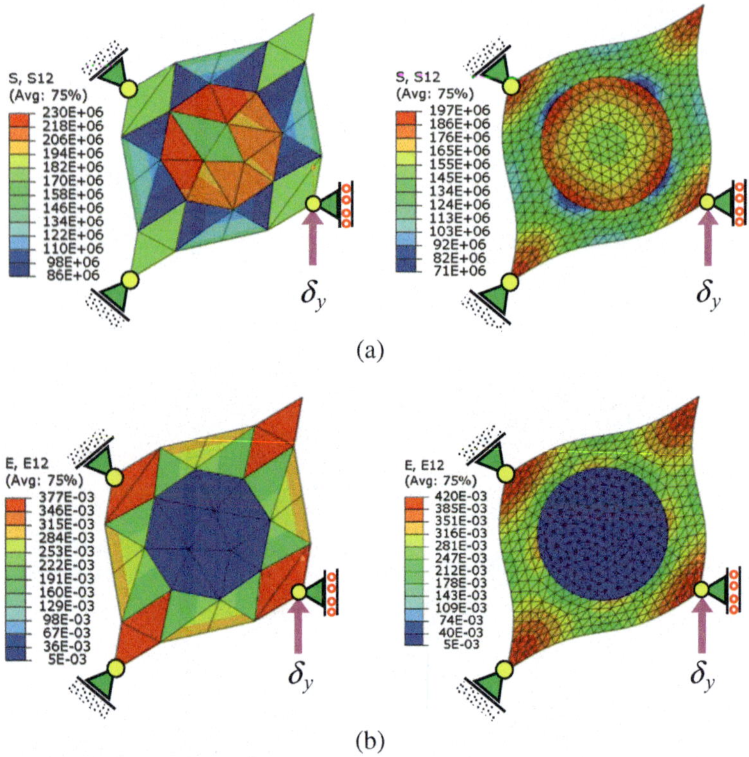

Fig. 8.21 Contour plots for *shear deformation on the xy-plane* for coarse and fine meshes. The plots are for: (**a**) stress, $\tau_{12} \equiv \tau_{XY}$; and, (**b**) strain, $\gamma_{12} \equiv \gamma_{XY}$

Table 8.3 Comparison of effective properties for two mesh densities of a UD composite (for Example 8.4)

Effective property	Load case	Coarse mesh (GPa)	Fine mesh (GPa)	Percentage error (%)
Young's modulus, E_{11}	Uniaxial X	1.1473	1.1586	1.0
Young's modulus, E_{22}	Uniaxial Y	1.1472	1.1586	1.0
Shear modulus, G_{12}	Shear XY	0.4774	0.4843	1.5

study in order that a mesh-independent mechanical response can be reached before we can deem the FEM solution acceptable.

(e) **Stress-Strain plots, effective properties and the effect of mesh density:** The **PBC2DGEN** code was also used to generate the effective properties for all three load cases. These were derived after the simulation had been completed. The results from these simulations are given in Table 8.3.

Comments on effective properties: The results in Table 8.3 show that there is comparatively little difference between the fine and coarse meshes. This may seem surprising considering that the coarse mesh is of quite poor mesh quality compared with the fine mesh.

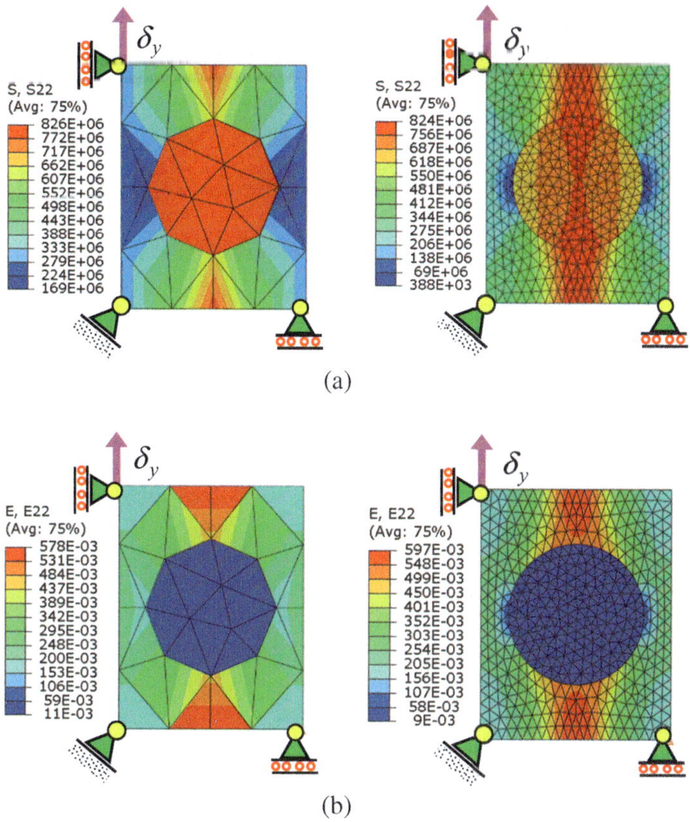

Fig. 8.22 Contour plots for *uniaxial deformation along the y-axis* for coarse and fine meshes. The plots are for: (**a**) stress, $\sigma_{22} \equiv \sigma_{YY}$; and, (**b**) strain, $\varepsilon_{22} \equiv \varepsilon_{YY}$

However, this is not unexpected, because the use of periodic boundary conditions significantly improves the convergence of the simulation to the desired solution. In fact, it is known in the micromechanics of heterogeneous material literature that a convergent solution is quickly achieved using periodic BCs.

Example 8.5 A solder thermal interface material (**STIM**) is used to ensure adequate heat dissipation in the heat sink compartment of an electronic circuit board. The STIM is made from Sn-Ag-Cu solder alloy called SAC305 with Young's Modulus, $E = 38.7\,\text{GPa}$ and Poisson ratio, $\nu = 0.35$ [15].

During a routine investigation, the STIM was observed to have several voids with a void area fraction, $A_f = 35\%$. A 2-D RVE window of $45 \times 45\,\mu\text{m}^2$ was isolated for the numerical study of the effect of the voids. Each void was initially idealized to be circular with an average void diameter, $D_v = 10\,\mu\text{m}$.

(a) Calculate the total number of voids that should fill the 2-D RVE for the area fraction, $A_f = 35\%$ to be achieved.
(b) Develop the virtual domain for the voided STIM. Assume the voids to be air-filled (empty space) and of random arrangement within the 2-D RVE window.
(c) As a result of the thermal loads on the STIM, residual stresses build up in the material causing a shear distortion/deformation of the right edge by $\Delta L = 0.45\,\mu\text{m}$:

  (i) Assume linear elastic behaviour of the STIM, determine its shear contour plot.
  (ii) What is the maximum critical stress in the material? Do you think the STIM can fail due to this stress, assuming the yield stress of steel, $\sigma_Y = 50\,\text{MPa}$?

(d) Run the model again considering elasto-plastic behaviour of the STIM with onset of yield at $\sigma_Y = 50\,\text{MPa}$. Compare the stress-strain and contour plots from the elastic and elasto-plastic simulations.

Solution

(a) **Total number of voids in 2-D RVE window:** Assume the 2-D RVE has length, L, width, W, void diameter, D_v and the number of voids enclosed completely by the RVE, N. The void area fraction, A_f, becomes:

$$A_f = \frac{\text{Total area of void}}{\text{Total area of RVE}} \quad \Longrightarrow \quad A_f = \frac{N\pi D_v^2}{4LW}$$

Re-writing the expression, making N the subject of the formula, gives:

$$\text{Number of voids, } N = \frac{4LWA_f}{\pi D_v^2} = \frac{4 \times 45 \times 45 \times 0.35}{\pi \times 10^2} = 9.02 \cong 9.00$$

(b) **Virtual domain of voided STIM:** `ABAQUS` was used to generated the voided STIM with random arrangement of voids within the square geometry that represents the STIM. Also, the meshed geometry is given in Fig. 8.23.
(c) **Shear Contour plot (linear elastic): `PBC2DGEN`** was used to impose periodic BCs on the meshed STIM and also to impose a strain: $\gamma_{xy} = \Delta L/L = 0.45/45 = 0.010$. The STIM is treated here as a linear elastic material with $E = 38.7\,\text{GPa}$ and $\nu = 0.35$. The shear stress and strain contour plot is shown in Fig. 8.24.

  Given a yield stress of STIM of $\sigma_Y = 50\,\text{MPa}$, the contour plot of elastic deformation gave a critical shear stress, $\tau_{12,crit}^{elastic} = 239\,\text{MPa}$. Since, $\tau_{12,crit}^{elastic} \gg \sigma_Y$, we conclude that the STIM would have failed at this strain level. Even a consideration of the von Mises stress, $\sigma_{vonMises} = 577\,\text{MPa}$ supports this view.
(d) **Elastic and elasto-plastic shear response compared:** However, the failure mode of STIM is not elastic, but a plastic deformation. Therefore, the above conclusion is not valid. To further determine the exact critical failure stress

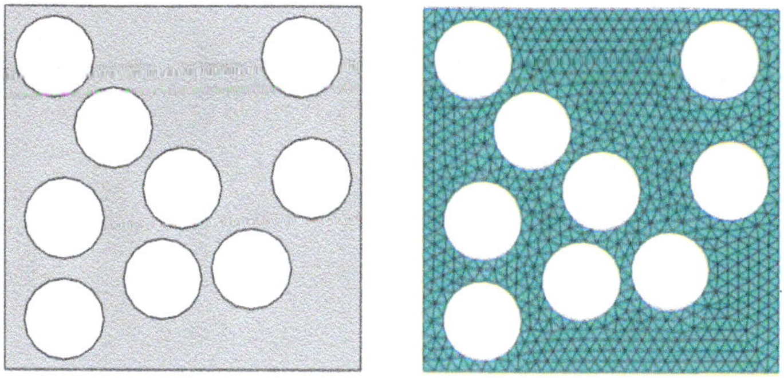

Fig. 8.23 The virtual domain and mesh of the voided solder thermal interface material

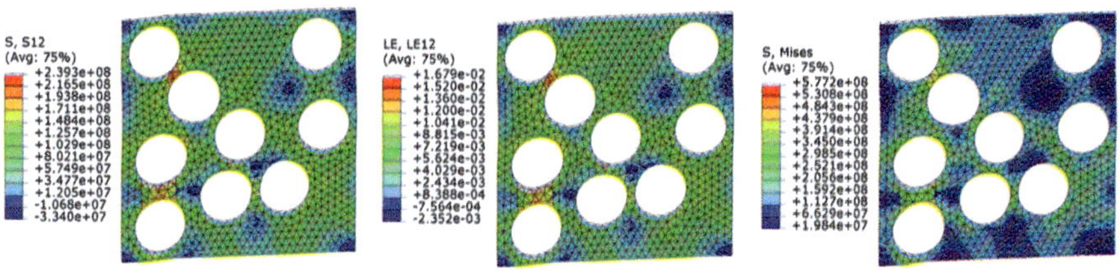

Fig. 8.24 Contour plots of shear stress and strain for the STIM model. The test material is considered a linear elastic material

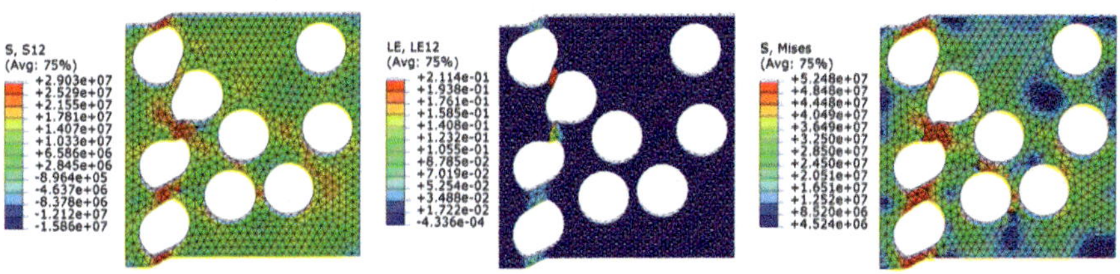

Fig. 8.25 Contour plots of shear stress and strain for the STIM model. The test material is considered a elasto-plastic material

allowing for plastic deformation, we will re-run the model assuming elasto-plastic behaviour of the STIM. The apparent yield stress is $\sigma_Y = 50\,\text{MPa}$, as specified in the question.

The shear stress and strain contour plots for this second scenario is given in Fig. 8.25. This showed a critical shear stress, $\tau_{12,crit}^{plastic} = 29\,\text{MPa}$. Notice that $\tau_{12,crit}^{plastic} \ll \sigma_Y$, hence the STIM did not fail in shear. We need to consider other stress measures, especially the von Mises stress, to ascertain the exact effective stress state to determine if the STIM has failed.

Figure 8.25 shows that the critical von Mises stress, $\sigma_{vonMises}^{critical} = 55\,\text{MPa} > \sigma_Y = 50\,\text{MPa}$, which confirms that the STIM has failed. Also, the comparison of elastic and elasto-plastic displacement contour plots of the STIM model is given in Fig. 8.26. The distinct deformation profiles shows the significance of allowing for plastic deformation in this type of study.

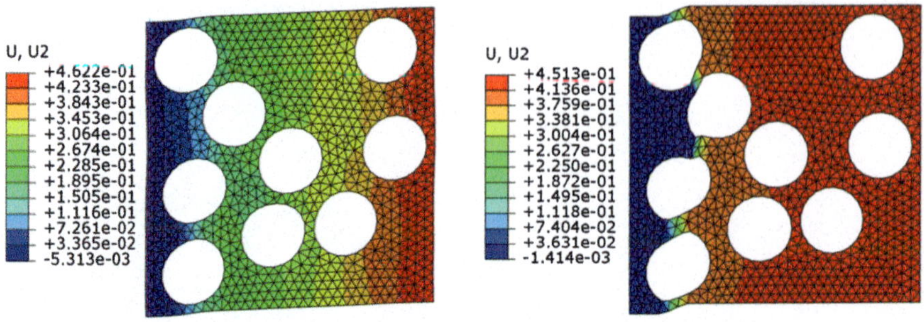

Fig. 8.26 Displacement contour plots for (*left*) elastic (*right*) elasto-plastic behaviour of the STIM model

Table 8.4 Material properties for aluminum-SiC PRAC material [16]

Material	Young's modulus, E (GPa)	Poisson ratio, ν
Silicon carbide (SiC)	450	0.15
6061 Aluminum	70	0.33

Example 8.6 Consider a particle reinforced aluminium alloy composite (**PRAC**), required for a study on the effect of particle shape on the critical stress and effective properties of a 6061 Aluminium matrix reinforced by silica carbide (SiC). The study is to be carried out using computational mechanics. The volume fraction of the silica is 15% and a probable RVE for the study was chosen to be $120 \times 120\,\mu m^2$. A reference diameter of a circular shaped SiC particle was $D_c = 14\,\mu m$. Properties of the composite constituents are given in Table 8.4.

(a) Determine the number of circular shaped particles needed here.
(b) To investigate the influence of other shaped particles on the stress and strain state, hexagonal and elliptical shapes were considered too. To make these shapes comparable with the circular shapes, assume each of the new shapes are circumscribed in a circle with same diameters as the circular shapes. Keeping the areas of the three shapes constant, and based on the circumscribed argument, determine the number of hexagonal and elliptical silica required for this study.
(c) Generate the virtual domain of the test composite for the three particle types.
(d) Impose uniaxial deformation along $x-$ and $y-$ axes, as well as shear deformation on the $xy-$plane for the three material types. Run the simulations to obtain the contour plots of stresses and strains for all three cases.
e) Determine the effective properties, as well as critical stress, for all three cases.

Solution

(a) **Total number of circular voids in 2-D RVE window:** Let us define the RVE to consist of length, L, and width, W, with circular shaped particles of average diameter, $D_c = 14\,\mu m$. The number of particles to result in an inclusion area fraction of 15% is:

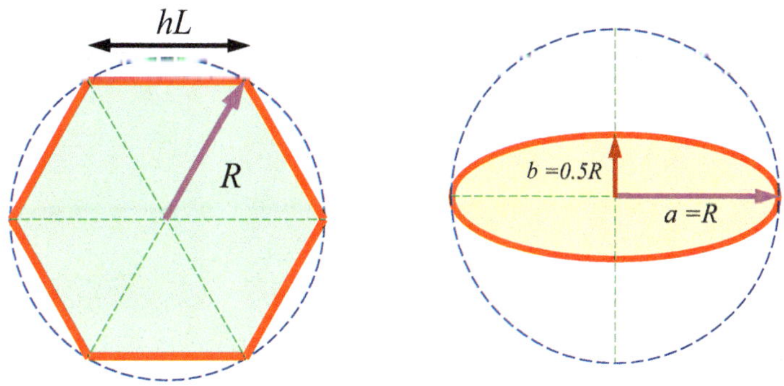

Fig. 8.27 Circumscribed particles of: (**a**) hexagonal and (**b**) elliptical shapes

$$\text{Number of circular particles, } N = \frac{4LWA_f}{\pi D^2}$$
$$= \frac{4 \times 120 \times 120 \times 0.15}{\pi \times 14^2}$$
$$= 14.03 \cong 14 \text{ particles}$$

(b) **Total number of other shaped particles:** Here, we consider hexagonal and elliptical shaped particle; both circumscribed inside a circle as shown in Fig. 8.27.

Regular hexagonal particles: If the hexagonal particle is circumscribed inside a circle of radius, R, and each edge length is hL, then hL must be equal to the radius, R, of the circumscribed circle ($hL = R$). The following conclusions are thus true:

$$\text{Area of hexagon, } hA = 1.5\sqrt{3}R^2$$
$$\text{Number of hexagonal particles, } N = \frac{LWA_f}{hA}$$
$$= \frac{120 \times 120 \times 0.15}{1.5\sqrt{3}R^2}$$
$$= 16.96 \cong 17 \text{ particles.}$$

Elliptical particles: For the elliptical inclusion circumscribed inside a circle, we make the assumption that the minor axis is half the major axis. We also see that the major axis is equal to the radius of the circle. If b and a are the lengths of the major and minor axes respectively, then $b = R$ and $a = 0.5R$.

$$\text{Area of ellipse, } eA = \pi ab = 0.5\pi R^2$$

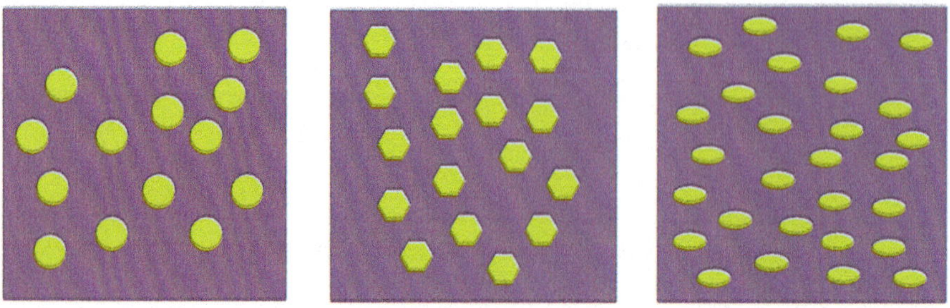

Fig. 8.28 Virtual domain of the particulate composites for all three particle shapes

Fig. 8.29 Periodic meshes of the particulate composites for all three particle shapes

$$\begin{aligned}\text{Number of elliptical inclusions, } N &= \frac{LWA_f}{eA} = \frac{LWA_f}{0.5\pi R^2} \\ &= \frac{120 \times 120 \times 0.15}{0.5\pi \times 7^2} \\ &= 28.06 \cong 28 \text{ particles}\end{aligned}$$

(c) **Virtual domains of the three composite types:** ABAQUS was used to generate the three composite types with the different particle geometries. The SiC particles were randomly distributed within the aluminium matrix as shown in Fig. 8.28. In all cases, the area fractions, A_f, were kept at approximately 15% so that subsequent comparison of effective properties can be deemed acceptable. The meshed representations for the three composite types are shown in Fig. 8.29 and **PBC2DGEN** was used to impose periodic BCs.

(d) **Contour plots for all three composite types:** Figs. 8.30, 8.31, 8.32, 8.33, 8.34, and 8.35 show the contour plots of the logarithmic (true) strain (LE) and (true) stress (S) for the three load cases and the three composite types.

(e) **Comparison of effective elastic properties and critical stresses:** By considering the simulation results shown in Figs. 8.30, 8.31, 8.32, 8.33, 8.34, and 8.35, together with **PBC2DGEN**, effective elastic properties were determined for all three composite types. Also, the critical stresses, σ_{max}, for the given mesh

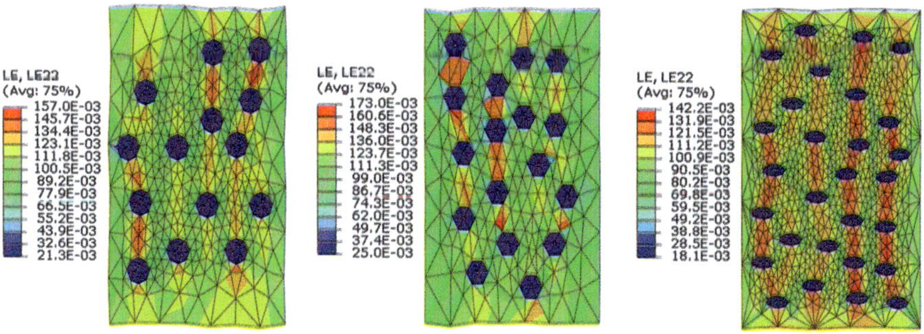

Fig. 8.30 True strain for *uniaxial* deformation along *y-axis* for the three composites

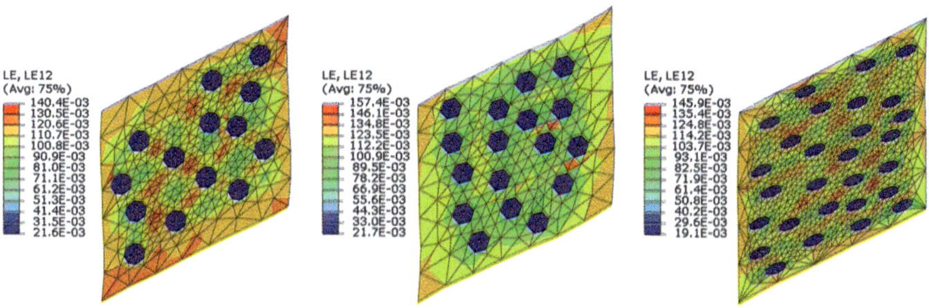

Fig. 8.31 True strain for *shear* deformation on *xy-plane* for the three composites

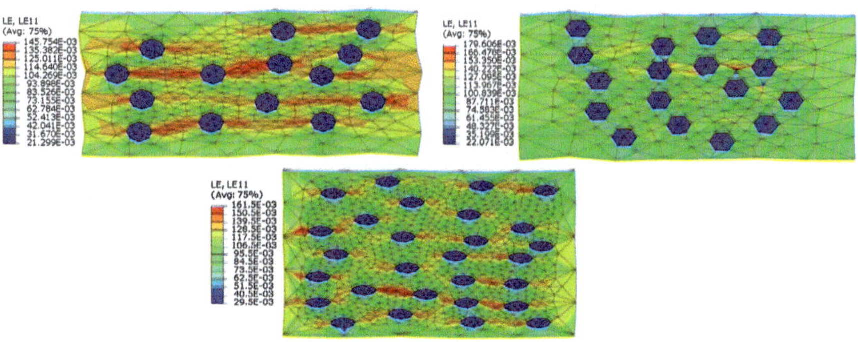

Fig. 8.32 True strain for *uniaxial* deformation along *x-axis* for the three composites

type for all three composites, were also determined. These are compiled for comparison in Table 8.5.

Comments on effect of inclusion shape on effective properties: The above results show that there is an influence of inclusion type shape on both elastic and maximum stress values for all load cases considered here. It is an important observation as it will help in optimal design and manufacture of this type of composites i.e. **PRAC**. Of course, the influence of RVE size and mesh densities,

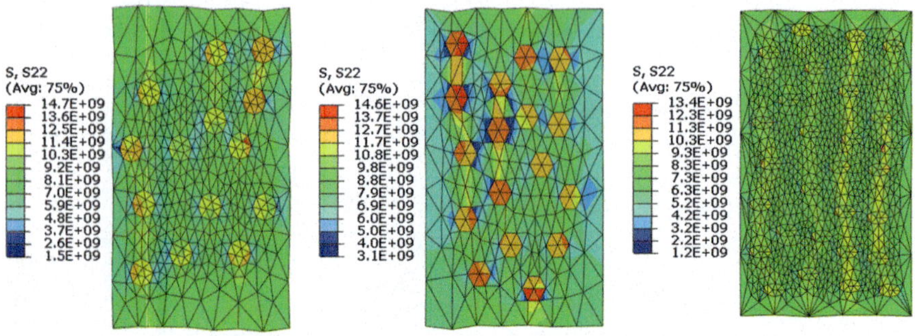

Fig. 8.33 True stress for *uniaxial* deformation along *y-axis* for the three composites

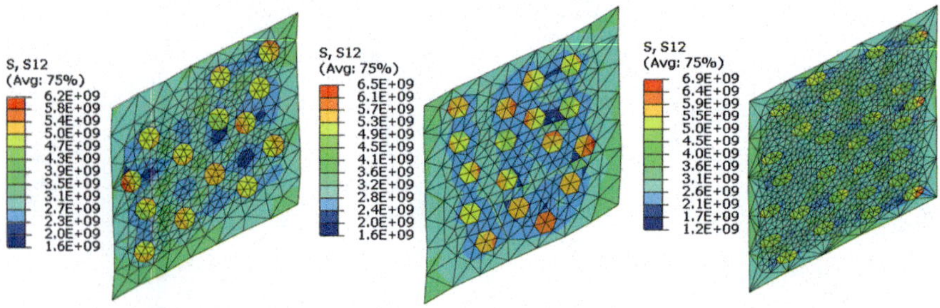

Fig. 8.34 True stress for the *shear* deformation on *xy-plane* for the three composites

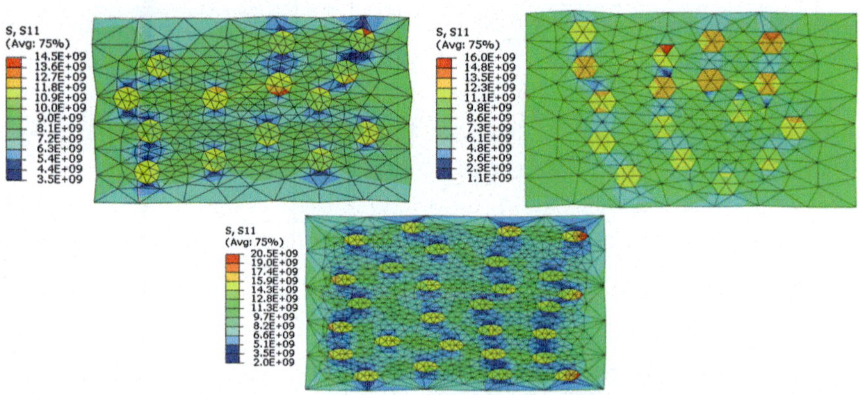

Fig. 8.35 True stress for the *uniaxial* deformation along *x-axis* for the three composites

on the effective properties have not been investigated in this study. These will have clear consequences on the effective properties determined through this computational method. For a definitive numerically determined Young's and shear moduli and maximum stresses, then the FEM user must include the effect of mesh density and RVE size on numerically predicted effective properties.

Table 8.5 Comparison of properties of the three composite types

Model variables	Load case	Circular (GPa)	Hexagonal (GPa)	Elliptical (GPa)
Young's modulus, E_{11}	Uniaxial X	79.04	81.97	83.41
Young's modulus, E_{22}	Uniaxial Y	79.16	82.83	77.72
Shear modulus, G_{12}	Shear XY	2.90	2.97	2.86
Maximum stress, $\sigma_{11,max}$	Uniaxial X	14.5	16.0	20.5
Maximum stress, $\sigma_{22,max}$	Uniaxial Y	14.7	14.6	13.4
Maximum stress, $\tau_{12,max}$	Shear XY	6.2	6.5	6.9

8.7 Periodic Boundary Conditions for Non-periodic Meshes

The above discussions on periodic BCs was focussed on periodic meshes and the approach used is also referred to as the *node-coupling* approach, where the kinematic coupling equations of node pairs were used to enforce the periodic BCs. However, there are specific categories of physical problems where such node-coupling constraints are not always possible. These specifically involve non-periodic meshes.

For example, if we are considering the microstructure of a trabeculae bone, foam structure, multi-component engine part, etc., it is hard to create periodic meshes for such structures. There are no clear boundary edges that can be enforced with periodic meshing arguments and hence only non-periodic meshes are possible. Such challenging mesh types have led to studies of alternative approaches to imposing periodic BCs on such virtual domains without having to meet the node-coupling requirement. Some of these alternative approaches are discussed in the following sections.

8.7.1 Gusev Constraint Elimination Technique

One of the key authors who started exploring a solution to the node-coupling bypass conundrum is Andrei Gusev [5]. He used what was described as a ***constraint elimination technique***. This method works by relating the degrees of freedom (dof) of a particular node, P^{-}, on a particular edge/surface to a corresponding node, P^{+}, located on an opposite surface of the RVE domain without specifying explicitly i.e. linear constraint equations. The requirement for node-coupling is not enforced however the two corresponding nodes have to share some geometric linearity (directly opposite each other).

8.7.2 Tyrus Polynomial Interpolation Technique

Other researchers have improved on the above Gusev approach. Of note is the work of Tyrus et al. [19], who used a ***polynomial interpolation technique*** to impose periodic boundary conditions on what was described as an *arbitrary 2-D mesh*, in order words, a non-periodic mesh. To do this, the displacement field along several boundaries of the RVE was interpolated.

Two distinguished regions were specified explicitly around the boundary upon which the interpolation was to be carried out. One region consisted of the location of stiff fibres along the boundary and the other region consisted of the matrix along the boundary. A first order polynomial interpolation (i.e. linear interpolation) was used to interpolate the displacement field along a stiff fibre edge whilst a polynomial function of order greater than 1 was used to interpolate the displacement field along the matrix edge.

The motivation behind the distinction between the interpolation order for the fibre and the matrix are:

(a) The fibre was considered as a very stiff member whose behaviour is essentially elastic; thus its displacement could always be described by a simple linear relationship.
(b) The displacement field around the matrix was observed to be nonlinear. Therefore, a cubic spline interpolant (Hermite shape function) was regarded as being adequate to interpolate the displacement field along the matrix region of the RVE edge.

After the polynomial interpolation had been carried out on predetermined edges, the displacement fields at these boundaries were transposed to the opposite boundaries of an RVE via explicitly specified mathematical relations. Although this method does not require identical meshes on the opposite edges of an RVE, the drawback is that the locations of nodes that lie on fibres and matrices along the edges of an RVE have to be known a priori. This adds to the computational cost of sorting the nodal positions along fibre-only and matrix-only boundary domains.

Figure 8.36 shows a comparison contour plot for a 2-D square RVE of a unidirectional composites with the Tyrus-style implementation on both periodic and non-periodic meshes. The contour plot shows no noticeable difference between the mesh types, hence confirming the viability of the approach for such micromechanical studies.

8.7.3 Nguyen Polynomial Interpolation Technique

The original Tyrus polynomial interpolation technique was further extended by Nguyen et al. [13]. Their approach still relies on ***polynomial interpolation shape functions***, however the requirement of prior knowledge of the location of fibres

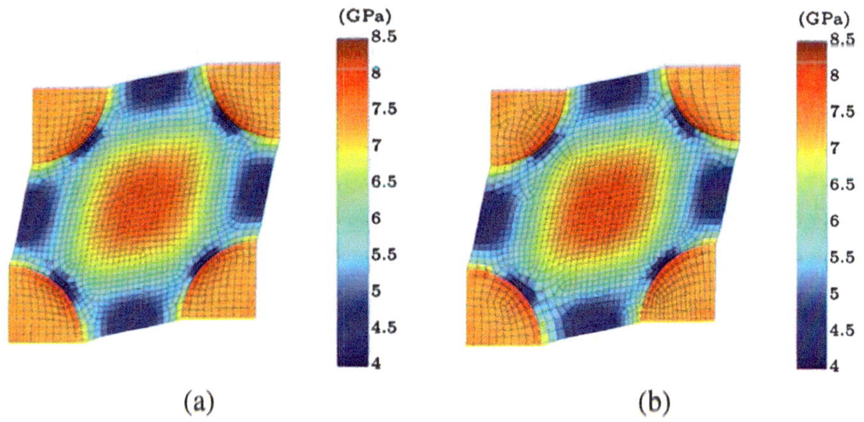

Fig. 8.36 Contour plots illustrating the local implementation of the polynomial interpolation technique on: (**a**) periodic; and, (**b**) non-periodic meshes (Images reproduced with kind permission of Tyrus et al. [19])

and matrices along the edges of an RVE was eliminated. In addition, the proposed method was also extended to 3-D RVEs. In their method, Lagrangian and cubic spline polynomial shape functions were used to interpolate the displacement fields along the edges of the RVE.

Furthermore, a surface interpolation function, referred to as a Coons patch, was used to interpolate the displacement field over the surfaces of the 3-D RVEs. They concluded that as the order of the polynomial shape function approached infinity, the accuracy of the prediction from the imposed PBC approached the exact solution. Comparison contour plots for 2-D and 3-D RVEs illustrating the Nguyen-approach are given in Figs. 8.37 and 8.38 respectively.

8.7.4 Akpoyomare Polynomial Interpolation Technique

Akpoyomare et al. [1] proposed yet another polynomial interpolation technique for imposing periodic BCs on arbitrary meshes. Their work was designed to be easily integrated within a virtual testing scheme for heterogeneous materials. The implementation was validated using an angle interlock textile composite: a significantly complex microstructure with 3-D RVE. The Akpoyomare approach was an extension of the Nguyen approach discussed in Sect. 8.7.3. Here, the authors used different interpolation functions for 2-D RVE edges and 3-D RVE surfaces. A ***piecewise cubic hermite interpolation function*** was used to interpolate edge displacements for a 2-D RVE.

On the other hand, the authors used a ***piecewise linear triangulation interpolation function*** to interpolate boundary displacements between one surface of the

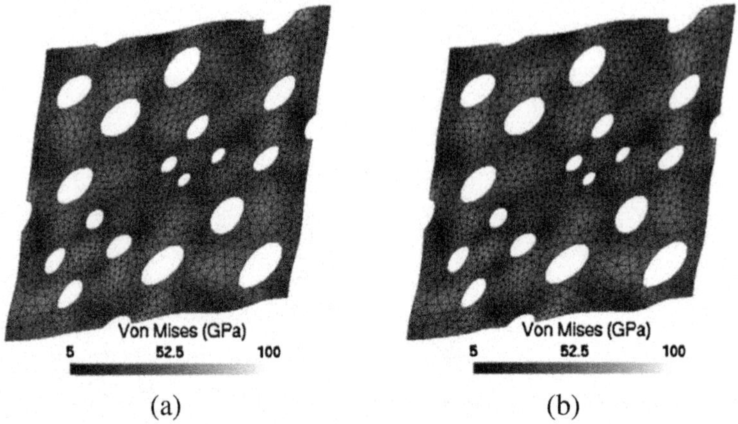

Fig. 8.37 Von mises stress contour plots on a non-periodic microstructure derived using the Nguyen approach with: (**a**) Lagrange; and, (**b**) the cubic spline interpolation polynomials (Images reproduced with kind permission of Nguyen et al. [13])

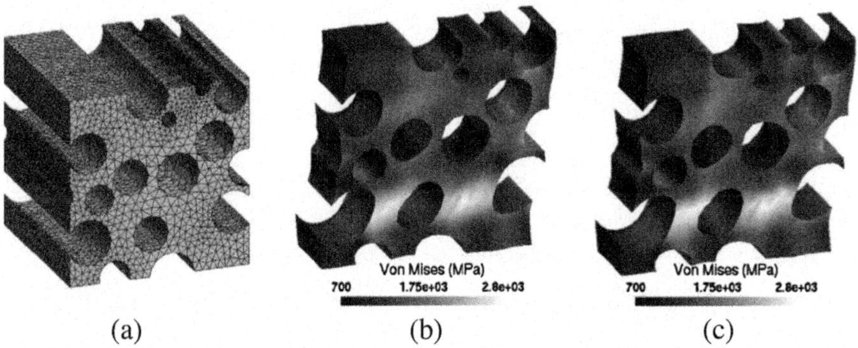

Fig. 8.38 Von Mises stress contour plots for a 3-D RVE based on the Nguyen approach: (**a**) non-periodic mesh (**b**) the Lagrange interpolation; and, (**c**) the cubic spline interpolation (Images reproduced with kind permission of Nguyen et al. [13])

3-D RVE and a directly parallel surface of same RVE. The Delaunay triangulation technique was employed by the authors to produce optimally shaped triangles needed for the piecewise linear triangulation.

8.7.5 Jacques Grid-Coupling Technique

This approach is similar to the node-coupling approach except that instead of individual nodes, an ***Euclidean grid structure*** is used. It was proposed and implemented by Jacques et al. [9]. This grid consists of the set of nodes that have

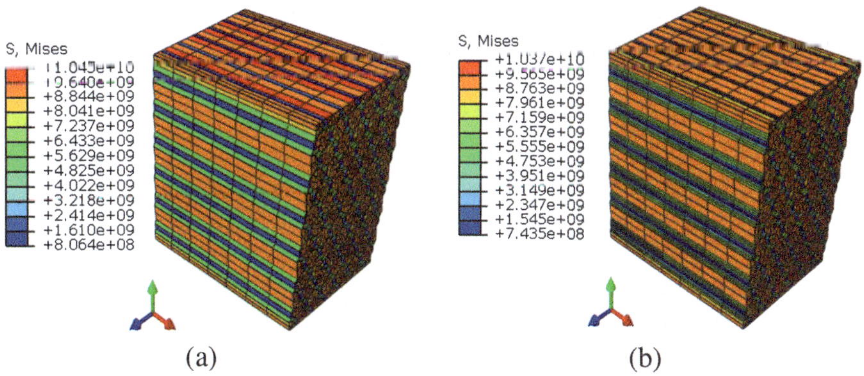

Fig. 8.39 Implementation of the Akpoyomare approach: illustration of shear deformation on the xy−plane for a UD composite using: (**a**) periodic; and, (**b**) non-periodic meshes (Images reproduced with kind permission of Akpoyomare et al. [1])

to be kinematically tied to corresponding nodes of the directly opposite face of the RVE. The sizes of the grid can vary, and the finer the grid size, the more the solution approaches the node-coupling solution.

Since the mesh structure is arbitrary, there may arise a case where the nodes on the corresponding face are not equal to the nodes on the initial grid structure. For this case, *reference nodes* are created artificially for the kinematic tying. The location of the reference node is defined by a what the authors called the ***Laplacian spatial averaging*** of the extra nodes that need to be coupled. This implies finding a position for the reference node that is equal to the average spatial location of the extra nodes. The authors used this approach to simulate the mechanical behaviour of a woven textile composites.

8.8 Conclusions

This chapter has presented a discussion on effect of boundary conditions on FEM models. In particular, this chapter has focussed on boundary condition types with specific attention given to periodic boundary conditions. The other types of boundary conditions are relatively easy to implement in FEM, however the periodic BCs present interesting challenges.

Existing FEM solvers have in-built methods for implementing multi-freedom constraint (MFC) equations. This approach has been demonstrated for implementing periodic BCs for periodic meshes. In doing this, the chapter has presented a user-defined MATLAB script, called **PBC2DGEN**, for automatically imposing periodic BCs on models with periodic meshes. Example problems drawn from micromechanics were presented to highlight the robustness of the **PBC2DGEN** algorithm.

The chapter concluded by discussing the existing methods for imposing periodic BCs on non-periodic meshes.

At the end of this chapter, the reader should be able to:

(a) Discuss the importance of boundary conditions in FEM.
(b) Explain the different types of boundary conditions.
(c) Analyze the mathematics of periodic boundary conditions.
(d) Demonstrate the use of **PBC2DGEN** within ABAQUS to solve simple micromechanics problems involving heterogeneous microstructure.
(e) Discuss ways of imposing periodic BCs on non-periodic meshes.

8.9 Problems: Boundary Conditions

Problem 8.1 A fracture test specimen is shown in Fig. 8.40. It is made of steel of Young's Modulus, $E_s = 210\,\text{GPa}$, and Poisson ratio, $\nu = 0.33$. It has dimensions of $40 \times 40 \times 5$ units^3. A crack is introduced on the left edge about the centreline with dimensions as shown.

The steel plate is subjected to a Dirichlet deformation on the front edges with a tensile test loading configuration. The displacement imposed is equivalent to a nominal strain of $\varepsilon_n = 10\%$. The back end of the plate has zero displacements and zero rotations in all directions. Initially treat the specimen as linear elastic.

(a) Carry out a mesh dependency study of maximum von Mises, σ_{11} and σ_{22} stresses. Consider up to 5 mesh densities in your study.
(b) For the convergent mesh, determine the volume averaged stresses and strains in the model and show this as a stress-strain plot.

Fig. 8.40 Dimensions, loads and boundary conditions of a plate with holes

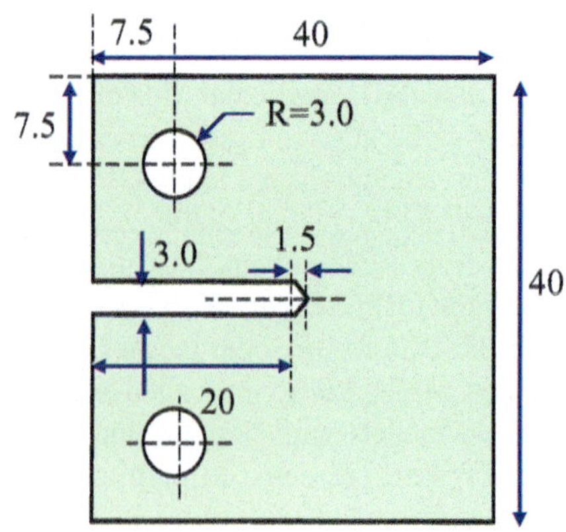

(c) Based on the volume-averaged stress-strain plot, determine the Young's moduli, E_{11} and E_{22}, for steel.
(d) Modify the model to capture the expected fracture behaviour by modelling the steel using the *plastic* and the *ductile damage* material models in ABAQUS.
(e) Determine the stress-strain plot for the failed specimen based on the von Mises stress. What is the failure stress from your plot?

Problem 8.2 A pipeline is made from steel with inside diameter, $D_o = 300$ mm and thickness, $t = 10$ mm. It sustains a gaseous pressure, $P_i = 5$ kPa. Steel properties are: Young's Modulus, $E = 210$ GPa and Poisson ratio, $\nu = 0.3$. Consider the steel material as an elasto-plastic material with yield stress, $\sigma_y = 1000$ MPa, and a plastic strain, $\varepsilon_p = 0.0$ at yield.

(a) Create an FEM model of the problem choosing an appropriate virtual domain to represent your model, using a Cartesian coordinate system.
(b) Determine the stress through-the-thickness of the tube along the $y-$ direction by extracting the values along a path specified through the thickness
(c) Consider revising the coordinate system to a cylindrical coordinate system with the origin at the centre. Display the contour plot of the model to obtain radial and tangential stresses
(d) Follow a linear path along the thickness of the tube, determine the plot of radial stress against position along the tube.
(e) Normally, the radial stress is zero on the outer wall. Does your plot from above confirm this?

Problem 8.3 An Arcan test is a special experiment used to investigate the shear response of brittle materials. It requires a butterfly shaped specimen. The dimensions of a typical specimen is given in Fig. 8.41 with holes representing gripping points.

(a) Develop the 2-D virtual domain of the butterfly shaped specimen.

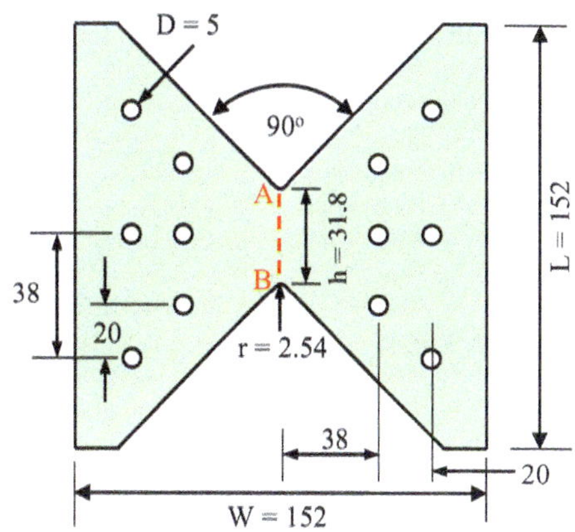

Fig. 8.41 The dimensions of an Arcan shear test butterfly specimen. Note: all dimensions are in millimetres

(b) If the test specimen is assumed to behave linear elastically with Young's Modulus, $E = 3.1$ GPa, and Poisson ratio, $\nu = 0.38$, choose an appropriate Dirichlet BC that will re-create the loading type, which prescribes shear stress within the gauge length of the specimen.
(c) Obtain the stress-strain plots of the material using elements only in the gauge-lengths region.
(d) Determine the elastic shear modulus of the material using simulation outputs.
(e) Assuming the test material is an elasto-plastic material, re-run the simulation (with the finest mesh of your choice to limit mesh effects) and obtain the stress-strain plot for this new case. The yield stress of the material is $\sigma_y = 60$ MPa.

Problem 8.4 Using a similar butterfly specimen as in Problem 8.3, investigate the effect of different stress concentrators to the shear response.

(a) Adopt the virtual domain, boundary conditions and results from Problem 8.3. This represents a reference/control case for this investigation.
(b) Develop other butterfly-shaped designs incorporating different types of stress concentrators: *circles, squares, hexagons, ellipses, triangles*. The radius of the circular stress concentrator, $R = 10$ mm. In all these cases, ensure that the areas of the different shapes are exactly the same, for the sake of comparison. The stress concentrator will be at the midpoint of the gauge length, as shown in Fig. 8.42.
(c) Consider the test material to be an elasto-plastic material with ductile failure where Young's Modulus, $E = 25$ GPa, and Poisson ratio, $\nu = 0.26$. Table 8.6 gives the yield stress as a function of plastic strain for the material. Run simulations for all the stress concentrator types.
(d) Using the simulation results above, generate the stress-strain plots for all the stress concentrators, and compare them with the reference case.
(e) To investigate the effect of increasing stress concentrator size on shear response, run simulations based on virtual domains that have circular stress concentrators but with increasing diameters: 3, 4, 5, 6, 7 and 8. Plot the resulting stress-strain graphs and comment on the observations you have made.

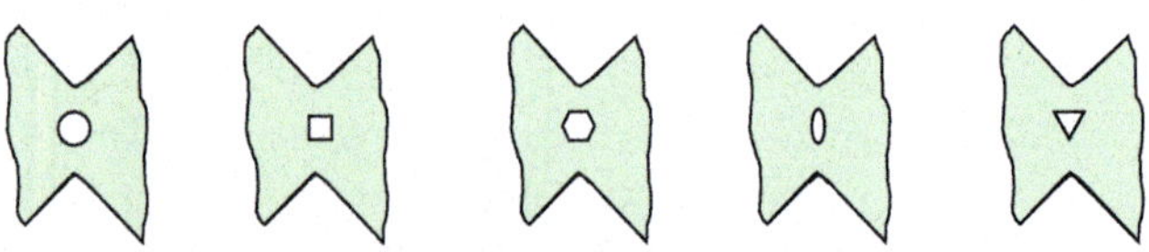

Fig. 8.42 Stress concentrators in the gauge section of Arcan shear specimens

Table 8.6 Elasto-plastic properties of tested material specimen

Stress, σ [MPa]	400	402	406	410	415
Plastic strain, ε_p	0.00	0.01	0.02	0.03	0.06

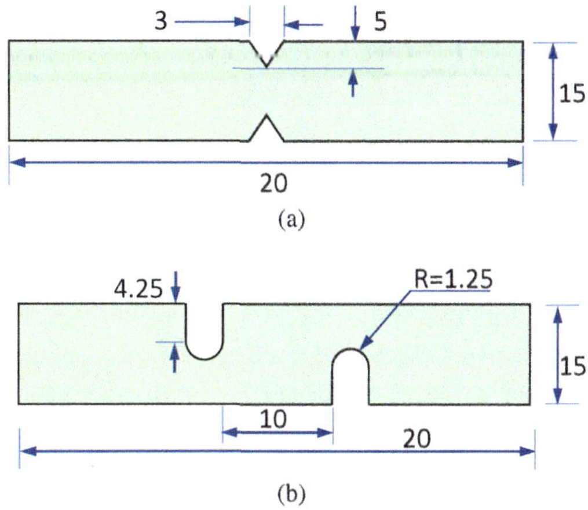

Fig. 8.43 Schematic representations of two double-notched shear specimens comprising of: (**a**) V- and (**b**) U-notches. Note: all dimensions are in millimetres

(f) To assess the nonlinear stress profiles around the stress concentrators, obtain the stress versus position plot along a path that runs through A–B for a stress concentrator of diameter 8 mm.

Problem 8.5 Two configurations of *double-notched shear* test specimens, with dimensions shown in Fig. 8.43, are to be used for this study. The first of the configurations has two directly opposite V-notches whilst the second has two U-notches at a staggered distance apart of 10 mm.

(a) Create a 2-D virtual domain for each of the two configurations.
(b) Isolate the gauge section and mesh it finely. Create five representative mesh densities.
(c) Determine the appropriate boundary conditions to be enforced on the model to cause the material to experience shear deformation.
(d) Run simulations assuming linear elastic behaviour of the test material with Young's modulus, $E = 28\,\text{GPa}$, and Poisson ratio, $\nu = 0.28$.
(e) Obtain stress-strain plots and shear modulus for all five mesh densities.

Problem 8.6 The following are a set of multi-freedom constraints (MFC) equations that define the boundary nodal interactions for a typical virtual domain. Write the ABAQUS-style linear constraint equation that will be used to update the ABAQUS input file to incorporate these MFC equations.

(a) $2u_3^{N_{10}} - 3u_3^{N_{11}} + 6u_3^{N_{12}} = 0$
(b) $5u_3^{N_{60}} - 3u_2^{N_{78}} = 2u_1^{N_{15}} - 3u_3^{N_{15}} + 3u_1^{N_6} - 10u_2^{N_{54}}$
(c) $\mathbf{u}^{N_1} - 5\mathbf{u}^{N_2} = 3\mathbf{u}^{N_3} - a\mathbf{u}^{N_4}$ where $\mathbf{u}$ = 2-D nodal displacement vector
(d) $u_2^{N_4} - 3u_1^{N_2} = 2u_1^{N_5} - 3u_2^{N_{16}}$ where $u_2^{N_{16}} = ce^2u_3^{N_3} - bu_1^{N_4}$ with b and c being material constants
(e) $u_2^{N_2} - au_3^{N_5} = \left(1 + a^2\right) u_3^{N_8} - (ab)^2\, u_1^{N_9}$ where a and b are kinematic constants

(f) $u_1^{N_6} - \left(\dfrac{b}{a}\right)^3 u_2^{N_7} + b\,(ab)^3\, u_2^{N_7} - (1 - b^2)\left(c - \sqrt[1/3]{b^4}\right) u_1^{N_6} = 0$ where a, b, and c are geometric constants

Problem 8.7 A 2-D RVE window of a particulate composite, shown in Fig. 8.44, shows different particle geometries dispersed within a polymeric square window. This represents a heterogeneous material requiring therefore the use of periodic BCs for obtaining the effective properties. For the given system as shown:

(a) Write the Multi-freedom constraint (MFC) equations for boundary nodes.
(b) Convert all the MFC equations of all boundary nodes (excluding retained and slave nodes) to ABAQUS-style linear constraint equations.

Problem 8.8 To determine the effective elastic properties of continuous fibre reinforced composite, a microscale 2-D $50 \times 50\,\mu m^2$ RVE window of the test composite was chosen, as shown in Fig. 8.45.

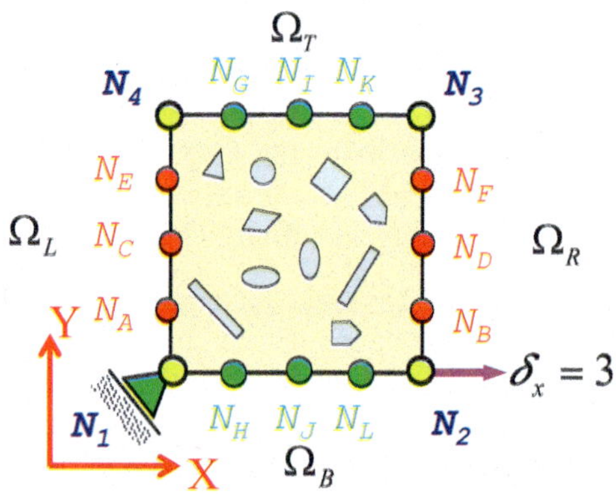

Fig. 8.44 A 2-D RVE window of a particulate composite

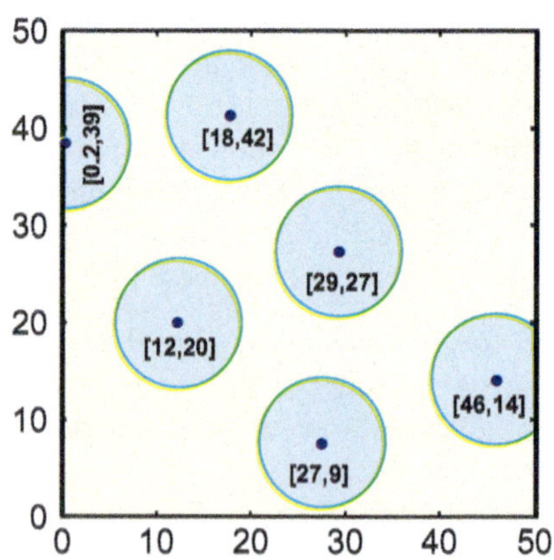

Fig. 8.45 A 2-D RVE of a unidirectional composite where the *circles* are fibres and the *square* is the matrix. *Coordinates* indicate *centre coordinates* of the circular fibres

Table 8.7 The elastic properties of the constituents of a unidirectional composite

Material	Young's Modulus, E (GPa)	Poisson ratio
E-glass fibre	75	0.25
Polypropylene	1.4	0.40

Table 8.8 The elastic properties of the constituents of a unidirectional composite

Material	Y. Modulus, E (GPa)	P. ratio, ν	σ_y (MPa)	Pl. strain ε_p
Silicon Carbide (SiC)	450	0.15		
6061 aluminium	1.4	0.33	280	0.00
			285	0.05
			282	0.15

The fibre reinforcement is E-glass fibre with a volume fraction of the fibre, $V_f = 35\%$, and the matrix is polypropylene. Elastic properties of the constituents are given in Table 8.7. The average diameter of the fibre, $D_f = 14\,\mu\text{m}$.

(a) Create the 2-D virtual domain of the unidirectional composite. *Hint: If using ABAQUS, use the* `Extrude cut` *option to trim off the overlapping fibres.*
(b) Run **PBC2DGEN** for these load cases: *uniaxial deformation* along the $x-$ and $y-$ axes as well as *shear deformation* on the $xy-$plane.
1. Determine the in-plane effective properties of the test composite.

Problem 8.9 A $70 \times 70\,\mu\text{m}^2$ particulate composite material is made up of aluminium inclusions with SiC particles. The properties of the inclusions are given in Table 8.8. The volume fraction of the SiC particles is $V_p = 15\%$. Assume that the aluminium shows an elasto-plastic mechanical response with ductile failure, while the SiC particles shows a brittle linear elastic response.

To assess the effect of particle shapes on the effective properties as well as yield behaviour of the test composite, the following six particle shapes were modelled: *circles, hexagons, triangles, squares, ellipses,* and *multi-shape particle types*, all within a single RVE window.

The particle shapes and the coordinate positions of the centres of the circumscribed circle, enclosing the particles is given in Fig. 8.46.

(a) Create the 2-D virtual domains of all six-particle types of the composite.
(b) With the six composite types, run **PBC2DGEN** for these load cases: *uniaxial deformation* along the $x-$ and $y-$ axes, as well as *shear deformation* on the $xy-$plane.
(c) Determine the volume-average stress-strain plots per load type showing results from all six composite types.
(d) Determine the in-plane effective properties (E_{11}, E_{22} and G_{12}) of the composite.
(e) Plot the histogram of effective properties (on the $y-$ axis) and particle types (on the $x-$ axis) for this composite. Comment on your findings.

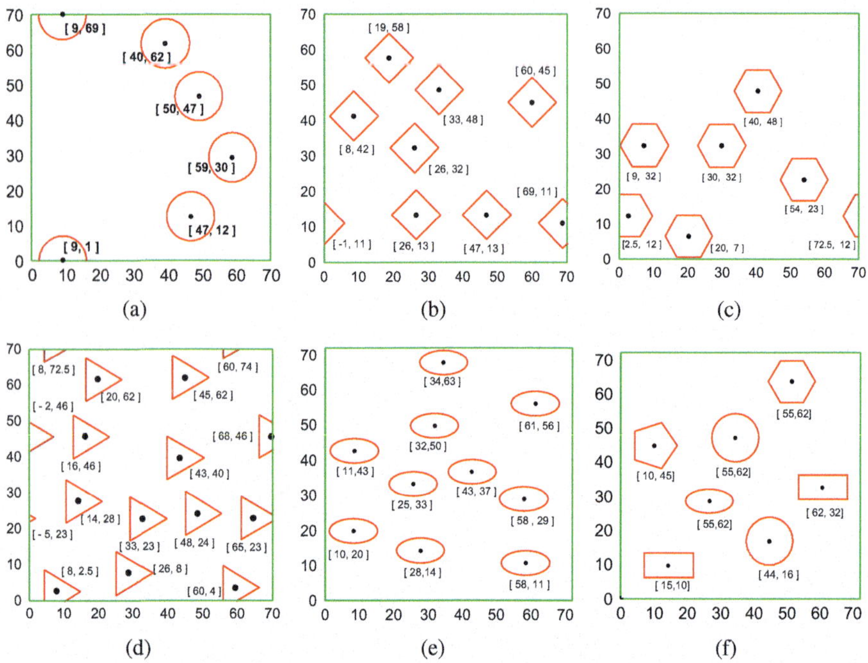

Fig. 8.46 An aluminium-SiC particulate composite with SiC represented by: (**a**) circular; (**b**) square; (**c**) hexagonal; (**d**) triangular; (**e**) elliptical; and, (**f**) multi-shaped particles

Note: For the effective properties to be comparable across all particles, the area fractions of all particles must be kept constant, similar to Example 8.6.

References

1. Akpoyomare, A., Okereke, M., Bingley, M.: Virtual testing of composites: imposing periodic boundary conditions on general finite element meshes. Compos. Struct. **160**, 983–994 (2017). http://dx.doi.org/10.1016/j.compstruct.2016.10.114, http://www.sciencedirect.com/science/article/pii/S0263822316322711
2. Al Kassem, G., Weichert, D.: Micromechanical material models for polymer composites through advanced numerical simulation techniques. PAMM **9**(1), 413–414 (2009). http://dx.doi.org/10.1002/pamm.200910180
3. de Morais, A.B.: Transverse moduli of continuous-fibre-reinforced polymers. Compos. Sci. Technol. **60**(7), 997–1002 (2000). http://dx.doi.org/10.1016/S0266-3538(99)00195-5, http://www.sciencedirect.com/science/article/pii/S0266353899001955
4. Gitman, I., Askes, H., Sluys, L.: Representative volume: existence and size determination. Eng. Fract. Mech. **74**(16), 2518–2534 (2007). http://dx.doi.org/10.1016/j.engfracmech.2006.12.021, http://www.sciencedirect.com/science/article/pii/S0013794406004772
5. Gusev, A.A.: Numerical identification of the potential of whisker- and platelet-filled polymers. Macromolecules **34**(9), 3081–3093 (2001). http://dx.doi.org/10.1021/ma001979b
6. He, Q.C.: Effects of size and boundary conditions on the yield strength of heterogeneous materials. J. Mech. Phys. Solids **49**(11), 2557–2575 (2001). http://dx.doi.org/10.1016/S0022-5096(01)00068-0, http://www.sciencedirect.com/science/article/pii/S0022509601000680. The Jean-Paul Boehler Memorial Volume

7. Hibbeler, R., Sekar, K.: Mechanics of Materials. Always Learning. Pearson Education Canada (2013). https://books.google.co.uk/books?id=Vu12ngEACAAJ
8. Huet, C.: Coupled size and boundary-condition effects in viscoelastic heterogeneous and composite bodies. Mech. Mater. **31**(12), 787–829 (1999). http://dx.doi.org/10.1016/S0167-6636(99)00038-1, http://www.sciencedirect.com/science/article/pii/S0167663699000381
9. Jacques, S., Baere, I.D., Paepegem, W.V.: Application of periodic boundary conditions on multiple part finite element meshes for the meso-scale homogenization of textile fabric composites. Compos. Sci. Technol. **92**, 41–54 (2014). http://dx.doi.org/10.1016/j.compscitech.2013.11.023, http://www.sciencedirect.com/science/article/pii/S0266353813004594
10. Kanit, T., Forest, S., Galliet, I., Mounoury, V., Jeulin, D.: Determination of the size of the representative volume element for random composites: statistical and numerical approach. Int. J. Solids Struct. **40**(13–14), 3647–3679 (2003). http://dx.doi.org/10.1016/S0020-7683(03)00143-4, http://www.sciencedirect.com/science/article/pii/S0020768303001434
11. Kouznetsova, V., Brekelmans, W.A.M., Baaijens, F.P.T.: An approach to micro-macro modeling of heterogeneous materials. Comput. Mech. **27**(1), 37–48 (2001). http://dx.doi.org/10.1007/s004660000212
12. Melro, A., Camanho, P., Pinho, S.: Generation of random distribution of fibres in long-fibre reinforced composites. Compos. Sci. Technol. **68**(9), 2092–2102 (2008). http://dx.doi.org/10.1016/j.compscitech.2008.03.013, http://www.sciencedirect.com/science/article/pii/S0266353808001048
13. Nguyen, V.D., Béchet, E., Geuzaine, C., Noels, L.: Imposing periodic boundary condition on arbitrary meshes by polynomial interpolation. Comput. Mater. Sci. **55**, 390–406 (2012). http://dx.doi.org/10.1016/j.commatsci.2011.10.017, http://www.sciencedirect.com/science/article/pii/S0927025611005866
14. Okereke, M., Akpoyomare, A.: A virtual framework for prediction of full-field elastic response of unidirectional composites. Comput. Mater. Sci. **70**, 82–99 (2013). http://dx.doi.org/10.1016/j.commatsci.2012.12.036, http://www.sciencedirect.com/science/article/pii/S0927025612007744
15. Otiaba, K.C., Okereke, M., Bhatti, R.: Numerical assessment of the effect of void morphology on thermo-mechanical performance of solder thermal interface material. Appl. Thermal Eng. **64**(1), 51–63 (2014)
16. Qin, S., Chen, C., Zhang, G., Wang, W., Wang, Z.: The effect of particle shape on ductility of sicp reinforced 6061 al matrix composites. Mater. Sci. Eng.: A **272**(2), 363–370 (1999)
17. Sun, C., Vaidya, R.: Prediction of composite properties from a representative volume element. Compos. Sci. Technol. **56**(2), 171–179 (1996). http://dx.doi.org/10.1016/0266-3538(95)00141-7, http://www.sciencedirect.com/science/article/pii/0266353895001417
18. Terada, K., Hori, M., Kyoya, T., Kikuchi, N.: Simulation of the multi-scale convergence in computational homogenization approaches. Int. J. Solids Struct. **37**(16), 2285–2311 (2000). http://dx.doi.org/10.1016/S0020-7683(98)00341-2, http://www.sciencedirect.com/science/article/pii/S0020768398003412
19. Tyrus, J., Gosz, M., DeSantiago, E.: A local finite element implementation for imposing periodic boundary conditions on composite micromechanical models. Int. J. Solids Struct. **44**(9), 2972–2989 (2007). http://dx.doi.org/10.1016/j.ijsolstr.2006.08.040, http://www.sciencedirect.com/science/article/pii/S0020768306003489
20. van der Sluis, O., Schreurs, P., Brekelmans, W., Meijer, H.: Overall behaviour of heterogeneous elastoviscoplastic materials: effect of microstructural modelling. Mech. Mater. **32**(8), 449–462 (2000). http://dx.doi.org/10.1016/S0167-6636(00)00019-3, http://www.sciencedirect.com/science/article/pii/S0167663600000193
21. Xia, Z., Zhang, Y., Ellyin, F.: A unified periodical boundary conditions for representative volume elements of composites and applications. Int. J. Solids Struct. **40**(8), 1907–1921 (2003). http://dx.doi.org/10.1016/S0020-7683(03)00024-6, http://www.sciencedirect.com/science/article/pii/S0020768303000246

Chapter 9
Material Response: Measures of Stress and Strain

Abstract Amongst the many pillars upon which the FEM solutions stand is the pillar of *material response*. This defines the physical behaviour of the material type under investigation in the FEM problem. It is under this pillar that one distinguishes between rubbery, elastic, nonlinear, fracture and interface failure mechanisms. It is a key component of the FEM problem and must be correctly defined if one is to obtain reliable solutions. A common theme for this pillar of the FEM process is what is described as *predictive modelling*, which is the use of computational methods to determine the material behaviour of a given material. In this chapter, we have presented the principles of the material response module of an FEM scheme. The focus here is exploiting the principles of continuum mechanics in defining the response of a material body. Specifically, this chapter introduces the kinematics of finite deformation of a material body; measures of strains and stresses; and concludes with the practical formulations of stresses needed by engineers during the design process. Such practical stress formulations include: principal stresses, von Mises stresses, etc. This chapter lays down the theory needed to understand the material model implementations in the finite element solver.

Keywords Material response • Finite deformation • Deformation gradient tensor • Strain tensor • Strain tensor • Stress measures

9.1 Introduction

Consider a virtual domain, shown in Fig. 9.1, bounded by Ω_{before} which represents the boundaries of a given material system *before* deformation. Let us impose a distributed load, $\boldsymbol{F}$, on the left hand side of the structure such that we can identify three possible deformation profiles. Whilst *Materials 1* and *2* are dominated by a form of compressive deformation, *Material 3* shows a dominant shear deformation. The material system is bounded by a virtual domain *after* the deformation represented by Ω_{after}. This deformed domain is different for all three material types. The answer to why this is the case lies in an understanding of the material composition of all three material systems.

The differences in deformation profiles are quantified by constructing a material model to capture the individual composition of the three materials. It is the goal

M. Okereke, S. Keates, *Finite Element Applications*, Springer Tracts in Mechanical Engineering, https://doi.org/10.1007/978-3-319-67125-3_9

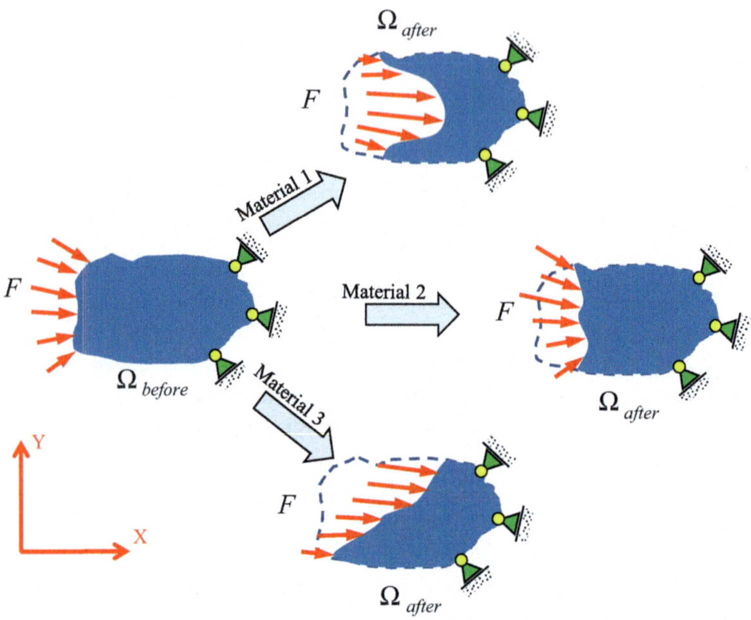

Fig. 9.1 Schematic representations of different material responses under the effect of a distributed load

of the FEM user to choose the correct material response to capture the expected material behaviour. The study of *material responses* here focuses on the numerical representation of the material behaviours that define how a structure can deform under the effect of load(s).

To describe the material response, we will seek to provide answers to the following questions.

- What are the reference frames that are required to describe the observed material response?
- What are the measures that describe the observed material response?
- What are the constitutive formulations that described the observed material response?
- How do FEM solvers implement the constitutive formulations into a material mode?
- What are the in-built material models in a typical FEM solver like `ABAQUS`?
- How can one extend the material models available in an FEM solver like `ABAQUS`?

In this chapter and the next, we will provide answers to these questions. The current chapter addresses the question of defining parameters needed to describe the deforming body. In the next chapter, the constitutive material models and their implementation within an FEM solver will be presented.

9.2 Chapter Objectives

At the end of this chapter, the reader should be able to:

(a) Define the kinematics of deformation needed in describing material response within the FEM process.
(b) Understand the different measures of representing the material response in an FEM scheme.
(c) Explore the different measures of strain needed to describe material response.
(d) Derive the different measures of stress that describe finite deforming systems.
(e) Introduce the different practical formulations of stresses needed by design in engineering to aid the design process for material systems subjected to multiple load histories.

9.3 Kinematics of Deformation

Continuum mechanics provides the fundamental principles for describing the material behaviours of different materials. Let us first isolate a *material body*, shown in Fig. 9.2, which represents either a form of solid, liquid or gas system. Consider the material body is bounded by a material domain, Ω, such that its microstructure consists of sets of elements, $\boldsymbol{X}$, called *particles* or *material points*.

On a reference frame of choice, *XYZ*, each of the material points can be related to a coordinate position on defined regions of the physical space. Each of these coordinate positions are measured with respect to the origin of the *XYZ* reference frame. The *kinematics of deformation* introduces the principles for describing the changes observed in material systems under the effect of loads.

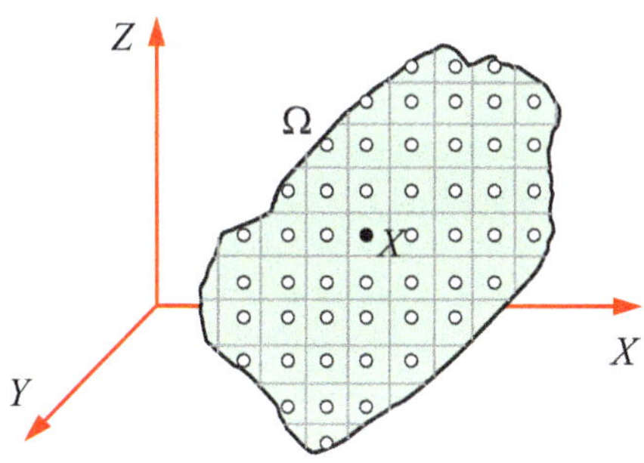

Fig. 9.2 A material body consisting of material points, $\boldsymbol{X}$, at defined coordinate positions

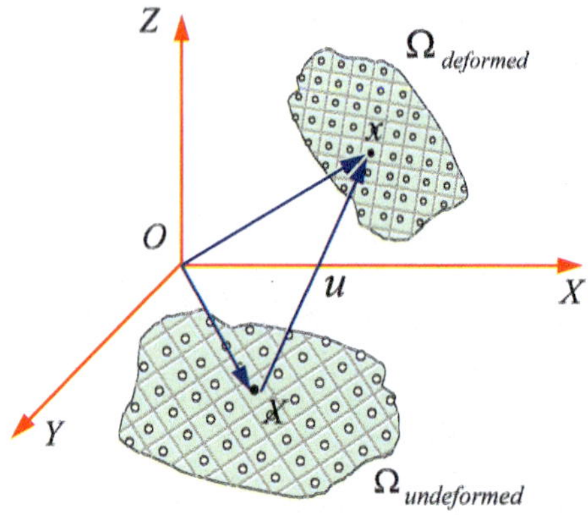

Fig. 9.3 A material body subjected to a displacement vector, $\boldsymbol{u}$, displacing it from its original (*material*) configuration, $\boldsymbol{X}$, to a new *spatial* configuration, $\boldsymbol{x}$

9.3.1 Material Configuration and Displacement

If we consider the material body of Fig. 9.3 to be a *rigid body*, the material points remain *locked* in their *original* coordinate positions during the deformation. However, for the purposes of defining material behaviour, we consider the material body to be *deformable*, i.e. it changes shape and size under the effect of loads. For a deformable body, the coordinate positions of the material points in Fig. 9.2 change in space and in time. At any given time, under the effect of external loads, the assembly of material points of the material body defines the *material configuration* of the body.

If the new configuration is defined by a new set of material points, $\boldsymbol{x}$, distinct from the initial reference configuration, $\boldsymbol{X}$, it is possible to mathematically relate the $\boldsymbol{x}$ material points to the original $\boldsymbol{X}$ material points according to the equations:

$$\boldsymbol{x} = \boldsymbol{\kappa}(\boldsymbol{X}) \qquad \text{and} \qquad \boldsymbol{X} = \boldsymbol{\kappa}^{-1}(\boldsymbol{x}) \tag{9.1}$$

where $\boldsymbol{\kappa}$ is the vector function that maps the $\boldsymbol{x}$ positions to the $\boldsymbol{X}$ positions. The above assumes that the mapping function is *invertible*. The above equation implies that the material points on a deformed material body can be related to their original positions and vice versa. It is this principle that drives the formulation of *stress* and *strain* measures that will be presented later.

Consider the material body experiences a change from its reference position, $\boldsymbol{X}$, to a new configuration, $\boldsymbol{x}$, as shown in Fig. 9.3. The change in configuration is caused by a *displacement vector*, $\boldsymbol{u}$, which consists of simultaneous *translation* and *rotation* of the material points relative to the origin of the fixed reference frame.

For a given displacement vector, $\boldsymbol{u}$, as shown in Fig. 9.3, the map of the material points at both original and deformed configurations becomes:

$$\boldsymbol{x} = \boldsymbol{\kappa}(\boldsymbol{X}) + \boldsymbol{u} \tag{9.2}$$

We will now extend the above discussion to account for the fact that the material points in the new configuration are a function not only of coordinate positions, but also time. Thus, Eq. 9.1 of the material body can be extended to incorporate time-dependence of the deformation, such that we write:

$$\boldsymbol{x} = \boldsymbol{\kappa}\,(\boldsymbol{X}, t) \qquad \text{and} \qquad \boldsymbol{X} = \boldsymbol{\kappa}^{-1}\,(\boldsymbol{x}, t) \tag{9.3}$$

9.3.2 Material and Spatial Reference Frames

In Continuum mechanics, two material configurations are essential in describing the kinematics of deformation of a material body. The first is the original configuration of the body before applying a displacement vector, called the *reference configuration*. It is a baseline set of material points of the material body from which subsequent measures of deformation are taken. From Eq. 9.3, this configuration coincides with time, $t = 0$. It is also the reference at which $\boldsymbol{u} = 0$, from Eq. 9.2, in which case: $\boldsymbol{x} = \boldsymbol{X}$.

Once time has elapsed, we define a new material configuration called the *current configuration*. From Fig. 9.3, the reference and current configurations are bounded by the $\Omega_{undeformed}$ and $\Omega_{deformed}$ boundary domains. It is customary in continuum mechanics to associate two coordinate systems to the two material configurations to quantitatively describe the kinematics of motion of the material body.

In the reference or original configuration, the *material reference frame* is used to describe the positions of the material points within the physical space. It is denoted by upper case letters. For example, for a Cartesian representation, the material reference frame is the XYZ–axes. The set of positions that describe a material body is represented in this frame by the vector, $\boldsymbol{X}$. Since the material coordinate is static and does not follow changing material configurations of the material body, it is said to be *time-independent*.

On the other hand, the *spatial* or *current reference frame* is a coordinate system that describes the current coordinate positions of a material body under the effect of external or internal loads. It is the cumulative location of the particles that make up the material body after each of the particles has been acted upon by a displacement vector, $\boldsymbol{u}$. It is represented by lower case letters. For the Cartesian coordinate representation, the spatial reference frame for a set of particles, $\boldsymbol{x}$, is the xyz–axes. The spatial coordinates are *time-dependent* as their values change with changing configurations of the material body.

We can illustrate these two reference frames by considering the fixed axis rotation of the bar shown in Fig. 9.4, rotating with angular velocity, ω. The bar is kept rotating by a perpendicular force vector, $\boldsymbol{F}$. The material reference frame is the XY-axes attached to the bar such that at all stages during the rotation, the axes rotate with the material. The material points of the body, with respect to the XY-axes, will remain the same at all three states of motion. The spatial reference frame is the xy-axes. It is similar to the global reference frame. With changing positions during motion of the bar, the material points coordinate positions will be changing with respect to this spatial (xy-axes) coordinate system.

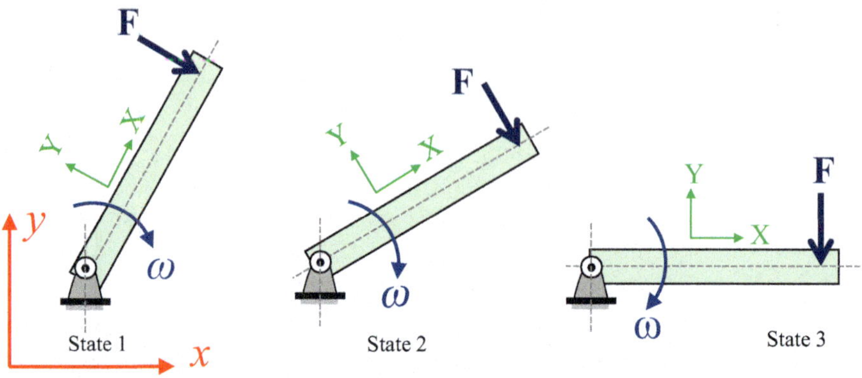

Fig. 9.4 An illustration of material (XY-) and spatial (xy-) reference frames for a fixed axis rotation of a slender bar during three states of the motion

When a physical property of a material body is expressed in terms of material coordinates, $\boldsymbol{X}$, and time, t, it is said to be given by *material*, or *Lagrangian, description*. In this description, the observer follows the material point as it moves along space and time. On the other hand, the *Eulerian*, or *spatial, description* describes the physical property in terms of the spatial coordinate system. This means the observer sits at a fixed point and assesses/observes the changes in configuration of the material body in space and time.

9.3.3 Deformation Gradient

Consider the material body of Fig. 9.2 being subjected to a displacement vector, $\boldsymbol{u}$. The material will experience deformation that comprises a change in shape and size. To quantify the deformation, let us isolate a line segment, $\overline{AB}$ of length, $\mathrm{d}\boldsymbol{X}$ in the material reference frame and $\overline{A'B'}$ of length $\mathrm{d}\boldsymbol{x}$ in the spatial reference frame, as shown in Fig. 9.5. The line $\overline{AB}$ is deformed, resulting in line $\overline{A'B'}$ such that all the material points that make up the material body will deform according to the equation: $\boldsymbol{x} = \boldsymbol{X} + \boldsymbol{u}$.

We introduce the *deformation gradient tensor* (or, for short, *deformation gradient*), $\boldsymbol{F}$, to represent the mapping function that links the deformation of the material reference frame to that of the spatial reference frame. This implies that any infinitesimal deformation, $\mathrm{d}\boldsymbol{X}$ on the material reference frame is transformed to its equivalent spatial reference frame deformed state, $\mathrm{d}\boldsymbol{x}$, by the *deformation gradient*, and defined according to the equation:

$$\mathrm{d}\boldsymbol{x} = \boldsymbol{F}\mathrm{d}\boldsymbol{X} \qquad \Longrightarrow \qquad \boldsymbol{F} = \frac{\partial \boldsymbol{x}}{\partial \boldsymbol{X}} = \frac{\partial (\boldsymbol{X} + \boldsymbol{u})}{\partial \boldsymbol{X}} = \boldsymbol{I} + \frac{\partial \boldsymbol{u}}{\partial \boldsymbol{X}} \tag{9.4}$$

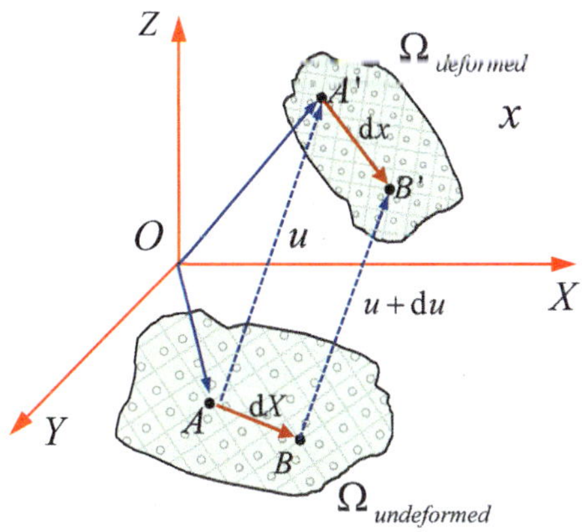

Fig. 9.5 An illustration of the deformation of a line segment, $\overline{AB}$ from material to spatial reference frames

where $\boldsymbol{I}$ is an identity matrix. The deformation gradient, $\boldsymbol{F}$, is the key parameter that defines the deformation of a material body. In its most general form (for a 3D system), $\boldsymbol{F}$ consists of nine components defined in terms of time, t. It is a second-order tensor and according to Eq. 9.4, the mapping function, $\boldsymbol{F}$, is a *linear* transformation function that maps $\mathrm{d}\boldsymbol{X}$ deformation to $\mathrm{d}\boldsymbol{x}$.

Using the left-hand side expression of Eq. 9.4, we can re-write it in component form for a Cartesian coordinate system representation in 3D of the material body thus:

$$\begin{bmatrix} \mathrm{d}x \\ \mathrm{d}y \\ \mathrm{d}z \end{bmatrix} = \begin{bmatrix} F_{xx} & F_{xy} & F_{xz} \\ F_{yx} & F_{yy} & F_{yz} \\ F_{zx} & F_{zy} & F_{zz} \end{bmatrix} \begin{bmatrix} \mathrm{d}X \\ \mathrm{d}Y \\ \mathrm{d}Z \end{bmatrix} \tag{9.5}$$

The deformation gradient represents the behaviour of a material point with respect to another neighbourhood point, hence some authors describe it as a *two-point tensor* that links material points in two separate material configurations [5]. The deformation gradient can also be expressed in terms of the partial derivatives of the terms of the material points in the spatial reference frame thus:

$$\boldsymbol{F} = \frac{\partial \boldsymbol{x}}{\partial \boldsymbol{X}} \qquad \Longrightarrow \qquad \boldsymbol{F} = \nabla \boldsymbol{x} \tag{9.6}$$

where $\nabla = \dfrac{\partial}{\partial \boldsymbol{X}}$ is the transpose of the vector differential operator. The component-wise expansion of Eq. 9.5 thus becomes:

$$\begin{bmatrix} \mathrm{d}x \\ \mathrm{d}y \\ \mathrm{d}z \end{bmatrix} = \begin{bmatrix} \frac{\partial x}{\partial X} & \frac{\partial x}{\partial Y} & \frac{\partial x}{\partial Z} \\ \frac{\partial y}{\partial X} & \frac{\partial y}{\partial Y} & \frac{\partial y}{\partial Z} \\ \frac{\partial z}{\partial X} & \frac{\partial z}{\partial Y} & \frac{\partial z}{\partial Z} \end{bmatrix} \begin{bmatrix} \mathrm{d}X \\ \mathrm{d}Y \\ \mathrm{d}Z \end{bmatrix} \tag{9.7}$$

Based on Eq. 9.3, let us assume mapping function, $\boldsymbol{\kappa}$, exists and also is invertible. The same philosophy will also hold for the mapping deformation gradient tensor, the material points have been displaced by a displacement vector, $\boldsymbol{u}$. As a result, we would expect there to exist an *inverse deformation gradient tensor*, $\boldsymbol{F}^{-1}$, which maps material points at the material reference frame, $\boldsymbol{X}$, to the spatial reference frame, $\boldsymbol{x}$. In other words, the inverse deformation gradient tensor transforms the spatial line segment, $\overline{A'B'}$ of Fig. 9.5 into the material line segment, $\overline{AB}$. Hence, the tensorial expression of the inverse deformation gradient is:

$$\boldsymbol{F}^{-1} = \frac{\partial \boldsymbol{X}}{\partial \boldsymbol{x}} \quad \Longrightarrow \quad \boldsymbol{F}^{-1} = \nabla^{-1}\boldsymbol{x} \quad \text{or} \quad \mathrm{d}\boldsymbol{X} = \boldsymbol{F}^{-1}\mathrm{d}\boldsymbol{x} \tag{9.8}$$

9.3.4 Rotation and Stretch Tensors

Its important at this point to introduce two new terms associated with kinematics of deformation of a material body: *rotation* and *stretch*. Rotation is a measure of rigid body deformation of the body such that only angular changes occur during the deformation of the body without any change in size or shape. Stretch on the other hand refers to a change in size of the body independent of any changes in orientation. Generally, when a body deforms, both stretch and rotation measures are required to describe comprehensively the deformation of the body. These two terms arise due to a local motion of material points within a material body.

Consider a material body bounded by the $\Omega \subset \Re^n$ such that in its undeformed state, the boundary domain is bounded by $\Omega_{undeformed}$, while in the deformed state, the boundary domain is bounded by $\Omega_{deformed}$, as shown in Fig. 9.6. If the undeformed material body is subjected to a deformation gradient, $\boldsymbol{F}$, the resulting deformed configuration is given as a cumulative effect of *rotation* and *stretch*. The principle of *polar decomposition* is introduced to separate the contributions of the stretch and rotation tensors. The mapping deformation gradient, $\boldsymbol{F}(\boldsymbol{X}, t)$, can be decomposed into *pure stretch*, $\boldsymbol{U}$ or $\boldsymbol{V}$, and *pure rotation*, $\boldsymbol{R}$, components.

The first form of polar decomposition requires that the undeformed material body is first subjected to an orthogonal rotation tensor, $\boldsymbol{R}$, so that there is a resultant change of *local orientation*. $\boldsymbol{R}$ is a measure of this *local change in orientation*. After operating on the material body with the rotation tensor, the *intermediate deformed*

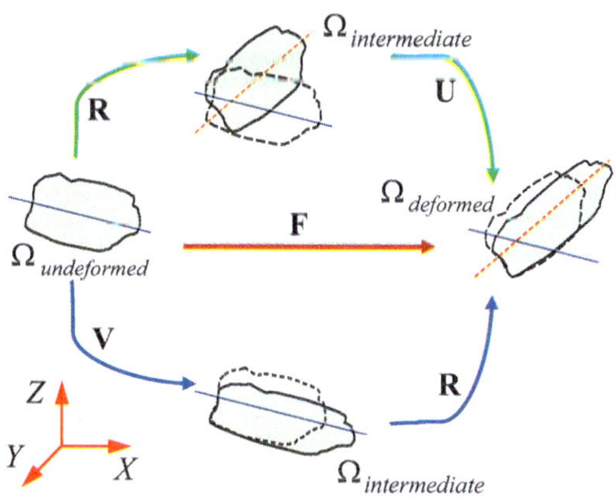

Fig. 9.6 An illustration of polar decomposition of the deformation gradient, $\boldsymbol{F}$, into rotation, $\boldsymbol{R}$, and stretch, $\boldsymbol{U}$ or $\boldsymbol{V}$, tensors

domain, $\Omega_{intermediate}$, is further operated by a *right* (or *material*) *stretch tensor*, $\boldsymbol{U}$, which introduces a local stretch (elongation or contraction) of $\Omega_{intermediate}$ along its main axial direction. This results in a *local change in size* without any change in orientation. Mathematically, the first form of polar decomposition is written as:

$$\boldsymbol{F} = \boldsymbol{R}\boldsymbol{U} \qquad \text{such that} \qquad \boldsymbol{R}^T\boldsymbol{R} = \boldsymbol{I}, \quad \boldsymbol{U} = \boldsymbol{U}^T \quad \text{and} \quad \det(\boldsymbol{R}) = 1 \tag{9.9}$$

Similar to the first form of polar decomposition, we can also achieve the deformed configuration by first operating on $\Omega_{undeformed}$ with a *left* (or *spatial*) *stretch tensor*, $\boldsymbol{V}$, which causes a change in size of the material body resulting into the $\Omega_{intermediate}$ domain, as shown in Fig. 9.6. After that, the rotation tensor is operated on the intermediate domain, $\Omega_{intermediate}$, to achieve the deformed domain, $\Omega_{deformed}$. The mathematical expression on this second form of polar decomposition becomes:

$$\boldsymbol{F} = \boldsymbol{V}\boldsymbol{R} \qquad \text{such that} \qquad \boldsymbol{R}^T\boldsymbol{R} = \boldsymbol{I} \quad \text{and} \quad \boldsymbol{V} = \boldsymbol{V}^T \tag{9.10}$$

9.3.5 *Velocity Gradient*

We established previously that the deformation of a material body is a function of not only displacement, but also time. The previous discussions of deformation gradient, stretch, rotation and strain measures considered a time-independent material body. It is important that for exhaustive an understanding of material responses within finite element modelling, the considerations of effect of time on the deformation need to be made. Plasticity formulations and viscoelasticity are ready examples to time-dependent material responses, hence need to be defined as rate quantities. The constitutive laws that describe material behaviour are often implemented within finite element modelling in their rate forms, hence the necessity of this section. We will now define time-dependent kinematics of deformation quantities.

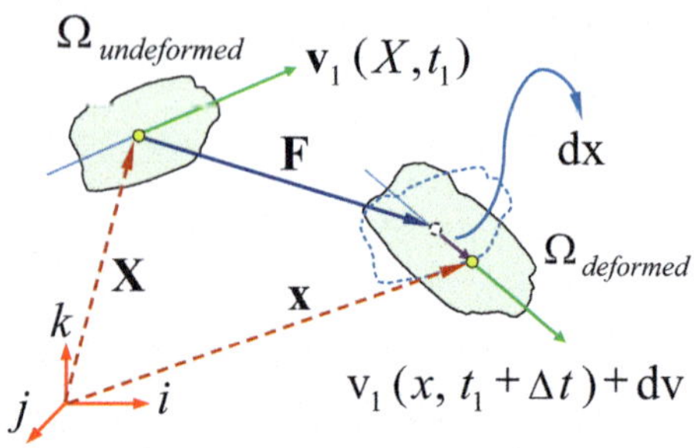

Fig. 9.7 An illustration of a material body with spatial varying velocity field

Consider a material body, $\Omega \subset \Re^n$, bounded by $\Omega_{undeformed}$ and $\Omega_{deformed}$ in the undeformed and deformed material configurations respectively and shown in Fig. 9.7. Let us assume that material point located at position $\boldsymbol{X}$ within $\Omega_{undeformed}$ has a velocity field, $\boldsymbol{v}_1$. If the material body is subjected to a deformation gradient, $\boldsymbol{F}$, for duration of Δt, the body rotates and elongates and the original material point now occupies a new position, $\boldsymbol{x}$, and the velocity field becomes $\boldsymbol{v}_1 + \mathrm{d}\boldsymbol{v}$.

Across the change of position, $\mathrm{d}\boldsymbol{x}$, of the material points in the deformed domain, the velocity field has changed by $\mathrm{d}\boldsymbol{v}$. The *velocity gradient*, $\boldsymbol{l}$, is the spatial rate of change of velocity field over the length, $\mathrm{d}\boldsymbol{x}$. The velocity gradient is derived by:

$$\mathrm{d}\boldsymbol{v} = \frac{\partial \boldsymbol{v}}{\partial \boldsymbol{x}}\mathrm{d}\boldsymbol{x} \qquad \text{where} \qquad \boldsymbol{l} = \frac{\partial \boldsymbol{v}}{\partial \boldsymbol{x}} = \begin{bmatrix} \frac{\partial v_x}{\partial x} & \frac{\partial v_x}{\partial y} & \frac{\partial v_x}{\partial z} \\ \frac{\partial v_y}{\partial x} & \frac{\partial v_y}{\partial y} & \frac{\partial v_y}{\partial z} \\ \frac{\partial v_z}{\partial x} & \frac{\partial v_z}{\partial y} & \frac{\partial v_z}{\partial z} \end{bmatrix} \tag{9.11}$$

Let us also relate the velocity gradient, $\boldsymbol{l}$, to the deformation gradient, $\boldsymbol{F}$, that caused the deformation of the material domain. Here, we consider the *rate of change of deformation gradient*, $\dot{\boldsymbol{F}}$.

$$\dot{\boldsymbol{F}} = \frac{\boldsymbol{F}}{\Delta t} \quad \equiv \quad \dot{\boldsymbol{F}} = \frac{\partial}{\partial t}\left(\frac{\partial \boldsymbol{x}}{\partial \boldsymbol{X}}\right) \quad \longrightarrow \quad \dot{\boldsymbol{F}} = \frac{\partial \boldsymbol{v}}{\partial \boldsymbol{X}} = \frac{\partial \boldsymbol{v}}{\partial \boldsymbol{x}}\frac{\partial \boldsymbol{x}}{\partial \boldsymbol{X}} \quad \longrightarrow \quad \dot{\boldsymbol{F}} = \boldsymbol{l}\boldsymbol{F}$$

We can re-write the above thus:

$$\dot{\boldsymbol{F}} = \boldsymbol{l}\boldsymbol{F} \quad \longrightarrow \quad \boldsymbol{l} = \dot{\boldsymbol{F}}\boldsymbol{F}^{-1} \tag{9.12}$$

which suggests that the velocity gradient is the mapping function between the deformation gradient and the rate of change term of $\boldsymbol{F}$. It is acting on the current configuration of the material domain.

Just like the deformation gradient can be multiplicatively decomposed into its *stretch* and *rotation* tensors, the velocity gradient can likewise be decomposed additively according to Eq. 9.13:

$$\boldsymbol{l} = \boldsymbol{d}(\boldsymbol{x}, t) + \boldsymbol{w}(\boldsymbol{x}, t) \quad \longrightarrow \quad \boldsymbol{l} = \text{sym}(\boldsymbol{l}) + \text{asym}(\boldsymbol{l}), \tag{9.13}$$

where:

- $\boldsymbol{d}$ = ***rate of deformation tensor***, which is a *stretch-related symmetric* part of the velocity gradient. It is also commonly described as the *rate of strain tensor* or *strain rate*; and,
- $\boldsymbol{w}$ = ***continuum spin***, which is a *rotation-related antisymmetric/skew* part of the velocity gradient. It is also commonly called the *rate of rotation tensor* or *vorticity tensor*.

Mathematically, the $\boldsymbol{d}$ and $\boldsymbol{w}$ tensors can be written as:

$$\begin{aligned} &\text{Rate of deformation tensor:} \quad && \boldsymbol{d} = \frac{1}{2}\left(\boldsymbol{l} + \boldsymbol{l}^T\right) = \boldsymbol{d}^T \\ &\text{Continuum spin tensor:} \quad && \boldsymbol{w} = \frac{1}{2}\left(\boldsymbol{l} - \boldsymbol{l}^T\right) = -\boldsymbol{w}^T \end{aligned} \tag{9.14}$$

9.4 Measures of Strain

In this section, we present the different measures of strain used in computational mechanics. The discussion considers the classic expressions of normal and shear strain as usually presented in engineering mechanics before discussing the less common strain measures used in finite deformation literature such as the Cauchy-Green deformation tensors, and such like.

9.4.1 Normal and Shear Strain Measures

An important measure of material response is *strain*. Strain, in undergraduate textbooks, is defined as a measure of the relative change in size or shape of a body with respect to a reference configuration. For a material body shown in Fig. 9.8, we recall the following strain measures:

$$\text{Normal strain} \quad \varepsilon_N = \frac{\Delta L}{L_0} \tag{9.15}$$

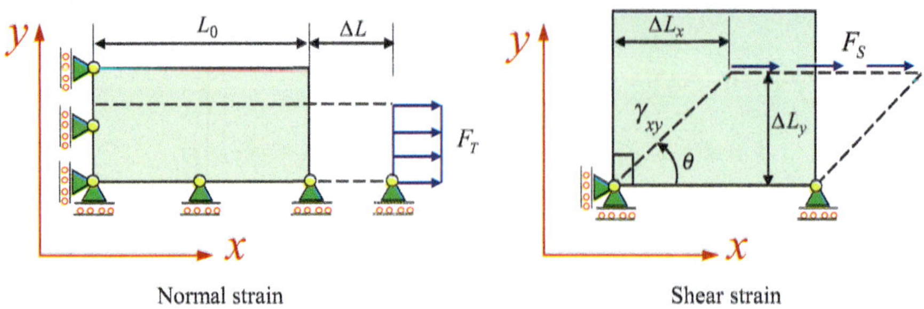

Fig. 9.8 Schematics of tensile and shear deformations

$$\text{Shear strain} \qquad \gamma_{xy} = \frac{\pi}{2} - \theta \quad \equiv \quad \tan^{-1}\left(\frac{\Delta L_x}{\Delta L_y}\right) \tag{9.16}$$

where L_0 is original length, and ΔL is the extension of the tensile test specimen. $\boldsymbol{F}_T$ and $\boldsymbol{F}_S$ are tensile and shear forces respectively. However, for a finite element process, the measures of strain adopted is based on the deformation gradient tensor derived previously. This way, not only small strain, but also finite strain, deformations can be accounted for within the FEM process.

9.4.2 *Right Cauchy-Green Deformation Tensor*

Consider again the material body of Fig. 9.5 and now let us define ds as the length of the line segment, $\overline{A'B'}$, in the deformed state. Therefore:

$$\begin{aligned} \mathrm{d}s^2 = \mathrm{d}\boldsymbol{x} \cdot \mathrm{d}\boldsymbol{x} \qquad \longrightarrow \qquad \mathrm{d}s^2 &= (\boldsymbol{F}\mathrm{d}\boldsymbol{X}) \cdot (\boldsymbol{F}\mathrm{d}\boldsymbol{X}) \\ \mathrm{d}s^2 &= \mathrm{d}\boldsymbol{X}^T \boldsymbol{F}^T \boldsymbol{F} \mathrm{d}\boldsymbol{X} \\ \mathrm{d}s^2 &= \mathrm{d}\boldsymbol{X}^T \boldsymbol{C} \mathrm{d}\boldsymbol{X}. \end{aligned}$$

In the above, we defined $\boldsymbol{C}$, called the *right Cauchy-Green deformation tensor*, as:

$$\boldsymbol{C} = \boldsymbol{F}^T \boldsymbol{F} \tag{9.17}$$

It is the first measure of strain that we can deduce based on the deformation gradient tensor. It is described as *right*, because the deformation gradient, $\boldsymbol{F}$, is on the right hand side of the transpose of $\boldsymbol{F}$ equation. It is also sometimes referred to as the *Green deformation tensor*. As shown above, the physical meaning of $\boldsymbol{C}$ is that it gives a measure of the square of the distances (ds^2) due to the deformation of a material body in the material reference frame.

9.4.3 Left Cauchy-Green Deformation Tensor

Similar to the approach taken above, we can also determine the length, dS, of the material body in the spatial reference frame. Recall that in the spatial reference frame, d$\boldsymbol{X} = \boldsymbol{F}^{-1}d\boldsymbol{x}$. The distance d$S$ can be determined according to the following:

$$\mathrm{d}S^2 = \mathrm{d}\boldsymbol{X}^T \cdot \mathrm{d}\boldsymbol{X} \qquad \longrightarrow \qquad \begin{aligned} \mathrm{d}S^2 &= \left(\boldsymbol{F}^{-1}\mathrm{d}\boldsymbol{x}\right)^T \left(\boldsymbol{F}^{-1}\mathrm{d}\boldsymbol{x}\right) \\ \mathrm{d}S^2 &= \mathrm{d}\boldsymbol{x}^T \left(\boldsymbol{F}^{-1}\right)^T \boldsymbol{F}^{-1}\mathrm{d}\boldsymbol{x} \\ \mathrm{d}S^2 &= \mathrm{d}\boldsymbol{x}^T \boldsymbol{B}^{-1}\mathrm{d}\boldsymbol{x} \end{aligned}$$

Notice that we have determined another type of strain measure, the *left Cauchy-Green deformation tensor*, which is represented by $\boldsymbol{B}$ thus:

$$\boldsymbol{B}^{-1} = \left(\boldsymbol{F}^{-1}\right)^T \boldsymbol{F}^{-1} \qquad \text{or simply} \qquad \boldsymbol{B} = \boldsymbol{F}\boldsymbol{F}^T \tag{9.18}$$

The deformation gradient tensor, $\boldsymbol{F}$, is on the left hand side of the transpose of $\boldsymbol{F}$ hence the *left* Cauchy-green tensor. $\boldsymbol{B}$ is also called the *Finger deformation tensor*. Physically, the $\boldsymbol{B}$ tensor represents the square of the distances (dS^2) due to deformation of a material body in the spatial coordinates system.

9.4.4 Change in Length, ΔL Measure

Another common measure needed to determine the strain in a material body is the *change in length*. Having previously determined the distances in material (ds^2) and spatial (dS^2) reference frames, we will now go ahead and use them to determine the relative change in length for the material body.

Mathematically the change in length of a material body can be expressed as:

$$\Delta L^2 = \mathrm{d}s^2 - \mathrm{d}S^2 \qquad \longrightarrow \qquad \begin{aligned} \Delta L^2 &= \mathrm{d}\boldsymbol{x} \cdot \mathrm{d}\boldsymbol{x} - \mathrm{d}\boldsymbol{X} \cdot \mathrm{d}\boldsymbol{X} \\ \Delta L^2 &= \mathrm{d}\boldsymbol{x} \cdot \mathrm{d}\boldsymbol{x} - \mathrm{d}\boldsymbol{x} \cdot \boldsymbol{B}^{-1}\mathrm{d}\boldsymbol{x} \\ \Delta L^2 &= \mathrm{d}\boldsymbol{x} \cdot \left(\boldsymbol{I} - \boldsymbol{B}^{-1}\right)\mathrm{d}\boldsymbol{x}, \end{aligned}$$

where $\boldsymbol{I}$ is the identity tensor. The above demonstrates how the change in length i.e. *stretch* of a material body can be determined with respect to both material configurations. Under rigid body rotation, there is change in length ($\Delta L = 0$), therefore we note that:

$$\mathrm{d}\boldsymbol{x} \cdot \left(\boldsymbol{I} - \boldsymbol{B}^{-1}\right)\mathrm{d}\boldsymbol{x} = \boldsymbol{0} \qquad \longrightarrow \qquad \boldsymbol{B} = \boldsymbol{I}. \tag{9.19}$$

The above rigid-body motion condition implies that the mapping function between the material and spatial configuration (i.e. deformation gradient, $\boldsymbol{F}$) contains terms that do not cause a change in length of the material but rather a change in orientation.

9.4.5 Green Strain Tensor

It is vital that any strain measure used within a finite element scheme must return the same strain irrespective of orientation of the material body. This is a crucial requirement for formulation of strain in finite deformation systems. In particular, if the rotations arising from deformation of a material body grows, so does the errors arising in the strain formulation. It is a requirement therefore that an objective strain measure must not be polluted by rigid body rotations [3]. Hereafter, we introduce the next measure of strain which obeys these requirements.

Consider the material body of Fig. 9.5, we can derive the change of length in terms of material reference frame thus:

$$\begin{aligned}
\Delta L^2 = \mathrm{d}s^2 - \mathrm{d}S^2 \quad &\longrightarrow \quad \Delta L^2 = \mathrm{d}\boldsymbol{x} \cdot \mathrm{d}\boldsymbol{x} - \mathrm{d}\boldsymbol{X} \cdot \mathrm{d}\boldsymbol{X} \\
&\longrightarrow \quad \Delta L^2 = (\boldsymbol{F}\mathrm{d}\boldsymbol{X}) \cdot (\boldsymbol{F}\mathrm{d}\boldsymbol{X}) - \mathrm{d}\boldsymbol{X} \cdot \mathrm{d}\boldsymbol{X} \\
&\qquad\quad \Delta L^2 = \mathrm{d}\boldsymbol{X} \cdot \left(\boldsymbol{F}^T\boldsymbol{F}\right) \mathrm{d}\boldsymbol{X} - \mathrm{d}\boldsymbol{X} \cdot (\boldsymbol{I})\mathrm{d}\boldsymbol{X} \\
&\qquad\quad \Delta L^2 = \mathrm{d}\boldsymbol{X} \cdot \left(\boldsymbol{F}^T\boldsymbol{F} - \boldsymbol{I}\right) \mathrm{d}\boldsymbol{X}
\end{aligned}$$

In the above, we notice that we have defined a new term: $\boldsymbol{F}^T\boldsymbol{F} - \boldsymbol{I}$, which we will now equate to $2\boldsymbol{E}$ such that:

$$\boldsymbol{E} = \frac{1}{2}\left(\boldsymbol{F}^T\boldsymbol{F} - \boldsymbol{I}\right) \quad \longrightarrow \quad \boldsymbol{E} = \frac{1}{2}\left(\boldsymbol{C} - \boldsymbol{I}\right) \tag{9.20}$$

where the new term, $\boldsymbol{E}$, is called the *Green-Lagrange strain tensor* or *Green strain tensor* for short. Based on Eq. 9.20, we can observe that:

$$\boldsymbol{E} = \frac{1}{2}\left(\boldsymbol{F}^T\boldsymbol{F} - \boldsymbol{I}\right) \quad \longrightarrow \quad \boldsymbol{E} = \frac{1}{2}\frac{\Delta L^2}{\mathrm{d}\boldsymbol{X} \cdot \mathrm{d}\boldsymbol{X}} \tag{9.21}$$

Notice that Eq. 9.21 is similar to the engineering *normal* strain measure of Eq. 9.15 being a ratio of a change in length (ΔL^2) to the original length, $L_0^2 \equiv |\mathrm{d}\boldsymbol{X} \cdot \mathrm{d}\boldsymbol{X}|$, in the material reference frame. It is commonly regarded in continuum mechanics that this expression of normal strain (Eq. 9.20) is a more accurate measure of strain than that of Eq. 9.15.

Since any error due to rigid body rotation must be removed for a finite element strain measure, we will now assess if the strain, $\boldsymbol{E}$, is independent of rigid body motion. Recall the polar decomposition of the deformation gradient, $\boldsymbol{F} = \boldsymbol{R}\boldsymbol{U}$, thus we will now substitute these into Eq. 9.20.

$$\boldsymbol{C} = \boldsymbol{F}^T\boldsymbol{F} \qquad \longrightarrow \qquad \begin{aligned} \boldsymbol{C} &= (\boldsymbol{R}\boldsymbol{U})^T \boldsymbol{R}\boldsymbol{U} = \boldsymbol{U}^T \left(\boldsymbol{R}^T\boldsymbol{R}\right) \boldsymbol{U} \\ &= \boldsymbol{U}^T\,(\boldsymbol{I})\,\boldsymbol{U} = \boldsymbol{U}^T\boldsymbol{U} = \boldsymbol{U}^2 \end{aligned} \tag{9.22}$$

Since $\boldsymbol{R}$ is orthogonal (i.e. $\boldsymbol{R}^T\boldsymbol{R} = \boldsymbol{I}$) and $\boldsymbol{U}$ is symmetric (i.e. $\boldsymbol{U}^T\boldsymbol{U} = \boldsymbol{U}^2$), it can be seen that:

$$\boldsymbol{E} = \frac{1}{2}\left(\boldsymbol{C} - \boldsymbol{I}\right) \qquad \longrightarrow \qquad \boldsymbol{E} = \frac{1}{2}\left(\boldsymbol{U}^2 - \boldsymbol{I}\right) \tag{9.23}$$

which establishes that the Green-Lagrange strain tensor is independent of rigid body rotation as it is solely a function of the stretch tensor. It gives a reliable strain measure for use in describing the material response within a finite element modelling scheme.

9.4.6 *Almansi Strain Tensor*

Similar to the Green-Lagrange strain tensor, we will also introduce another strain measure whose definition is independent of the rigid body rotation associated with a material body subjected to a deformation gradient, $\boldsymbol{F}$. Consider again the material body of Fig. 9.5, we can derive the change of length in terms of spatial reference frame thus:

$$\Delta L^2 = \mathrm{d}s^2 - \mathrm{d}S^2 \qquad \longrightarrow \qquad \begin{aligned} \Delta L^2 &= \mathrm{d}\boldsymbol{x} \cdot \mathrm{d}\boldsymbol{x} - \mathrm{d}\boldsymbol{X} \cdot \mathrm{d}\boldsymbol{X} \\ \Delta L^2 &= \mathrm{d}\boldsymbol{x} \cdot \mathrm{d}\boldsymbol{x} - \left(\boldsymbol{F}^{-1}\mathrm{d}\boldsymbol{x}\right) \cdot \left(\boldsymbol{F}^{-1}\mathrm{d}\boldsymbol{x}\right) \\ \Delta L^2 &= \mathrm{d}\boldsymbol{x} \cdot (\boldsymbol{I})\mathrm{d}\boldsymbol{x} - \mathrm{d}\boldsymbol{x} \cdot \left(\boldsymbol{F}^{-T}\boldsymbol{F}^{-1}\right)\mathrm{d}\boldsymbol{x} \\ \Delta L^2 &= \mathrm{d}\boldsymbol{x} \cdot \left(\boldsymbol{I} - \boldsymbol{F}^{-T}\boldsymbol{F}^{-1}\right)\mathrm{d}\boldsymbol{x} \end{aligned}$$

We have now introduced another alternative strain measure called the *Almansi strain tensor*, $\boldsymbol{e}$ such that:

$$\boldsymbol{e} = \frac{1}{2}\left(\boldsymbol{I} - \boldsymbol{F}^{-T}\boldsymbol{F}^{-1}\right) \qquad \longrightarrow \qquad \boldsymbol{e} = \frac{1}{2}\left(\boldsymbol{I} - \boldsymbol{B}^{-1}\right) \tag{9.24}$$

This equation shares a similar definition as to the Green-Lagrange strain tensor except that the derivation is based on spatial material reference frame. It is the strain measure of a material body that is being subjected to a deformation gradient with the observer sat at a static location and observing the deformation of the body.

We will also assess if this strain is not affected by rigid body rotation. Again, recall the polar decomposition of $\boldsymbol{F}$ in the spatial reference frame thus: $\boldsymbol{F} = \boldsymbol{VR}$. If we substitute this into Eq. 9.24, we will obtain:

$$\boldsymbol{e} = \frac{1}{2}\left(\boldsymbol{I} - \boldsymbol{B}^{-1}\right) \quad \longrightarrow \quad \boldsymbol{e} = \frac{1}{2}\left\{\boldsymbol{I} - \left[\boldsymbol{F}\boldsymbol{F}^T\right]^{-1}\right\} = \frac{1}{2}\left\{\boldsymbol{I} - \left[(\boldsymbol{VR})(\boldsymbol{VR})^T\right]^{-1}\right\}$$

$$= \frac{1}{2}\left\{\boldsymbol{I} - \left[\boldsymbol{V}(\boldsymbol{R}^T\boldsymbol{R})\boldsymbol{V}^T\right]^{-1}\right\}$$

$$= \frac{1}{2}\left\{\boldsymbol{I} - \left[\boldsymbol{V}\boldsymbol{V}^T\right]^{-1}\right\} = \frac{1}{2}\left\{\boldsymbol{I} - \left[\boldsymbol{V}^2\right]^{-1}\right\}$$

which establishes that the Almansi strain tensor is solely a function of the left stretch tensor, $\boldsymbol{V}$, and so not affected by any rigid body rotation of the material body under the effect of $\boldsymbol{F}$. We can also observe that:

$$\boldsymbol{B}^{-1} = \left[\boldsymbol{V}^2\right]^{-1} \quad \longrightarrow \quad \boldsymbol{B}^{-1} = \left[\boldsymbol{V}^{-1}\right]^2 \tag{9.25}$$

9.4.7 Logarithmic Strain Tensor

Another commonly used strain measure, especially in reporting experimental work is the *logarithmic strain*. Let us consider an infinitesimally small line element of initial length, L_0, in its undeformed configuration. If it is subjected to a deformation gradient, $\boldsymbol{F}$, resulting in a final deformed length, l, having experienced a change in length, ΔL, we can define the *stretch* or *extension ratio*, λ thus:

$$\lambda = \frac{l}{L_0} \quad \longrightarrow \quad \lambda = \frac{L_0 + \Delta L}{L_0} = 1 + \frac{\Delta L}{L_o} = 1 + \varepsilon_N \tag{9.26}$$

where ε_N is *normal engineering strain* defined in Eq. 9.15. The above definition considers strain changes to be uniform over the line segment. However, it is common in plasticity for any strain increment to be nonlinear, in which case, the path of strain increment becomes important. We use the *logarithmic strain* to capture this type of strain increment.

If the infinitesimally small line element experiences a local change in length, δl, the resulting strain increment with respect to the final length, l, can be expressed:

$$\delta\varepsilon = \frac{\delta l}{l} \tag{9.27}$$

The sum of all strain-path-dependent strain increments becomes the logarithmic strain, which is the integration of the strain increments from the initial length, L_0, to the final length, l, thus:

$$\int \delta\varepsilon = \int_{l_0}^{l} \frac{1}{l}\delta l \quad \longrightarrow \quad \begin{aligned} \varepsilon &= \ln\left(\frac{l}{L_0}\right) \\ &= \ln(\lambda) \\ &= \ln(1+\varepsilon_N) \\ &= \varepsilon_N - \frac{1}{2}(\varepsilon_N)^2 + \frac{1}{3}(\varepsilon_N)^3 - \cdots \end{aligned} \tag{9.28}$$

For a three-dimensional material body, we can define the same logarithmic strain in terms of the deformation gradient, $\boldsymbol{F}$, thus:

$$\boldsymbol{\varepsilon} = -\frac{1}{2}\ln \boldsymbol{B}^{-1} \quad \longrightarrow \quad \boldsymbol{\varepsilon} = -\frac{1}{2}\ln\left(\boldsymbol{F}^{-T}\boldsymbol{F}^{-1}\right) \tag{9.29}$$

The logarithmic strain tensor is also referred to as the *Hencky strain tensor*, named after H Hencky who, in 1928, made this derivation. It is necessary for finite element modelling studies that the strain tensor must not be affected by rigid body rotations, we will now show that the logarithmic strain tensor obeys this requirement.

We established previously in Eq. 9.25 that $\boldsymbol{B}$ can be related to the left stretch tensor, $\boldsymbol{V}$. We will exploit this relationship to re-define the logarithmic strain tensor thus:

$$\boldsymbol{\varepsilon} = -\frac{1}{2}\ln \boldsymbol{B}^{-1} \quad \longrightarrow \quad \boldsymbol{\varepsilon} = -\frac{1}{2}\ln\left[\boldsymbol{V}^{-1}\right]^2 = \ln \boldsymbol{V} \tag{9.30}$$

which shows that the logarithmic strain tensor is independent of the rotation tensor. It is thus a suitable measure of strain for finite element modelling studies.

9.4.8 Seth-Hill Family of Strain Tensors

Although previously, we have presented the different measures of strain, they can be combined into a single equation, which is generally regarded as the *Seth-Hill family of strain tensors*. For a material body, $\Omega \subset \Re^n$, subjected to a deformation gradient, $\boldsymbol{F}$, which can be decomposed into $\boldsymbol{F} = \boldsymbol{RU} = \boldsymbol{VR}$, we can define the Seth-Hill family of strains in terms of either the right stretch tensor, $\boldsymbol{U}$, or the left stretch tensor, $\boldsymbol{V}$, as:

$$\boldsymbol{\varepsilon} = \frac{1}{n}\left(\boldsymbol{U}^n - \boldsymbol{I}\right) \quad \text{or} \quad \boldsymbol{\varepsilon} = \frac{1}{n}\left(\boldsymbol{V}^n - \boldsymbol{I}\right) \tag{9.31}$$

where n is a real number and not necessarily an integer. The above equation is also referred to as the *generalized strain measures in Lagrangian and Eulerian descriptions*. For specific values of n, Eq. 9.31 shows any of the earlier strain measures as illustrated below.

$$
\begin{array}{llll}
\text{For } n = -2: & \text{Euler-Almansi strain,} & \boldsymbol{\varepsilon} = & 0.50\left(\boldsymbol{I} - \boldsymbol{V}^{-2}\right) \\
\text{For } n = -1: & \text{“True” strain,} & \boldsymbol{\varepsilon} = & \boldsymbol{I} - \ln \boldsymbol{U}^{-1} \\
\text{For } n \to \ 0: & \text{Logarithmic strain,} & \boldsymbol{\varepsilon} = & \ln \boldsymbol{U} \text{ or } \ln \boldsymbol{V} \\
\text{For } n = \ \ 1: & \text{Engineering strain,} & \boldsymbol{\varepsilon} = & \boldsymbol{U} - \boldsymbol{I} \text{ or } \boldsymbol{V} - \boldsymbol{I} \\
\text{For } n = \ \ 2: & \text{Green-Lagrange strain,} & \boldsymbol{\varepsilon} = & 0.50\left(\boldsymbol{U}^2 - \boldsymbol{I}\right)
\end{array}
$$

The Seth-Hill family of strains can be demonstrated for the elongation of a cylindrical bar of cross-sectional area, A, original length, L_0, undergoing a tensile deformation, ΔL, such that the final length, $L = L_0 + \Delta L$. The resulting stretch ratio becomes: $\lambda = L/L_0$.

The stretch ratio, λ, will always be positive, $\lambda > 0$ since lengths will always be positive. Therefore, the expressions of different strain measures, adapted from the Seth-Hill generalized family of strains, are given below while the comparison plots of the strain measures are shown in Fig. 9.9.

$$
\begin{array}{lllll}
\text{For } n = -2: & \text{Euler-Almansi strain,} & \varepsilon = & \frac{1}{2}\left(1 - \frac{1}{\lambda^2}\right) & = \frac{1}{2}\left(\frac{L^2 - L_0^2}{L^2}\right) \\
\text{For } n = -1: & \text{“True” strain,} & \varepsilon = & 1 - \frac{1}{\lambda} & = \frac{L - L_0}{L} \\
\text{For } n \to \ 0: & \text{Logarithmic strain,} & \varepsilon = & \ln \lambda & = \ln\left(\frac{L}{L_0}\right) \\
\text{For } n = \ \ 1: & \text{Engineering strain,} & \varepsilon = & \lambda - 1 & = \frac{L - L_0}{L_0} \\
\text{For } n = \ \ 2: & \text{Green-Lagrange strain,} & \varepsilon = & \frac{1}{2}\left(\lambda^2 - 1\right) & = \frac{1}{2}\left(\frac{L^2 - L_0^2}{L_0^2}\right)
\end{array}
\tag{9.32}
$$

Example 9.1 A 3×2 mm^2 rectangular plate shown in Fig. 9.10 is subjected to a uniaxial tensile deformation such that the plate undergoes a stretch of 50% along the x–axis and a contraction of 20% along the y–axis. The material configuration of the deformed the plate is given by:

$$
\begin{aligned}
x &= 1.5X + 0.0Y \\
y &= 0.0X + 0.8Y
\end{aligned}
\tag{9.33}
$$

(a) Determine the deformation gradient, $\boldsymbol{F}$, for the above problem.
(b) Determine the stretch, $\boldsymbol{U}$ and rotation, $\boldsymbol{R}$, tensors.
(c) Determine the Green-Lagrange strain tensor, $\boldsymbol{E}$.

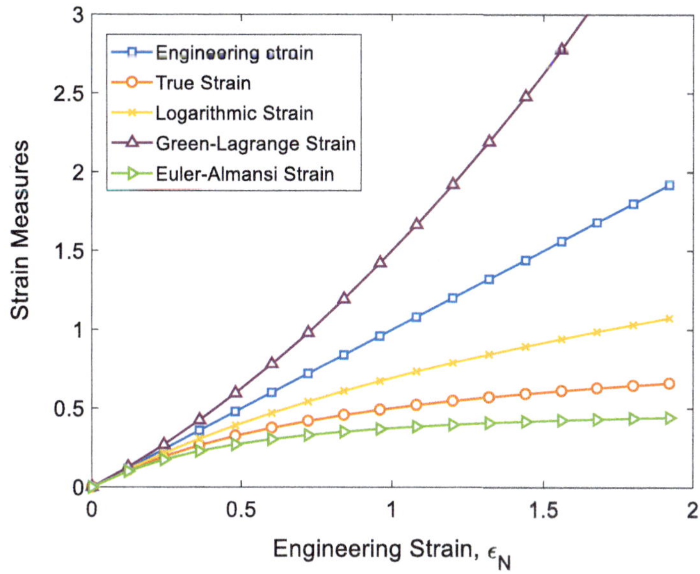

Fig. 9.9 A comparison of all the strain measures for uniaxial deformation of a rod

Fig. 9.10 Uniaxial deformation of a rectangular plate

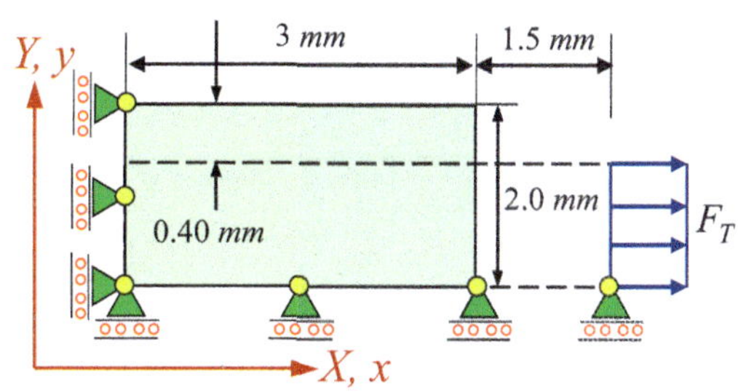

(d) Determine the Engineering strain, $\boldsymbol{\varepsilon}_{engineering}$, and prove that this yields the expected strains.

Solution

(a) **Deformation gradient tensor:** Based on Eq. 9.33, we get:

$$F = \frac{\partial \boldsymbol{x}}{\partial \boldsymbol{X}} \quad \longrightarrow \quad F = \begin{bmatrix} \frac{\partial x}{\partial X} & \frac{\partial x}{\partial Y} \\ \frac{\partial y}{\partial X} & \frac{\partial y}{\partial Y} \end{bmatrix} = \begin{bmatrix} 1.5 & 0.0 \\ 0.0 & 0.8 \end{bmatrix} \tag{9.34}$$

(b) **Stretch tensors**: We established previously that the stretch, $\boldsymbol{U}$, can be related to the left Cauchy-Green tensor, $\boldsymbol{C}$. We will use this to determine the stretch tensor only as given below:

$$\boldsymbol{C} = \boldsymbol{F}^T\boldsymbol{F} = \boldsymbol{U}^2 \quad \longrightarrow \quad \boldsymbol{U} = \sqrt{\boldsymbol{F}^T\boldsymbol{F}} \tag{9.35}$$

Substituting Eq. 9.34 into Eq. 9.35 becomes:

$$\boldsymbol{U} = \sqrt{\boldsymbol{F}^T\boldsymbol{F}} \quad \longrightarrow \quad \boldsymbol{U} = \sqrt{\boldsymbol{F}^2} = \boldsymbol{F} \quad \text{since} \quad \boldsymbol{F}^T = \boldsymbol{F}$$

$$\boldsymbol{U} = \begin{bmatrix} 1.5 & 0.0 \\ 0.0 & 0.8 \end{bmatrix} = \begin{bmatrix} U_{xx} & U_{xy} \\ U_{yx} & U_{yy} \end{bmatrix}$$

The above result shows the deformation is dominated by stretch along the $X-$ and $Y-$directions, hence the diagonal terms: $U_{xx} = \lambda_{xx} = 4.5/3.0 = 1.5$ and $U_{yy} = \lambda_{yy} = 1.6/2.0 = 0.8$, where $\lambda = L/L_0$ is the stretch ratio.

(c) **Rotation tensors**: Similar to the derivation of the stretch tensor, we will also calculate the rotation tensor based on $\boldsymbol{F}$ and $\boldsymbol{U}$. Recall,

$$\boldsymbol{F} = \boldsymbol{R}\boldsymbol{U} \quad \longrightarrow \quad \boldsymbol{R} = \frac{\boldsymbol{F}}{\boldsymbol{U}} = \begin{bmatrix} 1.0 & 0.0 \\ 0.0 & 1.0 \end{bmatrix}$$

Assuming the angle of rotation is φ, we also note that:

$$\boldsymbol{R} = \begin{bmatrix} \cos\varphi & -\sin\varphi \\ \sin\varphi & \cos\varphi \end{bmatrix} \quad \longrightarrow \quad \boldsymbol{R} = \begin{bmatrix} 1.0 & 0.0 \\ 0.0 & 1.0 \end{bmatrix} \tag{9.36}$$

The above implies that: $\cos\varphi = 1.0 \quad \longrightarrow \quad \varphi = \cos^{-1} 1.0 = 0^0$. Therefore, in the deformed configurations, the edges have not been rotated, hence rigid body rotation is excluded, as expected for a uniaxial test.

(d) **Green-Lagrange strain tensor**: This is determined as follows:

$$\boldsymbol{E} = \frac{1}{2}\left(\boldsymbol{F}^T\boldsymbol{F} - \boldsymbol{I}\right) \quad \longrightarrow \quad \boldsymbol{E} = \begin{bmatrix} 0.625 & 0.000 \\ 0.000 & -0.180 \end{bmatrix} \tag{9.37}$$

(e) **Engineering strain tensor**: This is determined using Eq. 9.32 thus:

$$\boldsymbol{\varepsilon}_{engineering} = \boldsymbol{U} - \boldsymbol{I} \quad \longrightarrow \quad \boldsymbol{\varepsilon}_{engineering} = \begin{bmatrix} 1.5 & 0.0 \\ 0.0 & 0.8 \end{bmatrix} - \begin{bmatrix} 1.0 & 0.0 \\ 0.0 & 1.0 \end{bmatrix}$$

$$\text{Therefore,} \quad \boldsymbol{\varepsilon}_{engineering} = \begin{bmatrix} 0.5 & 0.0 \\ 0.0 & -0.2 \end{bmatrix} = \begin{bmatrix} \varepsilon_{xx} & \varepsilon_{xy} \\ \varepsilon_{yx} & \varepsilon_{yy} \end{bmatrix} \tag{9.38}$$

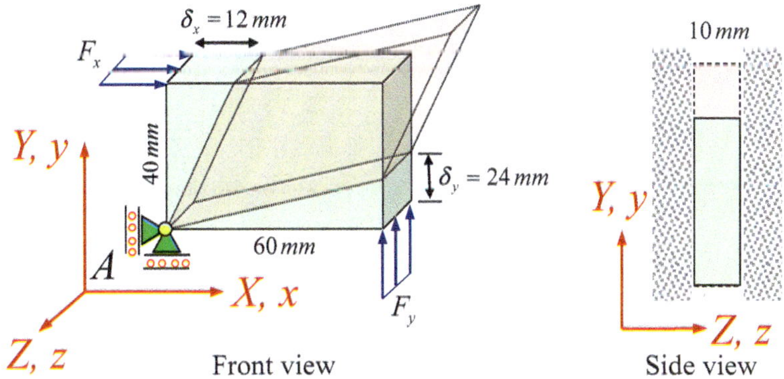

Fig. 9.11 Deformation of a PTFE plate subjected to two shear forces

Notice that along the x-axis, $\varepsilon_{xx} = 0.5$, which is a 50% increase in the X–edge and $\varepsilon_{yy} = -0.2$ is a 20% contraction of the Y–edge. This is the same load imposed on the plate thus demonstrating the calculations are correct.

Example 9.2 A PTFE sheet of dimensions 60 × 40 × 10 mm^3 shown in Fig. 9.11 is pinned securely at A and constrained along the z–axis. The plate is subjected to two shear forces, F_x and F_y, leading to edge deformations of $\delta_x = 12$ mm and $\delta_y = 24$ mm respectively. Assume that the deformation in the through-thickness direction (z–axis) is minimal.

(a) Determine the mapping function that relates material points in the material reference frame to those of the spatial reference frame.
(b) Determine the deformation gradient, $\boldsymbol{F}$, for the above problem.
(c) Determine the stretch, $\boldsymbol{U}$ and $\boldsymbol{V}$, and rotation, $\boldsymbol{R}$, tensors.
(d) Determine the Euler-Almansi strain tensor, $\boldsymbol{e}$.

Solution

(a) **Mapping function in spatial reference frame:** We develop the relationship between material and spatial configurations. Let us identify the material reference frame as: XYZ–axes and the spatial reference frame as: xyz–axes. Consider that the origin of both coordinate systems coincide as shown in Fig. 9.11. The mapping function along the x–axis should take the form:

$$x = \chi_x X + \chi_y Y + +\chi_z Z \tag{9.39}$$

where χ_x, χ_y and χ_z are the *coefficients/weighting terms* that describe the axis-specific mapping functions for X–, Y– and Z–axes respectively. We will now attempt to define what these coefficient/weighting terms are.

Recall that the deformation gradient tensor, $\boldsymbol{F}$, that caused the observed deformation is defined as:

$$F = \frac{\partial \boldsymbol{x}}{\partial \boldsymbol{X}} \quad \longrightarrow \quad F = \begin{bmatrix} \frac{\partial x}{\partial X} & \frac{\partial x}{\partial Y} & \frac{\partial x}{\partial Z} \\ \frac{\partial y}{\partial X} & \frac{\partial y}{\partial Y} & \frac{\partial y}{\partial Z} \\ \frac{\partial z}{\partial X} & \frac{\partial z}{\partial Y} & \frac{\partial z}{\partial Z} \end{bmatrix} \tag{9.40}$$

Considering the first row of the matrix of Eq. 9.39 and the coefficients of Eq. 9.40, we can conclude that for the x–axis:

$$\chi_x = \frac{\partial x}{\partial X} \qquad \chi_y = \frac{\partial y}{\partial X} \quad \text{and} \quad \chi_z = \frac{\partial z}{\partial X} \tag{9.41}$$

This implies that the first weighting term that describes the mapping function (i.e. χ_x) is essentially the ratio of the lengths: d$\boldsymbol{x}$ and d$\boldsymbol{X}$. Based on Fig. 9.11, along the x–axis, we notice that:

$$\chi_x = \frac{\partial x}{\partial X} \quad \equiv \quad \boldsymbol{\chi}_x = \frac{\mathrm{d}\boldsymbol{x}}{\mathrm{d}\boldsymbol{X}}$$

For the PTFE plate under shear loading, along the x–axis, material and spatial positions are coincident hence d$x = 40$ and d$X = 40$, so $\chi_x = \mathrm{d}x/\mathrm{d}X = 40/40 = 1.0$. Also, in the y–axis, $\chi_y = \frac{\partial x}{\partial Y} \equiv \chi_y = \frac{\mathrm{d}x}{\mathrm{d}Y} = \frac{12\text{ mm}}{60\text{ mm}} = 0.40$. There is no deformation in the z–axis, hence: dz = dZ (i.e. material and spatial reference frames are coincident).

Using this approach, we can determine all the weighting terms and the resulting mapping functions for the problem are:

$$x = \left(\frac{\mathrm{d}x}{\mathrm{d}X}\right) X + \left(\frac{\mathrm{d}x}{\mathrm{d}Y}\right) Y + \left(\frac{\mathrm{d}x}{\mathrm{d}Z}\right) Z \quad \longrightarrow \quad x = 1.0X + 0.4Y + 0.0Z$$

$$y = \left(\frac{\mathrm{d}y}{\mathrm{d}X}\right) X + \left(\frac{\mathrm{d}y}{\mathrm{d}Y}\right) Y + \left(\frac{\mathrm{d}y}{\mathrm{d}Z}\right) Z \quad \longrightarrow \quad y = 0.3X + 1.0Y + 0.0Z$$

$$z = \left(\frac{\mathrm{d}z}{\mathrm{d}X}\right) X + \left(\frac{\mathrm{d}z}{\mathrm{d}Y}\right) Y + \left(\frac{\mathrm{d}z}{\mathrm{d}Z}\right) Z \quad \longrightarrow \quad z = 0.0X + 0.0Y + 1.0Z$$

(b) **Deformation gradient tensor:** Based on the mapping functions and Eq. 9.40, $\boldsymbol{F}$ becomes:

$$\boldsymbol{F} = \begin{bmatrix} 1.0 & 0.4 & 0.0 \\ 0.3 & 1.0 & 0.0 \\ 0.0 & 0.0 & 1.0 \end{bmatrix} \tag{9.42}$$

(c) **Stretch and rotation tensors:** The right stretch tensor, $\boldsymbol{U}$ is determined using the deformation gradient tensor

$$\boldsymbol{U} = \sqrt{\boldsymbol{F}^T\boldsymbol{F}} \quad \longrightarrow \quad \boldsymbol{U} = \begin{bmatrix} 1.0440 & 0.8367 & 0.0000 \\ 0.8367 & 1.0770 & 0.0000 \\ 0.0000 & 0.0000 & 1.0000 \end{bmatrix}$$

Similarly, the left stretch tensor, $\boldsymbol{V}$ is given below.

$$\boldsymbol{V} = \sqrt{\boldsymbol{F}\boldsymbol{F}^T} \quad \longrightarrow \quad \boldsymbol{V} = \begin{bmatrix} 1.0770 & 0.8367 & 0.0000 \\ 0.8367 & 1.0440 & 0.0000 \\ 0.0000 & 0.0000 & 1.0000 \end{bmatrix}$$

The are two possible rotation tensors. Here, $\boldsymbol{R}_1$ can be determined from $\boldsymbol{U}$ while $\boldsymbol{R}_2$ is determined using $\boldsymbol{V}$ thus:

$$\boldsymbol{R}_1 = \boldsymbol{F}\boldsymbol{U}^{-1} \quad \longrightarrow \quad \boldsymbol{R}_1 = \begin{bmatrix} 1.7490 & -0.9873 & 0.0000 \\ -1.2099 & 1.8684 & 0.0000 \\ 0.0000 & 0.0000 & 1.0000 \end{bmatrix}$$

and

$$\boldsymbol{R}_2 = \boldsymbol{V}^{-1}\boldsymbol{F} \quad \longrightarrow \quad \boldsymbol{R}_2 = \begin{bmatrix} 1.8684 & -0.9873 & 0.0000 \\ -1.2099 & 1.7490 & 0.0000 \\ 0.0000 & 0.0000 & 1.0000 \end{bmatrix}$$

(d) **Euler-Almansi strain tensor:** This is derived using the left stretch tensor, $\boldsymbol{V}$:

$$\boldsymbol{e} = \frac{1}{2}\left\{\boldsymbol{I} - (\boldsymbol{V}^2)^{-1}\right\} \quad \longrightarrow \quad \boldsymbol{e} = \begin{bmatrix} -4.4677 & 4.9250 & 0.0000 \\ 4.9250 & -4.6620 & 0.0000 \\ 0.0000 & 0.0000 & 0.0000 \end{bmatrix}$$

9.5 Measures of Stress

The previous section introduced the kinematics of motion and deformation of a material body when subjected to a deforming stimulus. The deformation experienced can be quantified using the measures introduced previously. The consequence of a deforming body is the notion of *stress*. It can be also simply described as the *force per unit area.* The concept of internal forces is central to the discussion of stresses. This section discusses the formulation of *traction vectors, stress tensors* and diverse *stress measures* used in describing the state of stress of a material body undergoing finite deformation.

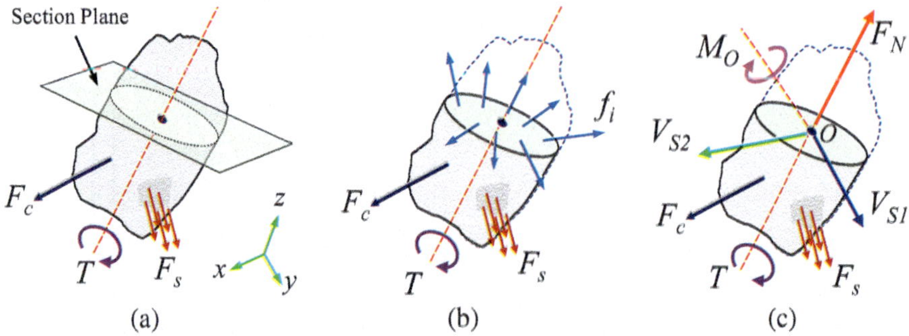

Fig. 9.12 An illustration of action of internal forces: (**a**) a material body under effect of external loads; (**b**) resulting internal forces acting on section plane; and, (**c**) resultant internal forces under static equilibrium conditions

9.5.1 The Concept of Internal Forces

In this section, we introduce the concept of *internal forces*, highlighting how crucial they are in describing the stress acting within a material body. Consider the material body shown in Fig. 9.12a under the effect of three *external loads* namely: concentrated force, F_c, surface forces, F_s and torque, T, acting about the longitudinal $z-$axis of the body.

The effect of the external forces is to excite an internal material response within the material body. To visualize this internal response, a *section plane* is introduced to 'cut through' the material body at a pre-defined location with respect to a chosen xyz reference frame.

Figure 9.12b shows the sectioned view of the material body revealing the set of internal forces, f_i where $i = 1, 2, \cdots, \infty$. On a given sectioned face, there can be very large numbers of internal forces, f_i, which result from the effect of the sectioned-off part on the remaining part of the material body. The size and orientation of these internal forces depend on the cumulative effect of the external loads. Material response derives from the effect of these internal forces.

If a typical internal force, f_i, is oriented with respect to the $x-$, $y-$and $z-$axes by the angles α_i, β_i and γ_i respectively, then the force vector system for this internal force becomes:

$$\boldsymbol{f}_i = f\cos\alpha_i\boldsymbol{i} + f\cos\beta_i\boldsymbol{j} + f\cos\gamma_i\boldsymbol{k} \quad \text{where} \quad f = \text{ magnitude of} f_i \tag{9.43}$$

The description of the material response will rely on a comprehensive analysis of the complete 3D set of force systems that are active on the sectioned surface. It is not always analytically convenient to determine this set of force systems hence it is common in continuum mechanics to make some assumptions about the relationship between the external loads and the complete set of internal force systems. The simplest assumption is that the set of internal forces must be in *static equilibrium* with the set of external forces. This requires that equations of static equilibrium are applied on the set on internal forces to reduce their effect to a few forces.

As shown in Fig. 9.12c, the resulting set of internal forces following application of equations of static equilibrium include: *axial normal force*, $\boldsymbol{F}_N$, two *shear forces*, $\boldsymbol{V}_{S1}$ and $\boldsymbol{V}_{S2}$, and *moment*, $\boldsymbol{M}_O$, about point O. Mathematically:

$$\begin{aligned}
&\text{Sum of forces along the z-axis, Axial force:} && \sum_{i=1}^{\infty} \boldsymbol{f}_i && = \boldsymbol{F}_N \\
&\text{Sum of forces along y-axis, shear force:} && \sum_{i=1}^{\infty} \boldsymbol{f}_i && = \boldsymbol{V}_{S1} \\
&\text{Sum of forces along the x-axis, shear force:} && \sum_{i=1}^{\infty} \boldsymbol{f}_i && = \boldsymbol{V}_{S2} \\
&\text{Sum of moment with respect to point O, moment:} && \sum_{i=1}^{\infty} \boldsymbol{r}_i \otimes \boldsymbol{f}_i && = \boldsymbol{M}_O
\end{aligned} \tag{9.44}$$

where $\boldsymbol{r}_i$ is the position vector of each internal force, $\boldsymbol{f}_i$. All the terms in Eq. 9.44 are internal quantities since they were derived by the internal force vector, $\boldsymbol{f}_i$. These internal forces are the basis upon which stresses, strains, and other continuum mechanics parameters associated with material response are derived.

Stress is the measure of the intensity of an internal force per unit area where the force acts. In other words, it is the ratio of the internal force at a point to the area over which that internal force is acting. High stress over the same unit implies high internal force and vice versa.

9.5.2 The Cauchy Stress Tensor

Let us now consider the concept of stress tensors and how these relate to the internal forces. Consider the material body shown in Fig. 9.13, but subject to the same external loads as Fig. 9.12a. Across the sectioned surface, there exist many internal force system, $\boldsymbol{f}$. We first isolate a reference configuration with an *infinitesimal surface spatial element*, $\mathrm{d}S$, located at position, $\boldsymbol{X}$, having a surface traction, $\boldsymbol{T}$. This surface element has an outward normal, $\boldsymbol{N}$ which coincides with the unit vector at $\boldsymbol{X}$. If we impose the equations of static equilibrium on the surface element, we will obtain an infinitesimal *resultant internal force*, $\mathrm{d}\boldsymbol{f}$ as shown in Fig. 9.13.

Similarly, the material body of Fig. 9.13 can be isolated in a *current configuration* after the *reference configuration* material points have been transformed using a time-dependent mapping function, $M(t)$. We can identify the current configuration surface element, $\mathrm{d}s$ which is located at $\boldsymbol{x}$ such that the resultant force, $\mathrm{d}\boldsymbol{f}$ exists with an outward normal vector, $\boldsymbol{n}$, and surface traction vector, $\boldsymbol{t}$. The infinitesimal resultant force can be defined analytically thus:

$$\mathrm{d}\boldsymbol{f} = \boldsymbol{T}\mathrm{d}S \quad \text{or} \quad \mathrm{d}\boldsymbol{f} = \boldsymbol{t}\mathrm{d}s \tag{9.45}$$

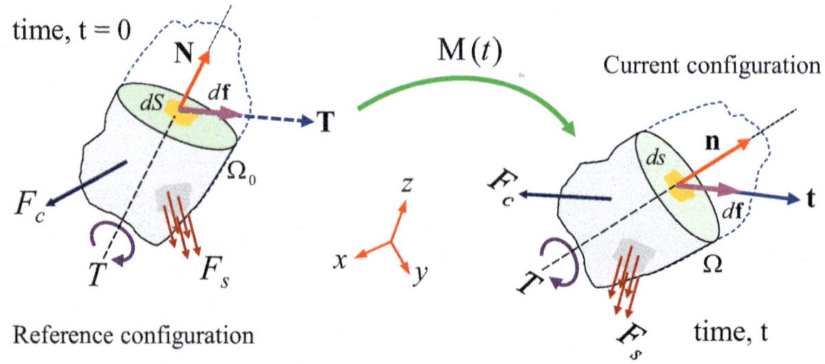

Fig. 9.13 An illustration of surface traction vectors on a material body

The *surface traction* (***T*** and ***t***) is the *force per unit surface area*. In the context of Eq. 9.45, ***t*** is called the *Cauchy traction vector*, which is the force per unit surface area, and d*s* defined on the current configuration. It is also regarded as the *true traction vector*.

On the other hand, the surface traction, ***T***, is called the *first Piola-Kirchoff traction vector*. It is the force per unit surface area, d*S* defined in the reference configuration. It is also called the *nominal traction vector* and is often described as a *pseudo traction vector* [5]. It is a virtual traction vector acting in the reference configuration, but with orientation coincident with the Cauchy traction vector, hence its representation by dash lines in Fig. 9.13. Examples of surface traction vectors include contact forces between contacting surfaces, and friction vectors, as well as hydrostatic forces. Even the kinetics of wind speed on a turbine blade constitute an example of a type of surface force.

According to the definitions of the surface traction vectors, i.e. force per unit surface area defined in a given configuration, one can make the projection that there must exist a relationship between surface traction vectors and a stress tensor since both quantities share the same unit. This extrapolation was made by Augustine-Louis Cauchy in what is now known as the *Cauchy stress theorem*.

The Cauchy stress theorem postulates that there must exist a second-order tensor field, $\boldsymbol{\sigma}$, in the current configuration or ***P***, in the reference configuration, such that:

$$\boldsymbol{t}(\boldsymbol{x}, t, \boldsymbol{n}) = \boldsymbol{\sigma}(\boldsymbol{x}, t)\boldsymbol{n} \qquad \text{or} \qquad \boldsymbol{T}(\boldsymbol{X}, t, \boldsymbol{N}) = \boldsymbol{P}(\boldsymbol{X}, t)\boldsymbol{N} \tag{9.46}$$

where $\boldsymbol{\sigma}$ is a *symmetric* spatial tensor field called the *Cauchy stress tensor*. It is also simply called the *Cauchy stress* or more appropriately, in line with undergraduate engineering mechanics, the true stress tensor. Also, ***P*** is called the *first Piola-Kirchoff stress tensor*. It is also regarded simply as the *Piola stress* or the *nominal stress tensor*.

In matrix notation, we re-write the current configuration representation of Eq. 9.46, for the *xyz*−reference frame, thus:

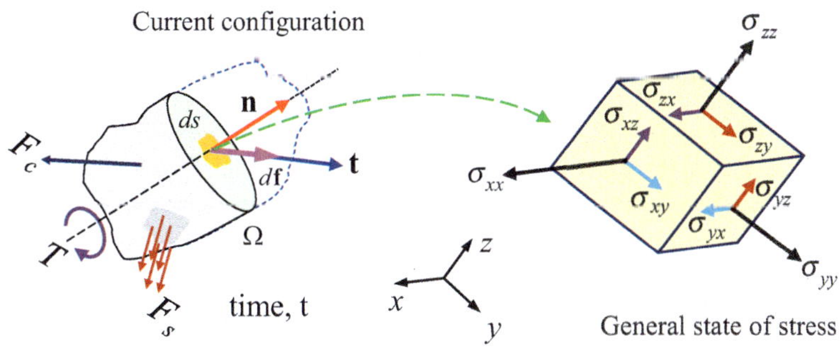

Fig. 9.14 The stress components that define the general state of stress within a material body

$$[t] = [\sigma]\,[n] \quad \longrightarrow \quad \begin{bmatrix} t_x \\ t_y \\ t_z \end{bmatrix} = \begin{bmatrix} \sigma_{xx} & \sigma_{xy} & \sigma_{xz} \\ \sigma_{yx} & \sigma_{yy} & \sigma_{yz} \\ \sigma_{zx} & \sigma_{zy} & \sigma_{zz} \end{bmatrix} \begin{bmatrix} n_x \\ n_y \\ n_z \end{bmatrix} \tag{9.47}$$

In the above, $[\boldsymbol{\sigma}]$ is the *Cauchy stress matrix* and it is a 3×3 matrix where the diagonal terms represents the normal stress terms and the off-diagonal terms are measures of shear stresses. The Cauchy stress matrix gives a comprehensive representation of a general state of stress for a point within a material body, as shown in Fig. 9.14.

As the material body continues to experience finite deformation, the material point of interest will change with position, time and orientation (through the outward normal vector). The Cauchy stress matrix must obey the principles of stress transformation discussed in undergraduate mechanics textbooks. The graphical representation of these orientation-dependent changes in magnitude of terms of the Cauchy stress matrix, is the *Mohr's circle of stress*.

Due to the requirement for static equilibrium and conservation of angular moment to be obeyed for a material body undergoing small deformations, we conclude that: $\sigma_{xy} \equiv \sigma_{yx}$, $\sigma_{xz} \equiv \sigma_{zx}$ and $\sigma_{yz} \equiv \sigma_{zy}$. Thus, the Cauchy stress therefore will have six independent terms. Since the Cauchy stress tensor is symmetric, it is fully defined by these six independent stress components instead of the nine terms specified in Eq. 9.47.

The Cauchy stress tensor, $\boldsymbol{\sigma}$, is described as an Eulerian stress formulation since its derivation is with respect to the current material configuration. It is very suitable for small deformation problems such as composite mechanics, viscoelasticity, etc. For problems where finite deformations dominate the material response, we recommend using other alternative measures of stress.

9.5.3 First Piola-Kirchoff Stress Tensor

Also, we can obtain the relationship between the Cauchy stress, $\boldsymbol{\sigma}$, and the first Piola-Kirchoff stress, $\boldsymbol{P}$. To do this, we have to map the area, dS defined in the reference configuration to the area, ds defined in the current configuration. The *Nanson's formula* is a useful relationship that accomplishes the mapping from one configuration to another. For the current area, ds, the Nanson's formula states:

$$\mathrm{d}s \quad = \quad J\boldsymbol{F}^{-T}\mathrm{d}S \quad \text{where} \quad J = \det\left[\boldsymbol{F}(\boldsymbol{X},t)\right] \tag{9.48}$$

where $\boldsymbol{F}$ is the deformation gradient tensor and J is the *volume ratio*. If $J = 1$, this suggests that det$\boldsymbol{F} = 1$ hence there is no motion of the material body. Such a deformation is defined as an *isochoric* or *volume-preserving* deformation.

Recall Eq. 9.45 which establishes that the infinitesimal internal force over an infinitesimal surface element can be defined equivalently in both reference and current configurations. Combining Eqs. 9.45 and 9.46 gives:

$$\boldsymbol{t}\left(\boldsymbol{x},\boldsymbol{n},t\right)\mathrm{d}s = \boldsymbol{T}\left(\boldsymbol{X},\boldsymbol{N},t\right)\mathrm{d}S \quad \longrightarrow \quad \boldsymbol{\sigma}\left(\boldsymbol{x},t\right)\boldsymbol{n}\mathrm{d}s = \boldsymbol{P}\left(\boldsymbol{X},t\right)\boldsymbol{N}\mathrm{d}S \tag{9.49}$$

Introducing the Nanson's formula of Eq. 9.48 into Eq. 9.49 yields:

$$\boldsymbol{\sigma}\left(\boldsymbol{x},t\right)\boldsymbol{n}\left[J\boldsymbol{F}^{-T}\mathrm{d}S\right] \quad = \quad \boldsymbol{P}\left(\boldsymbol{X},t\right)\boldsymbol{N}\mathrm{d}S \tag{9.50}$$

such that we obtain the expression of the first Piola-Kirchoff stress tensor as a function of the Cauchy stress:

$$\boldsymbol{P} = J\boldsymbol{\sigma}\boldsymbol{F}^{-T} \quad \text{or} \quad \boldsymbol{\sigma} = J^{-1}\boldsymbol{P}\boldsymbol{F}^{T} \tag{9.51}$$

The above is described as the *Piola transformation* with the stress in the current configuration transformed to its equivalent representation in the original/reference configuration. Since $\boldsymbol{\sigma}$ is a symmetric matrix, we note that: $\boldsymbol{\sigma} = \boldsymbol{\sigma}^T$, in which case we see that: $\boldsymbol{P}\boldsymbol{F}^T = \boldsymbol{F}\boldsymbol{P}^T$.

9.5.4 Kirchoff Stress Tensor

Apart from the Cauchy and first Piola-Kirchoff stress tensors, there exist other alternative measures of stress which are also used in continuum mechanics and the choice of a user depending on the type of problem under investigation. There are advantages and disadvantages associated with each measure of stress a user chooses to use. The *Kirchoff stress tensor*, $\boldsymbol{\tau}$ is one of such alternative stress tensor measures. It represents a multiple of the Cauchy stress and the volume ratio thus:

$$\boldsymbol{\tau} = J\boldsymbol{\sigma}. \tag{9.52}$$

9.5.5 Second Piola-Kirchoff Stress Tensor

A variant of the first Piola-Kirchoff stress tensor is introduced here, the *second Piola-Kirchoff stress tensor*, $\boldsymbol{S}$. It is obtained via the Kirchoff stress tensor according to the following equation:

$$\boldsymbol{S} = \boldsymbol{F}^{-1}\boldsymbol{\tau}\boldsymbol{F}^{-T} \quad \text{or also,} \quad \boldsymbol{S} = J\boldsymbol{F}^{-1}\boldsymbol{\sigma}\boldsymbol{F}^{-T} = \boldsymbol{F}^{-1}\boldsymbol{P} \tag{9.53}$$

where $\boldsymbol{P}$ is the first Piola-Kirchoff stress tensor, $\boldsymbol{\tau}$ is the Kirchoff stress tensor and $\boldsymbol{\sigma}$ is the Cauchy stress tensor. In continuum mechanics, it is widely accepted that $\boldsymbol{S}$ is the *pull-back* of $\boldsymbol{\tau}$ by $\boldsymbol{F}$ and $\boldsymbol{\tau}$ is the *push-forward* of $\boldsymbol{S}$.

The implication of Eq. 9.53 is that the second Piola-Kirchoff stress is symmetric i.e. $\boldsymbol{S} = \boldsymbol{F}^{-1}\boldsymbol{P} = \boldsymbol{S}^T$. A key relationship is therefore deduced between the first and second Piola-Kirchoff stress tensors:

$$\boldsymbol{P} = \boldsymbol{F}\boldsymbol{S} \tag{9.54}$$

which implies that the mapping function between the two stress measures is defined by the deformation gradient tensor.

9.5.6 Biot Stress Tensor

We define yet another material stress tensor called the *Biot stress tensor*, $\boldsymbol{T}_B$, which is defined in terms of the rotation tensor, $\boldsymbol{R}$, and the second Piola-Kirchoff stress tensor, $\boldsymbol{P}$, thus:

$$\boldsymbol{T}_B = \boldsymbol{R}^T\boldsymbol{P} \tag{9.55}$$

This is a *non-symmetric tensor* and not general ***positive-definite***. We can also re-derive $\boldsymbol{T}_B$ in terms of the right symmetric stretch tensor, $\boldsymbol{U}$. Recall that: $\boldsymbol{F} = \boldsymbol{R}\boldsymbol{U}$ and $\boldsymbol{P} = \boldsymbol{F}\boldsymbol{S}$ as derived previously. If we substitute these into Eq. 9.55, we obtain:

$$\boldsymbol{T}_B = \boldsymbol{R}^T\boldsymbol{P} \quad \longrightarrow \quad \boldsymbol{T}_B = \boldsymbol{R}^T\boldsymbol{F}\boldsymbol{S} = \boldsymbol{R}^T\boldsymbol{U}\boldsymbol{R}\boldsymbol{S} = \boldsymbol{U}\boldsymbol{S} \tag{9.56}$$

Also $\boldsymbol{R}^T\boldsymbol{P} \equiv \boldsymbol{R}\boldsymbol{P}^T$, so another expression of the Biot stress tensor becomes:

$$\boldsymbol{T}_B = \frac{1}{2}\left(\boldsymbol{R}^T\boldsymbol{P} + \boldsymbol{R}\boldsymbol{P}^T\right) \quad \text{or} \quad \boldsymbol{T}_B = \frac{\boldsymbol{R}^T\left(\boldsymbol{V}^{-1}\boldsymbol{\sigma} + \boldsymbol{\sigma}\boldsymbol{V}^{-1}\right)\boldsymbol{R}}{2} \tag{9.57}$$

If we multiply both parts of Eq. 9.55 by $\boldsymbol{R}$, we realize that: $\boldsymbol{P} = \boldsymbol{R}\boldsymbol{T}_B$, which implies that the second Piola-Kirchoff stress tensor, $\boldsymbol{P}$, is related to the Biot Stress tensor, $\boldsymbol{T}_B$, through a mapping function defined by the rotation tensor, $\boldsymbol{R}$. This stress measure is particularly suited for deformations in which finite rotations result during the deformation of the material body like hyperelasticity problems.

9.5.7 Corotated Cauchy Stress Tensor

We illustrated in Fig. 9.6 that for a material body subjected to a deformation gradient tensor, $\boldsymbol{F}$, its deformation comprises stretch tensors i.e. $\boldsymbol{U}$ or $\boldsymbol{V}$, and a rotation tensor, $\boldsymbol{R}$. We showed the polar decomposition of $\boldsymbol{F}$ to be: $\boldsymbol{F} = \boldsymbol{RU} = \boldsymbol{VR}$. In order to achieve the polar decomposition of $\boldsymbol{F} = \boldsymbol{RU}$, the material body was first *rotated* using $\boldsymbol{R}$ before it was stretched using $\boldsymbol{U}$. The result of the first rotation was a rotated material domain, which we called the *intermediate domain*. This intermediate domain is at the heart of a new stress tensor to be defined here.

The *Corotated Cauchy stress tensor*, $\boldsymbol{\sigma}_u$, is determined based on an intermediate domain that is neither the reference nor the current configuration. It is necessary for material responses in which a history of the immediate prior configuration is important.

To derive its values, let us consider Eq. 9.53 and make the Cauchy stress the subject of the formula:

$$\boldsymbol{S} = J\boldsymbol{F}^{-1}\boldsymbol{\sigma}\boldsymbol{F}^{-T} = \boldsymbol{F}^{-1}\boldsymbol{P} \quad \longrightarrow \quad \boldsymbol{\sigma} = J^{-1}\boldsymbol{FSF}^{T} \tag{9.58}$$

Equation 9.58 implies that the second Piola-Kirchoff stress, $\boldsymbol{S}$, has been operated by a deformation gradient, $\boldsymbol{F}$. If instead of using $\boldsymbol{F}$ to operate at this intermediate domain, we go ahead and operate on it using a stretch tensor according to Fig. 9.6, we end up determining the Corotated Cauchy stress tensor thus:

$$\boldsymbol{\sigma}_u = J^{-1}\boldsymbol{USU}^{T} \tag{9.59}$$

Recall also Eq. 9.53, which we now substitute into Eq. 9.60 such that we obtain:

$$\boldsymbol{\sigma}_u = J^{-1}\boldsymbol{U}\left[J\boldsymbol{F}^{-1}\boldsymbol{\sigma}\boldsymbol{F}^{-T}\right]\boldsymbol{U}^{T} \quad \longrightarrow \quad \boldsymbol{\sigma}_u = \boldsymbol{R}^{T}\boldsymbol{\sigma}\boldsymbol{R} \tag{9.60}$$

The above derivation of $\boldsymbol{\sigma}_u$ indicates that the original Cauchy stress has been rotated jointed (*corotated*) using the rotation tensor, $\boldsymbol{R}$. Therefore, this makes the formulation of the Cauchy stress now amenable to large deformation problems where finite deformations exists.

9.5.8 Mandel Stress Tensor

Another stress tensor especially for finite plasticity problems is called the *Mandel stress tensor*, Ξ. It is defined in terms of the right Cauchy-Green strain tensor, $\boldsymbol{C}$, and the second Piola-Kirchoff stress tensor, $\boldsymbol{S}$ according to:

$$\Xi = \frac{1}{2}(\boldsymbol{CS} + \boldsymbol{SC}) = \boldsymbol{CS} \quad \longrightarrow \quad \Xi = \boldsymbol{F}^{T}\boldsymbol{FS} \tag{9.61}$$

If we recall the expression of $\boldsymbol{S}$ in Eq. 9.53, we can re-derive the Mandel stress thus:

$$\boldsymbol{\Xi} = \boldsymbol{F}^T\boldsymbol{F}\boldsymbol{S} \quad \longrightarrow \quad \boldsymbol{\Xi} = \boldsymbol{F}^T\boldsymbol{F}\left[J\boldsymbol{F}^{-1}\boldsymbol{\sigma}\boldsymbol{F}^{-T}\right] \tag{9.62}$$

which demonstrates that the Mandel stress tensor can be related to the Cauchy stress tensor too via the deformation gradient, $\boldsymbol{F}$.

The Kirchoff, first and second Piola-Kirchoff, Biot, Corotated and Mandel stress tensors are all described as Lagrangian stress formulations since they are determined with respect to the reference configuration. They are comparable to nominal or engineering stresses. They are suitable for large deformation problems and hence these stress measures are commonly used in hyperelasticity problems.

Example 9.3 A steel plate of dimensions 40 × 40 mm², shown in Fig. 9.15, is subjected to three combined load cases such that edge $\overline{BC}$ is compressed a distance of 12 mm, edge $\overline{CD}$ compressed by 8 mm and also edge $\overline{CD}$ is sheared vertically upwards by 18 mm. Under this state of combined loading, the general state of stress at point, P within the plate is given by the Cauchy stress tensor, $\boldsymbol{\sigma}_P$ below.

$$\boldsymbol{\sigma}_P = \begin{bmatrix} -60 & 120 \\ 120 & -80 \end{bmatrix} \text{MPa}$$

(a) Determine the mapping function from material to spatial reference frame.
(b) Determine the deformation gradient, $\boldsymbol{F}$, for the above problem.
(c) Determine all the measures of stress tensors for this stress state.
(d) Plot a comparison bar chart for all stress measures for all the σ_{ij} terms.

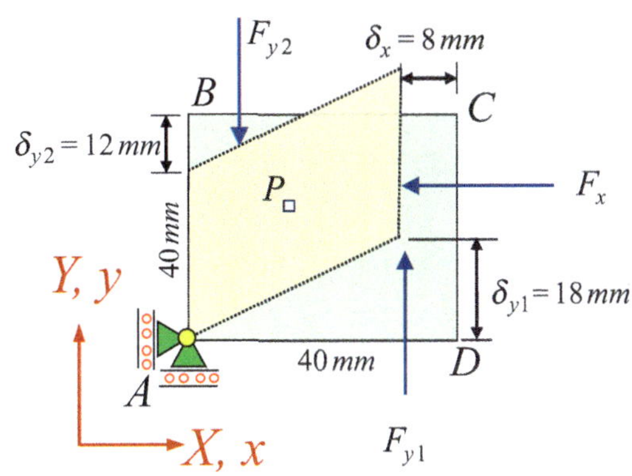

Fig. 9.15 The deformation of a steel plate subjected to three combined loads

Solution

(a) **Mapping function:** Similar to Example 9.2, the two-dimension mapping function should take the form:

$$x = \left(\frac{dx}{dX}\right)X + \left(\frac{dx}{dY}\right)Y \quad \longrightarrow \quad x = \left(\frac{32}{40}\right)X + \left(\frac{0}{40}\right)Y$$

$$y = \left(\frac{dy}{dX}\right)X + \left(\frac{dy}{dY}\right)Y \quad \longrightarrow \quad y = \left(\frac{18}{40}\right)X + \left(\frac{28}{40}\right)Y$$

The resulting mapping function becomes:

$$x = 0.80X + 0.00Y \tag{9.63}$$

$$y = 0.45X + 0.70Y \tag{9.64}$$

(b) **Deformation gradient tensor:** Based on mapping functions and Eqs. 9.63 and 9.64, the deformation gradient, $\boldsymbol{F}$ becomes:

$$\boldsymbol{F} = \begin{bmatrix} 0.80 & 0.00 \\ 0.45 & 0.70 \end{bmatrix} \tag{9.65}$$

(c) **Stress tensors:** We will determine the whole set of alternative measures of stress tensors discussed in Sect. 9.5 above.

First Piola-Kirchoff Stress Tensor, $\boldsymbol{P}$

$$\boldsymbol{P} = J\boldsymbol{\sigma}_P\boldsymbol{F}^{-T} \quad \longrightarrow \quad \boldsymbol{P} = \begin{bmatrix} -42 & 120 \\ 84 & -112 \end{bmatrix} \text{ MPa} \tag{9.66}$$

Kirchoff Stress Tensor, $\boldsymbol{\tau}$

$$\boldsymbol{\tau} = J\boldsymbol{\sigma}_P \quad \longrightarrow \quad \boldsymbol{\tau} = \begin{bmatrix} -33.6 & 67.2 \\ 67.2 & -44.8 \end{bmatrix} \text{ MPa} \tag{9.67}$$

Second Piola-Kirchoff Stress Tensor, $\boldsymbol{S}$

$$\boldsymbol{S} = J\boldsymbol{F}^{-1}\boldsymbol{\sigma}_P\boldsymbol{F}^{-T} \quad \longrightarrow \quad \boldsymbol{S} = \begin{bmatrix} -52.5 & 150.0 \\ 150.0 & -245.7 \end{bmatrix} \text{ MPa} \tag{9.68}$$

Biot Stress Tensor, $\boldsymbol{T}_B$

$$\boldsymbol{T}_B = \boldsymbol{R}^T J\boldsymbol{\sigma}_P\boldsymbol{F}^{-T} \quad \longrightarrow \quad \boldsymbol{T}_B = \begin{bmatrix} -88.9 & 223.4 \\ 152.0 & -280.9 \end{bmatrix} \text{ MPa} \tag{9.69}$$

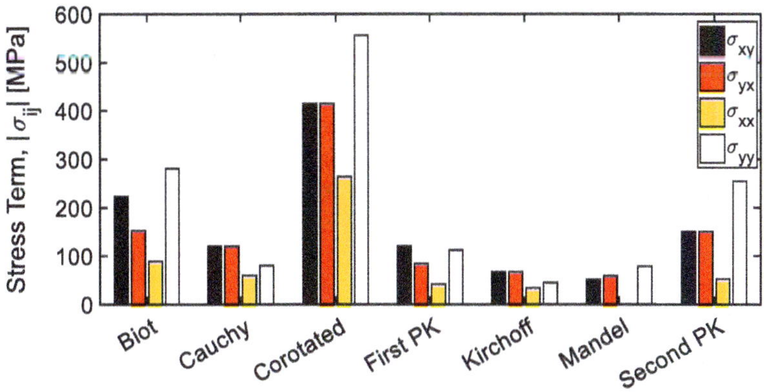

Fig. 9.16 A comparison bar chart of all stress measures for all terms of $\boldsymbol{\sigma}_P$

Corotated Stress Tensor, $\boldsymbol{\sigma}_u$

$$\boldsymbol{\sigma}_u = \boldsymbol{R}^T \boldsymbol{\sigma}_P \boldsymbol{R} \quad \longrightarrow \quad \boldsymbol{\sigma}_u = \begin{bmatrix} -264.0 & 414.6 \\ 414.6 & -556.0 \end{bmatrix} \text{ MPa} \tag{9.70}$$

Mandel Stress Tensor, $\boldsymbol{\Xi}$

$$\boldsymbol{\Xi} = \boldsymbol{F}^T \boldsymbol{F} J \boldsymbol{F}^{-1} \boldsymbol{\sigma}_P \boldsymbol{F}^{-T} \quad \longrightarrow \quad \boldsymbol{\Xi} = \begin{bmatrix} -0.0 & 51.2 \\ 58.8 & -78.4 \end{bmatrix} \text{ MPa} \tag{9.71}$$

(d) **Comparison bar chart for all stress measures:** Here, we consider the absolute magnitudes of the stress terms, $|\sigma_{ij}|$ where $i, j = x, y$ for all the different measures of stresses described previously. The result is shown in Fig. 9.16.

Remark 9.1 We observe from the comparison plots that there is a significant impact of a chosen stress measure in describing the stress state of a material point. For the problem considered, the *corotated stress tensor* gave consistently high values of the stress terms while the *Mandel stress tensor* gave the smallest stress term measure.

We can also observe that the *Cauchy stress* for all terms seems to be comparable with the *First Piola-Kirchoff stress tensor* terms. The FEM user should take great care in choosing the correct type of stress measure required for a given problem. The Corotated and Mandel stresses are particularly suited for problems in which significant deformation, and as a consequence, rotation of the problem domain, is experienced. Most problems with comparatively smaller deformations can be described sufficiently using the Cauchy stress and the Piola-Kirchoff stress tensors.

9.6 Practical Formulations of Stress

In the previous section, we identified classical formulations of the stress tensor generally used in nonlinear continuum mechanics. They describe comprehensively the general state of stress within a material body under deformation. Knowledge of them will help in material model development for not only the extreme stresses within the material body but also smaller stresses. This is why they have become quite useful in describing the material response of many materials in not only solid mechanics but geology, soil mechanics, and biomechanics.

However, in practice, it is important for design engineers to decide on the load bearing capacity of a chosen material based on a knowledge of the extreme stress values resulting from a chosen loading type. Also, *factors of safety* need to be chosen as part of the design process. When these two are known, choice of the right material for the design has to be made. Effective design processes within computational mechanics demand that the engineer gains knowledge of extreme values of stresses that act within the material domain.

Also, when a material body is subjected to a complex loading history, it is important for the engineer to determine the *size* and *direction* of the *effective stress* arising from the complex loading history. This effective stress is usually a single value, which can be related to material properties of candidate materials that can be used for the design.

All the above considerations call for definition of a new set of *practical formulations of stresses* that can be used to relate stress tensors to the mechanical properties of materials. Such formulations may include: *principal stresses, stress invariants, octahedral stresses, deviator stresses*, etc. In this section, we will present brief derivations of these practical formulations of stresses, with a view to using them in the next chapter to describe the material response within the finite element modelling process.

9.6.1 The Principal Stresses

The *principal stresses* are the first set of *extremal stress values* commonly used within the finite element scheme to describe the material response of a material body. It represents the *maximum* and *minimum normal stresses* as well as the *maximum shear stresses* acting in a material body. It also defines the orientation of the planes on which these extremal stress values act. In the following, we present the derivation of the principal stresses for a three-dimensional general state of stress tensor.

Consider a 3D material body bounded by domain, $\Omega \in \Re^n$, and subjected to multiple loads, as shown in Fig. 9.17. The key questions for a design engineer are:

- What is the *size* of the extremal stress value for the given load history?
- What is the *orientation* of the plane upon which this stress exists?

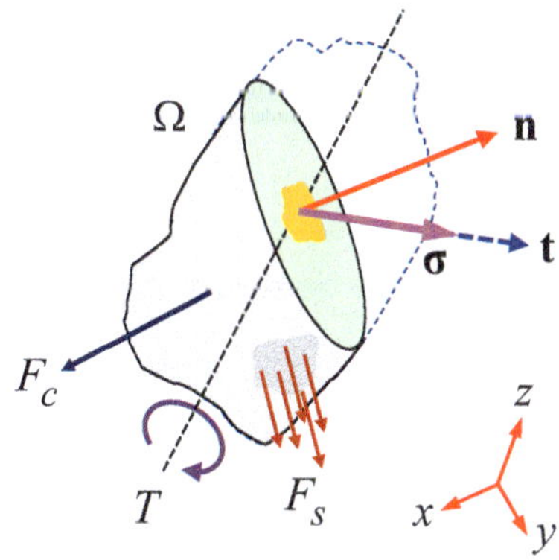

Fig. 9.17 An illustration of the principal stress plane, $\boldsymbol{n}$, for a sectioned material body

Let us assume that a material point in the 3D material body, located at position $\boldsymbol{x} \in \Omega$, and time, t, experiences a stress, $\boldsymbol{\sigma}$, such that the we can identify a plane shown in Fig. 9.17 with an outward pointing normal, $\boldsymbol{n}$, and Cauchy traction vector, $\boldsymbol{t}$. The principal stresses will exist on this plane, $\boldsymbol{n}$, if this plane is the plane on which the Cauchy traction vector, $\boldsymbol{t}$, is either a maximum or minimum, i.e. is an extremal stress value.

The Cauchy traction vector can be expressed in terms of the Cauchy stress thus:

$$\boldsymbol{t}(\boldsymbol{x}, t, \boldsymbol{n}) = \sigma(\boldsymbol{x}, t)\boldsymbol{n} \tag{9.72}$$

The unit vector, $\boldsymbol{n}$ at $\boldsymbol{x}$ is the direction of the *maximum* or *minimum* values of the *normal stress*, σ. The challenge here is to determine $\boldsymbol{n}$. In Continuum mechanics, the *Lagrange-multiplier method* is used to determine the principal stresses of σ.

Assume we are to determine the local maxima or minima of a function, $b(x, y)$, we describe how the *Lagrange-multiplier method* can be used to find these extremal values, subject to an *equality constraint*, say $d(x, y) = d_0$, where d_0 is a constant. The method defines a new *Lagrangian expression*, $\mathscr{L}(x, y, \lambda) = b(x, y) \pm \lambda\,[d(x, y) - d_0]$ where the new term, λ, called the *Lagrangian multiplier*, is introduced to link the original function with the equality constraint. The partial derivatives of $\mathscr{L}$ with respect to x and y must exist and have a value of zero at what is called a *stationary point*. local maximum or minimum of $b(x, y)$ exists at (x_0, y_0) if there also exists λ_0 such that the point (x_0, y_0, λ_0) of $\mathscr{L}$ is a stationary point. To obtain these extremal values, we have to solve the expression of the partial derivatives of $\mathscr{L}$.

Using the Lagrange-multiplier method [5], the Lagrange expression for Eq. 9.72 with respect to the normal stress, σ, and given an equality constraint of $|\boldsymbol{n}|^2 = 1$ becomes:

$$\mathscr{L}_\sigma(\boldsymbol{n}, \lambda) = \boldsymbol{n} \cdot \boldsymbol{\sigma n} - \lambda \left(|\boldsymbol{n}|^2 - 1 \right) \tag{9.73}$$

The equality constraint can be written in index notation as: $n_i n_i - 1 = 0$ (for i = 1,2,3). Taking λ as the *Lagrange multiplier*, let us re-write Eq. 9.73 in index notation thus:

$$\mathscr{L}_\sigma(n_i, \lambda) = n_i \sigma_{ij} n_j - \lambda (n_i n_i - 1) \tag{9.74}$$

The solution of the Lagrangian expression in Eq. 9.74 gives the extremal stress values needed to define the principal stresses and strain. First, we take the partial derivative of Eq. 9.74 with respect to the Lagrange multiplier, λ such that:

$$\frac{\partial \mathscr{L}_\sigma}{\partial \lambda} = n_i n_i - 1 = 0 \quad \longrightarrow \quad |n_i|^2 = 1 \quad \text{or} \quad |\boldsymbol{n}| = 1 \tag{9.75}$$

with the latter expression given in vector notation.

Next, we will also take the partial derivative of Eq. 9.74 with respect to n_i. However, for the 3-D material domain of Fig. 9.17, there is a dependence on the normal vectors i, j and k. Using the k-axis as the independent normal axis, with respect to Eq. 9.74, we will determine the partial derivatives in terms of the n_k normal axis, applying chain rule of differentiation, thus:

$$\frac{\partial \mathscr{L}_\sigma}{\partial n_k} = \frac{\partial \mathscr{L}_\sigma}{\partial n_i}\frac{\partial n_i}{\partial n_k} = 0 \tag{9.76}$$

We can now obtain the first term of Eq. 9.76 by taking the partial derivative with respect to n_i of Eq. 9.74 thus:

$$\frac{\partial \mathscr{L}_\sigma}{\partial n_i} = \sigma_{ij} n_j - 2\lambda n_i \tag{9.77}$$

However, to determine the partial derivative of the second term of Eq. 9.76, we introduce the *Kronecker delta*, δ_{ij}. Applying the Kronecker delta convention of Eq. 9.81 in terms of the i and k outward normal vectors, we determine (in index notation) that:

$$n_i = \delta_{ik} n_k \quad \longrightarrow \quad \frac{\partial n_i}{\partial n_k} = \delta_{ik} \tag{9.78}$$

We now combine Eqs. 9.77 and 9.78 according to Eq. 9.76 thus:

$$\begin{aligned} \frac{\partial \mathscr{L}_\sigma}{\partial n_k} &= \sigma_{ij} n_j \delta_{ik} - 2\lambda n_i \delta_{ik} &&= 0 \\ &= \sigma_{ij}\left(\delta_{jk} n_i + \delta_{ik} n_j\right) - 2\lambda n_i \delta_{ik} &&= 0 \\ &= 2\left(\sigma_{ki} n_i - \lambda n_k\right) &&= 0 \end{aligned} \tag{9.79}$$

Kronecker delta convention: Consider a 3-D material domain as in Fig. 9.17, we can define three orthonormal basis vectors, $\boldsymbol{e}_1, \boldsymbol{e}_2$ and $\boldsymbol{e}_3$ to define the 3-D coordinate system of the material domain. For the given *xyz*-reference frame, the orthonormal basis vectors reduce to a 3-D *Cartesian basis* defined by the ***ijk***−vectors. For such a system, it is true to state that:

$$\boldsymbol{e}_1 \cdot \boldsymbol{e}_2 = \boldsymbol{e}_2 \cdot \boldsymbol{e}_3 = \boldsymbol{e}_3 \cdot \boldsymbol{e}_1 = 0, \quad \text{and} \quad \boldsymbol{e}_1 \cdot \boldsymbol{e}_1 = \boldsymbol{e}_2 \cdot \boldsymbol{e}_2 = \boldsymbol{e}_3 \cdot \boldsymbol{e}_1 = 1$$

We can re-write the above expression by introducing the *Kronecker delta notation* thus:

$$\boldsymbol{e}_i \cdot \boldsymbol{e}_j = \delta_{ij} = \begin{cases} 1, & \text{if } i = j \\ 0, & \text{if } i \neq j \end{cases} \tag{9.80}$$

The following are implications of the Kronecker delta convention:

$$\delta_{ij}\delta_{jk} = \delta_{ik}, \quad \text{and} \quad \delta_{ij}\boldsymbol{n}_i = \boldsymbol{n}_j \tag{9.81}$$

where $\boldsymbol{n}_i$ and $\boldsymbol{n}_j$ are any given vectors which for our purposes here becomes the outward normal vectors.

The tensorial representation of Eq. 9.79 above becomes:

$$\boldsymbol{\sigma n} - \lambda \boldsymbol{n} = \boldsymbol{0} \quad \longrightarrow \quad (\boldsymbol{\sigma} - \lambda_i \boldsymbol{I})\,\boldsymbol{n}_i = \boldsymbol{0}, \text{ for } i = 1, 2, 3 \tag{9.82}$$

Remark 9.2 The solution of the Lagrangian expression reported in Eq. 9.82 represents an *eigenvalue problem* where the *eigenvalue* is the Lagrange multiplier, λ. For a stress tensor under investigation, there will exist three eigenvalues, λ_i, for $i = 1, 2, 3$. The unknown normal reaction vectors, $\boldsymbol{n}_i$, for $i = 1, 2, 3$ are the *eigenvectors*.

9.6.1.1 Stress Invariants

The physical nature of a tensor is defined by the eigenvalues and they are independent of chosen coordinates [5]. The eigenvalues of Eq. 9.82 are determined by obtaining the real roots of the characteristic equation of stress, $\boldsymbol{\sigma}$. The symmetry of $\boldsymbol{\sigma}$ ensures that the roots of the characteristic equation will always be real. The characteristic equation is generated by setting the determinant of Eq. 9.82 to zero. In matrix notation, the determinant expression of the stress tensor, is given as:

$$\det(\boldsymbol{\sigma} - \lambda \boldsymbol{I}) = \det\left(\begin{bmatrix} \sigma_{11} - \lambda & \sigma_{12} & \sigma_{13} \\ \sigma_{21} & \sigma_{22} - \lambda & \sigma_{23} \\ \sigma_{31} & \sigma_{32} & \sigma_{33} - \lambda \end{bmatrix}\right) = 0 \tag{9.83}$$

The resulting characteristic equation based on the determinant of the matrix of the above equation becomes:

$$\det(\boldsymbol{\sigma} - \lambda_i \boldsymbol{I}) = 0 \quad \longrightarrow \quad \lambda_i^3 - I_1\lambda_i^2 + I_2\lambda_i - I_3 = 0 \quad \text{for } i = 1,2,3 \tag{9.84}$$

In the above, we have introduced I_i, for $i = 1,2,3$, which are called *principal stress invariants* of the tensor, $\boldsymbol{\sigma}$. The specific expressions for all three stress invariants are:

$$\begin{aligned} I_1 &= \mathrm{tr}(\boldsymbol{\sigma}) &&= \lambda_1 + \lambda_2 + \lambda_3 \\ I_2 &= \frac{1}{2}\left[[\mathrm{tr}(\boldsymbol{\sigma})]^2 + \mathrm{tr}\left[\boldsymbol{\sigma}^2\right]\right] &&= \lambda_1\lambda_2 + \lambda_1\lambda_3 + \lambda_2\lambda_3 \\ I_3 &= \det(\boldsymbol{\sigma}) &&= \lambda_1\lambda_2\lambda_3 \end{aligned} \tag{9.85}$$

where $\mathrm{tr}(\boldsymbol{\sigma})$ is the *trace* of the stress tensor, σ. The above parameters are described as *stress invariants*, because these values of stresses do not depend on the chosen coordinate system of the stress tensor or any chosen area element upon which the stress tensor acts. They are independent quantities and hence can be comparable to the material properties for a given material.

9.6.1.2 Principal Normal Stresses

Principal normal stresses that we set out to define, are based on the three eigenvalues, λ_i for $i = 1,2,3$, of the stress tensor, $\boldsymbol{\sigma}$, such that:

- Maximum principal normal stress, $\sigma_1 = \max(\lambda_1, \lambda_2, \lambda_3)$
- Median principal normal stress, $\sigma_2 = \mathrm{median}(\lambda_1, \lambda_2, \lambda_3)$
- Minimum principal normal stress, $\sigma_3 = \min(\lambda_1, \lambda_2, \lambda_3)$
- Mean or average stress, $\sigma_m = \frac{1}{3}\mathrm{tr}\boldsymbol{\sigma} = \frac{1}{3}(\lambda_1 + \lambda_2 + \lambda_3)$

9.6.1.3 Principal Planes

Finally, the unknown normal reaction vector, $\boldsymbol{n}_i$ for $i = 1,2,3$, is the *eigenvector* of stress tensor, $\boldsymbol{\sigma}$. It represents the principal plane on which the principal stresses exist. For the three principal stresses, the orientation of their acting planes is defined by $\boldsymbol{n}_i$ for $i = 1,2,3$.

For example, let us say λ_1 is the corresponding eigenvalue for the maximum principal normal stress, σ_1. The principal plane of σ_1 can be determined thus:

$$(\boldsymbol{\sigma} - \lambda_1\boldsymbol{I})\,\boldsymbol{n}_1 = \boldsymbol{0} \longrightarrow \left(\begin{bmatrix} \sigma_{xx} & \sigma_{xy} & \sigma_{xz} \\ \sigma_{yx} & \sigma_{yy} & \sigma_{yz} \\ \sigma_{zx} & \sigma_{zy} & \sigma_{zz} \end{bmatrix} - \lambda_1 \begin{bmatrix} 1 & 0 & 0 \\ 0 & 1 & 0 \\ 0 & 0 & 1 \end{bmatrix}\right) \begin{bmatrix} n_x \\ n_y \\ n_z \end{bmatrix} = \begin{bmatrix} 0 \\ 0 \\ 0 \end{bmatrix} \tag{9.86}$$

Now that we have defined the eigenvalues and eigenvectors, we can understand their implications to the principal stresses discussions:

- The solution of the above Eq. 9.86 should yield the components of $\boldsymbol{n}$, i.e. n_x, n_y and n_y. Thus, in terms of the Cartesian orthonormal bases becomes: $\boldsymbol{n}_1 = n_x\boldsymbol{e}_1 + n_y\boldsymbol{e}_2 + n_z\boldsymbol{e}_3$. This also defines the principal plane in 3-D for the maximum principal stress, σ_1.
- Recall that the *direction cosines* of $\boldsymbol{n}_1$ for a typical 3-D system are the cosines of the angles α, β and γ which $|\boldsymbol{n}_1|$ makes with the $x, y-$ and $z-$axes respectively. The angular orientation of $\boldsymbol{n}_1$ with respect to the $x-$axis is:

$$\alpha = \cos^{-1}\left(\frac{n_x}{\sqrt{n_x^2 + n_y^2 + n_z^2}}\right) \tag{9.87}$$

- It also follows that: $\cos^2\alpha + \cos^2\beta + \cos^2\gamma = 1$. A similar process is repeated for the other principal stresses i.e. σ_2 and σ_3.
- For a detailed example of how to determine the eigenvectors, and hence the normal vector, $\boldsymbol{n}_i$ $(i = 1, 2, 3)$ that defines the principal normal stress plane, refer to [6].
- For our purposes here, we will use MATLAB™ to generate the eigenvalues and vectors. Please refer to Example 9.4 for a practical illustration of how this is done.
- The stress tensor, $\boldsymbol{\sigma}$, is **symmetric**, therefore the set of eigenvectors form a set of orthogonal basis whose directions coincide with the direction of action of the principal stresses, σ_i for $i = 1, 2, 3$. This new type of basis vector is described to be an *eigenbasis*.
- In the eigenbasis representation, the stress tensor $\boldsymbol{\sigma}$ will now consist solely of diagonal terms which are the principal stresses whilst other non-diagonal terms vanish (i.e. become zero). This is achieved using a *spectral decomposition* of the stress tensor, $\boldsymbol{\sigma}$, defined thus:

$$\boldsymbol{\sigma}(\boldsymbol{x}, t) = \sum_{i=1}^{3} \sigma_i \boldsymbol{n}_i \otimes \boldsymbol{n}_i \tag{9.88}$$

Example 9.4 A 2-D stress element, under multiple loading history, is shown below.

(a) Determine all the Lagrange multipliers, λ_i for $i = 1, 2$.
(b) Determine all the stress invariants.
(c) Determine the principal stresses for this system.
(d) Determine the principal normal stress plane, $\boldsymbol{n}_i$ for $i = 1, 2$.
(e) Determine the angles of the principal normal stress.

Solution

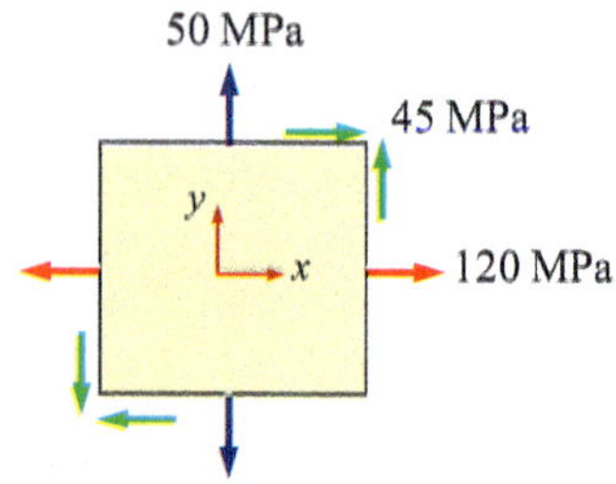

Fig. 9.18 A stress element subjected to multiple loads

(a) **Lagrange multipliers:** The stress tensor for the stress element shown in Fig. 9.18 is:

$$\boldsymbol{\sigma} = \begin{bmatrix} 120 & 45 \\ 45 & 50 \end{bmatrix} \quad \text{MPa} \tag{9.89}$$

The Lagrange multipliers, λ_i for $i = 1, 2$, is determined by obtaining the eigenvalues of $\boldsymbol{\sigma}$. We will use MATLAB™ here to obtain the eigenvalues and eigenvectors of $\boldsymbol{\sigma}$. The command for doing this is called `eig`. If N and L are the matrix of eigenvectors and eigenvalues of $\boldsymbol{\sigma}$ respectively, then the MATLAB™ implementation becomes:

```
stressTensor = [120e6 45e6; 45e6 50e6]

stressTensor =
   120000000    45000000
    45000000    50000000

[N,L] = eig(stressTensor)

N =
    0.4394   -0.8983
   -0.8983   -0.4394

L =
   1.0e+08 *
    0.2799         0
         0    1.4201

lambdaMultiplier = diag(L)

lambdaMultiplier =
   1.0e+08 *
    0.2799
    1.4201
```

From the above, we see that the Lagrange multipliers are:

$$L = [\lambda_1 \quad \lambda_2] \quad \longrightarrow \quad L = [27.99 \quad 142.01] \quad \text{MPa} \tag{9.90}$$

(b) **Stress invariants:** From Eq. 9.85, we get the stress invariants thus:

$$\begin{aligned} I_1 &= \text{tr}(\sigma) && \longrightarrow I_1 = 170 \times 10^6 \quad \text{Pa} \\ I_2 &= \frac{1}{2}\left\{[\text{tr}(\sigma)]^2 + \text{tr}\left[\sigma^2\right]\right\} && \longrightarrow I_2 = 2.4925 \times 10^{16} \quad \text{Pa}^2 \\ I_3 &= \det(\sigma) && \longrightarrow I_3 = 3.9750 \times 10^{15} \quad \text{Pa}^3 \end{aligned}$$

(c) **Principal stresses:** These are determined as follows:

$$\begin{aligned} &\text{Maximum: } \sigma_1 = \max(\lambda_1, \lambda_2, \lambda_3) && \longrightarrow \sigma_1 = 142.01 \ \text{MPa} \\ &\text{Minimum: } \sigma_2 = \min(\lambda_1, \lambda_2, \lambda_3) && \longrightarrow \sigma_2 = \ 27.99 \ \text{MPa} \end{aligned}$$

(d) **Principal normal stress plane:** The principal normal stress planes are determined based on the values of the eigenvectors. For the σ_2 stress, the plane is defined by the first column of the eigenvector, N, defined above. Therefore,

$$\begin{aligned} \text{Plane for } \sigma_1 : \boldsymbol{n}_1 &= -0.8983\boldsymbol{e}_1 - 0.4394\boldsymbol{e}_2 \\ \text{Plane for } \sigma_2 : \boldsymbol{n}_2 &= \ \ 0.4394\boldsymbol{e}_1 - 0.8983\boldsymbol{e}_2 \end{aligned}$$

where $\boldsymbol{e}_1$ and $\boldsymbol{e}_2$ are orthonormal bases that describe the problem domain.

(e) **Angle of principal normal plane for $\boldsymbol{n}_1$ and $\boldsymbol{n}_2$:** We use the principal normal plane, $\boldsymbol{n}_1$, and note as follows:

$$\begin{aligned} \cos\alpha_1 &= -0.8983 && \longrightarrow \alpha_1 = \cos^{-1}(-0.8983) = 153.94° \\ \cos\beta_1 &= -0.4394 && \longrightarrow \beta_1 = \cos^{-1}(-0.4394) = 116.06 \end{aligned}$$

Similar implementation applies in determining the principal normal plane angles for the second orthonormal vector, $\boldsymbol{n}_2$. The results are $\alpha_2 = 63.93°$ and $\beta_2 = 153.94°$ for angles of inclination with respect to the $\boldsymbol{e}_1$ and $\boldsymbol{e}_2$ orthonormal bases respectively.

9.6.1.4 Principal Shear Stresses

The next challenge is to establish a stress-tensor based representation of the *maximum* and *minimum shear stresses* acting on a material body under the effect of diverse loading history. We will work through the process of doing this, following similar approach used for determining the principal normal stress terms.

Consider the same material body of Fig. 9.17 subjected to three multiple loading types. Let us identify a location, $\boldsymbol{x}$, at time, t, such that it experiences a general Cauchy traction vector, $\boldsymbol{t}$, as shown in Fig. 9.19. The general traction vector can be decomposed into its *parallel* and perpendicular components. The parallel component of $\boldsymbol{t}$, defined by a *normal outward acting vector*, $\boldsymbol{n}$, leads to the normal Cauchy stress, $\boldsymbol{\sigma}$, treated previously [5]. The other perpendicular component of $\boldsymbol{t}$ acts parallel to the sectioned plane and defined by the outward pointing normal, $\boldsymbol{m}$.

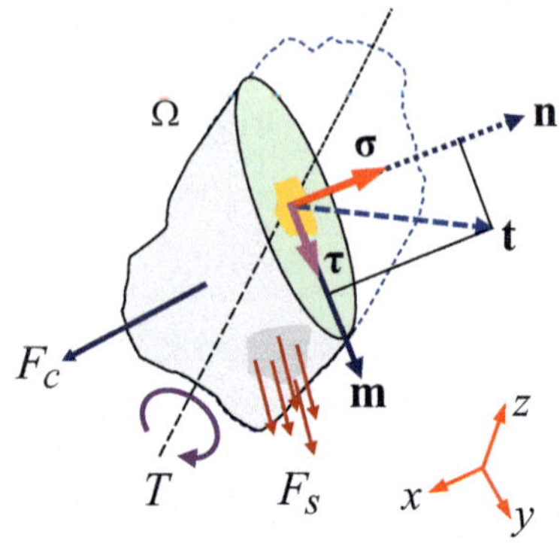

Fig. 9.19 An illustration of the principal shear and normal stresses on a sectioned plane

According to Fig. 9.19, we observe that:

$$\begin{array}{llll} \text{Normal stresses:} & \sigma = & \boldsymbol{n}\cdot\boldsymbol{t} & = \boldsymbol{n}\cdot\boldsymbol{\sigma}\boldsymbol{n} \\ \text{Shear stresses:} & \tau = & \boldsymbol{m}\cdot\boldsymbol{t} & = \boldsymbol{m}\cdot\boldsymbol{\sigma}\boldsymbol{n} \\ \text{Cauchy traction vector:.} & |\boldsymbol{t}|^2 = & \sigma^2+\tau^2 & \end{array} \tag{9.91}$$

If we switch to the eigenbasis representation, then the normal outward pointing vector, in orthogonal basis representation, becomes:

$$\boldsymbol{n} = n_1\boldsymbol{n}_1 + n_2\boldsymbol{n}_2 + n_3\boldsymbol{n}_3 \tag{9.92}$$

where $n_1 \equiv n_x, n_2 \equiv n_y$ and $n_3 \equiv n_z$ of Eq. 9.86. In the eigenbasis space, the stress tensor, $\boldsymbol{\sigma}$, is *diagonalized* with the diagonal elements being the principal normal stresses, σ_i for $i = 1, 2, 3$. Therefore, the representation of the general Cauchy traction vector, represented in terms of $\boldsymbol{n}$, becomes:

$$\boldsymbol{t} = \boldsymbol{\sigma}\boldsymbol{n} \quad \longrightarrow \quad \boldsymbol{t} = \sigma_1 n_1\boldsymbol{n}_1 + \sigma_2 n_2\boldsymbol{n}_2 + \sigma_3 n_3\boldsymbol{n}_3 \tag{9.93}$$

Using Eqs. 9.91_1 and 9.91_3, determine the function of the shear stress in terms of the Cauchy traction vector and the diagonalized (principal) stress tensor.

$$\tau^2 = |\boldsymbol{t}|^2 - \sigma^2 \quad \longrightarrow \quad \tau^2 = |\sigma_1 n_1\boldsymbol{n}_1 + \sigma_2 n_2\boldsymbol{n}_2 + \sigma_3 n_3\boldsymbol{n}_3|^2 - (\boldsymbol{n}\cdot\boldsymbol{\sigma}\boldsymbol{n})^2 \tag{9.94}$$

Evaluating Eq. 9.94, bearing in mind that: $\boldsymbol{n}_i \cdot \boldsymbol{n}_i = 1$ and $\boldsymbol{n}_i \cdot \boldsymbol{n}_j = 0$ for $i, j = 1, 2, 3$, we obtain:

$$\tau^2 = \left(\sigma_1^2 n_1^2 + \sigma_2^2 n_2^2 + \sigma_3^2 n_3^2\right) \quad - \quad \left[\sigma_1 n_1^2 + \sigma_2 n_2^2 + \sigma_3 n_3^2\right]^2 \tag{9.95}$$

The above is the expression of the variation of shear force on the arbitrary plane defined by the eigenbasis vector, $\boldsymbol{n}$. For the purposes of determining the *extremal shear stress values*, which are the principal shear stresses we set out at first to obtain, let us now implement again the *Lagrange-multiplier method*.

Using the *Lagrange multiplier*, λ, we define a new Lagrange expression, $\mathscr{L}_\tau$, for the shear stress expression above, thus:

$$\mathscr{L}_\tau = \tau^2 - \lambda^2(|\boldsymbol{n}|^2 - 1) \tag{9.96}$$

where $|\boldsymbol{n}|^2 = 1$ is the *equality constraint* for the Lagrangian expression, $\mathscr{L}_\tau$. Equation 9.96 in index notation becomes:

$$\mathscr{L}_\tau = \tau^2 - \lambda^2(n_1^2 + n_2^2 + n_3^3 - 1) \tag{9.97}$$

where τ^2 is according to Eq. 9.95. The extremal shear stress values will exist at the point where the partial derivative of $\mathscr{L}_\tau$ with respect to the orthogonal basis, n_i for $i = 1, 2, 3$ is equal to zero. For the first orthogonal basis, n_1, the partial derivative becomes:

$$\frac{\partial \mathscr{L}_\tau}{\partial n_1} = 2\sigma_1^2 n_1 - 2\left(\sigma_1 n_1^2 + \sigma_2 n_2^2 + \sigma_3 n_3^2\right)(2\sigma_1 n_1) - \lambda^2 n_1 = 0 \tag{9.98}$$

For all three orthogonal bases, the corresponding partial derivatives become:

$$\frac{\partial \mathscr{L}_\tau}{\partial n_1} = 0 \longrightarrow \sigma_1^2 n_1 - \left(\sigma_1 n_1^2 + \sigma_2 n_2^2 + \sigma_3 n_3^2\right)(2\sigma_1 n_1) - \lambda^2 n_1 = 0 \tag{9.99}$$

$$\frac{\partial \mathscr{L}_\tau}{\partial n_2} = 0 \longrightarrow \sigma_2^2 n_2 - \left(\sigma_1 n_1^2 + \sigma_2 n_2^2 + \sigma_3 n_3^2\right)(2\sigma_2 n_2) - \lambda^2 n_2 = 0 \tag{9.100}$$

$$\frac{\partial \mathscr{L}_\tau}{\partial n_3} = 0 \longrightarrow \sigma_3^2 n_3 - \left(\sigma_1 n_1^2 + \sigma_2 n_2^2 + \sigma_3 n_3^2\right)(2\sigma_2 n_3) - \lambda^2 n_3 = 0 \tag{9.101}$$

Equations 9.99, 9.100 and 9.101 satisfy a *necessary* and *sufficient* condition for $\mathscr{L}_\tau$ to take stationary values. For us to obtain the extremal values of $\mathscr{L}_\tau$, we established that $\mathscr{L}_\tau$ must have an equality constraint condition $|\boldsymbol{n}|^2 = 1$ or in expanded index nodation: $n_1^2 + n_2^2 + n_3^2 = 1$.

For the equality constraint to be obeyed, we identify three combinations of n_1, n_2 and n_3 thus:

- Consider the instance where $n_1 = 0$, $n_2 = 0$ and $n_3 = \pm 1$. Substituting these into Eqs. 9.99, 9.100, 9.101 and 9.95 results in: $\lambda = \sigma_3$ and $\tau = 0$.
- Similarly, if $n_1 = 0, n_3 = 0$ and $n_2 = \pm 1$, then: $\lambda = \sigma_2$ and $\tau = 0$.
- Also, when $n_2 = 0, n_3 = 0$ and $n_1 = \pm 1$, hence: $\lambda = \sigma_1$ and $\tau = 0$.
- If only $n_1 = 0$, then we will have: $n_2^2 + n_3^2 = 1$. However, we established previously that $n_2^2 = n_3^2$, which implies that we can conclude that: $n_2 = \pm \frac{1}{\sqrt{2}}$ and $n_3 = \pm \frac{1}{\sqrt{2}}$. When we substitute these values of n_1, n_2 and n_3 into Eq. 9.97, we obtain:

$$\tau^2 = \left[0 + \frac{\sigma_2^2}{2} + \frac{\sigma_3^2}{2}\right] - \left[0 + \frac{\sigma_2}{2} + \frac{\sigma_3}{2}\right]^2$$

- The simplification of the above equation yields the first expression of the extremal shear stress value, $\tau_{(2,3)}$ determined in terms of σ_2 and σ_3 principal normal stress values. Hence:

$$\tau_{(2,3)} = \pm\frac{1}{2}\left(\sigma_2 - \sigma_3\right) \tag{9.102}$$

- Similar evaluation of Eq. 9.97 using $n_1 = n_2 = \frac{1}{\sqrt{2}}$, $n_3 = 0$ and $n_1 = n_3 = \frac{1}{\sqrt{2}}$, $n_2 = 0$ will yield the following expressions of extremal shear stress values:

$$\tau_{(1,2)} = \pm\frac{1}{2}\left(\sigma_1 - \sigma_2\right) \quad \text{and} \quad \tau_{(1,3)} = \pm\frac{1}{2}\left(\sigma_1 - \sigma_3\right) \tag{9.103}$$

respectively. In the above, $\tau_{(i,j)}$ for $i, j = 1, 2, 3$, is the extremal shear stress value associated with σ_i and σ_j principal normal stresses.

The *maximum shear stress*, τ_{max}, is usually the largest of the three extremal values given in Eqs. 9.102 and 9.103. In other words, if *maximum principal normal stress* is σ_{max} and the *minimal principal normal stress* is σ_{min}, we can express the *maximum shear stress*, τ_{max} as:

$$\tau_{max} = \frac{1}{2}\left|\sigma_{max} - \sigma_{min}\right| \tag{9.104}$$

9.6.2 The Octahedral Stresses

We introduce the next practical formulation of stresses called the *octahedral stresses*. It is another measure of stress that can be used to define failure or yield criteria for a given type of material. The octahedral stresses are generally *octahedral normal stresses*, $\boldsymbol{\sigma}_{oct}$, and the *octahedral shear stresses*, $\boldsymbol{\tau}_{oct}$. We will now describe their representation and formulation.

Consider a sectioned 3-D view of material body shown in Fig. 9.20, which is subjected to diverse loading histories. We can isolate a set of eight planes within the material body that combine to form an *octahedron*. These planes are described as *octahedral planes*. The octahedron has its coordinate axes coincident with the three principal normal stress directions described previously.

The octahedral plane is also the plane defined by its outward acting normal vector, $\boldsymbol{n}_{oct}$, and parallel-to-octahedral-plane vector, $\boldsymbol{m}_{oct}$, as shown in Fig. 9.21. Each of the set of octahedral planes forms equal angles with the principal normal

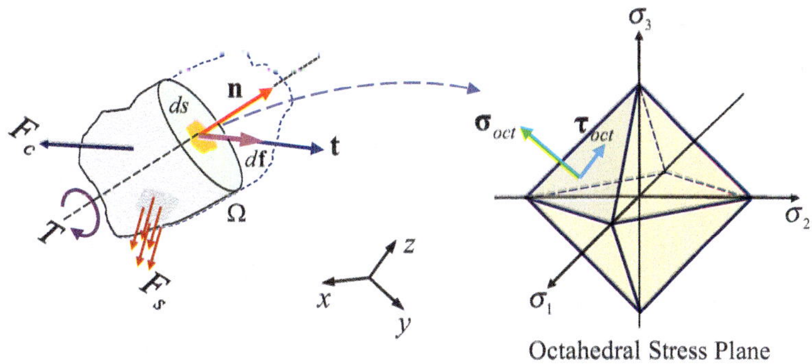

Fig. 9.20 The octahedral planes, showing octahedral normal and shear stresses

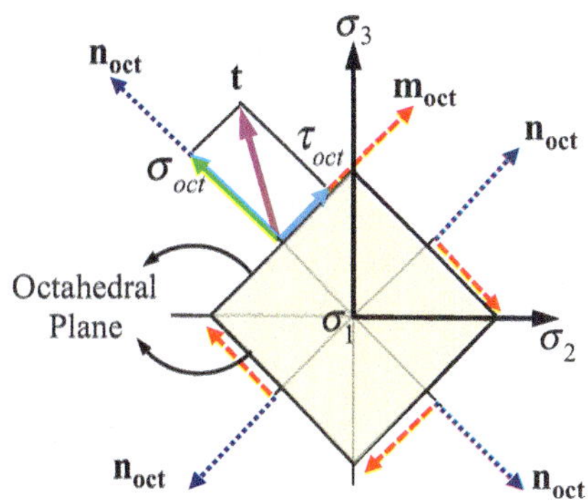

Fig. 9.21 A 2-D view of the octahedral stresses octahedron showing octahedral normal, $\boldsymbol{\sigma}_{oct}$, and shear, $\boldsymbol{\tau}_{oct}$, stresses and traction vector, $\boldsymbol{t}$. Note that each plane is defined by a normal outward acting vector, $\boldsymbol{n}_{oct}$, and a parallel-to-plane vector, $\boldsymbol{m}_{oct}$

stress directions. If the material body is stressed by a Cauchy vector, $\boldsymbol{\sigma}$, and we isolate the octahedral planes, the octahedral normal stress, $\boldsymbol{\sigma}_{oct}$, acts along the $\boldsymbol{n}_{oct}$ direction whilst the octahedral shear stress, $\boldsymbol{\tau}_{oct}$, acts along the $\boldsymbol{m}_{oct}$ direction.

It is a *necessary requirement* that the octahedral plane makes equal angles with the principal normal directions. If we define the angles of inclinations the outward normal acting vector, $\boldsymbol{n}_{oct}$, makes with the principal axes (1,2,3 – directions) as α, β and γ, then we note that since $\alpha = \beta = \gamma$, hence $\cos^2\alpha + \cos^2\beta + \cos^2\gamma = 1$. Therefore, $\cos\alpha = \dfrac{1}{\sqrt{3}}$ and $\alpha = 54.74°$, is the angle $\boldsymbol{n}_{oct}$ makes with the principal axes directions. Therefore, the components of $\boldsymbol{n}$ become: $n_1 = n_2 = n_3 = \dfrac{1}{\sqrt{3}}$.

Hence, the vector that describes a given octahedral plane is:

$$\boldsymbol{n}_{oct} = \frac{1}{\sqrt{3}}(\boldsymbol{e}_1 + \boldsymbol{e}_2 + \boldsymbol{e}_3) \tag{9.105}$$

where $\boldsymbol{e}_i$ for $i = 1, 2, 3$ are the orthonormal bases for the principal stress space. If the Cauchy stress acting on the sectioned plane is defined as $\boldsymbol{\sigma}$, then in the principal stress space, the applicable stress will be its equivalent *diagonalized* form, $\boldsymbol{\sigma}_{diag}$ comprising the principal stress terms thus:

$$\boldsymbol{\sigma}_{diag} = \begin{bmatrix} \sigma_1 & 0 & 0 \\ 0 & \sigma_2 & 0 \\ 0 & 0 & \sigma_3 \end{bmatrix} \tag{9.106}$$

where σ_1, σ_2 and σ_3 are the principal normal stresses of $\boldsymbol{\sigma}$. The component of the traction vector, $\boldsymbol{t}$, that acts in direction of the octahedral normal stress is $\boldsymbol{t}_{oct}$, and it is determined thus:

$$\boldsymbol{t}_{oct} = \sigma_{diag}\boldsymbol{n}, \quad \longrightarrow \quad \boldsymbol{t}_{oct} = \sigma_{ij}n_j\boldsymbol{e}_j = \frac{1}{\sqrt{3}}\left(\sigma_1\boldsymbol{e}_1 + \sigma_2\boldsymbol{e}_2 + \sigma_3\boldsymbol{e}_3\right) \tag{9.107}$$

for $i, j = 1, 2, 3$. The octahedral normal stress, σ_{oct}, is derived accordingly:

$$\begin{aligned} \sigma_{oct} = \boldsymbol{n} \cdot \boldsymbol{\sigma}_{diag}\boldsymbol{n} \quad \longrightarrow \sigma_{oct} &= t_{i,oct}n_i = \sigma_{ij}n_in_j \\ &= \sigma_1 n_1^2 + \sigma_2 n_2^2 + \sigma_3 n_3^2 \\ &= \tfrac{1}{3}\left(\sigma_1 + \sigma_2 + \sigma_3\right) \end{aligned} \tag{9.108}$$

The octahedral normal stress can be expressed in terms of the stress invariants (refer to Eq. 9.109), where I_1 is the first stress invariant of $\boldsymbol{\sigma}$, i.e. $\lambda_1 \equiv \sigma_1$.

$$\sigma_{oct} = \tfrac{1}{3}I_1 \tag{9.109}$$

Similarly, we can deduce the octahedral shear stress as follows:

$$\tau_{oct}^2 = |\boldsymbol{t}_{oct}|^2 - \sigma_{oct}^2 \quad \longrightarrow \tau_{oct} = \sqrt{\left(t_{i,oct}\right)^2 - \sigma_{oct}^2} \tag{9.110}$$

where, based on Eq. 9.107, we deduce that:

$$\left(t_{i,oct}\right)^2 = \tfrac{1}{3}\left(\sigma_1^2 + \sigma_2^2 + \sigma_3^3\right) \tag{9.111}$$

and using Eq. 9.108, the square of octahedral normal stress becomes:

$$\begin{aligned} \sigma_{oct}^2 &= \tfrac{1}{9}\left(\sigma_1 + \sigma_2 + \sigma_3\right)^2 \\ &= \tfrac{1}{9}\left(\sigma_1^2 + \sigma_2^2 + \sigma_3^2 + 2\sigma_1\sigma_2 + 2\sigma_1\sigma_3 + 2\sigma_2\sigma_3\right) \end{aligned} \tag{9.112}$$

Now, if we evaluate Eq. 9.110 by combining Eqs. 9.111 and 9.109, we obtain:

$$\begin{aligned} \tau_{oct} &= \sqrt{\left(t_{i,oct}\right)^2 - \sigma_{oct}^2} \\ &= \tfrac{1}{3}\left[\left(\sigma_1^2 + \sigma_2^2 + \sigma_3^2\right) - 2\left(\sigma_1\sigma_2 + \sigma_1\sigma_3 + \sigma_2\sigma_3\right)\right]^{\frac{1}{2}} \\ &= \tfrac{1}{3}\left[\left(\sigma_1 - \sigma_2\right)^2 + \left(\sigma_2 - \sigma_3\right)^2 + \left(\sigma_3 - \sigma_1\right)^2\right]^{\frac{1}{2}} \end{aligned} \tag{9.113}$$

We can show from Eqs. 9.85, 9.108 and 9.111 that:

$$\left(t_{i,oct}\right)^2 = \tfrac{1}{3}\left(I_1^2 - 2I_2\right) \quad \text{and} \quad \sigma_{oct}^2 = \tfrac{1}{3}I_1^2 \tag{9.114}$$

such that we can express Eq. 9.110 in terms of the stress invariants thus:

$$\tau_{oct} = \tfrac{1}{3}\sqrt{2I_1^2 - 6I_2} \tag{9.115}$$

9.6.3 The Deviatoric and Hydrostatic Stresses

We present two practical formulations of stresses called *hydrostatic* and *deviatoric stresses*. Previously, we defined deformation as *change of shape and size of a material body under the effect of a deformation gradient*. When a material body is subjected to a deformation gradient, stresses build up in such a body. This results in a *stress tensor*, as discussed previously in Sect. 9.5.2.

The stress tensor can be decomposed into a component that leads to a *size-change* herein referred to a *hydrostatic stress* while the *deviatoric stress* is the stress that leads to a *shape-change*. It is important to extract the components of the original stress tensor that lead to these two distinct modes of deformation, especially during the development of constitutive models for such a material body. Here, we present the mathematics that describe these two types of stresses.

The ***hydrostatic stress*** is the mean of the three normal stress components of a given stress tensor. Consider a stress tensor, $\boldsymbol{\sigma}$, with normal components defined by: σ_{xx}, σ_{yy} and σ_{zz} acting along an outward pointing normal vector along the $x-$, $y-$and $z-$axes respectively. The hydrostatic stress becomes:

$$\sigma_h = \left(\frac{\sigma_{xx} + \sigma_{yy} + \sigma_{zz}}{3}\right) \tag{9.116}$$

In terms of stress invariants, we can also describe the hydrostatic stress as:

$$\sigma_h = \tfrac{1}{3}I_1 = \tfrac{1}{3}\left(\sigma_{xx} + \sigma_{yy} + \sigma_{zz}\right) = \tfrac{1}{3}\text{tr}(\boldsymbol{\sigma}) \tag{9.117}$$

where I_1 is the first stress invariants and $\text{tr}(\cdots)$ is the *trace* of the tensor $(\cdots)$.

The hydrostatic stress is a scalar quantity and also equivalent to the *mean stress*, σ_m. It is also described as a *confining stress*, for example in geomechanics, which has to be determined when designing foundations of buildings.

The hydrostatic stress always acts normal to the planes that make up a material body. It is often also comparable to *pressure loads* as it enforces the same negative hydrostatic loading on a material body, i.e. $p = -\frac{1}{3}I_1$, where p is the magnitude of the pressure load. In constitutive modelling the hydrostatic stress is given in tensorial form thus:

$$\boldsymbol{\sigma}_h = \begin{bmatrix} \frac{1}{3}I_1 & 0 & 0 \\ 0 & \frac{1}{3}I_1 & 0 \\ 0 & 0 & \frac{1}{3}I_1 \end{bmatrix} = -p\boldsymbol{I} \tag{9.118}$$

where $\boldsymbol{I}$ is a 3×3 identity matrix and p is the magnitude of the hydrostatic pressure.

Given a stress tensor, $\boldsymbol{\sigma}$, the hydrostatic stress is purely the *volumetric stress* contribution of $\boldsymbol{\sigma}$, while the deviatoric stress leads to an *isochoric* stress contribution such that:

$$\boldsymbol{\sigma} = \boldsymbol{\sigma}_{vol} + \boldsymbol{\sigma}_{iso} \tag{9.119}$$

following an additive decomposition of the stress tensor, $\boldsymbol{\sigma}$, hence $\boldsymbol{\sigma}_{vol} = -p\boldsymbol{I}$. Therefore, hydrostatic stresses lead to volumetric changes in materials. Zhou and Song [9] reported that the volumetric distortion at grain boundaries of metals lead to *hydrogen embrittlement*: a premature failure of materials in the presence of hydrogen.

On the other hand, the ***deviatoric stresses*** are responsible for shape changes (or distortions) in the material body. The deviatoric stress component of the stress tensor is also referred to as the *stress deviator tensor* and commonly symbolized by $\boldsymbol{S}$ in vector form and S_{ij} in index format.

Let us define a stress tensor, $\boldsymbol{\sigma}$, and its corresponding hydrostatic stress tensor, $\boldsymbol{\sigma}_h$ we can then define the deviator stress, $\boldsymbol{S}$, as the difference of the stress tensor and its hydrostatic component thus:

$$\boldsymbol{S} = \begin{bmatrix} S_{xx} & S_{xy} & S_{xz} \\ S_{yx} & S_{yy} & S_{yz} \\ S_{zx} & S_{zy} & S_{zz} \end{bmatrix} = \begin{bmatrix} \sigma_{xx} & \sigma_{xy} & \sigma_{xz} \\ \sigma_{yx} & \sigma_{yy} & \sigma_{yz} \\ \sigma_{zx} & \sigma_{zy} & \sigma_{zz} \end{bmatrix} - \begin{bmatrix} \sigma_m & 0 & 0 \\ 0 & \sigma_m & 0 \\ 0 & 0 & \sigma_m \end{bmatrix} \tag{9.120}$$

where $\sigma_m = \frac{1}{3}\left(\sigma_{xx} + \sigma_{yy} + \sigma_{zz}\right) = \frac{1}{3}I_1$ is the mean stress of the stress tensor, $\boldsymbol{\sigma}$. This deviatoric stress, $\boldsymbol{S}$, is similar to the stress tensor, except that it exclusively enforces a distortion of the material body.

Shear deformations, characterized by a shape change, are dominated by the deviatoric stress. Expanding Eq. 9.120, we obtain the expression of the deviatoric stress:

$$\boldsymbol{S} = \begin{bmatrix} S_{xx} & S_{xy} & S_{xz} \\ S_{yx} & S_{yy} & S_{yz} \\ S_{zx} & S_{zy} & S_{zz} \end{bmatrix} = \begin{bmatrix} \sigma_{xx} - \sigma_m & \sigma_{xy} & \sigma_{xz} \\ \sigma_{yx} & \sigma_{yy} - \sigma_m & \sigma_{yz} \\ \sigma_{zx} & \sigma_{zy} & \sigma_{zz} - \sigma_m \end{bmatrix} \tag{9.121}$$

We note that $\boldsymbol{S}$ has similar structure as the *stress tensor*, which led us to solve an eigenvalue problem. The same characteristics apply for the deviatoric stress.

Now, let us evaluate the determinant of the deviator stress tensor based on Eq. 9.121. We obtain the characteristic equation of the deviatoric stress tensor as:

$$\det(\boldsymbol{S}) = |S_{ij} - \lambda\delta_{ij}| = -\lambda^3 + J_1\lambda_2 - J_2\lambda + J_3 = 0 \tag{9.122}$$

where λ is the *eigenvalues* of the deviatoric stress tensor, while J_1, J_2, and J_3 are three invariants of the deviatoric stress. The λ values are determined by obtaining the roots of the above characteristic equation (refer to Eq. 9.122). Its values are the three principal deviatoric stresses, namely S_1, S_2 and S_3 for *maximum, median* and *minimum principal deviatoric stresses* respectively.

Therefore, the ***deviatoric stress invariants*** can be represented as follows:

- ***First deviatoric stress invariant,*** J_1:

$$J_1 = S_1 + S_2 + S_3 = 0 \quad \longrightarrow \quad J_1 = \text{tr}(\boldsymbol{S}) = 0 \tag{9.123}$$

  The implication of $J_1 = 0$ is that all the normal deviatoric stress terms are zero, leaving only shear terms. Therefore, the deviator stress tensor is in a state of ***pure shear***. This is why its an important measure of stress for developing constitutive models for many material systems.

- ***Second deviatoric stress invariant,*** J_2:

$$\begin{aligned} J_2 = \tfrac{1}{2}\text{tr}(\boldsymbol{S}^2) \quad &= \tfrac{1}{2}\left(S_1^2 + S_2^2 + S_3^2\right) = \tfrac{1}{2}(\boldsymbol{S} : \boldsymbol{S}) \\ = \tfrac{1}{3}\left(I_1^2 - 3I_2\right) \quad &= \tfrac{1}{2}\left[\text{tr}(\sigma^2) - \tfrac{1}{3}\text{tr}\{(\sigma)^2\}\right] \end{aligned} \tag{9.124}$$

- ***Third deviatoric stress invariant,*** J_3:

$$\begin{aligned} J_3 = \det(\boldsymbol{S}) \quad &= \tfrac{1}{3}\left(S_1^3 + S_2^3 + S_3^3\right) \\ = S_1 S_2 S_3 \quad &= \tfrac{2}{27}I_1^3 - \tfrac{1}{3}I_1 I_2 + I_3 \end{aligned} \tag{9.125}$$

We can also express the *octahedral shear stress*, derived previously in terms of the second invariant of deviatoric stress, thus:

$$\tau_{oct} = \sqrt{\frac{2}{3}J_2} \quad \longrightarrow \quad \tau_{oct} = \sqrt{\tfrac{1}{3}\boldsymbol{S} : \boldsymbol{S}} \tag{9.126}$$

with the latter expression being a tensorial form of the octahedral shear stress expressed in terms of the deviator stress tensor, $\boldsymbol{S}$.

9.6.4 The von Mises Stresses

The next practical formulation of stress is the *von Mises stress*, σ_{vm} which is named after von Mises [8] who proposed the current formulation of the stress in 1913. However, the von Mises stress was originally suggested by Huber in 1904 [1]. The von Mises stresses proposed by the authors were simply mathematical equations and they did not make any connection of the von Mises stresses to the physics of material response. It was in 1924 when Hencky [4] identified the link between von Mises stress and the *deviatoric strain energy* of a material body. We will describe below the deviatoric strain energy.

The von Mises stress is the most common practical stress formulation used in many FEM solvers. This is because it has, over the years, provided the direct measure between laboratory-derived strength measures and the effective stresses for a body subjected to a complex combined loading history. The von Mises stress is a single stress measure that can be compared directly to such quantities as the ultimate tensile strength, σ_{UTS}, of a material body. With a von Mises stress, the design engineer now has a single value which can be compared directly with attributes such as tensile strength, compressive yield stress and shear strength of a given material. This stress formulation is usually recommended for materials that experience significant ductility.

As regards the deviatoric strain energy, let us recall that the *strain energy density*, W, is the energy experienced in a material body per unit of its volume. This is expressed analytically, for given stress, $\boldsymbol{\sigma}$, and strain, $\boldsymbol{\varepsilon}$, tensors as:

$$W = \int \sigma : d\varepsilon \quad \longrightarrow \quad W = \frac{1}{2}\boldsymbol{\sigma} : \boldsymbol{\varepsilon} \quad \text{(for linear elastic materials)} \tag{9.127}$$

The stress and strain tensors can be decomposed into their hydrostatic and deviatoric components such that the strain energy density can be expressed thus:

$$W = \frac{1}{2}\boldsymbol{\sigma} : \boldsymbol{\varepsilon} = \frac{1}{2}\left(\boldsymbol{\sigma}_h + \boldsymbol{S}\right) : \left(\boldsymbol{\varepsilon}_h + \boldsymbol{\varepsilon}'\right) \tag{9.128}$$

where $\boldsymbol{\varepsilon}'$ is the deviatoric strain tensor. We can now expand Eq. 9.128, bearing in mind that $\boldsymbol{\sigma}_h : \boldsymbol{\varepsilon}' = \mathbf{0}$ and $\boldsymbol{S} : \boldsymbol{\varepsilon}_h = \mathbf{0}$, to obtain:

$$W = \frac{1}{2}\boldsymbol{\sigma} : \boldsymbol{\varepsilon} \equiv W = W_h + W' \quad \longrightarrow \quad W = \underbrace{\frac{1}{2}\boldsymbol{\sigma}_h : \boldsymbol{\varepsilon}_h}_{\text{hydrostatic}} + \underbrace{\frac{1}{2}\boldsymbol{S} : \boldsymbol{\varepsilon}'}_{\text{deviatoric}} \tag{9.129}$$

where W_h and W' are the hydrostatic and deviatoric components of the strain energy density respectively.

We can further show that under Hooke's law conditions, the *deviatoric strain tensor*, $\boldsymbol{\varepsilon}'$, becomes:

$$\boldsymbol{\varepsilon}' = \tfrac{1}{2G}\boldsymbol{S} \tag{9.130}$$

where G is the shear modulus of the material body. The above equation demonstrates that the strain energy density can also be decomposed into its *hydrostatic* and *deviatoric* components. It is the deviatoric strain energy that Hencky used to establish the physical definition of the von Mises stress as will be shown later. Therefore, the expression of the deviatoric strain energy density in terms of deviatoric stress tensor, becomes:

$$W' = \frac{1}{2}\boldsymbol{S} : \boldsymbol{\varepsilon}' \quad \longrightarrow \quad W' = \frac{1}{4G}\boldsymbol{S} : \boldsymbol{S}. \tag{9.131}$$

Equation 9.131 also implies that the deviatoric strain energy is proportional to the double dot product of the deviatoric strain tensor: $W' \propto \boldsymbol{S} : \boldsymbol{S}$. To evaluate the von Mises stress, we need to determine the double dot product of the deviatoric stress tensor.

Let us consider again a material body subjected to multiple loads such that the stress tensor at a given point is $\boldsymbol{\sigma}$ for a three-dimensional system. It is important to identify the *effective stress* acting on the body under the effect of combined loading. We can define, for now, this effective stress as the von Mises stress, σ_{vm}, and it is expressed thus:

$$\sigma_{vm} = \sqrt{\tfrac{1}{2}\left[(\sigma_{xx} - \sigma_{yy})^2 + (\sigma_{yy} - \sigma_{zz})^2 + (\sigma_{zz} - \sigma_{xx})^2\right] + 3\left(\tau_{xy}^2 + \tau_{yz}^2 + \tau_{xz}^2\right)}$$

If we expand the above equation, we will obtain the following:

$$\sigma_{vm} = \sqrt{\left(\sigma_{xx}^2 + \sigma_{yy}^2 + \sigma_{zz}^2 - \sigma_{xx}\sigma_{yy} - \sigma_{yy}\sigma_{zz} - \sigma_{zz}\sigma_{xx}\right) + 3\left(\tau_{xy}^2 + \tau_{yz}^2 + \tau_{xz}^2\right)}$$

If we consider a two-dimensional system, where the through-thickness direction is the z–axis, then $\sigma_{zz} = \tau_{xz} = \tau_{yz} = 0$, such that the 2-D von Mises stress becomes:

$$\sigma_{vm} = \sqrt{\sigma_{xx}^2 + \sigma_{yy}^2 - \sigma_{xx}\sigma_{yy} + 3\tau_{xy}^2} \tag{9.132}$$

It is customary in continuum mechanics to compare the von Mises stress directly with the uniaxial tensile strength of the test material, as a reference case. The stress tensor, $\boldsymbol{\sigma}_{tensile}$, associated with uniaxial tensile deformation and the mean stress, σ_{mean}, are:

$$\boldsymbol{\sigma}_{tensile} = \begin{bmatrix} \sigma & 0 & 0 \\ 0 & 0 & 0 \\ 0 & 0 & 0 \end{bmatrix} \quad \text{and} \quad \sigma_{mean} = \tfrac{1}{3}\mathrm{tr}(\boldsymbol{\sigma}_{tensile}) = \tfrac{1}{3}\sigma$$

Thus, the deviatoric stress for the uniaxial tensile load case becomes:

$$\boldsymbol{S} = \boldsymbol{\sigma}_{tensile} - \sigma_{mean}\boldsymbol{I} \quad \longrightarrow \quad \boldsymbol{S} = \begin{bmatrix} \frac{2}{3}\sigma & 0 & 0 \\ 0 & -\frac{1}{3}\sigma & 0 \\ 0 & 0 & -\frac{1}{3}\sigma \end{bmatrix}$$

The double dot product of the deviatoric stress tensor for this uniaxial case becomes:

$$\sqrt{\boldsymbol{S} : \boldsymbol{S}} = \sqrt{\tfrac{4}{9}\sigma^2 + \tfrac{1}{9}\sigma^2 + \tfrac{1}{9}\sigma^2} = \sqrt{\tfrac{2}{3}}\,\sigma$$

If we define the tensile strength following a uniaxial tensile test as σ_{UTS}, it is clear that when tensile failure occurs: $\sigma_{tensile} = \sigma_{UTS}$. Since the von Mises stress is set to be equal to the tensile strength of the tensile load case. At the incidence of failure, we notice that Eq. 9.133 becomes: $\sqrt{\boldsymbol{S}:\boldsymbol{S}} = \sqrt{\frac{2}{3}}\sigma_{UTS}$. This represents a fraction of the ultimate tensile strength.

For the expected equivalence with σ_{UTS} to exist, we have to multiply the square root of the double dot product of the deviatoric stress tensor by $\sqrt{\frac{3}{2}}$ such that the expression of the von Mises stress in terms of the deviatoric stress tensor becomes:

$$\sigma_{vm} = \sqrt{\tfrac{3}{2}\boldsymbol{S}:\boldsymbol{S}} \quad \longrightarrow \quad \sigma_{vm} = \sqrt{\tfrac{3}{2} \times \tfrac{2}{3}}\, \sigma_{UTS} = \sigma_{UTS}$$

Consequently, the von Mises Stress is equivalent to the uniaxial tensile strength and proportional to the deviatoric strain energy density of the material body.

The von Mises stress derived above will always be a positive number since it is the square root of sum of squared stresses. However, FEM solvers often report negative von Mises stress, for example in a compression test. Let us consider how this is possible for a von Mises stress. In fact, in fracture and fatigue mechanics[7], a quantity called the *signed von Mises stress*, $\sigma_{signed-vm}$ is commonly used to determine the stress range in the vicinity of a crack, under effect of multiaxial loading. The signed von Mises stress has the same magnitude as a typical von Mises stress except that its sign is defined by the sign of the largest principal stress in the system i.e.

$$\sigma_{signed-vm} = \sigma_{vm} \times \mathrm{sign}(\sigma_1) \tag{9.133}$$

where $\mathrm{sign}(\sigma_1)$ indicates the sign of the maximum principal normal stress, σ_1.

FEM solver developers have adopted this approach in assigning signs to calculated von Mises stress [2]. In some other instances, the FEM developer might use the sign of the hydrostatic stress rather than the sign of the maximum principal normal stress. Whichever approach is adopted, it must be noted that these are artificial constructs to distinguish between tensile and compressive dominant stress state to help the design engineer make conclusions on the practical stresses acting in the material body. The signed von Mises stress has no physical relevance except that it is related to the classic von Mises stress value.

If a material body is subjected to a uniaxial compressive stress state,

$$\boldsymbol{\sigma}_{comp} = \begin{bmatrix} -\sigma & 0 & 0 \\ 0 & 0 & 0 \\ 0 & 0 & 0 \end{bmatrix}, \quad \text{and} \quad \sigma_h = \tfrac{1}{3}\mathrm{tr}(\boldsymbol{\sigma}_{comp}) = -\tfrac{1}{3}\sigma$$

where σ_h is the hydrostatic stress value. The computed von Mises stress, $\sigma_{vm} = +\sigma_{comp}$, and is positive, but the sign of the maximum principal stress will be negative i.e $\sigma_1 = -\sigma$, as well as the sign of σ_h is also negative. As a consequence, the

FEM solver will report a negative signed von Mises stress for the compression test. This is acceptable to design engineers thus aiding them make design decisions, and compare model predictions with, for example, compressive strength of the test material.

9.6.5 *The Tresca Stress*

Similar to the von Mises stress described above, we introduce another practical formulation of stresses called the *Tresca stress*, σ_{tresca}. Previously, the von Mises stress was set as equivalent to the tensile strength of the material body, where material failure or yielding initiates. The Tresca stress is a similar stress, but defines the yielding/failure condition for a material body under the effect of multiple loading conditions. The *Tresca yield criterion* states that the material body begins to yield when the *maximum shearing stress* reaches a critical value called the *Tresca stress*. Therefore, the effective stress in the material body must equal the maximum shear stress, τ_{max}, for us to define the effective stress to be a Tresca stress.

Mathematically, we define the tresca stress is defined thus:

$$\sigma_{tresca} = \tau_{max} = \frac{1}{2}\left|\sigma_{max} - \sigma_{min}\right| \tag{9.134}$$

where σ_{max} and σ_{min} are the maximum and minimum normal stresses experienced by the material body. In terms of principal stresses, if $\sigma_1 \leq \sigma_2 \leq \sigma_3$, the Tresca stress becomes:

$$\sigma_{tresca} = \frac{1}{2}\left|\sigma_1 - \sigma_3\right| \tag{9.135}$$

9.7 Conclusions

To understand the material response of material bodies modelled within a finite element scheme, its important to understand the parameters drawn from continuum mechanics required for describing the material response observed. This chapter has introduced the kinetics of deformation of a material body, highlighting the relationship between the deformed material configuration and the original material configuration.

This chapter has also introduced the role played by deformation gradient tensors in describing the observed changes in size and shape (i.e. deformation) of the material body. The different measures of strain have also been presented in this chapter with a special focus on how these strain measures result from the deformation gradient tensor.

Stress was defined as the intensity of the internal force system on a given plane. The different measures of stresses needed to describe a material's response were also introduced ranging from the Cauchy stress to the co-rotated and Mandel stresses. The sensitivity of these stress measures to rotation of the body were discussed and it was established that, for defining constitutive models, the stress measure should not be affected by significant rotation of the material body.

At the end of the chapter, we presented different practical formulations of stresses needed to relate stress state of a material body to the mechanical properties of materials. The latter is essential for design engineers to make design choices for any given complex stress state history. In the next chapter, we will present commonly used material models in FEM solvers.

At the end of this chapter, the reader should be able to:

(a) Describe the different kinematics of deformation used within the FEM process to describe material response.
(b) Operate on the different measures of strain and how these relate to mechanical properties of materials.
(c) Determine the different measures of stress and explore their sensitivity to rotation tensors.
(d) Appreciate and query the suitability of the different practical formulations of stresses needed by design engineers during the design process.

9.8 Problems: Material Response – Measures of Stress and Strain

Problem 9.1 A $6 \times 6\,\text{mm}^2$ square plate shown in Fig. 9.22 is subjected to a uniaxial tensile deformation such that the plate undergoes a stretch of 75% along the x–axis and a contraction of 45% along the y–axis. The mapping function that defines the material configuration of the deformed the plate is given by (where X, Y are material coordinates and x, y are spatial coordinates):

$$\begin{aligned} x &= 1.75X + 0.00Y \\ y &= 0.00X + 0.55Y \end{aligned} \tag{9.136}$$

(a) Determine the deformation gradient tensor, $\boldsymbol{F}$, for the above problem.
(b) Determine the right stretch, $\boldsymbol{U}$, and rotation, $\boldsymbol{R}$, tensors.
(c) Determine the Green-Lagrange strain tensor, $\boldsymbol{E}$.

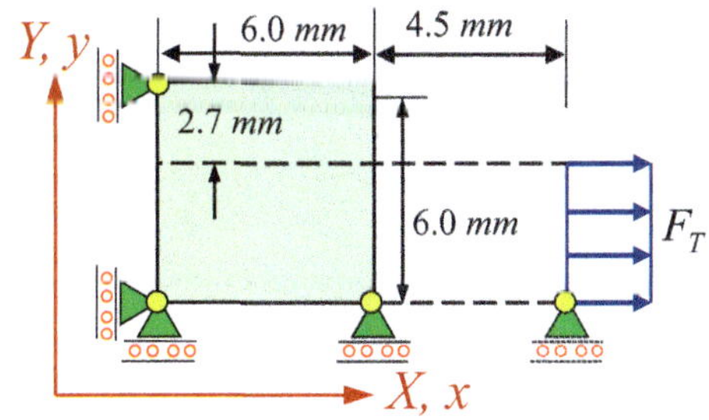

Fig. 9.22 Uniaxial deformation of a square plate

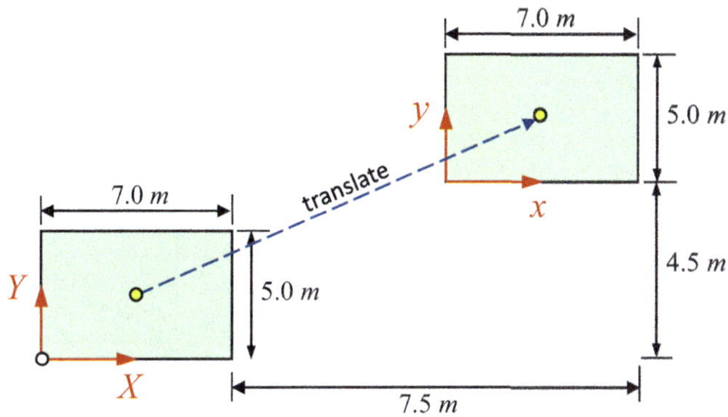

Fig. 9.23 Translation of a rectangular board

(d) Determine the Engineering strain, $\boldsymbol{\varepsilon}_{engineering}$, and prove that this yields the expected strains.

Problem 9.2 An information display steel board of in-plane dimensions $7 \times 5\,\mathrm{m}^2$ is located above a production floor. The board is connected to a lifting mechanism such that it is moved to different positions to give information to staff. Consider the instance where the board is translated horizontally by 7.5 m and vertically by 4.5 m as shown in Fig. 9.23. The mapping function that describes the material configuration of the board is given by (where X, Y are material coordinates and x, y are spatial coordinates):

$$\begin{aligned} x &= X + 7.5 \\ y &= Y + 4.5 \end{aligned} \tag{9.137}$$

(a) Determine the deformation gradient tensor, $\boldsymbol{F}$.
(b) Determine the right stretch, $\boldsymbol{U}$, and rotation, $\boldsymbol{R}$, tensors.
(c) Comment on the state of deformation of the board.
(d) What boundary conditions defines the mechanical response of the board?

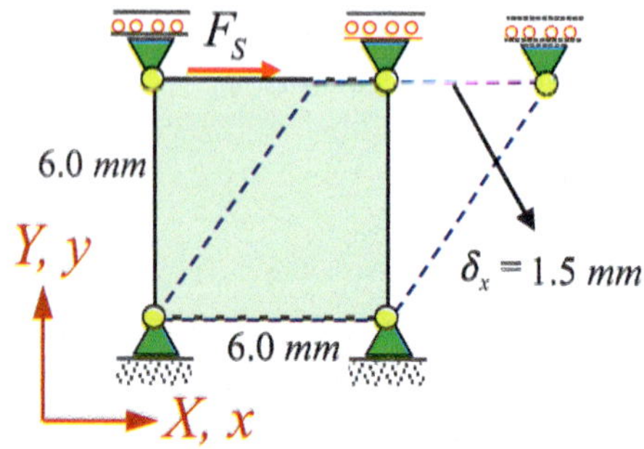

Fig. 9.24 Simple shear deformation of a square plate

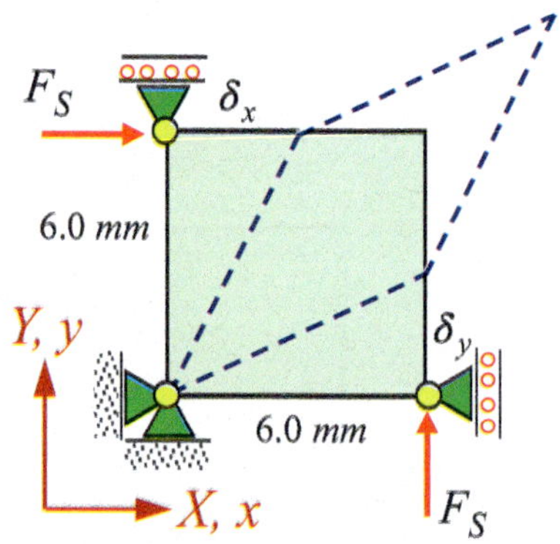

Fig. 9.25 Pure shear deformation of a square plate

Problem 9.3 A silicon rubber plate, of dimensions 6×6 mm^2, is subjected to a shear force, F_S, such that the top edge is displaced by 1.5 mm as shown in Fig. 9.24.

(a) Determine the function that maps the material to spatial reference frames.
(b) Determine the deformation gradient tensor, $\boldsymbol{F}$.
(c) Determine the left stretch, $\boldsymbol{V}$, and rotation, $\boldsymbol{R}$, tensors.
(d) Determine the Almansi strain tensor, $\boldsymbol{e}$.

Problem 9.4 Assume the silicon rubber plate of Problem 9.3 is subjected to a pure shear deformation as shown in Fig. 9.25 such that two of the corners are displaced equally by $\delta_x = \delta_y = 1.5$ mm using equal shear forces, F_S.

(a) Determine the function that maps the material to spatial reference frames.
(b) Determine the deformation gradient tensor, $\boldsymbol{F}$.
(c) Determine the left, $\boldsymbol{V}$, and right, $\boldsymbol{U}$, stretch tensors.
(d) Determine the engineering and logarithmic strain tensors.
(e) Compare the rotation, $\boldsymbol{R}$, tensors for the simple (Problem 9.3) and this pure shear deformation. Comment on your results.

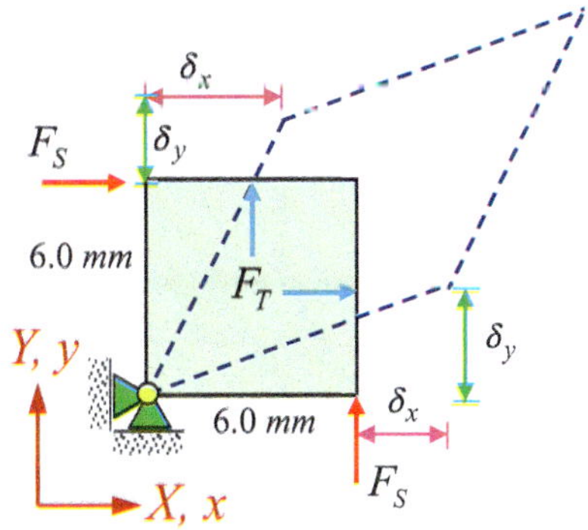

Fig. 9.26 Combined tensile and shear loading of a square plate

Problem 9.5 A steel plate is subjected to a complex loading history such that its edges are sheared by F_S and stretched by F_T, as shown in Fig. 9.26. If $\delta_x = \delta_y = 1.5$ mm, then:

(a) Determine the function that maps the material to spatial reference frames.
(b) Determine the deformation gradient tensor, $\boldsymbol{F}$.
(c) Determine the Almansi and Green strain tensors.

Problem 9.6 A three-dimensional structure is subjected to a shear force, F_s, as shown in Fig. 9.27. The expressions that maps its material configuration to the spatial reference frame are:

$$\begin{aligned} x &= 1.00X + 0.25Y + 0.00Z \\ y &= 0.00X + 1.00Y + 0.00Z \\ z &= 0.00X + 0.00Y + 1.00Z \end{aligned} \tag{9.138}$$

(a) Determine the deformation gradient tensor, $\boldsymbol{F}$.
(b) Determine the left, $\boldsymbol{V}$, and right, $\boldsymbol{U}$, stretch and rotation, $\boldsymbol{R}$, tensors.
(c) Determine the Green strain tensors, $\boldsymbol{E}$.

Problem 9.7 A cubic test specimen of dimensions $10 \times 10 \times 10\,\text{mm}^3$ is subjected to out-of-plane shear testing, as shown in Fig. 9.28. The expressions that maps its material configuration to the spatial reference frame are:

$$\begin{aligned} x &= 1.00X + 0.25Y + 0.00Z \\ y &= 0.00X + 1.00Y + 0.00Z \\ z &= 0.00X + 0.00Y + 1.00Z \end{aligned} \tag{9.139}$$

(a) Determine the deformation gradient tensor, $\boldsymbol{F}$.
(b) Determine the right stretch, $\boldsymbol{U}$, and rotation, $\boldsymbol{R}$, tensors.

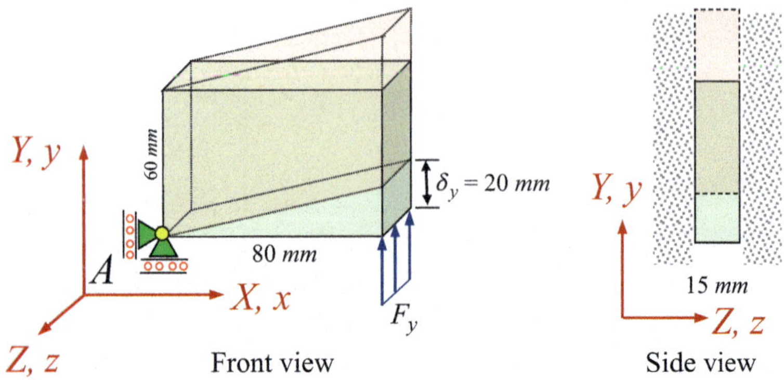

Fig. 9.27 Shear deformation of a three-dimensional structure

Fig. 9.28 Out-of-plane shear deformation of a cubic test specimen

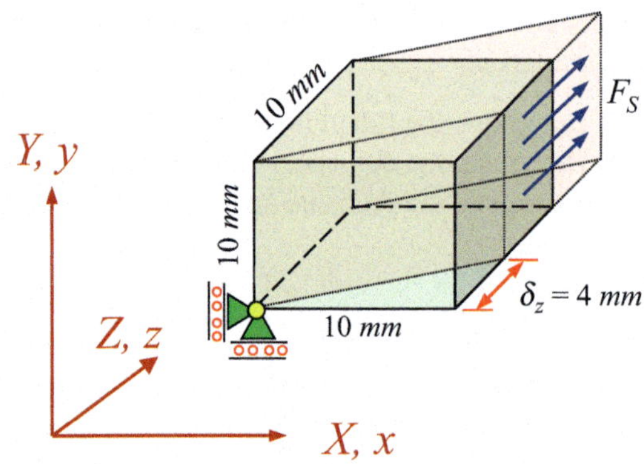

(c) Determine the engineering strain tensor, $\boldsymbol{\varepsilon}_{engineering}$, and deduce the magnitude of the out-of-plane shear strain on the $xz-$ and $yz-$planes.

Problem 9.8 The cubic test specimen of Problem 9.7 is subjected to two out-of-plane and one in-plane shear stresses, as shown in Fig. 9.29 such that it experiences these displacements: $\delta_x = 2$ mm, $\delta_y = 5$ mm, and $\delta_z = 4$ mm.

(a) Determine the expression that maps the material to the spatial reference frames.
(b) Determine the deformation gradient tensor, $\boldsymbol{F}$.
(c) Determine the two stretch ($\boldsymbol{U}$, and $\boldsymbol{V}$,) and rotation, $\boldsymbol{R}$, tensors.
(d) Determine the left, $\boldsymbol{B}$, and right, $\boldsymbol{C}$, Cauchy-Green tensors.

Problem 9.9 A polypropylene test specimen of dimensions $50 \times 40\,\text{mm}^2$, shown in Fig. 9.30, is subjected to biaxial compressive loading such that edge $\overline{BC}$ is compressed a distance of 8 mm and edge $\overline{CD}$ is compressed by 20 mm.

Under this state of combined loading, the general state of stress at point, Q, within the plate is given by the Cauchy stress tensor, $\boldsymbol{\sigma}_Q$ below.

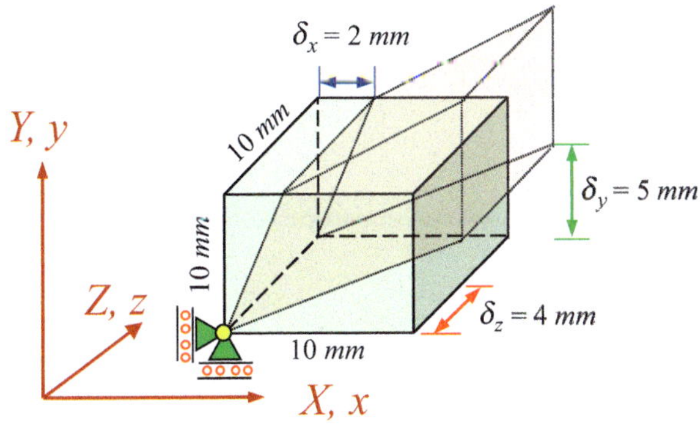

Fig. 9.29 Combined shear loading imposed on a cubic test specimen

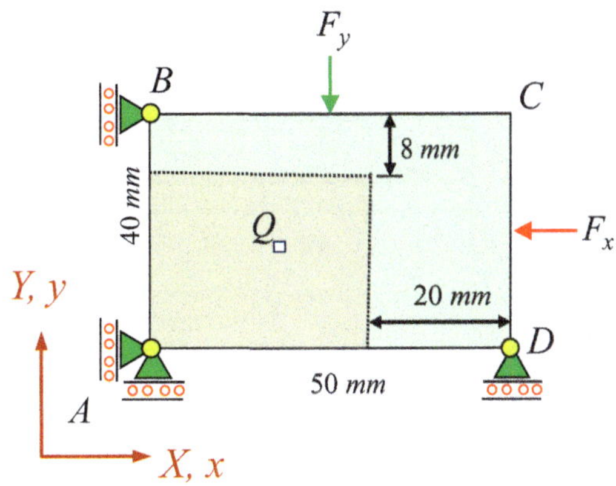

Fig. 9.30 A polypropylene plate under biaxial compressive loading

$$\boldsymbol{\sigma}_Q = \begin{bmatrix} -100 & 0 \\ 0 & -20 \end{bmatrix} \text{ MPa}$$

(a) Determine the mapping function from material to spatial reference frame.
(b) Determine the deformation gradient, $\boldsymbol{F}$, for the above problem.
(c) Determine the logarithmic strain tensor, $\boldsymbol{\varepsilon}_{logarithmic}$.
(d) Show that the ε_{xx} and ε_{yy} terms of $\boldsymbol{\varepsilon}_{engineering}$ are equal to the normal strain values along the x–and y–axes.

Problem 9.10 Using the material body of Problem 9.9 which is subjected to biaxial compressive loading:

(a) Determine all the measures of stress tensors defined in Sect. 9.5, for this stress state.
(b) Plot a comparison bar chart for all stress measures for all the σ_{ij} terms for $i, j = x, y$. *Hint: Adopt the approach of Example* 9.3 *above.*

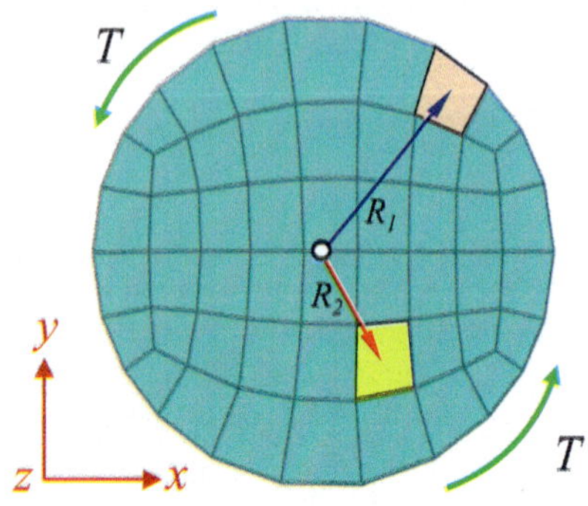

Fig. 9.31 A 2D cross-section view of a cylindrical solid shaft showing two selected quadrilateral elements

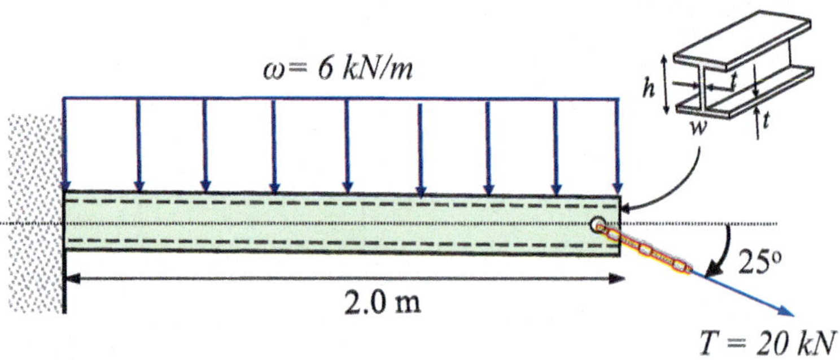

Fig. 9.32 An I-section beam, supported by a tension cable

Problem 9.11 A solid cylindrical shaft of diameter, $\phi = 50$ mm, is used to transmit 150 kW power and torque, T, when driven by a 1800 rpm electric motor.

If the shaft is meshed using quadrilateral elements as shown in Fig. 9.31:

(a) Determine the shear stresses for two elements located at radial distances, $R_1 = 22.5$ mm and $R_2 = 12.5$ mm.
(b) Determine the Cauchy stress tensor for the two elements.
(c) Determine the principal normal and shear stresses on these two stress elements.

Problem 9.12 An UB203 × 102 × 23 wide flange I-section beam, of length 2.0 m supports a uniformly distributed load, $\omega = 6$ kN/m.

To limit deflection of the beam, its right end is supported by cable of tension, $T = 20$ kN, as shown in Fig. 9.32. *Note: width,* $w = 203$ mm, *height,* $h = 102$ mm *and thickness,* $t = 23$ mm.

(a) Determine the Cauchy stress tensor for the three stress elements located on the outermost upper and bottom layers as well as on the neutral axis of the beam.
(b) Determine the principal normal and shear stresses on these three stress elements.
(c) Determine the principal normal and shear stresses for a stress element located on the neutral axis but closest to the wall.

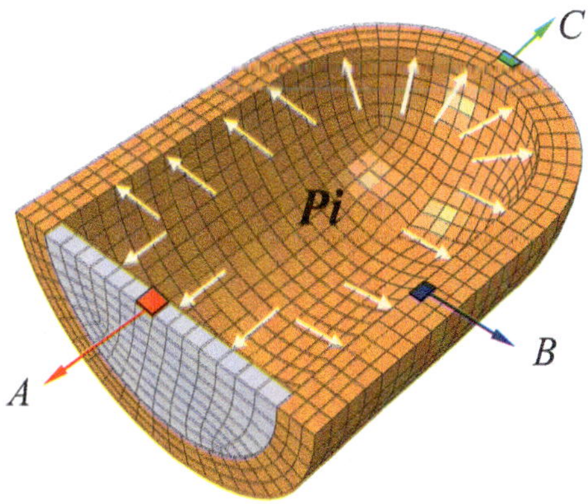

Fig. 9.33 Sectioned view of a cylindrical pressurized cylinder, showing stress elements A (on shut-out valve), B (on outer wall of cylinder) and C (on outer wall of hemispherical end)

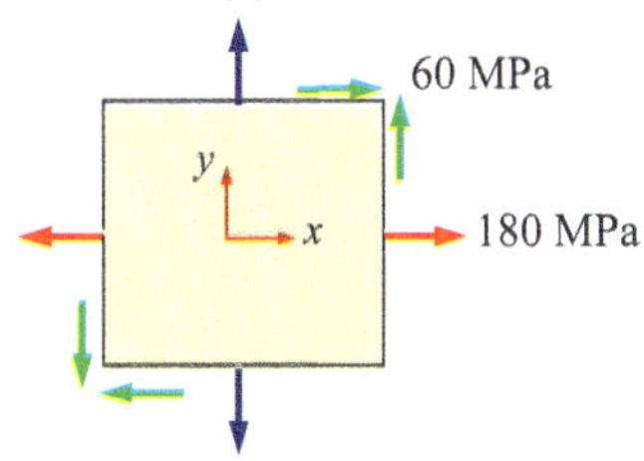

Fig. 9.34 A 2D stress element under effect of multiple stresses

Problem 9.13 A pressurized cylindrical vessel of internal diameter, $\phi_1 = 2.8\,\text{m}$, and external diameter, $\phi_2 = 3.6\,\text{m}$, is carrying a pressure, $P_i = 100\,\text{kPa}$, over its length, $L = 6\,\text{m}$. The cylinder has a hemispherical shape at the right end and a shut-out valve at the left end, as shown in Fig. 9.33.

The cylinder is discretized using hexahedral elements. To assess the state of stress within the pressure vessels, three elements located at A, B, and C were chosen.

(a) Determine the cauchy stress tensor for the three stress elements.
(b) Determine the principal normal and shear stresses on these three stress elements.
(c) Determine the three stress invariants for the element located at *B*.

Problem 9.14 A 2D stress element is subjected to a multiple loading history as shown in (Fig. 9.34).

(a) Determine all the Lagrange multipliers, λ_i for $i = 1, 2$.
(b) Determine all the stress invariants.
(c) Determine the principal stresses for this system.
(d) Determine the principal normal stress plane, $\boldsymbol{n}_i$ for $i = 1, 2$.
(e) Determine the angles of the principal normal stress.

Problem 9.15 The root of a metallic fan blade of a commercial aircraft engine is subjected to interface failure arising from the effect of a complex three-dimensional stress state.

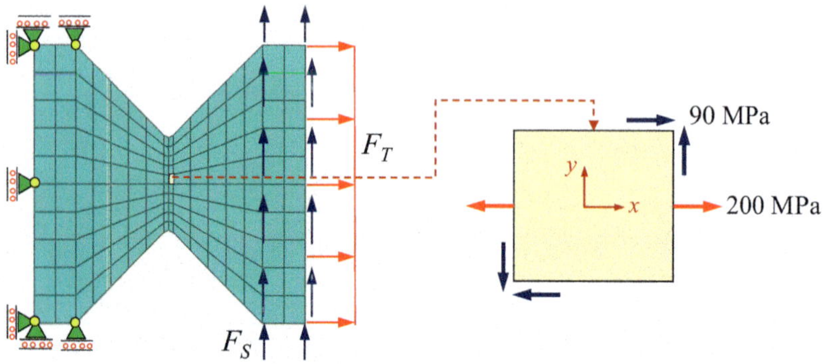

Fig. 9.35 A typical Arcan shear butterfly specimen, with applicable combined shear and tensile loading. Inset stress element shows the stress element in the gauge section of the Arcan specimen

A typical Cauchy stress tensor for a fan blade is given as:

$$\boldsymbol{\sigma} = \begin{bmatrix} 220 & 110 & -80 \\ 110 & -350 & 45 \\ -80 & 45 & 120 \end{bmatrix} \quad \text{MPa}$$

(a) Determine the principal stresses for this fan blade.
(b) Determine the principal normal stress plane, $\boldsymbol{n}_i$ for $i = 1, 2, 3$.
(c) Determine the angles of the principal normal stress.
(d) Determine the principal shear stresses, τ_{max}, on the fan blade.

Problem 9.16 An arcan shear test specimen is subjected to a combined tensile, F_T, and shear, F_S, forces, shown in Fig. 9.35. The specimen is discretized using a quadrilateral element.

Using a stress element taken from the gauge section, with tensile stress, $\sigma_{xx} = 200$ MPa, and shear stress, $\tau_{xy} = 90$ MPa:

(a) Determine all the Lagrange multipliers, λ_i for $i = 1, 2$.
(b) Determine all the stress invariants.
(c) Determine the principal stresses for this system.
(d) Determine the principal normal stress plane, $\boldsymbol{n}_i$ for $i = 1, 2$.
(e) Determine the angles of the principal normal stress.

Problem 9.17 A design engineer is considering choosing the correct material for a re-design of the metallic fan blade of Problem 9.15.

The engineer decides to explore different practical formulations of effective stresses that can be related to known mechanical properties of materials.

(a) Determine the octahedral normal, σ_{oct}, and shear, τ_{oct}, stresses.
(b) Determine the hydrostatic stress, σ_h, experienced by the fan blade.
(c) Determine the von Mises stress, σ_{vm}, arising from to the above stress state.

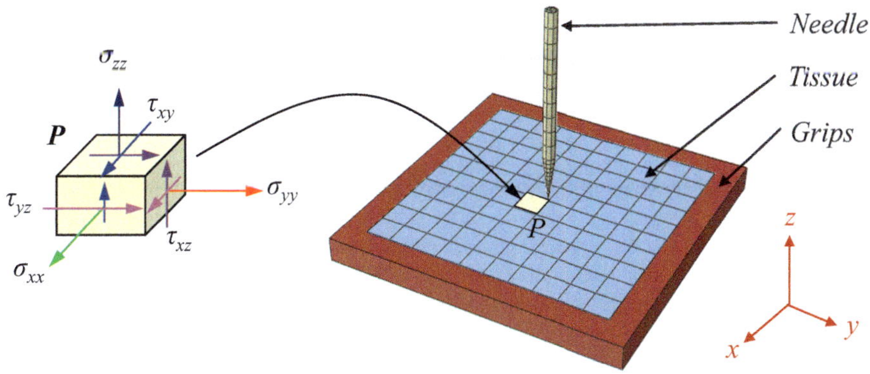

Fig. 9.36 Setup of a needle puncture test of a tissue material, showing the 3-D stress state at point, P

(d) Determine the signed von Mises stress, $\sigma_{vm-signed}$.
(e) If the engineer settles on a grade of titanium alloy of tensile yield strength, $\sigma_y = 950\,\text{MPa}$, show whether the above stress state will lead to yielding of the fan blade.

Problem 9.18 Biomedical engineers regularly undertake an ASTM F2132 needle pucture test on a stretched material, representing a human tissue, to assess the resistance to puncture of the test material. A needle is securely held in position and the test material is clamped securely on its four sides.

To optimize the test setup, finite element studies were recommended for investigating effects of tissue thicknesses, needle sizes and clamping configurations on the puncture resistance. A typical setup is shown in Fig. 9.36.

A stress element, P, was chosen for exploring the material response of the tissue material and the stress state is given as:

$$\boldsymbol{\sigma} = \begin{bmatrix} 20 & 8 & 5 \\ 8 & 20 & 8 \\ 5 & 8 & -10 \end{bmatrix} \text{ MPa}$$

(a) Determine the octahedral normal, σ_{oct}, and shear, τ_{oct}, stresses.
(b) Determine the hydrostatic stress, σ_h, experienced by the tissue material.
(c) Determine the von Mises stress, σ_{vm}, of point, P.
(d) Determine the signed von Mises stress, $\sigma_{vm-signed}$.
(e) Determine the Tresca stress associated with point, P.

References

1. Anandarajah, A.: Computational Methods in Elasticity and Plasticity: Solids and Porous Media. Springer, New York (2011). SpringerLink : Bücher. https://books.google.co.uk/books?id=Hsrj1snaYacC
2. Bishop, N., Sherratt, F.: Finite Element Based Fatigue Calculations. NAFEMS (2000). https://books.google.co.uk/books?id=Kaxi07Ou2jwC
3. Bonet, J., Gil, A., Wood, R.: Nonlinear Solid Mechanics for Finite Element Analysis: Statics. Cambridge University Press (2016). https://books.google.co.uk/books?id=2zxyDAAAQBAJ
4. Hencky, H.: On the theory of plastic deformations and the subsequent stresses induced in the material/zur theorie plastischer deformationen und der hierdurch im material hervorgerufenen nachspannungen. ZAMM – J. Appl. Math. Mech./Zeitschrift für Angewandte Mathematik und Mechanik **4**(4), 323–334 (1924). http://dx.doi.org/10.1002/zamm.19240040405
5. Holzapfel, G.: Nonlinear Solid Mechanics: A Continuum Approach for Engineering. Wiley (2000). https://books.google.co.uk/books?id=_ZkeAQAAIAAJ
6. Krishnamachari, S., Broutman, L.: Applied Stress Analysis of Plastics: A Mechanical Engineering Approach. Springer US (2013). https://books.google.co.uk/books?id=IiQRBwAAQBAJ
7. Lemaitre, J., Desmorat, R.: Engineering Damage Mechanics: Ductile, Creep, Fatigue and Brittle Failures. Springer, Berlin/Heidelberg (2005). https://books.google.co.uk/books?id=WgeaXK6-ZgwC
8. Mises, R.V.: Mechanics of solid bodies in the plastically-deformable state With 4 figures in the text. Nachr. d. Kgl. Ges. Wiss. Göttingen Math.-phys. Klasse **4**, 582–592 (1913)
9. Zhou, X., Song, J.: Effect of local stress on hydrogen segregation at grain boundaries in metals. Mater. Lett. **196**, 123–127 (2017). http://doi.org/10.1016/j.matlet.2017.03.035, http://www.sciencedirect.com/science/article/pii/S0167577X17303695

Chapter 10
Material Response: Constitutive Models and Their Implementation

Abstract The material response of a defined virtual domain within FEM is defined using carefully formulated constitutive models. This chapter describes some of the most common in-built material models in commercially available FEM solvers. It is important to gain this theoretical, as well as practical, understanding in order for the user to interpret FEM outputs. The chapter concludes by walking the reader through the principles of developing *user-defined material (UMAT) models*. This is very important as not all known material behaviours have been modelled and validated as in-built material models within common FEM solvers. Where new materials are discovered, or even modifications and improvements desired for existing in-built material models, the FEM user usually resorts to describing computationally their versions of the constitutive mathematics to reflect such changes. The UMAT subroutine within FEM solvers helps the reader to do so. This chapter is a necessary guide for reliable interpretation of FEM results as the knowledge gained here will help the reader query the predictive fidelity of the FEM framework in comparison with the known material response of the test material.

Keywords Constitutive models • Hyperelasticity • Viscoelasticity • Linear elasticity • User-defined subroutines

10.1 Introduction

Previously, we introduced the different principles of an FEM setup, namely: the virtual domain; creation of an appropriate mesh for the problem; implementation of representative boundary conditions to simulate loading history, etc. In the preceding chapter, we introduced the stress and strain measures needed to define the material response. Material response was considered as another essential principle for any FEM framework. In fact, the material modules of FEM solvers continue to present the most challenges to FEM users.

There is a need for the FEM user to understand how material models work and how they can also be created. The predictive fidelity of an FEM solution relies strongly on creating the most robust material model that captures experimentally observed mechanical behaviour. This starts with formulating the constitutive mathematics that define the experimentally observed behaviour before deploying the

M. Okereke, S. Keates, *Finite Element Applications*, Springer Tracts in Mechanical Engineering, https://doi.org/10.1007/978-3-319-67125-3_10

formulations into a numerical scheme, herein called the *material model*, which is then used within the FEM process to undertake simulations. Consequently, this chapter addresses two vital requirements for creating material models and these are:

- Development of the constitutive mathematics of the chosen test material; and,
- Numerical implementation of constitutive mathematics into a material model for use within the FEM framework.

It is the objective of this chapter to present some thoughts on how these two can be done not only in commercially available FEM solvers, but also, for instance, where the user decides to develop bespoke material models. This latter requirement leads us later in this chapter to discuss the *user-defined material model sub-routine* (UMAT). An introduction is given of the salient steps needed to develop UMATs.

10.2 Chapter Objectives

At the end of this chapter, the reader should be able to:

(a) Gain a quick overview of common constitutive models for describing material response.
(b) Appreciate special cases where material response predictions do not always match FEM predictions.
(c) Introduce the user-defined material subroutine within an FEM solver.
(d) Understand and illustrate steps required for creating a simple user-defined material models, especially in ABAQUS.

10.3 An Introduction to Constitutive Models

Constitutive models are buzz words in computational mechanics that represent a numerical material model. The keyword *constitutive* requires that the numerical models must capture the whole gamut of mechanical behaviour that constitutes known experimentally observed responses of the chosen material. It is therefore, a laudable objective for the *constitutive modeller*[1] to create a model that represents comprehensively the behaviour of the test material. Although most material models do not necessarily meet this requirement, it remains the goal amongst model developers to continue to develop models that are truly 'constitutive.'

[1]A constitutive modeller is someone who develops material models for use in Computational Mechanics. They work within the area of Solid Mechanics and use measures of stress and strain to describe the material response of the chosen material.

All constitutive models are developed based on *constitutive mathematics*, which are a set of analytical formulations that describe the mechanics of the chosen material. The constitutive mathematics are based on some of the principles and parameters presented in Chap. 9. In this section, we will present some of the common constitutive models and their underlining constitutive mathematics. This offers users the key formulations that are implemented in commercial FEM solvers. Users are also encouraged, once they can develop their own UMATs, to begin to create UMATs based on these described constitutive mathematics.

For the purposes of development of constitutive models in this chapter, we will be considering the relationship between stress and strain. The stress-strain diagram, in engineering mechanics, gives the most reliable window into the mechanics of material behaviour, hence establishing that the formulation of stress and strain is at the core of the constitutive mathematics.

Although analytical formulations of stresses and strains are given as one-dimensional expressions, finite element simulations require that the stress and strain are tensorial expressions leading to the development of triaxial constitutive models. It is essential within the FEM scheme that every element within a discretized virtual domain has a *complete* state of stress so that the simulation will be realistic and thus convergence to appropriate solution obtained more quickly. In formulating the constitutive models in this section, we will use the Cauchy triaxial stress tensor and also the strain tensors defined previously (see Sects. 9.4 and 9.5.2), which for a material point, $\boldsymbol{x}$, and time, t, can be expressed:

$$\boldsymbol{\sigma}(\boldsymbol{x},t) = \begin{bmatrix} \sigma_{11} & \sigma_{12} & \sigma_{13} \\ \sigma_{21} & \sigma_{22} & \sigma_{23} \\ \sigma_{31} & \sigma_{32} & \sigma_{33} \end{bmatrix} \quad \boldsymbol{\varepsilon}(\boldsymbol{x},t) = \begin{bmatrix} \varepsilon_{11} & \varepsilon_{12} & \varepsilon_{13} \\ \varepsilon_{21} & \varepsilon_{22} & \varepsilon_{23} \\ \varepsilon_{31} & \varepsilon_{32} & \varepsilon_{33} \end{bmatrix} \tag{10.1}$$

The above stress and strain tensors are **symmetric**. It is also common in the constitutive modelling community to use the *Voigt (contracted) notation* of these tensors and they are given thus:

$$\begin{aligned} \boldsymbol{\sigma}(\boldsymbol{x},t) &= \begin{bmatrix} \sigma_{11} & \sigma_{22} & \sigma_{33} & \tau_{12} & \tau_{23} & \tau_{31} \end{bmatrix}^T \\ &= \begin{bmatrix} \sigma_1 & \sigma_2 & \sigma_3 & \sigma_4 & \sigma_5 & \sigma_6 \end{bmatrix}^T \end{aligned} \tag{10.2}$$

and

$$\begin{aligned} \boldsymbol{\varepsilon}(\boldsymbol{x},t) &= \begin{bmatrix} \varepsilon_{11} & \varepsilon_{22} & \varepsilon_{33} & \gamma_{12} & \gamma_{23} & \gamma_{31} \end{bmatrix}^T \\ &= \begin{bmatrix} \varepsilon_1 & \varepsilon_2 & \varepsilon_3 & \varepsilon_4 & \varepsilon_5 & \varepsilon_6 \end{bmatrix}^T \end{aligned} \tag{10.3}$$

for the stress and strain expressions respectively, with T being the transpose of the column matrix. Also, note that $\tau_{ij} = \sigma_{ij}$ and $\gamma_{ij} = 2\varepsilon_{ij}$ for $\forall\ i \neq j$. Constitutive models presented hereafter will be based on vector forms of Eqs. 10.2 and 10.3.

10.4 Linear Elasticity

Linear ***elasticity*** *models* are the simplest and most common material models in use for many practical applications. They are the key (practical) material models used for solids mechanics and structural analyses. At the core of the models is **Hooke's Law** which established the dependence of deformation of an elasticity spring with the extension force. However, it was Thomas Young (1773–1829) who made the extrapolation from Hooke's Law of linear strings to that of *stress*, σ, and *strain*, ε, by introducing a modulus of elasticity, E, such that the expression of one-dimensional linear elasticity becomes:

$$\sigma = E\varepsilon \tag{10.4}$$

where E = **Young's Modulus**, also called the *modulus of elasticity.* The above expression is a one-dimensional representation and usually used for uniaxial representation of experimental data for common engineering materials.

In order to incorporate the one-dimensional linear elastic Hooke's law as a material model formulation for the FEM process, Eq. 10.4 has to be generalized into a triaxial stress and strain state where the proportionality constant is a fourth-order tensor. The most general vectorized expression of linear elasticity, implemented in many FEM solvers is given thus:

$$\boldsymbol{\sigma} = \boldsymbol{D} : \boldsymbol{\varepsilon}$$
$$\text{or in index notation: } \sigma_{ij} = D_{ijkl}\varepsilon_{kl}$$

where $\boldsymbol{\sigma}$ and $\boldsymbol{\varepsilon}$ are second-order stress and strain tensors,[2] which can be presented in the Voigt contracted notation, while $\boldsymbol{D}$ is a fourth-order tensor called the ***stiffness tensor***, or in some texts, the *elasticity tensor*. Each term of the stiffness tensor represents a modulus of elasticity in form and structure according to the one-dimensional Hooke's Law of Eq. 10.4. In its most comprehensive form, the stiffness tensor, $\boldsymbol{D}$, has 81 components.

Consider for example, the expansion of the first term of the stress tensor:

$$\sigma_{11} = D_{1111}\varepsilon_{11} + D_{1112}\varepsilon_{12} + D_{1113}\varepsilon_{13} + D_{1121}\varepsilon_{21} + \tag{10.5}$$

$$D_{1122}\varepsilon_{22} + D_{1123}\varepsilon_{23} + D_{1131}\varepsilon_{31} + D_{1132}\varepsilon_{32} + D_{1133}\varepsilon_{33} \tag{10.6}$$

In practice, the set of 81 terms of the $\boldsymbol{D}$ tensor is never used as not all of the terms are independent. It was established previously that the Cauchy stress is symmetric and its conjugate strain tensor should also be symmetric. This introduces simplifications

[2]Refer to Chap. 9 for an extensive discussion of stress and strain and their tensors. It is recommended that the reader goes through that chapter first before this one.

to the formulation of the $\boldsymbol{D}$ tensor, for example: $D_{ijkl} = D_{ijlk} = D_{jikl} = D_{jilk}$, and so on. Consequently, using the simplifications, the stiffness tensor reduces from 81 terms to 36 modulus of elasticity terms.

Also, the stiffness tensor has symmetric properties: $\boldsymbol{D} = \boldsymbol{D}^t$, which further reduces its number of terms to 21 elastic constants. In the next section, we consider special cases of the stiffness tensor according to specific classes of symmetry within the problem domain.

10.4.1 Fully Anisotropic Linear Elasticity

In the area of mechanical properties of materials, *isotropy* refers to uniformity of material properties (i.e. elastic moduli) in all directions. For example, polypropylene will have the same modulus of elasticity irrespective of the direction it is tested in, hence it is said to be an **isotropic** material. Most metals and polymers show isotropic properties.

In fact, possessing a homogeneous microstructure usually results in isotropic material properties. Such materials are easier to predict since only a few elastic properties are required to describe their material response. When a material is not isotropic, i.e. its properties change with the changing orientation of loading, it is said to be ***anisotropic***. In this section, we will define the constitutive relations for such anisotropic materials.

Given a stress tensor, $\boldsymbol{\sigma}$, and its conjugate strain tensor, $\boldsymbol{\varepsilon}$, the fully anisotropic linear elasticity constitutive relationship becomes:

$$\begin{bmatrix} \sigma_{11} \\ \sigma_{22} \\ \sigma_{33} \\ \sigma_{12} \\ \sigma_{23} \\ \sigma_{13} \end{bmatrix} = \begin{bmatrix} D_{1111} & D_{1122} & D_{1133} & D_{1112} & D_{1123} & D_{1113} \\ D_{2211} & D_{2222} & D_{2233} & D_{2212} & D_{2223} & D_{2213} \\ D_{3311} & D_{3322} & D_{3333} & D_{3312} & D_{3323} & D_{3313} \\ D_{1211} & D_{1222} & D_{1233} & D_{1212} & D_{1223} & D_{1213} \\ D_{2311} & D_{2322} & D_{2333} & D_{2312} & D_{2323} & D_{2313} \\ D_{1311} & D_{1322} & D_{1333} & D_{1312} & D_{1323} & D_{1313} \end{bmatrix} \begin{bmatrix} \varepsilon_{11} \\ \varepsilon_{22} \\ \varepsilon_{33} \\ \gamma_{12} \\ \gamma_{23} \\ \gamma_{13} \end{bmatrix} \tag{10.7}$$

In material modelling, it may be essential that the resulting strain tensor, $\boldsymbol{\varepsilon}$, is determined based on a certain applied stress tensor. For such a case, the inverse of the stiffness tensor is generated and this new parameter is called the ***compliance tensor***, $\boldsymbol{C}$, which leads to this constitutive relationship:

$$\begin{bmatrix} \varepsilon_{11} \\ \varepsilon_{22} \\ \varepsilon_{33} \\ \gamma_{12} \\ \gamma_{23} \\ \gamma_{13} \end{bmatrix} = \begin{bmatrix} C_{1111} & C_{1122} & C_{1133} & C_{1112} & C_{1123} & C_{1113} \\ C_{2211} & C_{2222} & C_{2233} & C_{2212} & C_{2223} & C_{2213} \\ C_{3311} & C_{3322} & C_{3333} & C_{3312} & C_{3323} & C_{3313} \\ C_{1211} & C_{1222} & C_{1233} & C_{1212} & C_{1223} & C_{1213} \\ C_{2311} & C_{2322} & C_{2333} & C_{2312} & C_{2323} & C_{2313} \\ C_{1311} & C_{1322} & C_{1333} & C_{1312} & C_{1323} & C_{1313} \end{bmatrix} \begin{bmatrix} \sigma_{11} \\ \sigma_{22} \\ \sigma_{33} \\ \sigma_{12} \\ \sigma_{23} \\ \sigma_{13} \end{bmatrix} \tag{10.8}$$

10.4.2 Orthotropic Linear Elasticity

Orthotropic *materials* have elastic properties that depend on three orthogonal planes associated with the test material. For example, consider a given a textile composite material oriented in *XYZ* coordinate system, it will show different moduli of elasticity with respect to the $X-$, $Y-$ and $Z-$axes leading to: E_{11}, E_{22} and E_{33}, for example, given that $1 \equiv X, 2 \equiv Y$ and $3 \equiv Z$. Under this scenario, every other term that is not located on these orthogonal planes will have zero values but values associated with 12, 23 and 13 planes will not be zero. The resulting constitutive relation for orthotropic materials is given in Eq. 10.9. The same implementation applies for the compliance tensor, requiring the inversion of the stiffness tensor.

$$\begin{bmatrix} \sigma_{11} \\ \sigma_{22} \\ \sigma_{33} \\ \sigma_{12} \\ \sigma_{23} \\ \sigma_{13} \end{bmatrix} = \begin{bmatrix} D_{1111} & D_{1122} & D_{1133} & 0 & 0 & 0 \\ D_{1122} & D_{2222} & D_{2233} & 0 & 0 & 0 \\ D_{1133} & D_{2233} & D_{3333} & 0 & 0 & 0 \\ 0 & 0 & 0 & D_{1212} & 0 & 0 \\ 0 & 0 & 0 & 0 & D_{2323} & D_{2313} \\ 0 & 0 & 0 & 0 & D_{1323} & D_{1313} \end{bmatrix} \begin{bmatrix} \varepsilon_{11} \\ \varepsilon_{22} \\ \varepsilon_{33} \\ \gamma_{12} \\ \gamma_{23} \\ \gamma_{13} \end{bmatrix} \quad (10.9)$$

10.4.3 Transversely Isotropic Linear Elasticity

Transverse isotropic *materials* are materials that possess a plane of symmetry such that all properties that are *transverse* or *at right angles* to this plane will have the same value. Consider, for example, a cylinder with the longitudinal axis oriented in, say, the $Z-$axis. All properties in the $X-$ and $Y-$directions will be the same. A fibre reinforced unidirectional composite shows this property with the modulus in the main fibre direction being, say, E_{33} for the $Z-$ or 3−axis. The properties in the two transverse-to-the-fibre axes (E_{22} and E_{33}, for example) will be the same.

The constitutive formulation for the linear elasticity of transversely isotropic materials is given by Eq. 10.10. We should note that for the in-plane properties: $D_{1212} = \frac{1}{2}(D_{1111} - D_{1122})$ and $D_{1313} = D_{2323}$ and for the normal axis properties: $D_{1111} = D_{2222}$ and $D_{1133} = D_{2233}$.

$$\begin{bmatrix} \sigma_{11} \\ \sigma_{22} \\ \sigma_{33} \\ \sigma_{12} \\ \sigma_{23} \\ \sigma_{13} \end{bmatrix} = \begin{bmatrix} D_{1111} & D_{1122} & D_{1133} & 0 & 0 & 0 \\ D_{1122} & D_{1111} & D_{1133} & 0 & 0 & 0 \\ D_{1133} & D_{1133} & D_{3333} & 0 & 0 & 0 \\ 0 & 0 & 0 & \frac{1}{2}(D_{1111} - D_{1122}) & 0 & 0 \\ 0 & 0 & 0 & 0 & D_{1313} & 0 \\ 0 & 0 & 0 & 0 & 0 & D_{1313} \end{bmatrix} \begin{bmatrix} \varepsilon_{11} \\ \varepsilon_{22} \\ \varepsilon_{33} \\ \gamma_{12} \\ \gamma_{23} \\ \gamma_{13} \end{bmatrix} \quad (10.10)$$

10.4.4 Isotropic Linear Elasticity

Finally, ***isotropic** materials* are materials whose properties remain the same even with changing orientation. We have previously described this type of material, but here we will now formulate its linear elastic constitutive relations, as given in Eq. 10.12. We have to also note that due to the symmetry of this type of material, $D_{1133} = D_{1122}$ and $D_{1111} = D_{3333}$ for the normal properties while for the in-plane properties:

$$D_{in-plane} \equiv D_{1212} = \frac{1}{2}(D_{1111} - D_{1122}) \tag{10.11}$$

Also note that as a consequence of the symmetry, $D_{in-plane} \equiv D_{1313} \equiv D_{2323}$.

$$\begin{bmatrix} \sigma_{11} \\ \sigma_{22} \\ \sigma_{33} \\ \sigma_{12} \\ \sigma_{23} \\ \sigma_{13} \end{bmatrix} = \begin{bmatrix} D_{1111} & D_{1122} & D_{1122} & 0 & 0 & 0 \\ D_{1122} & D_{1111} & D_{1122} & 0 & 0 & 0 \\ D_{1122} & D_{1122} & D_{1111} & 0 & 0 & 0 \\ 0 & 0 & 0 & D_{in-plane} & 0 & 0 \\ 0 & 0 & 0 & 0 & D_{in-plane} & 0 \\ 0 & 0 & 0 & 0 & 0 & D_{in-plane} \end{bmatrix} \begin{bmatrix} \varepsilon_{11} \\ \varepsilon_{22} \\ \varepsilon_{33} \\ \gamma_{12} \\ \gamma_{23} \\ \gamma_{13} \end{bmatrix} \tag{10.12}$$

The above is not easy to deploy within a finite element framework in this form without relating it to material properties such as **Young's Modulus**, E, and **Poisson's ratio**, ν, obtained from experiments. It is known from engineering mechanics that the E and ν parameters are sufficient to describe isotropic linear elasticity without needing any other parameters. Therefore, we have to extend the stiffness tensor, $\boldsymbol{D}$, of Eq. 10.12 to contain the E and ν parameters.

The above requirement is achieved by describing the constitutive mathematics of a homogeneous isotropic material model (under isothermal deformation) in terms of the ***Lamé constants*** – G and λ – named after Gabriel Lamé (1795–1870). The constitutive formulation becomes:

$$\boldsymbol{\sigma} = 2G\boldsymbol{\varepsilon} + \lambda \text{tr}(\boldsymbol{\varepsilon})\boldsymbol{I} \tag{10.13}$$

where $\boldsymbol{I}$ is the identity matrix and tr(·) is the trace of the (·) matrix expression.

The linear elasticity formulation can be extended via the Lamé constants to include the Young's modulus, E, and the Poisson ratio, ν. As a result, the first Lamé constant, λ, is called the *Cross Modulus* and expressed as:

$$\lambda = \frac{E\nu}{(1+\nu)(1-2\nu)} \tag{10.14}$$

while the second Lamé constant, G is the *Shear Modulus* and expressed as:

$$G = \frac{E}{2(1+\nu)} \tag{10.15}$$

The two Lamé constants can be related via the *Bulk Modulus*, K thus:

$$\lambda = K - \frac{2}{3}G \quad \text{where} \quad K = \frac{E}{3\,(1-2\nu)} \tag{10.16}$$

Equation 10.12 and its *stiffness tensor*, can be re-defined as:

$$\begin{bmatrix} \sigma_{11} \\ \sigma_{22} \\ \sigma_{33} \\ \sigma_{12} \\ \sigma_{23} \\ \sigma_{13} \end{bmatrix} = \begin{bmatrix} \lambda + 2G & \lambda & \lambda & 0 & 0 & 0 \\ \lambda & \lambda + 2G & \lambda & 0 & 0 & 0 \\ \lambda & \lambda & \lambda + 2G & 0 & 0 & 0 \\ 0 & 0 & 0 & G & 0 & 0 \\ 0 & 0 & 0 & 0 & G & 0 \\ 0 & 0 & 0 & 0 & 0 & G \end{bmatrix} \begin{bmatrix} \varepsilon_{11} \\ \varepsilon_{22} \\ \varepsilon_{33} \\ \gamma_{12} \\ \gamma_{23} \\ \gamma_{13} \end{bmatrix} \tag{10.17}$$

Again, the constitutive formulation described in terms of the *compliance tensor* expressed according to the Lamé constants can be written as follows:

$$\begin{bmatrix} \varepsilon_{11} \\ \varepsilon_{22} \\ \varepsilon_{33} \\ \gamma_{12} \\ \gamma_{23} \\ \gamma_{13} \end{bmatrix} = \frac{1}{E} \begin{bmatrix} 1 & -\nu & -\nu & 0 & 0 & 0 \\ -\nu & 1 & -\nu & 0 & 0 & 0 \\ -\nu & -\nu & 1 & 0 & 0 & 0 \\ 0 & 0 & 0 & 2(1+\nu) & 0 & 0 \\ 0 & 0 & 0 & 0 & 2(1+\nu) & 0 \\ 0 & 0 & 0 & 0 & 0 & 2(1+\nu) \end{bmatrix} \begin{bmatrix} \sigma_{11} \\ \sigma_{22} \\ \sigma_{33} \\ \sigma_{12} \\ \sigma_{23} \\ \sigma_{13} \end{bmatrix} \tag{10.18}$$

10.5 Plasticity Models

The next set of constitutive models that we will discuss under this chapter are the ***plasticity** models*. Plasticity sets in once the material begins to experience deviation from linear elastic behaviour. If the test material undergoes a load-unload history, plasticity occurs if the material experiences permanent deformation. This section will explore some of the plasticity models implemented in common finite element solvers, so that the reader will understand the constitutive formulation that describe *plastic deformations*.

In this section, we will not concern ourselves with the underlying crystalline, molecular and/or microstructural mechanisms that drive permanent deformation of a material body. There exist excellent undergraduate textbooks that deal with these issues very well. It is our objective simply to present the constitutive mathematics of plastic deformation. It is our hope that this will serve as a reference resource for the finite element user when choosing an appropriate plasticity model for a given application. This will also help the reader understand the stress formulations that have been implemented within commercial finite element solvers.

10.5.1 Elastoplastic Material Model

The simplest plasticity model is the ***elastoplastic*** *material model*. Let us consider an isotropic material volume subjected to a **uniaxial** stress history. If the material experiences an elastoplastic response, the material will initially experience a linear elastic deformation characterized by a predominanly **Hookean** material response. The slope of the linear elastic behaviour is the Young's modulus, E. Beyond a certain point in stress, the material begins to deviate from linearity. This point is the yield stress, σ_y, and beyond it, the material can take many possible mechanical responses.

Once yield is attained, there are three possible post-yield material responses that a plasticity model needs to capture. These are:

- **Perfect plasticity**: The simplest post-yield behaviour of an elastoplastic material is the *perfect plasticity* where the body experiences increasing strain without any significant change in stress, as shown in Fig. 10.1. The slope of the line is zero hence: $E = 0$.
- **Isotropic strain hardening:** This type of plasticity is common with many metals and evident in increasing load-bearing capacity of the material volume after the yield point is exceeded. For such plasticity, the amount of stress required to cause further plastic deformation rises and it is usually a function of accumulated plastic strain [11]. As shown in Fig. 10.1, the slope of the linear hardening plot is positive, i.e. $E_h > 0$.
- **Isotropic stress softening:** This is the converse to the strain hardening where after yielding, the material experiences a loss of load-bearing capacity leading to reduction of stress with increasing strain. Its usually a consequence of microstructural post-yield mechanics as an accumulation of damage, structural rejuvenation or even adiabatic heating resulting from, for example, impact rates loading history [8]. The slope of a linear segment of strain softening arm of the stress-strain plot is less than zero, i.e. $E_s < 0$.

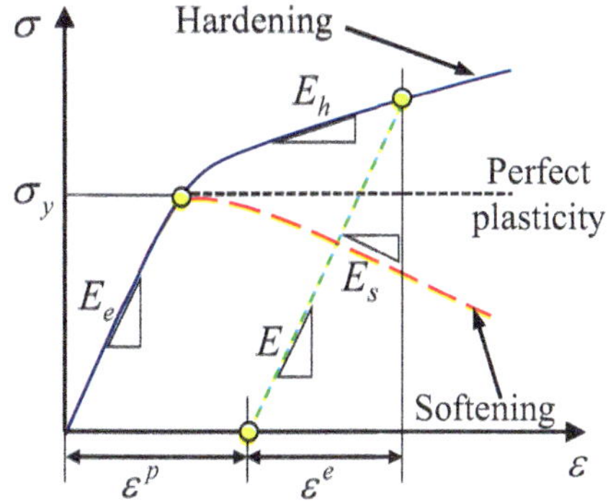

Fig. 10.1 A typical elastoplastic stress-strain profile showing total strain decomposed into elastic and plastic components, as well as three post-yield plasticity responses

10.5.2 The Constitutive Mathematics of Elasto-Plasticity

An important tool for formulating a constitutive model for an elastoplastic material model is the mechanical analogue shown in Fig. 10.2. It consists of a serial arrangement of an elastic spring, which describes the elastic behaviour of the material and a *friction element*, which defines the onset of yield and subsequent post-yield mechanical response. Its common in continuum mechanics to describe the behaviour of the friction element using a ***flow theory*** *of plasticity*. The flow theory describes a gradual transition from elastic behaviour to a plastic behaviour.

The total strain, ε, experienced by a material volume undergoing elastoplastic response can be additively decomposed into an elastic contripution, ε^e, and a plastic contribution, ε^p. The plastic strain, as shown in Fig. 10.1, is the irrecoverable strain stored within the material volume following removal of the load history. It is this plastic strain that drives the plasticity and the basis of formulation of plasticity models.

The additive decomposition of the total strain rate becomes:

$$\dot{\varepsilon} = \dot{\varepsilon}^e + \dot{\varepsilon}^p \tag{10.19}$$

where $\dot{\varepsilon} = \dfrac{d\varepsilon}{dt}$, $\dot{\varepsilon}^e$ is the strain rate due to the elastic strain and, finally, $\dot{\varepsilon}^p$ is the plastic strain rate. The elastic contribution is defined by a Hookean mechanical response such that for an isotropic material system:

$$\dot{\varepsilon} = \frac{\dot{\sigma}}{E} \tag{10.20}$$

and this equation is applicable when $\sigma \leq \sigma_y$.

Experimentally generated stress-strain data of elastoplasticity, according to Fig. 10.3, can also be decomposed into plasticity-driven response, i.e. a post-yield stress-strain data, and an elastic response. The stress due to linear elasticity arguments becomes: $\sigma = E(\varepsilon - \varepsilon^p)$. If the elastic strain is quite small, then $\varepsilon \approx \varepsilon^p$, especially for finite deforming materials or during a superplastic forming. In this case, the stress-strain data will take the form shown in Fig. 10.3 for all three types of post-yield plastic behaviour. It is this type of material response that the aforementioned flow theory of plasticity seeks to describe.

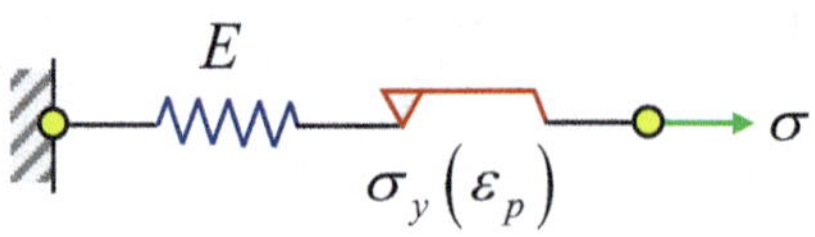

Fig. 10.2 A typical mechanical analogue of an elastoplastic model showing an elastic spring of stiffness, E, and a friction element, which fails once a yield stress, σ_y, is reached

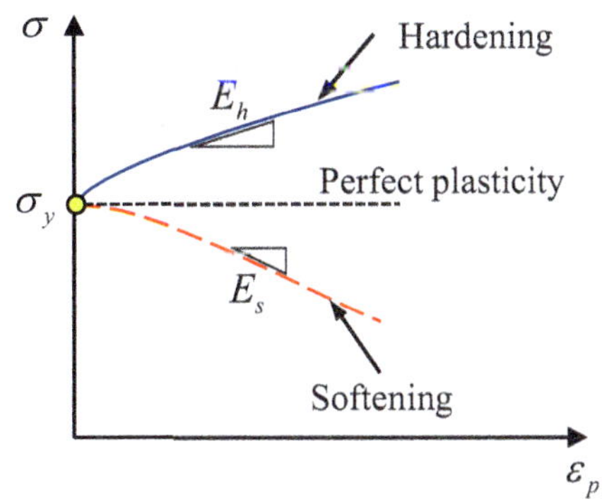

Fig. 10.3 Stress-strain data for a plasticity-only mechanical response, with three typical post-yield responses shown

10.5.2.1 Onset of Yield

The flow theory of plasticity is introduced into the constitutive mathematics once a yield point is reached during the loading history of the material volume. Correctly identifying the yield point is essential for correct representation of the underlying plasticity.

- *Uniaxial loading history:* For a uniaxial tests with monotonically increasing loading history, the yield point is approached once the stress state in the material becomes equal to the yield stress, σ_y, of the test material i.e.

$$\sigma_{ij} \equiv \sigma_y \quad \text{for} \quad i,j = 1,2,3 \quad \text{and} \quad \forall\, i = j \tag{10.21}$$

  The strain that corresponds to the yield stress is the yield strain, ε_y, and it is the strain at which the yield stress exists.
- *Multi-axial loading history*: When the material volume is subjected to a complex loading history, it is not trivial to isolate the yield point. In this instance, the effective stress state due to the combined and complex loading history needs to be determined. The effective stresses are defined commonly using the **von Mises stress**[3] or any of the other practical stress formulations[4] (see Sect. 9.6). The corresponding strain at which the effective stress exists is the effective plastic strain.

[3]For extensive discussion of the von Mises stress, the reader should refer to Sect. 9.6.4. It is one of many practical stress formulations used to generate a single stress measure for a body under complex loading history.

[4]Depending on the type of problem under consideration, the FEM user will determine which of the practical stress formulations best suits the yield behaviour shown by the material volume.

The representation of the effective plastic strain rate, $\dot{\varepsilon}^p$ becomes:

$$\dot{\varepsilon}^p = \frac{\sqrt{2}}{3}\left[\left(\dot{\varepsilon}_1^p - \dot{\varepsilon}_2^p\right)^2 + \left(\dot{\varepsilon}_2^p - \dot{\varepsilon}_3^p\right)^2 + \left(\dot{\varepsilon}_3^p - \dot{\varepsilon}_1^p\right)^2\right]^{1/2} \tag{10.22}$$

where $\dot{\varepsilon}_i^p$ for $i = 1, 2, 3$ is the i-th plastic principal strain rate.

10.5.2.2 Yield Criterion

Once the yield point is identified, increasing the load will lead to plastic flow. The analytical condition that has to be obeyed to establish the point of departure from linearity, and hence the initiation of plastic deformation, is described as the *Yield criterion*. It is usually defined by the *yield function*, f. There exist many yield criteria for the plastic deformation of metals. For example, there is the *Von Mises Yield Criterion*, which is based on the Von Mises stress. Details of this yield condition are given in Sect. 9.6.4.

The yield criterion, f, does not depend on the hydrostatic or mean stress,[5] σ_m. This implies it is purely driven by a the deviatoric stress-state, σ', of the material volume. For isotropic materials, the yield function is such that: $f = 0$. The yield function/criterion, f, is an *even* and *symmetric* function of the deviatoric stress, σ_i', where $i = 1, \cdots, 3$.

We can then re-write the Von Mises yield criterion thus:

$$f = \sigma_e - \sigma_y \longrightarrow f = \left(\frac{3}{2}\boldsymbol{S} : \boldsymbol{S}\right)^{1/2} - \sigma_y \tag{10.23}$$

where $\boldsymbol{S}$ is the deviatoric stress tensor. If $f(\boldsymbol{\sigma}) < 0$, then elastic deformation dominates the material behaviour. Yielding, and the consequent onset of plastic deformation, occur when $f = 0$ and finally the different *plastic flow* profiles occur when $f(\boldsymbol{\sigma}, \boldsymbol{\varepsilon}^p) \geq 0$.

For a two-dimensional principal stress space, where the principal stress in the 3-axis is set as: $\sigma_3 = 0$, the analysis is said to be carried out under a *plane stress* condition. The geometric representation of the *Von Mises yield criterion* for such stress space is shown in Fig. 10.4 as a skewed ellipse. The *yield surface* is the set of σ_1 and σ_2 that corresponds to $f(\boldsymbol{\sigma}, \boldsymbol{\varepsilon}^p) = 0$. It defines the limit beyond which plastic deformation initiates.

The *Associated flow rule* is required to define the direction of flow of plasticity following the initiation of yield. Let us assume that there is an increase in the load imposed on the material body, the direction of flow of plasticity is thus defined by what in plasticity literature is termed as the *normality hypothesis of plasticity*. The hypothesis states that once yielding has occurred, the 'direction' of flow is

[5] See Sect. 9.6 for description of hydrostatic, mean, and deviatoric stresses.

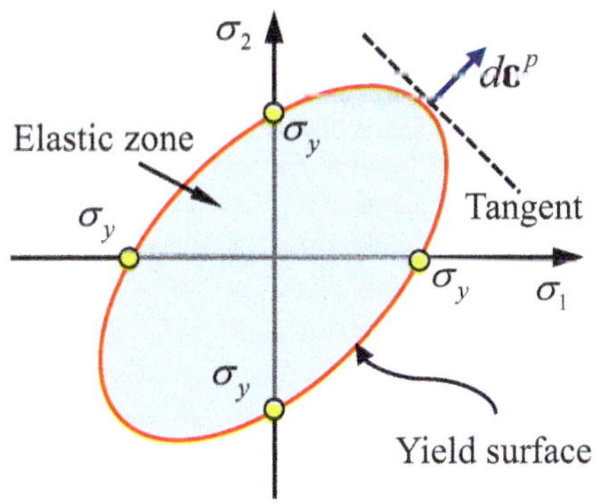

Fig. 10.4 The Von Mises yield surface under plane stress conditions, showing the direction of the increment of plastic strain, $d\boldsymbol{\varepsilon}^p$

determined by the *increment of the plastic strain tensor*, $\boldsymbol{\varepsilon}^p$. This increment of the plastic strain tensor is in a direction relative to the principal stress directions, which is normal to the tangent to the yield surface as the load point. This is illustrated in Fig. 10.4 and defined mathematically in terms of the yield function, f, thus:

$$\mathrm{d}\boldsymbol{\varepsilon}^p = d\lambda \frac{\partial f}{\partial \boldsymbol{\sigma}} \quad \longrightarrow \quad \dot{\boldsymbol{\varepsilon}}^p = \dot{\lambda} \frac{\partial f}{\partial \boldsymbol{\sigma}} \quad \text{in terms of strain rate} \tag{10.24}$$

With respect to the normality hypothesis, the direction of flow is defined by $\partial f / \partial \boldsymbol{\sigma}$, whilst its magnitude is determined by the plastic multiplier, λ. Also, the parameter $\mathrm{d}\lambda$ is described as the *hardening parameter* and usually is $\mathrm{d}\lambda > 0$. This type of plasticity is described by the *associated flow rule*.

10.5.3 The Perfect Plasticity

Consider the associated flow rule formulation of Eq. 10.24. The perfect plasticity formulation requires that following onset of yield, the flow evolves at a constant yield stress, σ_y. For this type of material behaviour, the change in stress is zero, i.e. $\mathrm{d}\boldsymbol{\sigma} = \mathbf{0}$, although there is an increment of plastic strain, i.e. $\mathrm{d}\boldsymbol{\varepsilon}^p > 0$.

Similarly, if the perfectly plastic material is subjected to a cyclic loading as shown in Fig. 10.5, the elastic component of the total strain, ε^e, will be constant for all the cycles such that $d\varepsilon^e = 0$. For such a system, the yield surface does not vary with increasing plastic strain hence a clearly defined yield surface for such material exists.

This type of material model is introduced in `ABAQUS` via the `Plasticity` material module, which requires the user to stipulate the yield stress, σ_y, and a plastic strain,[6] ε^p. In FEM solvers, this type of plasticity is a commonly implemented

[6]The plastic strain must begin at zero.

Fig. 10.5 A sequence of load-unload cyclic response of a perfectly plastic material showing equivalent elastic strain, ε^e

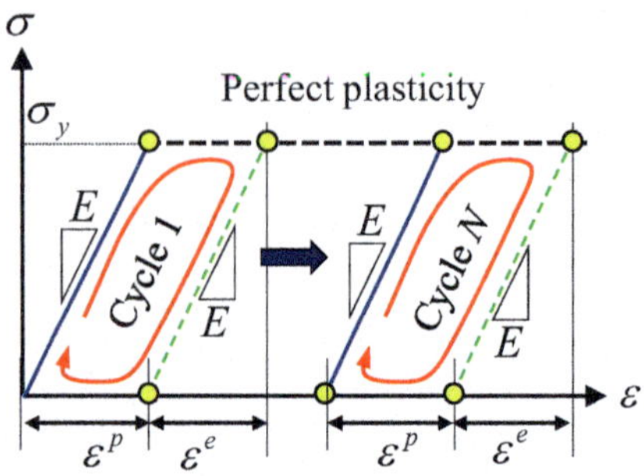

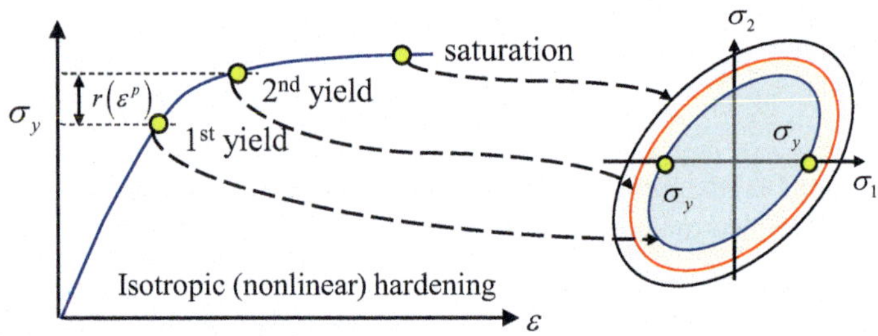

Fig. 10.6 A uniaxial plot of an elastoplastic material showing *nonlinear isotropic hardening*. The insert image is a multiple envelop yield surface

in-built material model. It is mostly used to describe metal plasticity hence some textbooks describe this type of plasticity as the *classic plasticity of metals*. Table 8.8 shows typical elastoplastic material properties for 6061 aluminium.

10.5.4 The Isotropic Hardening

To illustrate the second type of elastoplastic post-yield response, let us subject a material volume to a uniaxial test under plane stress conditions. An example stress-strain profile for such a body is shown in Fig. 10.6, with the inserted image, being the yield surface of the material body.

The material response is initially elastic until a yield point, σ_y, is reached and identified as the *first yield*. With increasing loading such that $\mathrm{d}\boldsymbol{\sigma} > 0$, a subsequent yield point, called *second yield*, is reached. With subsequent loading, the stress-strain profile reaches a *saturation point* where there is no significant increase in stress ($d\boldsymbol{\sigma} \approx 0$). This is also identified as a third yield envelope: the largest of the three envelopes.

The hardening is described as *nonlinear* since the increase of stress, $\mathrm{d}\boldsymbol{\sigma}$, has a nonlinear dependence with plastic strain, $\boldsymbol{\varepsilon}^p$. As a consequence, the yield surfaces *expand* concentrically outwards, as shown in Fig. 10.6, until a saturation point is reached. The nonlinear hardening is described as *isotropic* if the expansion of the yield surface is uniform in all directions of the stress space. For the hardening to be termed isotropic, the change of yield surface must be characterized by a homogeneous distortion of the surface in all directions.

The resulting *isotropic hardening*, for the example under consideration, is given as $r(\varepsilon^p)$. Thus, the new yield stress accounting for the isotropic hardening becomes:

$$\sigma\,(\varepsilon^p) = \sigma_{y0} + r(\varepsilon^p) \tag{10.25}$$

where σ_{y0} is the first/initial yield stress and $r(\varepsilon^p)$ is the measure of isotropic hardening. The formulation of $r(\varepsilon^p)$ determines whether the isotropic hardening is described as linear or not.

10.5.5 The Kinematic Hardening

Previously, we introduced isotropic hardening to represent a sequence of yield surfaces resulting from a plasticity flow rule involving homogeneous distortion (expansion or contraction) of the initial yield surface in all directions. This is not always the case with many materials. Here, we introduce the *kinematic hardening* which represents a *translation of the yield surfaces*. In such an instant, the nonlinear plastic flow is characterized by a set of constant hardening modulus which corresponds to a defined yield surface. The subsequent yield surfaces results from a *kinematic translation* of the initial centre of yield surface as shown in Fig. 10.7.

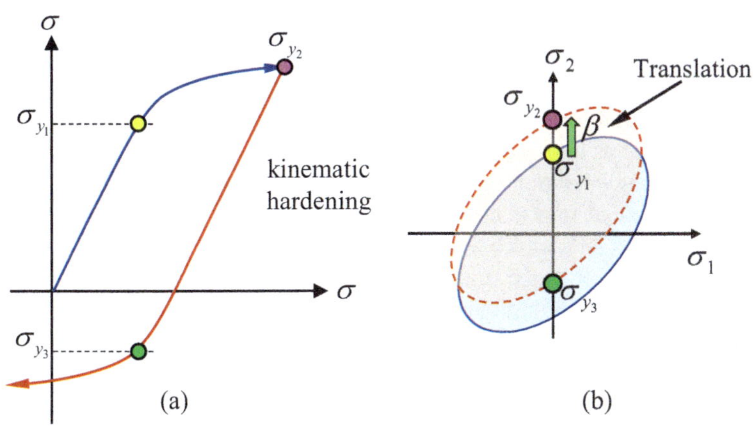

Fig. 10.7 (**a**) A load-unload uniaxial plot of an elastoplastic material showing *kinematic hardening*. (**b**) A sequence of yield envelopes illustrating effect of kinematic hardening with original envelope translated by back stress, β

The yield function that results from a kinematic hardening system is expressed in terms of the deviatoric stress, $\boldsymbol{S}$ as:

$$f = \left[\frac{3}{2}(\boldsymbol{S} - \boldsymbol{\beta}') : (\boldsymbol{S} - \boldsymbol{\beta}')\right]^{\frac{1}{2}} - \sigma_y \tag{10.26}$$

where σ_y is the uniaxial yield stress. The term $\boldsymbol{\beta}'$ is the deviatoric form of the *kinematic hardening variable*, $\boldsymbol{\beta}$, which is also called the *back stress*.

10.5.6 Johnson-Cook Plasticity

Prior to this point, we assumed the metal plasticity to be rate-independent, hence formulations of yield and hardening were all rate- and temperature-insensitive. Actual experimental data for many metals show significant temperature- and rate-dependence. In this section, we will introduce the *Johnson-Cook plasticity model*: one of the most popular nonlinear hardening formulations of plasticity for modelling material responses that show temperature- and rate-dependence. Although there are other competing models, the model introduced here remains very popular amongst researchers. It has also been implemented as an in-built material model within the ABAQUS FEM framework.

The Johnson-Cook plasticity model was developed jointly by Gordon R. Johnson and William H. Cook [18] in 1985. It is a type of Von Mises plasticity model, however it comprises explicit analytical formulations of the hardening behaviour and rate-dependence effects. Their original paper was aimed at modelling the fracture characteristics of three metals (4340 Steel, Armco Iron and OFHC copper). The effects of strain rates, temperature and pressure on the strain to failure of these metals were assessed.

Current uses of the Johnson-Cook plasticity model continue to be based around the same conditions as the original paper with the model commonly used for 'dynamic' studies as high velocity impact, blast dynamics, detonation of explosives and metal forming operations. Predictive modelling of experimental data from these studies must account for the high strain rates, the temperature-dependence and the effect of high pressure waves associated with them. The Johnson-Cook plasticity model is well suited for such studies. The Johnson-Cook plasticity model presented here will be divided into the Johnson-Cook (a) hardening formulation and (b) strain rate dependence formulations.

10.5.6.1 Johnson-Cook Hardening Formulation

The Johnson-Cook hardening formulation is a type of isotropic hardening where the first yield stress, σ_{y0} (also called the *static yield stress*), is represented as:

$$\sigma_{y0} = [A + B(\bar{\varepsilon}^p)^n]\left(1 - \hat{T}^m\right) \tag{10.27}$$

where $\bar{\varepsilon}^p$ is the *equivalent plastic strain*. Also, A = *Johnson-Cook (J-C) yield strength*, B = *J-C hardening coefficient*, n = *J-C strain hardening exponent*, m = *J-C softening exponent*. All these are also material parameters obtained at or below a transition temperature, T_{tr}.

The parameter, $\hat{T}$, is called the *homologous temperature* and it is a non-dimensional ratio of current temperature, T, to the melting temperature, T_m, and it is defined as:

$$\hat{T} \equiv \begin{Bmatrix} 0 & \text{for } T < T_{tr} \\ \dfrac{T - T_{tr}}{T_m - T_{tr}} & \text{for } T_{tr} \leq T \leq T_m \\ 1 & \text{for } T > T_m \end{Bmatrix} \tag{10.28}$$

The transition temperature, T_{tr}, defines the limit below which temperature dependence of material properties are neglected and above which temperature-dependence has to be accounted for in the expression of the yield stress. With respect to Eq. 10.27, if $T \geq T_m$, then $\hat{T} = 1$ and the term $\left(1 - \hat{T}^m\right) \longrightarrow 0$ such that $\sigma_{y0} = 0$. Under this condition, the material begins to flow as a fluid and isotropic hardening is precluded by forcing the equivalent plastic strain to be zero: $\bar{\varepsilon}^p = 0$.

10.5.6.2 Johnson-Cook Rate Dependence

Having accounted for the hardening behaviour of the Johnson-Cook plasticity, here we demonstrate how strain rate dependence can be incorporated with the original static yield stress formulation. According to the authors [18], the expression of yield stress that incorporates rate dependence is:

$$\sigma_y = [A + B(\varepsilon^p)^n]\left[1 + C\ln(\dot{\varepsilon}^*)\right]\left(1 - \hat{T}^m\right) \tag{10.29}$$

where $\dot{\varepsilon}^{p*}$ is the dimensionless plastic strain rate for $\dot{\varepsilon}_0 = 1.0\,\text{s}^{-1}$ and is defined as:

$$\dot{\varepsilon}^{p*} = \frac{\dot{\varepsilon}^p}{\dot{\varepsilon}_0} \tag{10.30}$$

For this model, $\dot{\varepsilon}^p$ = plastic strain rate, $\dot{\varepsilon}_0$ = reference strain rate and C = J-C strain constant. The C and $\dot{\varepsilon}_0$ parameters are also material constants and need to be specified as part of the model input parameters for the Johnson-Cook model. In the context of the original Johnson and Cook [18] paper, the model parameters for the three metals considered by the others are published in Table 10.1, along with a number of other materials.

Table 10.1 Johnson-Cook plasticity models parameters for some metals

Material	A	B	C	n	m	$\dot{\varepsilon}$	Ref.
–	[MPa]	[MPa]	–	–	–	[s^{-1}]	–
OFHC copper	90	292	0.025	0.310	1.090	1.000	[18]
Armco Iron	175	380	0.060	0.320	0.550	1.000	[18]
4340 steel	792	510	0.014	0.260	1.020	1.000	[18]
AISI 316L steel	305	1161	0.010	0.610	0.517	1.000	[10, 33]
Al 7050-T7451 alloy	491	530	0.005	0.580	-	-	[30]
Ti-6Al-4V titanium alloy	1098	1092	0.014	0.930	1.100	-	[34]

10.6 Viscoelasticity

The next set of material response to be considered is *viscoelasticity*. This is the material response shown by certain materials that shows a combination of *elastic* and *viscous* or *fluid-like* characteristics under the effect of a load. It is a typical property shown by polymers (both amorphous and semicrystalline), biopolymers (cellulose, starch, chitin, DNA, proteins, etc.), bitumen, biomaterials, and composite materials (laminates, particulates and nanocomposites). In this section, we will introduce the constitutive formulation of this type of material and the different mechanical analogues for describing the different features of viscoelastic response.

Viscoelastic responses are dominated by both rate- and temperature-sensitivity. The constitutive model for capturing viscoelasticity must reflect these dual characteristics. Consider a weight suspended on a polymeric film, as shown in Fig. 10.8a. If the material is treated as linearly elastic, it will experience an elastic strain, ε^e, which remains constant over the test duration of $t + \delta t$, as shown in Fig. 10.8b. The strain expression becomes: $\varepsilon_T = \varepsilon_0$, where ε_0 is the initial static strain experienced once the load, W, is released.

However, the polymer film is expected to show a viscoelastic behaviour since it a polymeric system. In effect, the polymer film extends gradually over the test duration of $t + \Delta t$. At first, the increase in strain is rapid before it approaches a saturation/threshold strain. The strain profile is shown in Fig. 10.8b. For the viscoelastic case: $\varepsilon_T = \varepsilon_0 + \Delta\varepsilon(t)$, where $\Delta\varepsilon(t)$ is the time-dependent accumulation of strain, which causes the gradual elongation of the polymeric film.

The dependence of strain on time, under constant load, is described as ***creep*** and shown in Fig. 10.9a. This behaviour is a manifestation of viscoelasticity and it is the effect of a macromolecular rearrangement of the solid under the effect of an induced load. A distinction between the mechanics of viscoelasticity and plasticity is that when a material is undergoing *creep* within the *linear viscoelasticity* range, once the load is removed, the material should return to its undeformed configuration. This is unlike plasticity where a permanent plastic deformation exists.

Consider another scenario where the polymer film is stretched suddenly to constant strain, γ_0. A measure of the resulting stress gives another manifestation of viscoelasticity called *stress relaxation*, as shown in Fig. 10.9b. For this type of

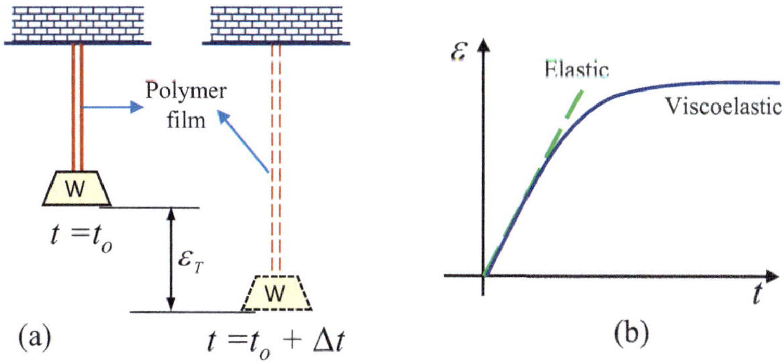

Fig. 10.8 (**a**) A material body, W, hanging at the end of a polymer film in undeformed and deformed configurations; and (**b**) the resulting strain-time profiles for *elastic* and *viscoelastic* responses

Fig. 10.9 The viscoelastic response for (**a**) creep and (**b**) stress relaxation experiments

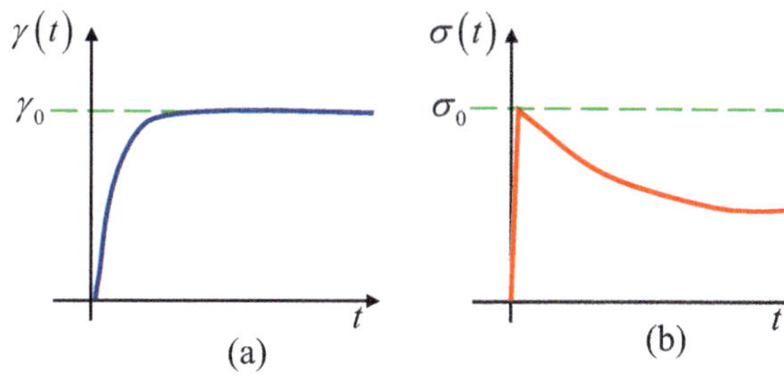

test, a constant strain, γ_0, is applied at time, $t = t_0$, such that instantly the stress, $\sigma_0(t)$, is obtained. This stress, however, does not remain constant over the duration of Δt, due to viscoelasticity effects. Instead, the stress continues to decrease with time.

In viscoelasticity, the rearrangement of the polymer macromolecules is a cumulative effect of *bond-stretching*, which is characterized by a dominant *elastic* response. The *viscous* deformation of the polymeric system is as a result of the diffusion of polymer macromolecules under the effect of a loading history.

10.6.1 A Theory of Viscoelasticity

The constitutive modelling of viscoelastic systems is based on phenomenological representations of the contributions of the elastic and viscous components, in a most suitable format to capture the experimentally observed behaviour. The approach is to use a certain combinations of *spring elements* to capture elastic behaviour and *dashpot elements* to capture viscous behaviour of polymer macromolecules. This is similar to *inductance* or *capacitance* and *resistance* used in electrical networks [22]. The elastic spring *stores energy* while the dashpot element *dissipates energy*.

10.6.1.1 Constitutive Models for Creep Response

The constitutive mathematics of a spring element is demonstrated here based on a creep experiment in which a viscoelastic material is subjected to a constant load, σ_0, such that it experiences a time-dependent strain, $\gamma_0(t)$. If a new material is subjected to a load, σ_1, a corresponding new time-dependent strain, $\gamma_1(t)$, results. A *creep compliance*, $J(t)$, is a material parameter that relates the stresses and the strains imposed on these two experiments (Fig. 10.10).

The definition of creep compliance at time, t, for the two experiments, is:

$$J(t) = \frac{\gamma_0(t)}{\sigma_0} = \frac{\gamma_1(t)}{\sigma_1} \quad \longrightarrow \quad \sigma = \frac{\gamma(t)}{J(t)} \tag{10.31}$$

The above equation shows that the creep compliance is a time-dependent quantity that changes depending on the decade of time in which the experiment/material is investigated.

When a polymeric material is tested at different strain rates, different creep compliances result and a plot of the creep compliance versus log t over a number of decades of time results in a *sigmoid-shaped curve*, which for a creep test is called a *master curve*. This curves gives a complete representation of the viscoelastic behaviour of the material tested from short to long time scales.

At short time scales, the polymer macromolecules do not have enough time to re-arrange, resulting in a creep compliance measure called the *unrelaxed creep compliance*, J_U, which is time-independent. At very long times, again the macromolecules would have re-arranged sufficiently leading to a *relaxed creep compliance*, J_R, which is also time-independent. In between these two states, the effect of the viscous element contributes significantly to the behaviour of the material and a strong sensitivity to time effects dominates the material response.

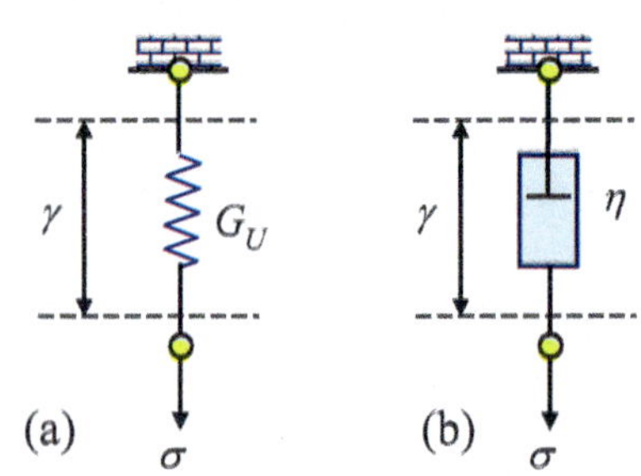

Fig. 10.10 Example mechanical analogues used for developing viscoelasticity constitutive models: (**a**) an elastic spring of unrelaxed shear modulus, G_U; and (**b**) viscous dashpot of viscosity, η. Both elements are subjected to a shear strain, γ

10.6.1.2 Constitutive Models for Stress Relaxation Response

The constitutive mathematics for a *stress relaxation* response is developed by again considering a material volume subjected to constant strain, γ_0, such the material experiences a time-dependent stress, $\sigma_0(t)$. If the material is unloaded and allowed to recover completely,[7] and deformed to a new strain, γ_1, the corresponding time-dependent stress is $\sigma_1(t)$. A *stress relaxation modulus*, $G(t)$, is a material parameter that connects the stress and strain obtained in the stress relaxation experiment according to the following:

$$G(t) = \frac{\sigma_0(t)}{\gamma_0} = \frac{\sigma_1(t)}{\gamma_1} \quad \longrightarrow \quad \sigma = G(t)\gamma(t) \tag{10.32}$$

Similarly, the stress relaxation modulus can also be measured over decades of time to generate a sigmoidal curve that represents the different time-dependent relaxation behaviour of the polymer macromolecules under the effects of a load history. At very short times, we obtain the *unrelaxed stress relaxation modulus*, G_U. At sufficiently long test durations, the polymer segments should have sufficiently relaxed leading to a *relaxed stress relaxation modulus*, G_R.

Both G_U and G_R are independent of time and result from the effect of the spring elements. In between these boundary stress relaxation moduli are the highly time-dependent, $G(t)$, which result from the effect of the viscous dashpot element of the mechanical analogue. We can also observe that there exists a relationship between the creep compliance and the stress relaxation modulus and is given as:

$$J(t) = \frac{1}{G(t)} \quad , \quad J_U = \frac{1}{G_U} \quad \text{and} \quad G_R = \frac{1}{J_R} \tag{10.33}$$

10.6.2 *Domain of Analysis of Viscoelasticity*

As established previously, viscoelasticity is a time-dependent evolution of stress and strain under the effect of either a load or strain history. The influence of time effects leads us to consider two possible domains of analysis:

(a) **Time domain**: This domain of analysis considers loads that are applied statically over a duration of time. In such analyses, the viscoelastic behaviour is thus also considered over a duration of time. The analyses of creep and stress relaxation above were based on assessing the viscoelasticity in a time domain. This was why the resulting master curve represents a history of the different relaxation stages of the polymer macromolecules in a time-scale albeit the representation was on a log t scale.

[7]This recovery time is usually at least three-times the initial loading duration.

(b) **Frequency domain**: This domain of analysis considers the viscoelastic behavior under the effect of a dynamic (reciprocatory) history. Under this, the material properties are not static quantities, but rather properties that change with the direction of the loading history. These properties are described as *dynamic properties*. Since a viscous dashpot is essentially a *damper*, the viscoelastic formulation will adopt the principles of engineering vibration of particles.

Consequently, in developing constitutive models of viscoelastic materials, it is important to a priori specify the domain of analysis one wants to use for the problem. Since we have already discussed the constitutive mathematics of a time-domain analysis problem, hereafter we will focus on the frequency domain of the analysis.

10.6.2.1 Dynamic Viscoelasticity

To understand dynamic viscoelasticity, the domain of analysis of the problem has to be set to a frequency mode. The classic nature of dynamic behaviour demands that the imposed load is oscillatory, with a known frequency. To illustrate this, consider a material body subjected to an oscillatory shear strain of frequency, ω. The formulations of the shear strain and shear stresses for a viscoelastic body become:

$$\gamma = \gamma_0 \sin(\omega t) \quad \text{and} \quad \sigma = \sigma_0 \sin(\omega t + \delta) \tag{10.34}$$

where δ is the *phase angle* by which the strain lags the stress. The expansion of the stress equation above gives:

$$\begin{aligned} \sigma &= (\sigma_0 \cos\delta)\sin(\omega t) + (\sigma_0 \sin\delta)\cos(\omega t) \\ &= \gamma_0 G' \sin(\omega t) + \gamma_0 G'' \cos(\omega t) \end{aligned} \tag{10.35}$$

Here, we have introduced two quantities, where $G' \sin(\omega t)$ is in phase with the dynamic strain and $G'' \cos(\omega t)$ is out of phase by 90°. The analytical definitions of these two new quantities are:

$$G' = \frac{\sigma_0}{\gamma_0}\cos\delta \quad \text{and} \quad G'' = \frac{\sigma_0}{\gamma_0}\sin\delta \tag{10.36}$$

These new quantities are part of a *complex shear modulus*, G^*, which, adopting electrical circuitry principles of current and voltage, can be represented as:

$$G^* = G' + iG'' \tag{10.37}$$

where G' is the *real shear modulus* and G'' is the *imaginary shear modulus*.

Similarly, the complex creep compliance[8] is given as:

$$J^* = J' - iJ'' \tag{10.38}$$

where $J' = \dfrac{\gamma_0 \cos\delta}{\sigma_0}$ and $J'' = \dfrac{\gamma_0 \sin\delta}{\sigma_0}$. An important material parameter that connects the two parts of the complex shear modulus or creep compliance is the *tangent of the phase angle*, tan δ, which is expressed as:

$$\tan\delta = \frac{J''}{J'} \quad \equiv \quad \tan\delta = \frac{G''}{G'} \tag{10.39}$$

Having laid down the theory of viscoelasticity and the parameters that feed into a constitutive model, the following section will focus on the different in-built material models within existing finite element solvers for modelling viscoelasticity. The objective is to generate a constitutive formulation: the basis of in-built viscoelastic material models.

10.6.3 The Standard Linear Solid Viscoelastic Model

The *standard linear solid model* is also called the *Zener model*. It is a model with two arrangements of elastic springs and viscous dashpots, as shown in Fig. 10.11. Both arrangements of the standard linear solid will lead to the same viscoelastic

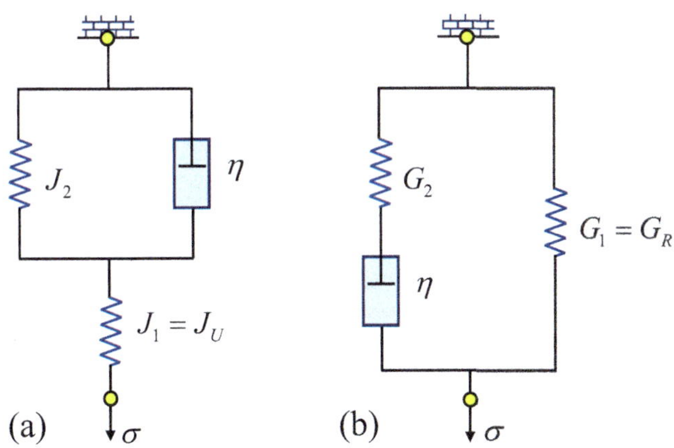

Fig. 10.11 Two versions of the mechanical analogue of a standard linear solid or Zener model

[8]Note that the complex creep compliance is also related to the complex shear modulus: $J^* = 1/G^*$.

response. McCrum and colleagues [22] gave an extensive undergraduate derivation of the creep compliance and the stress relaxation shear modulus.

In view of the *creep response* of a viscoelastic material, the Zener model expression of the creep compliance, $J(t)$, can be expressed as follows:

$$J(t) = J_U + \Delta J\left\{1 - \exp\left[-\tfrac{t}{\tau_\sigma}\right]\right\} \tag{10.40}$$

where

$$\Delta J = J_R - J_U \quad \text{and} \quad \tau_\sigma = J_2\eta \tag{10.41}$$

A new parameter called **relaxation time,** τ_σ is introduced. It is derived from a creep experiment in which a constant stress, σ, initiated the viscoelastic response. This is why the τ value has a stress, σ, subscript. To fully describe this type of viscoelastic response using the Zener model, the FEM user need only specify the J_U, J_R and τ_σ parameters. These can be determined from simple creep data, either over a long period of time or across a wide range of temperatures.

Similarly, the *stress relaxation* effect of the Zener model can be deduced carefully, based on Fig. 10.11b, such that the resulting stress relaxation shear modulus becomes:

$$G(t) = G_U - \Delta G\left\{1 - \exp\left[-\tfrac{t}{\tau_\gamma}\right]\right\} \tag{10.42}$$

where

$$\Delta G = G_U - G_R \quad \text{and} \quad \tau_\gamma = \tau_\sigma\frac{G_R}{G_U} \quad \longrightarrow \quad \frac{\tau_\sigma}{\tau_\gamma} = \frac{J_R}{J_U} = \frac{G_U}{G_R} \tag{10.43}$$

Also, the relaxation time from a stress relaxation experiment is τ_γ. Again, it is derived from a stress relaxation experiment.

Finally, in a frequency domain of analysis, we will assess the effect of the Zener model to demonstrate evidence of viscoelasticity of the dynamic mechanical properties. The complex creep compliance, J^*, of the Zener model, given a frequency, ω, is expressed as:

$$J^* = J_U + \frac{J_U - J_R}{1 + i\omega\tau_\sigma} \quad \longrightarrow \quad J^* = J' - iJ'' \tag{10.44}$$

such that:

$$J' = J_U + \frac{(J_R - J_U)}{1 + \omega^2\tau_\sigma^2} \quad \text{and} \quad J'' = \frac{(J_R - J_U)\omega\tau_\sigma}{1 + \omega^2\tau_\sigma^2} \tag{10.45}$$

Similarly, a formulation for the complex shear modulus, G^* becomes:

$$G^* = G_U - \frac{G_U - G_R}{1 + i\omega\tau_\gamma} \quad \longrightarrow \quad G^* = G' + iG'' \tag{10.46}$$

such that:

$$G' = J_U + \frac{(G_U - G_R)}{1 + \omega^2\tau_\gamma^2} \quad \text{and} \quad G'' = \frac{(G_U - G_R)\omega\tau_\gamma}{1 + \omega^2\tau_\gamma^2} \tag{10.47}$$

Let us consider the graphical representations of these viscoelastic parameters. Consider a piece of polypropylene subjected to a creep test where such a Zener-type model is fitted to the creep data. The viscoelastic parameters obtained are: unrelaxed creep compliance, $J_U = 1.25\,\text{GPa}^{-1}$, relaxed creep compliance, $J_R = 4.95\,\text{GPa}^{-1}$, and relaxation time, $\tau_\sigma = 12.2\,\text{s}$. The resulting $J(t)$ and $G(t)$ plot with respect to $\log_{10} t$ is given in Fig. 10.12.

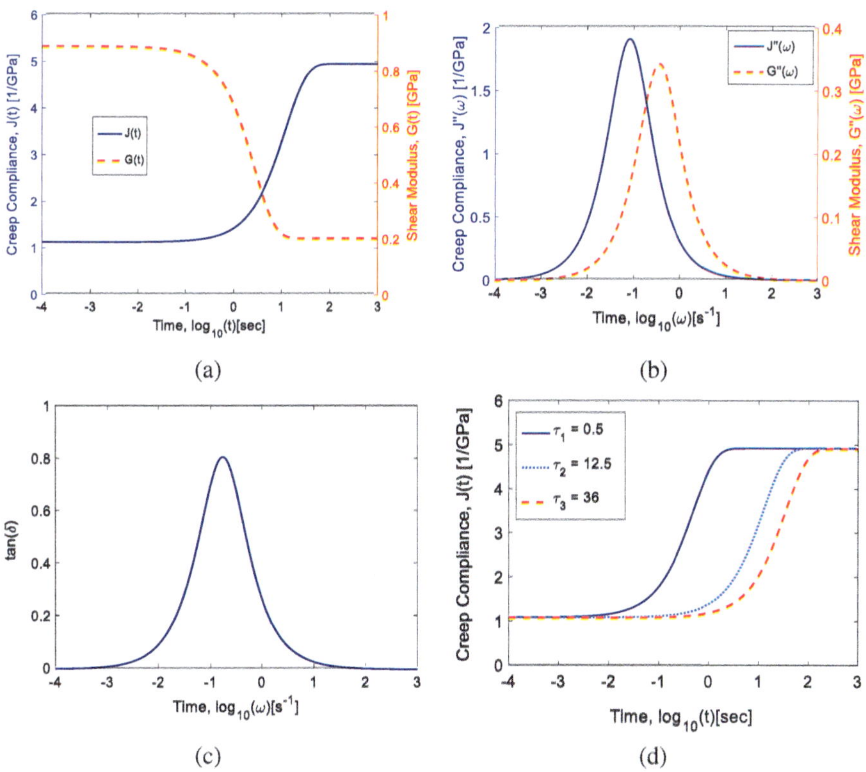

Fig. 10.12 Plots of Zener model parameters, showing: (**a**) creep compliance, $J(t)$, and stress relaxation shear modulus, $G(t)$; (**b**) imaginary part of complex creep compliance, $J''(\omega)$, and shear modulus, $G''(\omega)$; (**c**) tangent of phase angle, $\tan\delta$, profile; and, (**d**) the effect of multiple relaxation times, τ_i for $i = 1, \cdots, 3$ on the creep compliance of polypropylene

Similarly, we can explore the dynamic mechanical properties of polypropylene within the temperature window of the above creep test. Based on either a torsion pendulum's creep data or a dynamic mechanical analyzer's information, the J_U and J_R properties above were matched at a frequency, ω. For our purposes here, let us say $t \equiv \omega$ such that we can compare plots of J' and J'' with respect to $\log_{10} \omega$.

There is a problem with the standard linear solid model of Sect. 10.6.3 and that lies in its inability to represent the observed viscoelastic behaviour of actual polymers reliably. For example, the spread of $J(t)$ values in Fig. 10.12d was across six decades for $\tau_1 = 0.5$ s, and about five decades for $\tau_3 = 36$ s. In general, the key consideration is understanding the effect of changing the relaxation times such that a closely spaced relaxation time is used instead.

Figure 10.12d shows that the spread of decades of time for $J(t)$ and $G(t)$ will vary when relaxation times are changed. In fact, the differences in polymer segments between short and long polymer macromolecules will lead to different relaxation times. So, a simple Zener-type model will not accurately capture reliably the full relaxation times seen in real polymers. A Zener-type model can only cover three decades of time [22].

10.6.4 The Generalized Maxwell Model

It is quite simple to correct the shortcomings of the classic Zener model. This is by using a *distribution* or *spectrum of relaxation times* representation of the Zener model. Such a spectrum should cover all polymer macromolecular lengths and their associated relaxation times. This leads to the *generalized Maxwell model*, which is an extension of the Zener model of type shown in Fig. 10.11b. It consists of a spring element of stress relaxation modulus, G_R, and n Maxwell[9] model segments, arranged in parallel, as shown in Fig. 10.13.

The generalized model is based on stress relaxation data according to a parallel arrangement of N units of Maxwell's serial spring-dashpot segments. The expression of the stress relaxation modulus becomes:

$$G(t) = G_R + \sum_{i=1}^{N} G_i \exp\left[-\frac{t}{\tau_i}\right] \tag{10.48}$$

where G_R is the relaxed shear modulus, τ_i is i-th relaxation time of the N Maxwell units and t is the current time. The relaxation times have been found experimentally to be closely spaced [22], such that we can replace the summation with an integral as follows:

[9] A Maxwell model has a serial arrangement of spring and dashpot elements. The constitutive model is $\frac{d\gamma}{dt} = \frac{\sigma}{\eta} + \frac{1}{J}\frac{d\sigma}{dt}$ where J and η are spring's creep compliance and dashpot's viscosity.

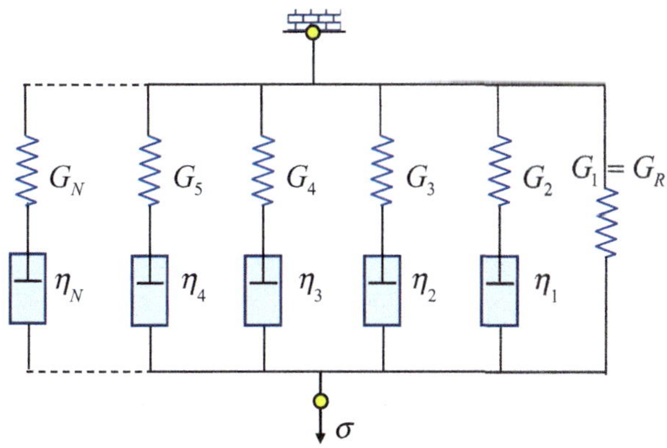

Fig. 10.13 The mechanical analogue of a generalized Maxwell model

$$G(t) = G_R + \int_0^\infty \exp\left[-\frac{t}{\tau}\right] g(\tau) d\tau$$

$$\text{where} \quad \int_0^\infty g(\tau) d\tau = (G_U - G_R) \tag{10.49}$$

Similarly, the creep compliance, $J(t)$, formulation that can account for multiple relaxation times can be obtained by adapting the Zener model of Fig. 10.11b such that N Kelvin-Voigt[10] segments are connect in series with an elastic spring. Following careful manipulation of the expressions of the strains from different parts of the model, we obtain the following expression of the creep compliance (in integral form):

$$J(t) = J_U + \int_0^\infty \left\{1 - \exp\left[-\frac{t}{\tau}\right]\right\} j(\tau) d\tau$$

$$\text{where} \quad \int_0^\infty j(\tau) d\tau = (J_R - J_U) \tag{10.50}$$

[10]The Kelvin-Voigt model is a representation of a viscoelastic material in which elastic spring and dashpot elements are arranged in parallel. For such a system, the stress-strain relationship is: $\sigma(t) = E\varepsilon(t) + \eta \frac{\mathrm{d}\varepsilon(t)}{\mathrm{d}t}$, where E and η are the spring's modulus of elasticity and the dashpot's viscosity, respectively.

10.6.5 Temperature Dependence and Viscoelasticity

Apart from loading history, the temperature history imposed on a viscoelastic material will affect its relaxation significantly. Thus, constitutive models for viscoelastic materials must account for the temperature-dependent response. In this section, we will explore some of the existing approaches that introduce temperature-dependence in constitutive models of viscoelastic materials.

For a typical creep test, the creep compliances, J_U and J_R, are not affected significantly by the changes in temperature in comparison with the creep relaxation times, τ_σ. From a molecular consideration, the effect of temperature on molecular motions is to increase the entropy of the molecular configuration, which will influence the relaxation times of the material. It is the influence of the temperature on the relaxation times of polymers (at a given temperature) that is used in accounting for temperature effects in viscoelastic models.

Let us consider a creep test carried out at two temperatures, T_{ref} and T, such that the corresponding relaxation times from these tests are: $\tau_{T_{ref}}$ and τ_T respectively. The effect of temperature on the creep response can be quantified by introducing a parameter, a_T, such that:

$$\tau_T = a_T \tau_{T_{ref}} \quad \longrightarrow \quad \log \tau_T = \log a_T + \log \tau_{T_{ref}} \tag{10.51}$$

The definition of the a_T parameter for a temperature dependent material is usually via the ***Arrhenius equation***, which establishes the relationship between relaxation times and temperature. It is analytically described thus:

$$\tau = \tau_\infty \exp\left(\frac{\Delta H}{RT}\right) \quad \text{where} \quad \tau_\infty = \frac{1}{\nu}\exp - \left(\frac{\Delta S}{R}\right) \tag{10.52}$$

where ν = frequency of excitation of a macromolecular segment required to move over an activation barrier; ΔH = change in activation enthalpy of the relaxation process; ΔS = change in entropy; R = universal gas constant and T = absolute temperature (in Kelvins). τ_∞ has an empirical meaning and corresponds to the relaxation time obtained from the intercept of the plot of ln τ versus T^{-1}, i.e. when $T^{-1} = 0$ [22].

Given a creep test carried out at temperatures, T and T_{ref}, the Arrhenius-type temperature shift factor, a_T, is:

$$a_T = \exp \frac{\Delta H}{RT}\left[\frac{1}{T} - \frac{1}{T_{ref}}\right] \quad \longrightarrow \quad \log a_T = \frac{\Delta H}{2.303R}\left[\frac{1}{T} - \frac{1}{T_{ref}}\right] \tag{10.53}$$

The above equation suggests that with increasing temperature, there is a commensurate reduction in the relaxation times, hence accelerating the creep process. As a result, one can assess the full range of viscoelastic behaviour of a material usingthis *accelerated creep process* through an array of different temperatures. The creep data from such temperature measurements are then subsequently *temperature-shifted* to form a creep *master curve*.

The Arrhenius-type temperature dependence is commonly used for describing secondary polymer transitions, but for temperatures around the glass-transition temperature, T_g, an alternative relationship between temperature shift factor and temperature is used. This is the *Williams-Landel-Ferry (WLF) equation*, used for temperature near and above the glass transition [37], and it is given as:

$$\log a_T = \frac{-C_1\left(T - T_{ref}\right)}{C_2 + \left(T - T_{ref}\right)} \tag{10.54}$$

where C_1 and C_2 are material constants whose values depend on material type and reference temperature, T_{ref}, and T is current temperature (in Celsius). The original paper by the authors [37] stated that if $T_{ref} = T_g$, then $C_1 = 17.4$ and $C_2 = 51.6$. These are regarded as 'universal' constants and have been shown to fit quite a wide range of polymers.

10.7 Nonlinear Elasticity

Only a few materials can be described appropriately using the linear **elasticity** arguments of Sect. 10.4. It was important to establish the foundations of linear elasticity, because other higher order material models will always have some linear elasticity response at quite small strains. Consequently, such higher order models will have formulations of linear elasticity as a base model and then deviations from that characterizes the test material's nonlinear behaviour.

Moreover, a large set of materials show some form of nonlinear mechanical response. Thus, this section is important in describing quite a large set of material models for practical uses in design, construction and fabrication. It is an extensive set, but we will only aim here to explore some of the commonly used nonlinear elasticity models.

Previously, it was shown that linear elasticity demands that when a load is removed, the material returns to its undeformed configuration without any permanent change of deformation. Similarly, *nonlinear elasticity* shows the same properties of non-permanent deformation following the removal of load. In other words, there is no dissipation of energy following the release of the disturbing load. The difference between linear and nonlinear elasticity is that the relationship between stress and strain is nonlinear. This is illustrated graphically in Fig. 10.14.

Given a Cauchy stress, $\boldsymbol{\sigma}$, the nonlinear elasticity models demand that $\boldsymbol{\sigma}$ has a nonlinear relationship with infinitesimal strain, $\boldsymbol{\varepsilon}$, thus:

$$\boldsymbol{\sigma}(\boldsymbol{x}, t) = f(\boldsymbol{x}, t, \boldsymbol{\varepsilon}) \tag{10.55}$$

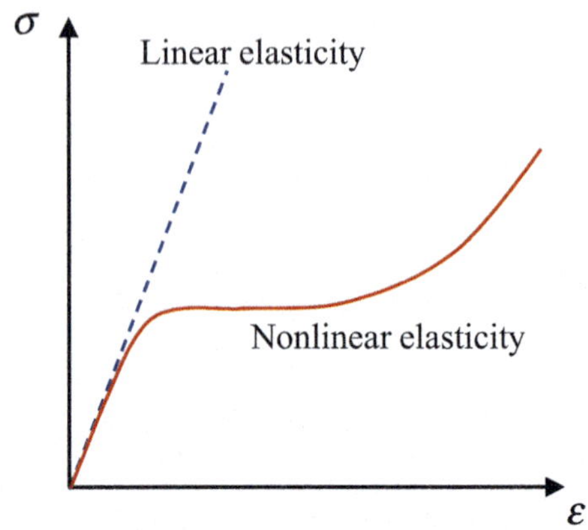

Fig. 10.14 A graphical representation of linear and nonlinear elasticity material behaviours

where f is the nonlinear function and $\boldsymbol{x}$ is the current/material configuration. In defining the nonlinear elasticity models, the FEM user must express the constitutive models such that the dependence of stress on strain must evolve in a nonlinear manner. The nonlinear elasticity models have to be distinguished from plasticity models where there is a permanent deformation following the removal of the load. A different set of formulations will be used to describe the plasticity laws and these will be presented subsequently.

10.7.1 An Introduction to Hyperelastic Material Models

The most common type of constitutive models for capturing the nonlinear elasticity of materials are the *hyperelastic material models* or *Green elasticity models*, named after the original work of George Green (1793–1841). The term is used for *ideally elastic*[11] materials whose stress-strain relationships can be derived strictly from the ***strain energy density function***. Rubber and soft tissues are commonly regarded as hyperelastic materials.

Just like linear elasticity models, the hyperelastic models are uniquely suited for describing the material response of rubber, etc. with nonlinear elastic relationship between stress and strain. Such materials also show isotropic material properties with the dependence of stress and strain being independent of changing strain rate. In finite element material models, several implementations of hyperelastic material models exist and these will be explored here.

[11]Ideally elastic materials have similar properties as linear elastic materials but the dependence of stress on current strain is nonlinear as well as always independent of history or rate of loading.

Consider a hyperelastic material bounded by domain, Ω subjected to a deformation gradient, $\boldsymbol{F}$ which results into a strain tensor, say $\boldsymbol{\varepsilon}$. Let us assume for this material, there exists a *Helmholtz free-energy function*,[12] Ψ, per unit volume of the material, and defined thus:

$$\Psi = \Psi(\boldsymbol{F}) \quad \text{or} \quad \Psi = \Psi(\boldsymbol{\varepsilon}) \tag{10.56}$$

which implies that the Helmholtz free energy function can be expressed interchangeably as functions of either the deformation gradient, $\boldsymbol{F}$, or the strain tensor, $\boldsymbol{\varepsilon}$. This condition is true if the system is *homogeneous*, in which case the material properties are independent of material directions. For *heterogeneous* material systems, the strain energy function is not solely dependent on $\boldsymbol{F}$, but also on material points, $\boldsymbol{x}$, within the material system. For example, a filled rubber will have a strain energy density function that is dependent on a cumulative effect of the rubber matrix and filler media.

The Helmholtz free energy function, Ψ is also generally regarded in hyperelasticity literature as the *stored-energy function* or *strain-energy function* [17]. Thus, hereafter, we will refer to this free energy function by the common terminologies of *strain energy* or *stored energy*. Note that the use of strain energy here is an short-form of the strain energy density function. It should not be confused with the '**strain energy**' which is the stored energy in a material undergoing deformation.

Given a strain energy function, $\Psi(\boldsymbol{\varepsilon})$, stress is determined by obtaining the differentiation of the strain energy with respect to the strain tensor thus:

$$\begin{aligned} \boldsymbol{\sigma} &= \frac{\partial \Psi(\boldsymbol{\varepsilon})}{\partial \boldsymbol{\varepsilon}} \\ \text{or in index notation,} \quad \sigma_{ij} &= \frac{\partial \Psi}{\partial \varepsilon_{ij}} \end{aligned} \tag{10.57}$$

Also, according to Holzapfel, the second Piola-Kirchoff stress tensor, $\boldsymbol{P}$ can be determine also from the strain energy function, $\Psi(\boldsymbol{F})$, according to the equation:

$$\begin{aligned} \boldsymbol{P} &= \frac{\partial \Psi(\boldsymbol{F})}{\partial \boldsymbol{F}} \\ \text{or in index notation,} \quad P_{ij} &= \frac{\partial \Psi}{\partial F_{ij}} \end{aligned} \tag{10.58}$$

[12]Helmholtz free energy is a thermodynamic quantity that captures the actual work derived from a closed thermodynamic system under constant temperature and volume conditions. Given an *internal energy*, U, *entropy*, S and *temperature*, T, the Helmholtz free energy, A can be expressed as: $A = U - TS$. It is a scalar-valued function.

Recall also, that the Cauchy stress tensor, $\boldsymbol{\sigma}$, can be related to the second Piola-Kirchoff stress tensor such that:

$$\sigma = J^{-1}\boldsymbol{P}\boldsymbol{F}^{T} \quad \longrightarrow \quad \sigma = J^{-1}\boldsymbol{F}\left(\frac{\partial\Psi(\boldsymbol{F})}{\partial\boldsymbol{F}}\right)^{T}$$
$$\text{or in index notation,} \quad \sigma_{ij} = J^{-1}F_{ij}\frac{\partial\Psi}{\partial F_{ij}} \tag{10.59}$$

where J is the volume ratio obtained thus: $J = \det(\boldsymbol{F})$.

If $J = 1$, the material is described as *incompressible*, which implies that the material keeps its volume throughout the deformation. However, if $J > 1$, the material is described as *compressible*, which requires that following finite deformation, the material volume changes with deformation. The constraint here is that the volume ratio has to be positive. In latter sections, we will derive both compressible and incompressible variants of the constitutive hyperelastic material models.

The exact formulation of the strain energy function distinguishes one type of hyperelastic model from another. At the core of derivation of hyperelastic models is the description of the mathematics of the given/applicable strain energy for the given type of model.

10.7.2 Classes Hyperelastic Material Models

The expression of the strain energy function is the key task in formulating an appropriate hyperelastic material model. The common examples of the hyperelastic material models can be classified as:

- **Phenomenological models**: These are created from empirical formulations of the dependence of stress on strain. They are informed solely by experimental information of the material system. Examples of phenomenological models include: Saint-Venant Kirchoff, Polynomial, Ogden, Mooney-Rivlin and Yeoh material models. Details of these models will be presented later.
- **Mechanistic models**: These models are derived based on information of the underlying structure of the material model. It is often based on the polymer macromolecular structure, for polymer-based systems or tissue microstructure like collagens, muscle fibres, etc. Some examples include: Arruda-Boyce and Edwards-Vilgris material models.

10.7.3 Saint-Venant Kirchoff Hyperelastic Material Model

If a hyperelastic material volume undergoes large displacements, but the material strain is quite small, the *Saint-Venant Kirchoff* hyperelastic material model is recommended. It is formulated similar to linear elasticity but represents a nonlinear elasticity response. The governing strain energy for this model is given in terms of the *Green strain tensor*, $\boldsymbol{E}$, thus:

$$\Psi(\boldsymbol{E}) = \frac{\lambda_{Lame}}{2}\left[\mathrm{tr}(\boldsymbol{E})\right]^2 + G_{Lame}\mathrm{tr}(\boldsymbol{E}^2) \quad (10.60)$$

where λ_{Lame} and G_{Lame} are the first and second **Lamé constants**, herein attached with the subscript 'Lame' to distinguish the constants from the stretch ratio, λ, which shall be referred to subsequently. It is essential that $\lambda_{Lame} > 0$ and $G_{Lame} > 0$. Taking the derivative of Eq. 10.60 with respect to $\boldsymbol{E}$ results in the second Piola-Kirchoff stress tensor:

$$\boldsymbol{S}(\boldsymbol{E}) = \lambda_{Lame}\mathrm{tr}(\boldsymbol{E})\boldsymbol{I} + 2G_{Lame}\mathrm{tr}(\boldsymbol{E}) \quad (10.61)$$

where $\boldsymbol{I}$ is a second order **identity matrix**. To use this model for simulation, the Lamé constants are the only parameters that need to be specified, given that the deformation gradient, $\boldsymbol{F}$, exists.

The classical definition of the Saint-Venant Kirchoff model is independent of the volume ratio, J, which means that one cannot distinguish between compressible and incompressible hyperelastic material response. A probable solution is to deduce a *modified Saint-Venant Kirchoff model* [17], where the strain energy function takes the form:

$$\Psi(\boldsymbol{E})_{\mathrm{mod}} = \frac{\kappa}{2}(\ln J)^2 + G_{Lame}\mathrm{tr}(\boldsymbol{E}^2) \quad (10.62)$$

where λ_{Lame} is replaced by κ = *Bulk Modulus* and $\mathrm{tr}(\boldsymbol{E})$ is replaced by $\ln J$, with $J = \det(\boldsymbol{F})$. This model also makes the Saint-Venant model adaptable for predictions of large compressive strains, a major weakness of the original model. The modified second Piola-Kirchoff stress tensor becomes:

$$\boldsymbol{S}(\boldsymbol{E})_{\mathrm{mod}} = 2\kappa(\ln J)\boldsymbol{I} + 2G_{Lame}\mathrm{tr}(\boldsymbol{E}) \quad (10.63)$$

10.7.4 Polynomial Hyperelastic Material Model

Another phenomenological hyperelastic material model is the *Polynomial material model*. It is also called the *Generalized Rivlin model*. The strain energy density function is formulated based on polynomial functions of two invariants, I_1 and I_2, of the left Cauchy-Green deformation tensor, $\boldsymbol{B}$.[13] It is also called the *Rivlin function* or *Rivlin series*. This formulation was made popular by Rivlin and Saunders [26] in 1951 and was originally developed for incompressible isotropic materials. The authors validated the model using vulcanized rubber test-pieces. In their paper, the strain energy density function for an incompressible material volume was defined as:

$$\Psi(I_1, I_2)_{\text{incomp}} = \sum_{i=0}^{m}\sum_{j=0}^{n} C_{ij}\,(I_1 - 3)^i\,(I_2 - 3)^j \tag{10.64}$$

where C_{ij} are material constants with the first term, $C_{00} = 0$. Also, m and n are the orders of the polynomials and it is common to set $m = n$. C_{ij} is related to the distortional response of the material volume under consideration. The left Cauchy-Green deformation tensor is defined as: $\boldsymbol{B} = \boldsymbol{F}\,\boldsymbol{F}^T$. Note that the invariants of $\boldsymbol{B}$ are defined as:

$$\begin{aligned} I_1 &= \text{tr}(\boldsymbol{B}) && \longrightarrow I_1 = \lambda_1^2 + \lambda_2^2 + \lambda_3^2 \\ I_2 &= \frac{1}{2}\left([\text{tr}(\boldsymbol{B})]^2 - \text{tr}(\boldsymbol{B}^2)\right) && \longrightarrow I_2 = \lambda_1^2\lambda_2^2 + \lambda_2^2\lambda_3^2 + \lambda_3^2\lambda_1^2 \end{aligned} \tag{10.65}$$

Some of the applications of this polynomial-type strain energy density functions have been particularly for rubber-like materials with characteristic nonlinear elasticity mechanics and incompressible behaviour [1, 3, 4, 15, 16].

If the material volume is *compressible*, in which case $J > 1$, then the *Rivlin function* for such systems is defined as polynomial functions of the deviatoric invariants of the left Cauchy-Green deformation tensor, thus:

$$\Psi(\bar{I}_1, \bar{I}_2)_{\text{comp}} = \sum_{i=0}^{m}\sum_{j=0}^{n} C_{ij}\,(\bar{I}_1 - 3)^i\,(\bar{I}_2 - 3)^j + \sum_{k=1}^{p} D_k\,(J-1)^{2k} \tag{10.66}$$

where D_k is a material constant, which describes the volumetric response of the material system. The volume ratio is defined by $J = \det(\boldsymbol{F})$ and p is the order of the extra polynomial function that accounts for compressibility. Notice that if the material is incompressible, in which case $J = 1$, then the extra polynomial term vanishes and reduces to Eq. 10.64. The deviatoric invariants are defined as follows:

$$\bar{I}_1 = J^{-2/3} I_1 \quad \text{and} \quad \bar{I}_2 = J^{-4/3} I_2 \tag{10.67}$$

The strain energy of the Rivlin function is dependent only on the first invariant, I_1, with minimal effects of the second invariant, I_2 [13, 14, 25, 39, 40]. Consequently,

[13] For a review of the definition and formulation of the $\boldsymbol{B}$ tensor, the reader should refer to Sect. 9.4.3.

the *Reduced polynomial hyperelastic material model* for compressible material volume becomes:

$$\Psi(\bar{I}_1, \bar{I}_2)_{\text{comp,Reduced}} = \sum_{i=0}^{n} C_{i0}\left(\bar{I}_1 - 3\right)^i + \sum_{k=1}^{p} D_k\,(J-1)^{2k} \tag{10.68}$$

where $j = 0$ in all cases, and n and p are orders of the polynomial, with n often set equal to p. If, say, $n = p = 2$, then the material constants required for this reduced polynomial function are: C_{01}, C_{02}, D_1 and D_2. These will be determined by adjusting the model prediction to the experimental data.

10.7.5 Mooney-Rivlin Hyperelastic Material Model

Prior to Rivlin deducing the *Generalized Rivlin model*, Mooney in 1940 had already proposed the strain energy density function of a hyperelastic (compressible) material model according to the equation:

$$\Psi(\bar{I}_1, \bar{I}_2) = C_1(\bar{I}_1 - 3) + C_2(\bar{I}_2 - 3) + D_1(J-1)^2, \tag{10.69}$$

where C_1, C_2 and D_1 are experimentally-generated material constants. Note $\bar{I}_1$ and $\bar{I}_2$ are respectively the first and second deviatoric invariants of the left Cauchy-Green deformation tensor, whose analytical expressions were given in Eq. 10.67.

According to the Polynomial strain energy function, if the order of the polynomial is set as $n = 1$ and $p = 1$, the strain energy function becomes:

$$\begin{aligned}\Psi(\bar{I}_1, \bar{I}_2) = C_{00}(\bar{I}_1 - 3)^0 + C_{01}(\bar{I}_2 - 3) + C_{10}(\bar{I}_1 - 3)\\ + C_{11}(\bar{I}_2 - 3) + D_1(J-1)^2\end{aligned} \tag{10.70}$$

Recall that $C_{00} = 0$, such that when we compare Eq. 10.69 with Eq. 10.70, we notice that: $C_{01} = C_2, C_{11} = 0, C_{10} = C_1$ and $p = 1$. As a consequence, the *Mooney-Rivlin strain energy function* is a degenerate form of the *Generalized Rivlin model* for $n = p = 1$.

10.7.6 Neo-Hookean Hyperelastic Material Model

The *Neo-hookean hyperelastic material model* is the simplest hyperelastic material model and, as the name would suggest, is similar to the Hookean linear elasticity model. It was also proposed by Rivlin in 1948. The strain energy function for a Neo-hookean solid depends solely on the first invariant of right Cauchy-Green deformation tensor, $\boldsymbol{C}$. For an incompressible material volume, the strain energy is defined analytically as:

$$\Psi(I_1)_{\text{incomp}} = C_1(I_1 - 3) \tag{10.71}$$

where C_1 is a material constant related to the distortional response of the material volume. The first invariant of the $\boldsymbol{B}$ tensor is derived according to Eq. 10.65 above. For a compressible material, the Neo-Hookean strain energy density function is given in terms of the deviatoric (first) invariant of $\boldsymbol{C}$, thus:

$$\Psi(\bar{I}_1)_{comp} = C_1(\bar{I}_1 - 3) + D_1(J - 1)^2 \tag{10.72}$$

where $\bar{I}_1 = J^{-2/3}I_1$, $J = \det(\boldsymbol{F})$ and D_1 is the material constant responsible for the volumetric response. Also, $\bar{I}$ represents the isochoric part of the right Cauchy-Green tensor, $\boldsymbol{C}$, such that: $\bar{C} = (\det(\boldsymbol{C}))^{-1/3}\boldsymbol{C} = J^{-2/3}\boldsymbol{C}$. The material constants can be deduced from comparing the Neo-Hookean response with the linear elasticity, such that:

$$C_1 = \frac{G_{Lame}}{2} \quad \text{and} \quad D_1 = \frac{\lambda_{Lame}}{2} \tag{10.73}$$

where G_{Lame} and λ_{Lame} are Lamé constants. These are equivalent to the initial shear and bulk moduli of the material volume.

10.7.7 Ogden Hyperelastic Material Model

The *Ogden hyperelastic material model* is similar to the polynomial material model discussed in Sect. 10.7.4 except that its strain energy function is described in terms of principal stretches, λ_i for $i = 1, 2, 3$. It is a very convenient model to use in computational mechanics, because the principal stretches are measurable from laboratory experiments such as *uniaxial tension, equi-biaxial tension* and *shear* tests. The model was developed by Raymond Ogden in 1972 [23]. The model describes the hyperelasticity of materials such as rubber, biological tissues and finite deforming polymer systems.

The Ogden strain energy function for incompressible material systems is given as:

$$\Psi(\lambda_1, \lambda_2, \lambda_3)_{\text{incomp}} = \sum_{i=0}^{\infty} \frac{\mu_i}{\alpha_i}\left(\lambda_1^{\alpha_i} + \lambda_2^{\alpha_i} + \lambda_3^{\alpha_i} - 3\right) \tag{10.74}$$

where each μ_i and α_i is a material constant. According to Ogden, the α_i constants do not necessarily need to be an integers [24]. It is advisable for practical purposes that the terms of the *Ogden function* should be limited to finite numbers rather than ∞.

Also, the material constants are related to the shear modulus, G, of the test material in the *stress free* configuration according to the equation:

$$\sum_{i=1}^{M} \mu_i \alpha_i = 2G \tag{10.75}$$

where M is a positive integer. Similar to the formulation of compressibility in Eq. 10.66, the *Ogden function* of Eq. 10.74 can be extended to account for compressible materials by adding the compressibility term: $\sum_{i=1}^{\infty} D_i (J-1)^{2i}$, where J is the volume ratio and $J > 1$.

In order to fit experimental data to the Ogden model, it is most commonly implemented with $M = 3$ or $M = 4$. Also, apart from the condition of Eq. 10.75, it is a necessary condition that $\mu_i \alpha_i > 0$. Different algorithms have been used to fit experimental data to the Ogden model with examples such as Twizzell and Ogden [32] and Ogden et al. [24].

10.7.8 Yeoh Hyperelastic Material Model

The last of the phenomenological hyperelastic material models treated here is the *Yeoh model* [39]. The strain energy function of the Yeoh model has similar attributes to the Generalized Rivlin model except that it is defined solely as a function of the first invariant of the Cauchy-Green deformation tensors, as well as setting the order of the polynomial to $n = 3$. Therefore, the *Yeoh strain energy function* is expressed analytically as:

$$\Psi(I_1)_{incomp} = \sum_{i=1}^{3} C_i (I_1 - 3)^i \tag{10.76}$$

with C_i being a material constant. In the limit of linear elasticity for an incompressible material, the C_i material constant equates to $2C_1$ of the Neo-Hookean model according to Eq. 10.73 above. Also, the Neo-Hookean model is a degenerate form of the Yeoh model where the power of the Yeoh model is set to $n = 1$. The Yeoh model can also be generalized beyond the three power terms to $n > 3$ for certain material types, for example, filled rubber and biological tissues. The higher the order, the more terms to be fitted but the better the fit.

Similarly, for a compressible material volume, the Yeoh strain energy function can be expressed as:

$$\Psi(\bar{I}_1)_{comp} = \sum_{i=1}^{3} C_i (\bar{I}_1 - 3)^i + \sum_{i=1}^{3} D_i (J-1)^{2i} \tag{10.77}$$

with C_i and D_i being material constants that relate to the distortional and volumetric response of the material volume respectively. Also, $\bar{I}_1 = J^{-2/3}I_1$. For a compressible material, C_{10} is the initial shear modulus and C_{11} is the initial bulk modulus.

10.7.9 Arruda-Boyce Hyperelastic Model

The phenomenological models described in Sects. 10.7.3, 10.7.4, 10.7.5, 10.7.6, 10.7.7, and 10.7.8 are very good in predicting a certain type of deformation mode, usually uniaxial tension with its characteristic 'S'-shaped load versus stretch plot for rubber deformation problems. Robust hyperelastic models should not only capture reliably uniaxial tension but also biaxial extension and shear deformation modes, up to very large stretches.

The Treloar [31] and Jones-Treloar [19] rubber data remain the classic experimental data for rubbery elasticity problems. The model parameters of the empirical models can only be derived by fitting these models to experimental data, often from multiple experiments. There is a need to create new sets of material models that are informed by the underlining micromechanics of the material volume. Such models are classified as the *mechanistic hyperelastic material models.*

The first mechanistic hyperelastic material model described here is the *Arruda-Boyce* model proposed jointly in 1993 by Ellen M. Arruda and Mary C. Boyce [5, 7] . They cited that their motivation for proposing the model was the desire for a constitutive relationship that "possesses mathematical simplicity, requires one test to characterize the material and has a limited number of parameters" [2].

The model is based on a cubic representation of the material volume where the rubbery elasticity is modelled using eight-chain segments that span eight diagonal spatial orientations within the cubic volume. This is why some literature refers to this model as the *eight-chain network model.* The authors assumed the material volume to be *incompressible.* The rubbery network statistics follows the *Langevin chain statistical mechanics.*

The strain energy density function of the Arruda-Boyce model, for an incompressible material, is given by:

$$\Psi(\lambda_{chain})_{incomp} = Nk_BT\sqrt{n}\left[\beta_{chain}\lambda_{chain} + \sqrt{n}\ln\left(\frac{\beta_{chain}}{\sinh\beta_{chain}}\right)\right] \tag{10.78}$$

where k_B is Boltzmann's constant; T is the absolute temperature; N is the number of chains in network of cross-linked polymer; and, n is the number of links per chain. The model parameter, β_{chain} depends on the *inverse Langevin function* of the principal stretches and defined as:

$$\beta_{chain} = \mathscr{L}^{-1}\left(\frac{\lambda_{chain}}{\sqrt{n}}\right) \tag{10.79}$$

where the inverse Langevin function is expressed as: $\mathscr{L}^{-1}(x) = \coth(x) - \frac{1}{x}$. Also, the stretch of each chain, λ_{chain}, is described by:

$$\lambda_{chain} = \left\{ \frac{1}{3} \sum_{i=1}^{3} \lambda_i \right\}^{\frac{1}{2}} \quad \longrightarrow \quad \lambda_{chain} = \sqrt{\tfrac{1}{3} I_1} \tag{10.80}$$

where I_1 is the first invariant of the left Cauchy-Green deformation tensor, $\boldsymbol{B}$.

Using a series expansion form of the inverse Langevin function, the expanded form of Eq. 10.78, up to the first five terms, is:

$$\Psi(\lambda_i)_{incomp} = nk_B T \left[\tfrac{1}{2}(I_1 - 3) + \tfrac{1}{20N}\left(I_1^2 - 9\right) + \tfrac{1050N^2}{2}\left(I_1^3 - 27\right) + \tfrac{19}{7000N^3}\left(I_1^4 - 81\right) + \tfrac{519}{673750N^4}\left(I_1^5 - 243\right) \right] + \cdots \tag{10.81}$$

The polynomial-hyperelasticity-type expression of the Arruda-Boyce model, with respect to Eq. 10.81, becomes:

$$\Psi(I_1)_{incomp} = nk_B T \sum_{i=1}^{5} \alpha_i \beta^{(i-1)} \left(I_1^i - 3^i\right) \tag{10.82}$$

where $\alpha_1 = \frac{1}{2}, \alpha_2 = \frac{1}{20}, \cdots, \alpha_5 = \frac{519}{673750}$ and $\beta = \frac{1}{N}$.

The Arruda-Boyce strain energy function is a nonlinear function in terms of I_1. The N parameter is a measure of the inextensibility limit of the molecular network. Beyond a certain value of N, the strain energy required to stretch the molecular network becomes excessive and hence the molecular entanglements are said to have 'locked'.

If the material volume is compressible, the polynomial-type representation of the Arruda-Boyce model is extended to incorporate a compressibility term and the complete formulation is given as [6]:

$$\Psi(I_1)_{comp} = nk_B T \sum_{i=1}^{5} \alpha_i \beta^{(i-1)} \left(\bar{I}_1^i - 3^i\right) + D_1 \left(\frac{J^2 - 1}{2} - \ln J\right) \tag{10.83}$$

where D_1 is a material constant, which reduces to $D_1 = \frac{\kappa}{2}$, where κ is Bulk Modulus, when linear elasticity constraints are imposed on the problem. Also, the deviatoric component of the first invariant of the left Cauchy-Green deformation tensor becomes: $\bar{I}_1 = J^{-2/3} I_1$.

10.7.10 Edwards-Vilgis Hyperelastic Model

Another mechanistic hyperelastic material model is the *Edwards-Vilgis model* which was proposed jointly by S. Edwards and T. Vilgis in 1986. It is an isotropic hyper-elasticity model that is typically used in the modelling of large deformation mechanics of homogeneous polymeric materials.

As a mechanistic model, the Edwards-Vilgis model formulation is inspired by macromolecular mechanics of, say, the polymer chains of the material volume under investigation. In fact, when a polymer material volume is subjected to a stress history such as stretch, λ_i for $i = 1, 2, 3$, this model accounts for the entropic elasticity of the polymer entanglements as they disentangle from their macromolecular confinements. At small strains, the authors proposed that the molecular mechanism of stretching is dominated by the slippage of the polymer chains [12].

The scalar-free *Edwards-Vilgis energy density function*, $\Psi(\lambda_i)$ depends on the principal stretches, λ_i for $i = 1, 2, 3$. It is an additive decomposition of the strain energy contributions from the network cross-links and the slip-links. The complete formulation of this strain energy is given thus:

$$\Psi(\lambda_i) = \Psi_c(\lambda_i) + \Psi_s(\lambda_i) \tag{10.84}$$

where strain energy due to cross-links becomes:

$$\Psi_c(\lambda_i) = \frac{N_c k_B T}{2} \left\{ \frac{\sum_{i=1}^{3} (1-\alpha^2)\lambda_i^2}{1 - \alpha^2 \sum_{i=1}^{3} \lambda_i^2} - \ln\left(1 - \alpha^2 \sum_{i=1}^{3} \lambda_i^2\right) \right\} \tag{10.85}$$

and the strain energy due to slip-links is:

$$\begin{aligned} \Psi_s(\lambda_i) = \frac{N_s k_B T}{2} \Bigg[\sum_{i=1}^{3} \left\{ \frac{(1-\eta)(1-\alpha^2)}{1 - \alpha^2 \sum_{i=1}^{3} \bar{\lambda}_i^{2}} \frac{\lambda_i^2}{1+\lambda_i^2} + \ln\left(1 + \eta\lambda_i^2\right) \right\} \\ - \ln\left(1 - \alpha^2 \sum_{i=1}^{3} \lambda_i^2\right) \Bigg] \end{aligned} \tag{10.86}$$

Here, N_s and N_c are the number density of *slip-links* (representing entanglements) and cross-links for the given polymer. Also, α, η and k_B are respectively, the inextensibility of chains parameter, an index describing ease of entanglements movement and Boltzmann's constant. η is also described as the *slipperiness factor* and can take values from infinity, for perfect sliding, to zero, for no sliding [35]. Finally, T and R are temperature and gas constants.

The Edwards-Vilgris formulation of $\Psi(\lambda_i)$ has been shown to be better than other competing polymer network models such as the Arruda-Boyce eight chain model [2] and the Wu and van der Giessen full network model [38]. Sweeney [29] showed that the Edwards-Vilgris function describes a broader range of material behaviour with the condition that finite extensibility of the chains is not approached too closely. If the user is interested in modelling thermosets where cross-links do not slip, the number of cross-links is non-zero, i.e. $N_c \neq 0$. However, for thermoplastics, the original Edwards-Vilgis function is reduced by assuming the density of cross-links, $N_c = 0$.

10.7.11 Stress Formulation for Hyperelastic Material Models

No matter which constitutive model one chooses to use for a given hyperelasticity model, it is necessary that a stress formulation is determined for implementation within a finite element scheme. In this section, we introduce briefly the approach taken to generate the stress formulation based on the strain energy formulations expressed previously. We show some of these stress formulation, for different scenarios.

- **Incompressible material with strain energy of form $\Psi = \hat{\Psi}(\lambda_i)$**: In formulating a constitutive model for an incompressible hyperelastic materials, the principal Cauchy (true) stress, σ_i, is found by differentiating the strain energy function, $\Psi(\lambda_i)$, with respect to the principal stretches, λ_i, thus:

$$\sigma_i = \lambda_i \frac{\partial \Psi(\lambda_i)}{\partial \lambda_i} + p \quad \text{(no sum on } i\text{)} \tag{10.87}$$

  where p is unknown hydrostatic pressure arising from satisfying certain boundary conditions, e.g. incompressibility constraints under pure deviatoric deformation (i.e. $\det(\boldsymbol{V}) = 1$), where $\boldsymbol{V}$ = left stretch tensor. In vector form, Eq. 10.87 can be expressed as:

$$\boldsymbol{\sigma} = \lambda_i \sum_{i=1}^{3} \frac{\partial \Psi(\lambda_i)}{\partial \lambda_i} \boldsymbol{n}_i \otimes \boldsymbol{n}_i + p\boldsymbol{I} \tag{10.88}$$

  where $\boldsymbol{n}$ is the unit eigenvector of the left Cauchy-Green deformation tensor, $\boldsymbol{B}$, and $\boldsymbol{I}$ is an identity matrix.
- **Incompressible material with strain energy of form $\Psi = \hat{\Psi}(I_1, I_2)$**: At times the strain energy function may be expressed in terms of the first and second invariants of $\boldsymbol{B}$ such that $\Psi = \hat{\Psi}(I_1, I_2)$. In this instance, the expression of the principal Cauchy stress becomes:

$$\sigma_i = \lambda_i \left[\frac{\partial \Psi}{\partial I_1} \frac{\partial I_1}{\partial \lambda_i} + \frac{\partial \Psi}{\partial I_2} \frac{\partial I_2}{\partial \lambda_i} \right] + p \quad \text{(no sum on } i\text{)} \tag{10.89}$$

- **Nominal stress for an incompressible material**: The expression for the nominal stress of a hyperelastic material is derived as a function of the true stress and the principal stretch, thus:

$$f_i = \frac{1}{\lambda_i} \sigma_i \quad \text{(no sum on } i\text{)} \tag{10.90}$$

where f_i is the nominal stress that corresponds to the true stress, σ_i.

- **Compressible material with strain energy of form $\Psi = \hat{\Psi}(I_1, I_2, J)$**: As shown above, the strain energy function for compressible material is expressed in terms of the invariants of the Cauchy-Green deformation tensor and the volume ratio. For this type of strain energy function, the expression of the Principal Cauchy stress becomes:

$$\sigma_i = \frac{\lambda_i}{J} \left[\frac{\partial \Psi}{\partial I_1} \frac{\partial I_1}{\partial \lambda_i} + \frac{\partial \Psi}{\partial I_2} \frac{\partial I_2}{\partial \lambda_i} + \frac{\partial \Psi}{\partial J} \frac{\partial J}{\partial \lambda_i} \right] \quad \text{(no sum on } i\text{)} \tag{10.91}$$

- **Nominal stress for an Compressible material**: The expression for the nominal stress for the compressible material becomes:

$$f_i = \frac{J}{\lambda_i} \sigma_i \quad \text{(no sum on } i\text{)} \tag{10.92}$$

where f_i is the nominal stress that corresponds to the true stress, σ_i.

10.8 Other Material Models

Apart from linear and nonlinear elasticity, as well as plasticity and viscoelasticity, there are other material models that the reader needs to be aware of when developing finite element modelling solutions to problems. The other models relate to those of *concrete*, *composite materials*, *foams*, *wood* and *biomaterials*, to name only a few. We consider these beyond the scope of this book and so will not go into their discussion. The objective here has been to define the constitutive models for common materials.

So far, the discussion has been based on linear elasticity, yield and onset of post-yield mechanics of the materials considered. These are just a subset of a range of material responses shown by even the materials we have considered here. The von Mises yield criterion, discussed previously in Sect. 10.5.2, is again one of the most common yield models for metals especially. Other yield models exist for both homogeneous and heterogeneous material systems. We have not discussed these in this text.

Finally, *damage* is a common feature of most materials. It typically refers to the loss of stiffness of a material due to microstructural failure within the material. The mechanics of damage and its evolution, which leads to complete failure of a material, have also not been discussed here. The interested reader is directed to fracture mechanics textbooks to see the computational mechanics implementation of damage. Every constitutive model can be implemented with damage criteria to capture the experimentally observed damage mechanics for materials modelled using such models. In commercial FEM solvers, there exist many damage models that can be used to capture either the ductile or the brittle failure seen in many real materials.

In the next section, we will conclude this chapter by introducing the reader to the principle of the *user-defined material model*. This is a bespoke design of a material model for describing unique material responses that are not currently implemented in most FEM solvers. It is both a technically and computationally intensive activity and not something to be undertaken lightly.

10.9 User-Defined Material Models

In this section, we will introduce in a systematic manner the computational implementation for extending the material libraries available to the FEM user. This section will present a justification for user-defined material models and discuss their classes and components before presenting the components of a user-defined material model. We will conclude by illustrating, with examples, the development of a user-defined material model with the ABAQUS FEM framework.

10.9.1 The Necessity of User-Defined Material Models

Existing finite element solvers have their own library of material models. The developers of such solvers would have taken painstaking efforts to code, validate and re-validate the material models before releasing it to public use as an *in-built material model*. These models are usually validated using a known set of experiments for a given type of material. The user expects that such models should be robust and accurate in replicating experimentally-known constitutive behaviour of the test material. As we stated previously, errors arising from the finite element simulation, can be due to the material model. What the FEM developer assures the user is that following extensive checks of their in-built material models, the released versions account for the least possible error arising from the material model.

However, in practice, especially with new materials being developed in the material science research community, the in-built material models will often fail to capture the observed material response comprehensively. The FEM user will therefore want a reliable material model to use for the new set of experimental data generated for the new material. This need demands development of a new set of material models, different from the in-built material models, capable of capturing numerically the new set of experimental data. It is this need that motivates the development of user-defined material models.

Finite element developers understand the need for users to create their own bespoke versions of material models. This will widen the usefulness of the finite element material modelling framework, beyond the narrow limits of the in-built material models. To help users do this, whilst still using the FEM solver material modelling framework, FEM developers will usually offer the user the **user-defined material subroutine** as part of the model release.

The *user-defined material subroutine* is a set of programming instructions, which the user can adapt and extend to describe a given material response. It is usually a complex tool and often written using FORTRAN77 or C languages. It is a powerful tool and can be used to model quite complex material responses. However, it is technically challenging and fraught with several difficulties, especially for the first-time user.

FEM users looking to become proficient in developing user-defined material models will need to familiarize themselves with the particular language, context and parameters that are available in their chosen FEM framework. This is vital if the user is to adapt and extend adequately the user-defined material subroutine for their specific type of material response. In Sect. 10.9.6, we will walk the reader through the different aspects of the ABAQUS subroutine. It is not intended to be an exhaustive discussion, but a brief introduction. The ABAQUS User's manual gives extensive information for the interested reader.

Finally, user-defined material models are only used, especially within ABAQUS, with *structural elements*. The reader should refer to Sect. 6.3.3 for a definition and the list of different structural elements. User-defined material models can also be described as *user-defined mechanical material models* since they are only used when describing the material response in which a mechanical loading history is involved. The user-defined material model is based on stress and strain formulations as well as the deformation gradient: all these are elements of mechanical loading.

10.9.2 Implicit Versus Explicit Methods

There is usually a confusion between *implicit* and *explicit* methods used in solutions of the finite element equation and the same methods used in development of user-defined material models. Generally, in finite element modelling, the *explicit methods* determine the solution state of model variables (e.g. displacement, velocity, temperature, stress) by making assumptions that the state of that solution equates the state at a later time.

On the other hand, *implicit methods* assume that the future state of a model variable is assumed to depend on both its current state and a later state. As a result, the implicit solution of a model variable is often very challenging to obtain as it requires extra computation and is thus harder to implement numerically. Let us look more closely at these two methods before illustrating their use in the finite element modelling problems.

10.9.2.1 Explicit Methods

Consider a model variable, Υ_t, with known values at current time, t. If we are interested in determining the value of this variable at a future time, $\Upsilon_{t+\Delta t}$, after a time step change, Δt, then explicit schemes makes the assumption that:

$$\Upsilon_{t+\Delta t} = f(\Upsilon_t) \tag{10.93}$$

such that the function, f, exists which makes the above equality possible. The challenge is, therefore, to determine objectively the value of $\Upsilon_{t+\Delta t}$. A numerical approach is commonly used to determine this and the most common used in finite element modelling is the *forward Euler method.*

To demonstrate the *forward Euler method*, consider, for example, a model variable, Υ, which has an ordinary differential equation: $\Upsilon'' = a + b\Upsilon^2$ with a and b being model constants. The explicit method solution, $\Upsilon_{t+\Delta t}$, at a future time can be obtained thus:

$$\Upsilon'' \approx \frac{\Upsilon_{t+\Delta t} - \Upsilon_t}{\Delta t} = a + \Upsilon_t^2 \quad \longrightarrow \quad \Upsilon_{t+\Delta t} = \Upsilon_t + \Delta t\left(a + b\Upsilon_t^2\right) \tag{10.94}$$

Consequently, the future state solution of Υ is dependent on the current state of the variable, as expected for the explicit method. The equation above also defines the function, f, required in Eq. 10.93.

10.9.2.2 Implicit Methods

On the other hand, an *implicit method* solution for Υ at a future time, takes the form:

$$\Upsilon_{t+\Delta t} = g(\Upsilon_{t+\Delta t}, \Upsilon_t) \tag{10.95}$$

where the function, g, must exist which makes the desired solution, $\Upsilon_{t+\Delta t}$, implicitly dependent on itself. Similar to the explicit method, the solution for $\Upsilon_{t+\Delta t}$ is best obtained using numerical methods and the *backward Euler method* is used here. To demonstrate the *backward Euler method* for the same ordinary differential equation above, the possible solution becomes:

$$\Upsilon'' \approx \frac{\Upsilon_{t+\Delta t} - \Upsilon_t}{\Delta t} = a + \Upsilon^2_{t+\Delta t} \longrightarrow \Delta t \Upsilon^2_{t+\Delta t} - \Upsilon_{t+\Delta t} + [\Upsilon_t + a\Delta t] = 0 \quad (10.96)$$

The resulting equation is a quadratic equation of $\Upsilon_{t+\Delta t}$. We need to determine the roots of the quadratic equation, such that the result becomes:

$$\Upsilon_{t+\Delta t} = \frac{-\Delta t \pm \sqrt{1 - 4\Delta t(\Upsilon_t + a\Delta t)}}{2\Delta t} \quad (10.97)$$

The above equation can be obtained easily using the root-finding algorithm for quadratic equations. Most roots of implicit functions are not always this straightforward, instead more robust root-finding algorithms are commonly required in finite element modelling. One of popular root-finding algorithm is the *Newton-Raphson method*. Other approaches that are commonly used in finite element studies to solve for momentum balance equation, constitutive and static equilibrium formulations are the *tangential stiffness method* and *initial tangential stiffness method* [11].

10.9.2.3 Newton-Raphson Method

Let us briefly illustrate the Newton-Raphson method for a constitutive formulation: $\boldsymbol{\sigma}(\boldsymbol{\varepsilon}) = 0$ with $\boldsymbol{\sigma}$ = stress and $\boldsymbol{\varepsilon}$ = strain. If the strain at i-th iteration is $\boldsymbol{\varepsilon}_i$, the predicted strain at the next iteration point, $\boldsymbol{\varepsilon}_{i+1}$, can be obtained using the Newton-Raphson approach thus:

$$\boldsymbol{\varepsilon}_{i+1} = \varepsilon_i - \frac{\boldsymbol{\sigma}(\boldsymbol{\varepsilon}_i)}{\boldsymbol{\sigma}'(\boldsymbol{\varepsilon}_i)} \quad (10.98)$$

where $\boldsymbol{\sigma}'(\boldsymbol{\varepsilon}_i)$ is the first-order partial derivative of the stress function.

The Newton-Raphson method is iterative such that the analysis continues to iterate steadily based on Eq. 10.98 until a certain condition is satisfied for a solution to be deemed acceptable. The iteration is stopped when this condition is satisfied:

$$\left| \frac{\boldsymbol{\varepsilon}_{i+1} - \boldsymbol{\varepsilon}_i}{\boldsymbol{\varepsilon}_i} \right| \leq \phi \quad (10.99)$$

where ϕ is a specified value called the *estimated relative error*. Alternatively, a tolerance can be set such that the solution ends when the absolute value of the stress function becomes smaller than a set value, called the *tolerance*, δ:

$$|\boldsymbol{\sigma}(\boldsymbol{\varepsilon}_i)| \leq \delta \quad (10.100)$$

The key observation in the Newton-Raphson Method is the requirement for the stress function (in this case) to be differentiable and continuous over the domain of strain being considered.

10.9.2.4 Implicit and Explicit FEM Solvers

The simulation engine of an FEM solver was introduced in Sect. 4.2. FEM solutions are approximate numerical solutions. The simulation engine operates on the finite element equation, $\boldsymbol{f} = \boldsymbol{k}\boldsymbol{u}$, where $\boldsymbol{f}$ = internal forces, $\boldsymbol{u}$ = nodal displacements and $\boldsymbol{k}$ = stiffness matrix. The objective is to determine the nodal forces given known nodal displacements or vice versa.

Thus, *implicit methods* can be applied to determine the unknown forces. The implicit methods, in preference to explicit methods, are very attractive to FEM developers due to the opportunity of obtaining the *residual force* at each step and iteration during the solution of the problem, until convergence is achieved within a set tolerance limit. As part of the solution, a *Jacobian*, $\boldsymbol{J}$, called the *tangent stiffness matrix*, needs to be determined to relate the internal forces to the nodal displacements. This Jacobian is different from the Jacobian for the material model discussed in Sect. 10.9.5.

The implicit and explicit classification of FEM solvers need not be confused with the implicit and explicit user-defined material models presented later. These solvers are available to users in ABAQUS. The implicit solvers are very robust and users can achieve good convergence to solutions with limited errors.

On the other hand, the explicit solvers yield rapid convergence to solutions, but users should be careful about their use as errors can easily arise from their use. The user should also be careful about choice of time-steps as their size can influence the solution significantly. This is due to the inherent definition of explicit methods, which necessitates equating the model variable at a current state to a future state. If the time step is too large, then errors arising from this *'equating condition'* will grow. In the following sections, we will discuss and distinguish between implicit and explicit method for developing user-defined material models.

10.9.3 Classes of User-Defined Subroutines

In Sect. 4.2, we explained that the *simulation engine* of a finite element solver is at the core of the analysis procedure that drive a given FEM solver. The *Implicit* and *Explicit* analysis methods were introduced as methods used in the simulation engine. The *implicit* method was suitable for problems in which time-dependence is not important such as static mechanics, modal analysis, etc. while the *explicit* method was important when rate and consequently inertial effects cannot be neglected during the analysis.

The explicit method is best suited for the following types of problems: impact engineering, dynamic behaviour of materials, blast loading, crash, and sometimes highly nonlinear finite deforming materials. These two analysis approaches define

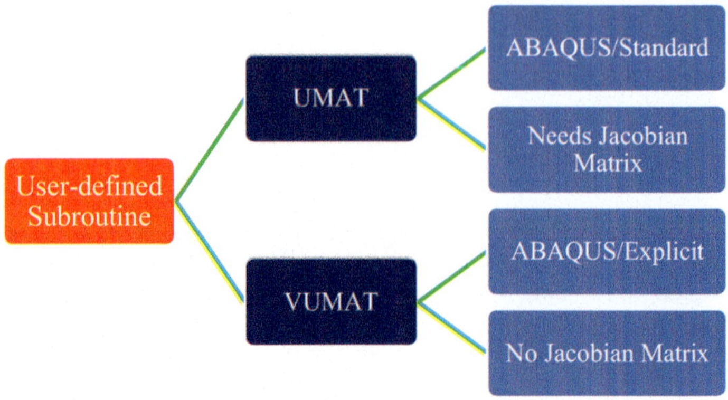

Fig. 10.15 Classes and unique features of a user-defined subroutine for ABAQUS

the classification of user-defined material models namely: *UMAT* and *VUMAT*, and their sub-divisions and uniqueness are illustrated in Fig. 10.15.

UMAT is used in the ABAQUS/Standard analysis interface. This is a simulation engine designed specifically for principally implicit analysis and hence the included user-defined material subroutine is a UMAT. It is strictly used for describing material responses where the analysis procedure comprises of a mechanical behaviour. As a result, it is not suitable for electromagnetic analysis. Other subroutines will need to be used for capturing those types of material behaviour.

The UMAT must have a *material* ***Jacobian matrix***. The FEM user looking to develop a UMAT, must determine the *Jacobian matrix* representation of the new material model that is to be implemented into a UMAT. It is also this requirement for a Jacobian matrix and the necessity for convergence that limits the usefulness of the UMAT especially for challenging analysis such as fracture, damage, etc.

The **VUMAT**, on the other hand, is used within the ABAQUS/Explicit analysis framework. Also, the UMAT in LS-DYNA (an explicit FEM code) is essentially a *vectorized UMAT*. The VUMAT is based on the explicit method. It frequently requires the user to define a set of solution-dependent state variables, which capture the analysis state during the numerical execution. Of course, the VUMAT isonly used if the in-built material models within ABAQUS/Explicit is not sufficient in describing the problem the user is solving. In the next section, the structures of header sections for UMAT and VUMAT are shown.

10.9.4 Components of a User-Defined Material Model

The interface available in different commercial finite element solvers are similar for both UMATs and VUMATs. The interface consists of, essentially, an information passing scheme between the actual finite element solver and a piece of computer code, often called a *subroutine*, which defines the user-defined constitutive material model. Although the computer code is written using the recommended programming language and syntax applicable to the finite element solver being used, the underlying philosophy is universal.

In general, there are three main sets of data required in implementing a user-defined constitutive material model, as shown in Fig. 10.16:

- **Unmodifiable data**: This set of data os *read-only* and, as a result, cannot be changed. They are essential variables needed for the running of the subroutine. For a UMAT, such unmodifiable data can include: `STRESS`, `STATEV`, `DDSDDE`, etc., while for a VUMAT example variables include: `NBLOCK`, `NSTATEV`, etc.[14] Algorithms 10.1 and 10.2 show header sections for the UMAT and VUMAT user subroutines respectively.

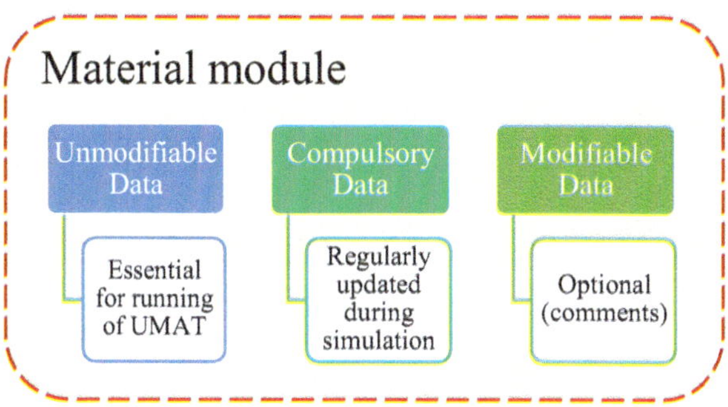

Fig. 10.16 Components of a typical ABAQUS UMAT subroutine

[14]These variables apply for the ABAQUS UMAT and VUMAT scripting interface. The interested reader should consult the ABAQUS User documentation for description of these variables.

Algorithm 10.1 A snippet of an ABAQUS UMAT header

```
*************************************************************
**         UMAT HEADER SECTION FOR ABAQUS/STANDARD        **
*************************************************************
          SUBROUTINE UMAT(STRESS,STATEV,DDSDDE,SSE,SPD,SCD,
     1 RPL,DDSDDT,DRPLDE,DRPLDT,
     2 STRAN,DSTRAN,TIME,DTIME,TEMP,DTEMP,PREDEF,DPRED,CMNAME,
     3 NDI,NSHR,NTENS,NSTATV,PROPS,NPROPS,COORDS,DROT,PNEWDT,
     4 CELENT,DFGRD0,DFGRD1,NOEL,NPT,LAYER,KSPT,KSTEP,KINC)
C
      INCLUDE 'ABA_PARAM.INC'
C
*************************************************************
```

Algorithm 10.2 A snippet of an ABAQUS VUMAT header

```
*************************************************************
**         VUMAT HEADER SECTION FOR ABAQUS/EXPLICIT        **
*************************************************************
        SUBROUTINE VUMAT(
C READ ONLY (UNMODIFIABLE VARIABLES) -
     1  NBLOCK, NDIR, NSHR, NSTATEV, NFIELDV, NPROPS, LANNEAL,
     2  STEPTIME, TOTALTIME, DT, CMNAME, COORDMP, CHARLENGTH,
     3  PROPS, DENSITY, STRAININC, RELSPININC,
     4  TEMPOLD, STRETCHOLD, DEFGRADOLD, FIELDOLD,
     5  STRESSOLD, STATEOLD, ENERINTERNOLD, ENERINELASOLD,
     6  TEMPNEW, STRETCHNEW, DEFGRADNEW, FIELDNEW,
C WRITE ONLY (MODIFIABLE) VARIABLES -
     7  STRESSNEW, STATENEW, ENERINTERNNEW, ENERINELASNEW )
C
      INCLUDE 'VABA_PARAM.INC'
C
*************************************************************
```

- **Compulsory data**: The compulsory data fall under the unmodifiable data and must exist for a given UMAT or VUMAT to run. For example, in developing a UMAT sub-routine, two tasks are essential: *update the stress state* and *determine the Jacobian matrix*. No UMAT can run without the variables that meet these requirements. As a result, both `STRESS` and `DDSDDE` are the some of the compulsory (unmodifiable) data needed for the UMAT. texttttDDSDDE is the *tangent stiffness*, which is equivalent to the slope of a stress-strain graph in the region of strain of interest. In terms of constitutive formulation, it also represents the material *Jacobian matrix*. On the other hand, the VUMAT must have the `NBLOCK` variable, because it defines the *number of blocks of material points* needed to be called for the subroutine to run. Also, the `NSTATEV` represents the number of *solution-dependent state variables*. Both are compulsory data that must be specified for a VUMAT subroutine to run.

- **Modifiable data**: This type of data can be changed by the user without affecting the user subroutine. They are usually optional for example comments that help the readability of the code.

Furthermore, the specifics of parameters comprising each data set depend on the finite element solver being used (i.e. whether an explicit or implicit solver). At a rudimentary level, the only compulsory data to modify in an explicit user-defined material code is the stress tensor. On the other hand, in an implicit solver the compulsory data is the stress tensor and the so-called Jacobian.

The reader should note that the `INCLUDE 'ABA_PARAM.INC'` and `INCLUDE 'VABA_PARAM.INC'` lines must appear in an `ABAQUS` UMAT and VUMAT respectively. These are `ABAQUS`-specific include files needed for the job to run. Although these are not variables, they are *compulsory* files that supply required information for setting up the user-defined subroutine environment. Other FEM solvers typically have similar variants of these files.

10.9.5 The Derivation of the Jacobian Matrix

As established previously, the Jacobian is very important for implicit methods. In developing a UMAT, the user will have to determine the expression of the Jacobian matrix for the constitutive formulation under consideration. There are two main types of Jacobian matrices commonly described in the literature and these are briefly described below.

- **Consistent Jacobian matrix**: This represents the exact expression of the Jacobian; derive from the constitutive equations of the material model. It is used for problems involving *small deformations* such as linear elasticity, or for *large deformation* problems with *small volume changes*, for example, metal plasticity. It is essential, especially for the ABAQUS/Standard FEM solver.

  Whether for such small-deformation or large deformation (with small volume changes) problems, the expression of the consistent material Jacobian matrix, $\boldsymbol{C}$, is defined as:

$$\boldsymbol{C} = \frac{\partial \Delta \boldsymbol{\sigma}}{\partial \Delta \boldsymbol{\varepsilon}} \tag{10.101}$$

  where $\Delta\boldsymbol{\sigma}$ and $\Delta\boldsymbol{\varepsilon}$ are the increments in the stress and strain tensors of the material respectively. If strain is very large, it is common to approximate $\boldsymbol{\varepsilon}$ by the logarithmic strain, ln $\boldsymbol{\varepsilon}$. The reader should note that Eq. 10.101 represents the tangent stiffness tensor of the constitutive model.

However, when the problem involves large deformations with large volume changes such as pressure-dependent plasticity problems, a modified version of the classic Jacobian is used instead. The exact form of this consistent Jacobian is written as:

$$\mathbf{C} = \frac{1}{J}\frac{\partial \Delta (J\boldsymbol{\sigma})}{\partial \Delta \boldsymbol{\varepsilon}} \tag{10.102}$$

where $J = \det(\boldsymbol{F})$ is the *volume ratio* and it represents the determinant of the deformation gradient tensor, $\boldsymbol{F}$. Using this form of the Jacobian, the influence of volume changes on the tangent stiffness tensor is accounted for. Thus, rapid **convergence** of the analysis to the desired solution is achieved.

The implicit methods, and their dependence on the highly iterative Newton-Raphson method, present a significant challenge in the development of consistent Jacobian matrices for complex material models. This is especially due to the need to determine first-order partial derivatives of whatever function the constitutive model is based on.

The evolution and widespread use of ***symbolic computation*** systems as Maple, Mathematica and MATLAB™, have made it easier for those partial derivatives to be developed with comparative ease. Increasingly, UMAT coders are now using to these symbolic mathematics toolboxes to tackle the challenge of obtaining partial derivatives of complex material models. However, there are still instances when the exact formulation of the Jacobian matrix is not possible, in which case, the approach explained next becomes acceptable.

- **Approximate Jacobian matrix**: It is often very challenging to derive the exact expressions of Jacobian matrix for complex material models. Thus, it is common practice to derive approximations in such cases. In adopting approximate forms of Eq. 10.101, it is instructive to note that this term is only necessary to help the rate of convergence of implicit solvers, not the accuracy of the actual solution to the constitutive model.

The determination of the approximate Jacobian matrix of the stress function, $\boldsymbol{\sigma}(\boldsymbol{\varepsilon})$, is based on a finite difference method for determining the partial derivatives associated with the stress function [9, 20, 36]. The finite deference expression for the Jacobian matrix is:

$$C_{ij}(\boldsymbol{\varepsilon}) = \frac{\partial \sigma_i}{\partial \varepsilon_j}(\boldsymbol{\varepsilon}) \quad \approx \quad \frac{\sigma_i(\boldsymbol{\varepsilon} + h\boldsymbol{e}_j) - \sigma_i(\boldsymbol{\varepsilon})}{h} \tag{10.103}$$

where h is small increment applied on the strain, $\boldsymbol{\varepsilon}$, and $\boldsymbol{e}_j$, is a unit vector in the direction of the strain tensor, $\boldsymbol{\varepsilon}$. If h is very small, the above is a satisfactory approximation of the partial derivatives. The choice of h is, however, left to the computer to determine for a given problem and when chosen correctly, it is possible to achieve rapid convergence to desired solution. Evaluating Eq. 10.103 is advantageous, because we do not have to analytically determine the partial derivative of $\boldsymbol{\sigma}(\boldsymbol{\varepsilon})$.

The above approximation is very useful in determining the convergent solution, because traditionally the Newton Raphson method is not very sensitive to poorly defined Jacobian matrix. As a result, the solution can still be obtained correctly even though we have used an approximate Jacobian matrices. This fact is exploited in many UMATs for complex material models where the user simply starts with a highly simplified Jacobian matrix or even a linear combination of constituent Jacobian matrices defined to capture different components of the material response. The latter approach has been used intelligently by Li and Buckley [21] in developing an ABAQUS UMAT for glass polymers, for example.

10.9.6 Steps for Developing User-Defined Material Models in ABAQUS

Here, we provide a step-by-step guide into the development of user-defined material subroutines using ABAQUS as an example FEM solver. Similar approaches will apply for other FEM solvers, but the reader is encouraged to consult the documentation that comes with such solvers to understand their specific instructions. The aim here is not to give an extensive list of what you need to do, but rather identifying essential considerations the user has to make when developing user-defined material models.

Step 1: Assemble Appropriate Software

The first step is to assemble appropriate software and the necessary documentation you will need when developing UMATs or VUMATs. For ABAQUS, the ABAQUS software must be installed on the PC you intend to use. The user must having the necessary permissions to be able to write to the disk drives. Also, the actual code development has to be done in Fortran77 or any of the C programming languages. Thus, the coder need to choose an appropriate application development environment. Traditionally, many users tend to work with Microsoft Visual Studio code editor, for example. The integrated development environment (IDE) of Visual Studio makes it possible for users to code in about 25 programming languages.

As well as the code editor, the user will need to also identify an appropriate compiler to compile the codes before deploying it within the ABAQUS UMAT framework. The most common compilers, recommended by 3DS Simulia, makers of ABAQUS, is the Intel C++ and Fortran compilers. The user will need a paid license, but note that the Intel compilers are well tested and used broadly in the UMAT community. There exist other open-source compilers, but the user should be careful when using these as the results may not always be correct. Extensive validation of the UMAT to known material response data is recommended if one is to use an unsupported compiler.

Step 2: Understand ABAQUS Utility Routines

Even before the user begins to code the UMAT, it is important to understand the available *ABAQUS utility routines*. These are snippets of programming code designed to accomplish specific actions. They support the code development, such that the coder can concentrate on coding the specific material model rather than secondary tasks, such as, for example, getting the *environment variables*, job directories, stress invariants, etc. ABAQUS has already implemented these snippets of codes and makes them available for the user to call upon when required.

The routines are activated within a given user-defined subroutine using the CALL command, which interrupts the current execution and enforces that the routine declared after the CALL command is executed before returning to the current user subroutine. It is common practice to identify earlier on these utility routines (at least the ones the user needs first) and place them at the bottom end of the user-defined subroutine development interface.

Three examples of the many utility routines are:

- **SPRINC**: This is one of the tensor operations utility routines and it *calculates the principal stresses* associated with a given problem. The reader should refer to Sect. 9.6.1 for a review of principal stresses. The call command is: CALL SPRINC(S,PS,LSTR,NDI,NSHR) where S is stress or strain tensor; LSTR is an identifer where if LSTR = 1 then S contains stress and if LSTR = 2, then S contains strain. NDI is number of direct or normal stress/strain components while NSHR is number of shear components in either the stress or strain tensor.
- **GETJOBNAME**: This routine helps the coder return the name of the current job. The format for calling it is: CALL GETJOBNAME(JOBNAME, LENJOBNAME) where LENJOBNAME is the length of the character string JOBNAME. This call is for ABAQUS/Standard, but the equivalent utility routine for ABAQUS/Explicit is VGETJOBNAME.
- **XIT**: This is an ABAQUS/Standard utility routine for terminating and analysis and is called thus: CALL XIT. The equivalent ABAQUS/Explicit routine is: XPLIB_EXIT. When this is used, instead of the STOP command, all files associated with the simulation are forced to close properly.

Step 3: Decide on Either UMAT or VUMAT

Early on, the coder will have to make a decision on the type of user-defined material sub-routine to develop. Both types of user-defined material models can always be developed for any material model, but it is always recommended for the coder to start with a UMAT, especially if it is a problem with small deformations or for large deformation studies, the constraint of small volume changes is enforced. The choice of the UMAT route will again depend on whether the user can determine either a consistent or approximate Jacobian matrix. If this is not possible, a VUMAT will have to be developed.

Also, problems that have high rate-dependence such as crash, impact, fracture and damage mechanics problems, are better addressed using a VUMAT. If the user is to use the user-defined material model to investigate the effect to impact loads or interface failure between two materials, it is better to start by writing a VUMAT.

Step 4: If UMAT, Derive Mathematics of the Jacobian Matrix

Once the decision has been made to develop a UMAT, the coder now focusses on deriving the expression of the Jacobian matrix: either a consistent or approximate Jacobian using the principles of Sect. 10.9.5. The key task here is to determine the partial derivatives of the constitutive model formulations using appropriate numerical methods. The use of symbolic computation software is recommended to help with the manipulations of the constitutive formulations.

It is important to know that even with the approximate form of the Jacobian, reliable solutions can still be achieved. Although the task of developing a Jacobian can be daunting, the coder should not waste time trying to determine the 'best' Jacobian, especially for complex material models. Well derived Jacobian matrices will clearly lead to faster convergence to the desired solution. Also, any improvements in computing power should have a synergistic effect on the improved 'convergence rate' of the FEM framework.

The Jacobian matrix, $\boldsymbol{C}$, can be determined by considering the incremental changes in stress, $\Delta\sigma$, versus similar changes in strain, $\Delta\boldsymbol{\varepsilon}$, so that $\boldsymbol{C} = \partial\Delta\boldsymbol{\sigma}/\partial\Delta\boldsymbol{\varepsilon}$. The model variable that defines the Jacobian matrix is `DDSDDE(i,j)`. This represents the ratio of change of the i-th component of stress caused by an *infinitesimal* change in the j-th component of strain i.e. `DDSDDE(i,j)` $= \Delta\sigma_i/\Delta\varepsilon_j$. When running a fully coupled thermal-stress analysis and the like, we use a similar type of Jacobian called `DDSDDT` that considers ratio of increments in stress, $\Delta\sigma$, to increments in temperature, ΔT.

Step 5: Define System and User Variables

The next challenge is to define the relevant system (or unmodifiable) variables that support the kind of UMAT or VUMAT that the coder wants to develop. The typical layouts for the UMATs and VUMATs are shown in Algorithms 10.1 and 10.2. If the developer has introduced a user variable that is not part of the unmodifiable variables, this is the stage where it is defined.

Some of the important variables and their meanings are described briefly here:

- **`STRESS`**: This is the array of the *stress tensor* passed into the UMAT at the beginning of the increment. It defines the stress state at the start of the solution and in the course of the solution, it will need to be updated to reflect a new stress state before the next time-step calculations. The full stress tensor array is given as `STRESS(NTENS)` where `NTENS` is the total number of components of the stress tensor and consists of the sum the of number of normal stress terms (`NDI`) and number of shear stress terms (`NSHR`).

- **STRAN**: This is the array of the total *strain tensors* obtained at the beginning of the increment. This value is essentially the elastic strain, STRAN = $\sum \boldsymbol{\varepsilon}^e$, for small strain problems, but the logarithmic strain, ln $\boldsymbol{\varepsilon}$ for finite strain problems. Similarly, the full array is defined as `STRAN (NTENS)`. Note this strain is strictly the mechanical strain, with the exclusion of any thermal strain from the total strain.
- **DSTRAN**: This is the array of *strain increments*. It is solely a mechanical strain value, which is the difference between total strains and thermal strains, where temperature changes, are involved in the analysis.
- **KINC**: This is the increment number for the given solution. It is an important parameter for developing the UMAT and needs to be declared and passed into the UMAT sub-routine at the start.
- **DFGRD0** and **DFGRD1**: This represents the array of the deformation gradient tensors at the start and end of the increment. It is always defined and stored as a three-dimensional tensor and hence is defined within the UMAT using `DFGRD0 (3,3)` or `DFGRD1 (3,3)` to account for its 3×3 matrix representation at the start and end of the increment respectively.

A variable that is always defined at the start is **CHARACTER*80 CMNAME** which defines the character string for the user material name (CMNAME). This particular definition enforces that the character spans a dimension of 80 storage spaces. The coder uses the resulting character string to store information about the user material name during the running of the user-defined sub-routine.

Also, the coder needs to declare the `DIMENSION` command at this stage. This passes information to the UMAT on the number of dimensions (lengths) for all the arrays of model data passed into the UMAT. A snippet of the dimension command is given in Algorithm 10.3.

Let us consider some of the variables defined in the `DIMENSIONS` command. For example, `TIME (2)` implies that the variable `TIME`, which is the value of the step time, has two terms/dimensions that describe it within the subroutine. The first, `TIME (1)`, defines the value of the step time at the start of the increment and `TIME (2)` is same step time, but now at the end of the increment.

Another declared variable is the *rotation increment matrix*, DROT, which is the increment of the rigid body rotation of a material volume (see Sect. 9.3.4). Dimensioning it as `DROT (3,3)` implies that the rotation matrix is a 3×3 matrix.

Algorithm 10.3 A snippet of ABAQUS UMAT `DIMENSIONS` command

```
DIMENSION STRESS(NTENS),STATEV(NSTATV), DDSDDE(NTENS, NTENS),
 1 DDSDDT(NTENS), DRPLDE(NTENS), STRAN(NTENS), DSTRAN(NTENS),
 2 TIME(2), PREDEF(1),DPRED(1), PROPS(NPROPS), COORDS(3),
 3 DROT(3,3), DFGRD0(3,3),DFGRD1(3,3)
```

Algorithm 10.4 Two examples uses of the ABAQUS UMAT PARAMETER command

```
*****************************************************
C   EXAMPLE 1
    PARAMETER( ZERO = 0.D0,   ONE = 1.D0,   TWO = 2.D0,
  1          THREE = 3.D0, FOUR = 4.D0,  HALF = 0.5D0)
*****************************************************
C   EXAMPLE 2
    PARAMETER(PI = 3.141593,  TOLER = 0.D-6, F = .FALSE,
  1     FLAG1 = 'DAMAGE INITIATED', EPSILON = 1.0E-6)
*****************************************************
```

Finally, as part of the definition of system and user variables, it is recommended the user uses the PARAMETER command to give a symbolic name to any constant or expression that is commonly used within the subroutine. Two example uses of the command are given in Algorithm 10.4.

Consider, for example, the parameter FOUR = 4.D0, which is a symbolic name that will be used every time the coder wants to use the **double precision** number[15] 4 within the subroutine. Similarly, for a user subroutine in which a certain analysis is undertaken once '*damage has initiated*,' a damage initiation flag might be needed at multiple times during the simulation. Using the PARAMETER command, the FLAG1 parameter with content 'DAMAGE INITIATED,' is now available for the coder to use within the subroutine.

Step 6: Define Material Properties

The next step in the development of a user-defined material subroutine is to declare the material properties or constants that are needed for the development of the constitutive formulation. For example, in a simple isotropic **Hookean** elasticity material model, material properties are simply *Young's Modulus*, E, and *Poisson ratio*, ν. These values need to be declared at the top end of the UMAT before further analysis.

There are two possible ways of specifying the material properties in the user subroutine. Most beginners will input these parameters directly into the code. This is acceptable if the parameters are not going to change often from simulation to simulation. For example, Boltzmann constant, $k_B = 1.381 \times 10^{-23}\,\text{m}^2\,\text{kg}\,\text{s}^{-2}\,\text{K}^{-1}$, can be 'hard-coded' in the user subroutine.

This way of specifying the material properties demands that the hard-coded properties cannot be changed without altering the code. It is an inconvenient way of specifying the material properties as it limits the flexibility and useability of the code for running other applications. For example, if a coder specifies the Young's Modulus of steel directly into the subroutine as $E = 210\,\text{GPa}$, the same code cannot be used to run another study for, say, aluminium with a Young's Modulus, $E = 70\,\text{GPa}$.

[15]The double precision indicator in Fortran is the .D0 suffix attached to the number 4.

Algorithm 10.5 Two ways of specifying material constants in an ABAQUS UMAT

```
*******************************************************
C   EXAMPLE 1
     E              = 210E9   !Youngs Modulus of steel: Pascal
     PRatio         = 0.3     !Poisson Ratio of steel: no units
*******************************************************
C   EXAMPLE 2
    E         = PROPS(1) !Youngs Modulus
    PRatio    = PROPS(2) ! Poisson Ratio
*******************************************************
```

The second method is via the PROPS and NPROPS variable in the ABAQUS user subroutine framework. The PROPS variable is used to supply user-specified material properties for the specific user-defined material subroutine under investigation. The NPROPS is the number of properties that the user intends to specify within the code. Within the DIMENSION call, PROPS(NPROPS) means NPROPS number of material constants are to be defined in the UMAT.

It is also important to provide inline commentary about the model parameters to help with reading and understanding the code. The exclamation sign after a Fortran statement implies that the rest of entries (after the sign) are comments. Also, use C, c, d, D, * and blank lines to indicate a comment.

Step 7: Define Model Formulation

This step depends on the kind of material model the user is looking to develop. This requires coding all the necessary relationships between stress and strain, as well as all other necessary aspects of the material formulation. It is also at this stage that the user might have to call some of the utility routines that are needed for the model development. Depending on the type of material model to be considered, the user might have to provide extensive code or just a snippet. We will not spend too much time on this aspect as it is really dependent on the material model formulation.

At this stage, at the first time step, it might be important to initialize all the solution-dependent state variables. At subsequent time-steps, the coder will simply have to load the solution-dependent state variables. If the user is using multiple UMATs, then at this stage, the user calls all the different UMATs to obtain different stress measures from the different component UMATs.

Step 8: Update the Stress

Once the stress values have been obtained at the given time-step, the next step is to update the stress. This will usually involve the equation: $\boldsymbol{\sigma}_{t+\Delta t} = \boldsymbol{\sigma}_t + \Delta\boldsymbol{\sigma}$, where $\Delta\boldsymbol{\sigma}$ is the increment in stress calculated based on the model formulation above. $\boldsymbol{\sigma}_t$ is the stress from the old time step. This updated stress for the current time is then passed onto the next time-step.

Step 9: Define the Coding for the Jacobian Matrix

This stage is aimed at the coding of the analytically derived Jacobian matrix of Step 4 above. This section is important only for ABAQUS/Standard, which relies on a Jacobian matrix to be in place. If not, the Jacobian coding is omitted and the iteration returns to the top for the next time step iteration.

Step 10: Write Results to the Database File, and End of UMAT Coding

At the end of the iteration, the solver exits from the user subroutine and continues with other aspects of the finite element analysis. It is advisable to write these results to the model database file, for post-processing and further detailed verification of the outputs. Extensive effort has to be given to checking, re-checking and editing the code to ensure the outputs agree with the expected results. Model outputs based on different loading conditions are used to assess the code is properly coded. A UMAT layout is given in Algorithm 10.6.

Algorithm 10.6 The typical layout of an ABAQUS UMAT

```
*****************************************************
**                 TYPICAL UMAT LAYOUT            **
*****************************************************
SUBROUTINE UMAT(STRESS,STATEV,DDSDDE,..., KINC)
   INCLUDE 'ABA_PARAM.INC' ! For ABAQUS/Standard only
   CHARACTER*80 CMNAME
C-----------------------------------------------------
*** STEP 5: DEFINE SYSTEM AND USER VARIABLES/
    PARAMETERS
C
   DIMENSION STRESS(NTENS),STATEV(NSTATV),...
   DIMENSION DSTRESS(4) ! Stress increment
   PARAMETER (ZERO  = 0.D0, ..., TEN = 10.D0)
C-----------------------------------------------------
*** STEP 6: DEFINE THE MATERIAL PROPERTIES
C
    PROP1 = PROPS(1) ! PROPS(1) - 1st property value
    PROP2 = PROPS(2) ! PROPS(2) - 2nd property value
C-----------------------------------------------------
*** STEP 7:  DEFINE MODEL FORMULATIONS
C  Codes that define the constitutive behaviour of
C  material appear here
C  It might be good to call other UMATs and routines
C
   CALL UMAT_1(STRESS1, ....) ! UMAT1 (Optional)
   CALL UMAT_2(STRESS2, ....) ! UMAT2 (Optional)
```

```
   CALL UTILITY_ROUTINE(...)   ! Use utility routine
C-----------------------------------------------------
*** STEP 8: UPDATE STRESS
C  The following is a typical stress update code
C  based on multiple UMATs yielding multiple
C  stresses e.g. STRESS1 and its increment, DSTRESS1
      DO K = 1,NTENS
      STRESS(K) = STRESS1(K) + STRESS2(K)
                + DSTRESS1(K) + DSTRESS2(K)
      END DO
C-----------------------------------------------------
*** STEP 9: CODING FOR THE JACOBIAN
C   All sets of codes that determine the Jacobian
C   matrix will appear here (only for ABAQUS/Standard
   )
     RETURN
     END
******************************************************
```

10.9.7 ABAQUS *Example of User-Defined Material Subroutines*

Here, we consider one of the most common and simplest material models that is used for modelling a wide range of materials. This is the *isothermal* ***Hookean*** *elastic material model*. The following shows the analysis that one has to make to create inputs for the UMAT.

- ***Constitutive model***: We have already introduced the constitutive formulation of this model in Sect. 10.4.4. We described this type of model as an *isotropic linear elasticity* material model. In that presentation, we did not consider the effect of temperature on the material properties, hence the limitation of ***isothermal*** conditions imposed on this material model.

  As an example, consider a test material whose constitutive mathematics can be represented as (please also refer to Eq. 10.13):

$$\boldsymbol{\sigma} = \lambda \mathrm{tr}(\boldsymbol{\varepsilon})\boldsymbol{I} + 2G\boldsymbol{\varepsilon} \tag{10.104}$$

  where λ and G are Lamé's constants, $\boldsymbol{\sigma}$ is the stress tensor and $\boldsymbol{\varepsilon}$ is the strain tensor. Note $\mathrm{tr}(\cdots)$ is the **trace** of a matrix. The incremental form of the above equation becomes:

$$\boldsymbol{\Delta\sigma} = \lambda \mathrm{tr}(\boldsymbol{\Delta\varepsilon})\boldsymbol{I} + 2G\boldsymbol{\Delta\varepsilon} \tag{10.105}$$

The incremental strain tensor, $\boldsymbol{\Delta\varepsilon}$, will be passed into the user sub-routine from the finite element analysis and using Eq. 10.105, the stress increment can be calculated at the current time-step.

- ***Material properties***: The material properties for an isotropic linear elastic material were defined in Sect. 10.4.4 as Young's Modulus, E, and Poisson ratio, ν. However, the constitutive model above indicates we need to determine the Lamé's constants: λ and G. The extra formulations to determine these are:

$$G = \frac{E}{2(1+\nu)} \quad \text{and} \quad \lambda = \frac{3K-3G}{3} \quad \text{where} \quad K = \frac{E}{3(1-2\nu)} \tag{10.106}$$

- ***Derivation of Stiffness Tensor***: In order to determine the stresses based on the strain using an isotropic linear elasticity formulation, the *stiffness tensor* has to be determined. According to Eq. 10.18, the plane strain constitutive formulation for this problem becomes:

$$\boldsymbol{\sigma} = \boldsymbol{D\varepsilon} \longrightarrow \begin{bmatrix} \sigma_{11} \\ \sigma_{22} \\ \sigma_{33} \\ \sigma_{12} \end{bmatrix} = \begin{bmatrix} \lambda+2G & \lambda & \lambda & 0 \\ \lambda & \lambda+2G & \lambda & 0 \\ \lambda & \lambda & \lambda+2G & 0 \\ 0 & 0 & 0 & 2G \end{bmatrix} \begin{bmatrix} \varepsilon_{11} \\ \varepsilon_{22} \\ \varepsilon_{33} \\ \varepsilon_{12} \end{bmatrix} \tag{10.107}$$

where $\boldsymbol{D}$ is the *elasticity stiffness tensor*. Notice that $\gamma_{12} = 2\varepsilon_{12}$, which is the format that is required for showing the shear strain within ABAQUS. A key parameter for the above is: $\lambda + 2G$, which has to be uniquely defined inside the UMAT during the development stage.

- ***Stress update***: A section of the UMAT has to be dedicated to the stress (tensor) update algorithm. Stress from a previous time step is given as $\boldsymbol{\sigma}_t$, and for the later stress to be evaluated after an elapse of time, Δt, the stress update formulation will simply be written as:

$$\boldsymbol{\sigma}_{t+\Delta t} = \boldsymbol{\sigma}_t + \boldsymbol{\Delta\sigma}_t \tag{10.108}$$

where the stress increment, $\boldsymbol{\Delta\sigma}_t$, is obtained according to Eq. 10.105, which is essentially the constitutive formulation of the material volume under consideration. It is this term that enforces the desired material response in the finite element analysis. This is computed inside the UMAT as a stress increment, which from one time-step to another continues to update until the expected materialresponse is generated. This is where the power of the user subroutine lies, as a means of introducing new material response to the finite element material modelling module.

- ***Derivation and coding of material Jacobian matrix***: The last essential part of the user defined coding framework is the derivation and coding of the material Jacobian matrix. According to the mathematics of the Jacobian matrix, we established based on Eq. 10.101 that the Jacobian matrix is the partial derivative of the stress tensor with respect to the strain tensor. Applying this requirement on the constitutive formulation of Eq. 10.107 shows that the Jacobian matrix is equivalent to the stiffness tensor (for this problem):

$$\mathbf{C} = \frac{\partial \Delta \sigma}{\partial \Delta \boldsymbol{\varepsilon}} \quad \longrightarrow \quad \mathbf{C} = \frac{\partial \sigma}{\partial \boldsymbol{\varepsilon}} = \frac{\partial (\boldsymbol{D}\boldsymbol{\varepsilon})}{\partial \boldsymbol{\varepsilon}} \approx \frac{\Delta \sigma}{\boldsymbol{\Delta \varepsilon}} = \boldsymbol{D} \tag{10.109}$$

The above conclusions are true only for small deformation problems or those with large deformations, but with small volume changes. This condition is obeyed classically by the isotropic elasticity problem.

The reader should note that this version of the isotropic linear elasticity UMAT has not considered the effect of temperature. This is because the development brief demands that we make the model isothermal in which case the temperature change is zero and all material properties are temperature insensitive.

The UMAT subroutine, shown in Algorithm 10.7, represents the implementation of the user-defined constitutive model given by Eq. 10.104 in ABAQUS/Standard, the ABAQUS implicit solver.

Algorithm 10.7 An ABAQUS UMAT for an isothermal hookean elasticity model

```
*****************************************************
**                                                 **
**   UMAT FOR ISOTHERMAL HOOKEAN ELASTICITY MODEL  **
**             FOR PLANE STRAIN ELEMENTS ONLY      **
**  (c) Copyright: AuthorName AuthorSurname Year   **
**                                                 **
*****************************************************
SUBROUTINE UMAT(STRESS,STATEV,DDSDDE,SSE,SPD,SCD, RPL,
  DDSDDT,DRPLDE,DRPLDT,STRAN,DSTRAN,TIME,DTIME,TEMP,
  DTEMP, PREDEF, DPRED,CMNAME,NDI,NSHR,NTENS,NSTATV,
  PROPS,NPROPS,COORDS,DROT, PNEWDT,CELENT,DFGRD0,
  DFGRD1,NOEL,NPT,LAYER,KSPT,JSTEP,KINC)
C
   INCLUDE 'ABA_PARAM.INC' ! For ABAQUS/Standard only
   CHARACTER*80 CMNAME
C
```

```
C-----------------------------------------------------------
*** STEP 5: DEFINE SYSTEM AND USER VARIABLES/PARAMETERS
   DIMENSION STRESS(NTENS),STATEV(NSTATV),
  1 DDSDDE(NTENS,NTENS),DDSDDT(NTENS),DRPLDE(NTENS),
  2 STRAN(NTENS),DSTRAN(NTENS),TIME(2),PREDEF(1),
  3 DPRED(1), PROPS(NPROPS),COORDS(3),DROT(3,3),
  4 DFGRD0(3,3),DFGRD1(3,3), JSTEP(4)
C
   DIMENSION DSTRESS(4), STIFFMAT(4,4)
   PARAMETER (ZERO  = 0.D0, ONE   = 1.D0, TWO   = 2.D0,
              THREE = 3.D0, FOUR = 4.D0, HALF = 0.5D0)
C
C
-----------------------------------------------------------
*** STEP 6: DEFINE AND CALCULATE ALL MATERIAL PROPERTIES
        EMODULUS  = PROPS(1) ! PROPS(1)-Elastic modulus, E
        PRATIO    = PROPS(2) ! PROPS(2)-Poissons ratio, NU

        KMODULUS  = EMODULUS/THREE*(ONE-TWO*PRATIO)
        GMODULUS  = EMODULUS/TWO*(ONE+PRATIO)
        GMODHALF  = GMODULUS*HALF
        LAMBDA    = (KMODULUS*THREE-GMODULUS*TWO)/THREE
C
C-----------------------------------------------------------
*** STEP 7:  DEFINE MODEL FORMULATIONS
C Define the Stiffness tensor - STIFFMAT
C (plane strain only)
      DO K1 = 1, NDI
         DO K2 = 1, NDI
           STIFFMAT(K2,K1) = LAMBDA
         END DO
        STIFFMAT(K1,K1) = GMODULUS*TWO + LAMBDA
      END DO
        STIFFMAT(4,4) = GMODHALF
***
C Define the Stress increment
    ETRACE=DSTRAN(1)+DSTRAN(2)+DSTRAN(3)!Trace(Strain)
    DO K=1,NDI
     DSTRESS(K) = 2*GMODHALF*DSTRAN(K)+LAMBDA*ETRACE
    END DO
    DSTRESS(4) = GMODHALF*DSTRAN(4)
C
C-----------------------------------------------------------
*** STEP 8: UPDATE STRESS
       DO K = 1,NTENS
       STRESS(K) = STRESS(K) + DSTRESS(K)
       END DO
C
C-----------------------------------------------------------
*** STEP 9: CODING FOR THE JACOBIAN
      DO I=1,NDI
       DO J=1,NDI
        DDSDDE(I,J) = STIFFMAT(I,J)
       END DO
```

```
      END DO
      DDSDDE(4,4)  = STIFFMAT(4,4)
C
      RETURN
      END
*****************************************************
```

10.10 Conclusions

One of the main strengths of the finite element framework in tackling a wide range of problem lies in the ability of the user to develop material models for a wide range of materials under different loading conditions. This chapter has explored the different constitutive formulations of materials that are readily modelled using the finite element process. Broadly, these include: *linear elasticity, classic metal plasticity, viscoelasticity* and *nonlinear (rubbery) elasticity*. The discussion has been limited to these as they cover quite a large set of test materials that are used in practice, namely: metals, polymers, metals, rubbery materials, elastomers as well as biological systems. Most of these materials have heterogeneous properties and need to be treated with isotropic considerations of their material properties.

There exist other material models, especially with heterogeneous microstructure, which have not been considered in this chapter. These include, for example, composite materials, sandwich structures, concrete, foams, etc. All these are valid considerations, but the authors have chosen not to include them as they will require extensive discussion to formulate their constitutive models. There are also other excellent textbooks to address these.

Finally, in this chapter, we have also presented the user-defined material models (UMAT). These help the user integrate new formulations of material response within the finite element framework. The need, and implementation of the UMAT principles, for the UMAT were presented. The authors have also used the ABAQUS UMAT framework to illustrate how a simple UMAT can be created. The ability to develop user-specific material models or make modifications to existing material models is a very useful skill and quite liberating to the constitutive modeller looking to capture quite extensive range of material responses.

At the end of this chapter, the reader should be able to:

(a) Describe the principles of common constitutive models available in commercial FEM solvers.
(b) Define the requirements needed to develop material models for FEM solvers.
(c) Justify the importance of the user-defined material subroutine within an FEM solver.

(continued)

(d) Appreciate the different components of a typical UMAT.
(e) Develop variants of user-defined subroutines for many material systems in ABAQUS.

10.11 Problems: Material Response – Constitutive Models and Their Implementation

Problem 10.1 The hyperelastic properties of a strip of carbon-black reinforced natural rubber is to be investigated using finite element methods. The mechanical properties of the material are given in Table 10.2.

(a) Use ABAQUS to generate the dogbone test specimen with details given in Fig. 10.17.
(b) Using the Neo-Hookean model parameters[16] given in Table 10.3, run a tensile simulation up to the strain limit of 100%.
(c) Generate the stress strain plot from your test simulation.

Table 10.2 Mechanical properties of carbon-black reinforced rubber [28]

Tensile strength [MPa]	Shear strength [MPa]	Shear modulus [GPa]	Elongation [%]
12	4.6	0.36	100

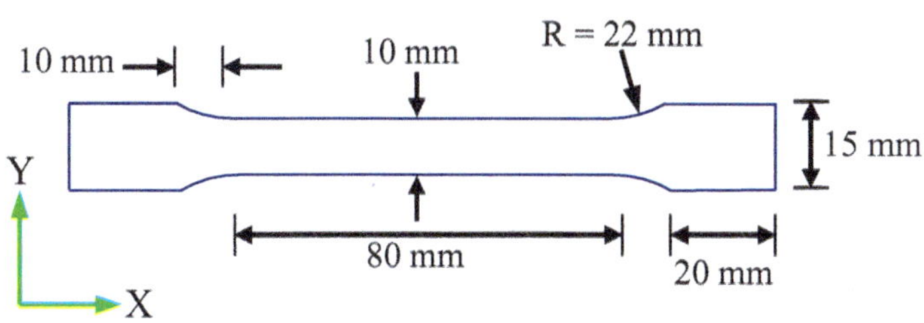

Fig. 10.17 Dimensions of a typical dogbone test specimen

Table 10.3 Neo-Hookean model constants for carbon-black filled natural rubber [28]

Coefficients	C_{10} [MPa]	D_1 [MPa]
Values	0.2587	1.5828×10^{-3}

[16]These coefficients were obtained by [28] using ABAQUS to fit experimental data.

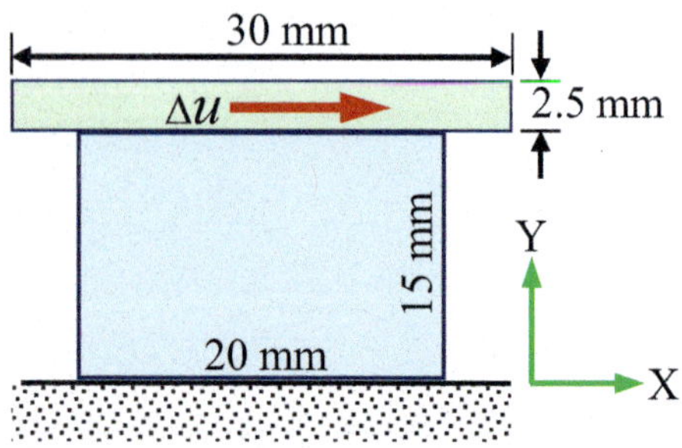

Fig. 10.18 A schematic diagram of a shear test specimen

Problem 10.2 The configuration of a shear test of a carbon-filled natural rubber is shown in Fig. 10.18. The rubber is a $20 \times 15\,\text{mm}^2$ block fixed at the base and sheared at the top by a displacement, $\Delta u = 20\,\text{mm}$, via a 2.5 mm thick steel plate with modulus of elasticity, $E = 210\,\text{GPa}$. The rubber material is to be modelled using the same Neo-Hookean material model as Problem 10.1, whose coefficients are given in Table 10.3.

(a) Use ABAQUS to generate the shear test configuration.
(b) Undertake a finite element investigation of the shear response assuming perfect contact between the metallic and rubber parts. The plate is constrained along the y-axis. *Hint: Treat the steel plates as linear elastic material with Poisson ratio,* $\nu = 0.30$.
(c) Generate the shear stress-strain plot from your test simulation.

Problem 10.3 A bulge test[17] is a type of test used to determine model constants for a hyperelastic material. In this problem, you are required to undertake a bulge test on the same carbon-black filled natural rubber from Problem 10.1. The rubber is a circular sheet of thickness, $t = 5\,\text{mm}$ and diameter, $\phi = 120\,\text{mm}$. It is firmly clamped on the outside edges of the sheet from a radius, $R_i = 55\,\text{mm}$ to $R_o = 60\,\text{mm}$, as shown in Fig. 10.19. A pressure load, $P_i = 80\,\text{kPa}$ is applied on one side of the rubber sheet. The rubber material is to be modelled using the same Neo-Hookean material model as Problem 10.1 with coefficients given in Table 10.3.

(a) Use ABAQUS to generate the bulge test configuration with the outer edges clamped securely in all directions.
(b) Carry out a finite element investigation of the bulge response .
(c) Generate the stress strain plot from your test simulation based on the element on the outer and bottom of the top of the bulge.

[17]The bulge test is an **equibiaxial** test with stress in x- and y-directions being equal: $\sigma_x = \sigma_y = \frac{P_i r}{2t} = \frac{pr}{2t_0\lambda_z}$ and $\sigma_z \approx 0$. Note here that P_i = pressure, r = radius of bulge, t = thickness, t_0 = thickness of the virgin material before testing and, finally, λ_z = stretch along the z-axis.

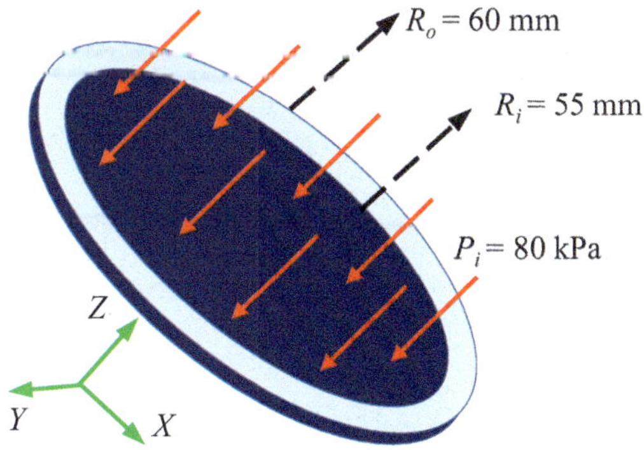

Fig. 10.19 A bulge test specimen (with dimensions)

Table 10.4 Bulge test data of carbon-black filled natural rubber [28]

Pressure [kPa]	0.0	0.5	2.7	5.0	10.8	20.1	35.1	49.1	64.8	79.9
Bulge height	0.0	7.5	12.2	15.2	19.3	23.6	30.1	36.8	44.5	55.0

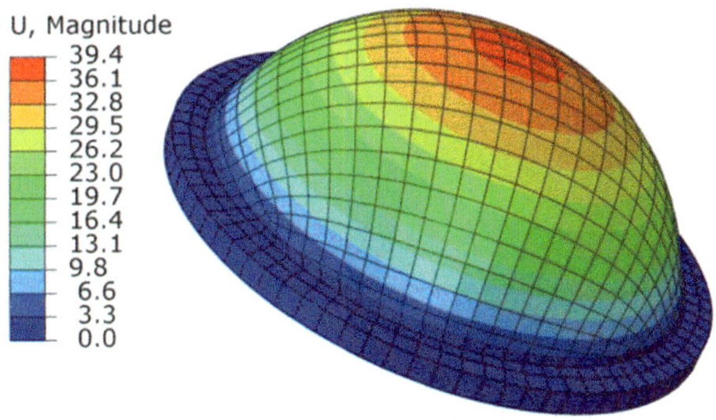

Fig. 10.20 Displacement contour plot from a bulge test with maximum height, $\Delta H_{max} = 39.4$ mm for $P_i = 80$ kPa. The model simulation is based on the Neo-Hookean material model of Problem 10.3

Problem 10.4 Experimental data on a bulge test of carbon-black filled natural rubber are given in Table 10.4. The data shown is the height, H of the bulge (i.e. the highest point of the bulge measured from the initial undeformed position) versus the corresponding deforming pressure, P_i.

(a) Carry out a finite element investigation of the bulge test response for the different pressures in Table 10.4 and determine numerically the corresponding heights based on the Neo-Hookean material model with parameters given in Table 10.3. Note that Fig. 10.20 shows a typical contour plot for $P_i = 80$ kPa.
(b) Plot a comparison graph of *Bulge height* against *Pressure* between the experimental data and the Neo-Hookean model predictions. Comment on the results you obtained.

Problem 10.5 Using the same experimental data shown in Table 10.4, undertake a finite element analysis based on the following conditions:

Table 10.5 Mooney-Rivlin 5-parameter model constants for carbon-black filled natural rubber [27]. *Assume incompressibility,* i.e. $D = 0$

Constants	C_{10} [MPa]	C_{01} [MPa]	C_{20} [MPa]	C_{11} [MPa]	C_{02} [MPa]
Values	0.58643	-3.8942×10^{-2}	-2.8359×10^{-3}	7.6169×10^{-3}	-7.7051×10^{-4}

Table 10.6 Ogden model constants ($N = 3$) for an incompressible natural rubber [27]

Constants	μ_1 [MPa]	α_1	μ_2 [MPa]	α_2	μ_3 [MPa]	α_3
Values	2.0336	-0.9900	-2.5731	-0.7749	1.6200	1.6910

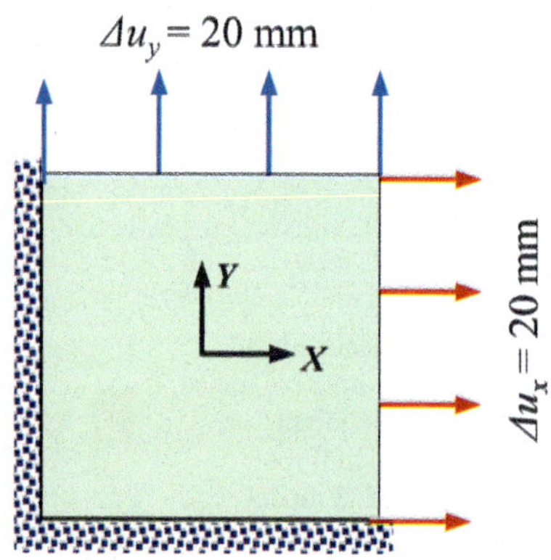

Fig. 10.21 An equibiaxial test specimen subjected to $x-$ and $y-$axes displacements

(a) Use the *five-parameter Mooney-Rivlin* material model with coefficients given in Table 10.5, determine the predicted bulge heights for the natural rubber.
(b) Repeat the analysis, but model the natural rubber using a *three-parameter Ogden* material model with coefficients given in Table 10.6.
(c) Generate a comparison *Bulge height* versus *Pressure* plot for the experimental data and the two material models.
(d) Quantify the cumulative error between the experimental data and the two models and determine which one of the two material models best fits experiment.

Problem 10.6 Model parameters of hyperelasticity can also be determined using an **equibiaxial** test where a test specimen is loaded equally in two perpendicular opposite directions. A test specimen, shown in Fig. 10.21, of dimensions $20 \times 20\,\text{mm}^2$ is subjected to an equibiaxial load of $\Delta u = 20\,\text{mm}$. The material is made from the same carbon-filled natural rubber of Problem 10.1 and modelled using same Neo-Hookean material model with coefficients given in Table 10.3.

(a) Use ABAQUS to design the equibiaxial test specimen and impose appropriate boundary conditions.
(b) Carry out a finite element investigation of the equibiaxial test response .
(c) Generate the stress-strain plots along the $x-$ and $y-$axes.

Table 10.7 Experimental data on carbon-black filled natural rubber [28]

Uniaxial		Biaxial		Planar extension		Volumetric	
Strain, ε	Stress, σ [MPa]	Strain, ε	Stress, σ [MPa]	Strain, ε	Stress, σ [MPa]	Pressure, P_i [MPa]	Volume ratio
0.0000	0.0000	0.0000	0.0000	0.0000	0.1477	111.05	0.9169
0.0860	0.1544	0.0428	0.1913	0.0663	0.2407	105.82	0.9203
0.3566	0.2335	0.1335	0.3715	0.1171	0.3444	100.49	0.9236
0.6194	0.3285	0.2100	0.5591	0.2031	0.4837	92.24	0.9284
0.8849	0.4591	0.2988	0.6583	0.3332	0.6202	88.34	0.9322
1.1400	0.5699	0.4143	0.7465	0.4831	0.7728	80.35	0.9377
1.3976	0.6609	0.4995	0.7464	0.7914	0.9800	73.31	0.9429
1.6240	0.7718	0.6380	0.8419	1.1439	1.1817	66.28	0.9481
1.8167	0.9222	0.8139	0.9852	1.5139	1.3671	59.25	0.9531
2.0275	1.0369	1.0413	1.2057	1.8223	1.4925	52.00	0.9588
2.3451	1.2784	1.2758	1.4041	2.0161	1.5906	44.54	0.9633
2.6603	1.6544	1.5121	1.6467	2.2099	1.7051	39.31	0.9671
3.0537	2.1728	1.7075	1.8378	2.4082	1.7950	33.35	0.9723
3.3690	2.6953	1.8994	2.0290	2.5844	1.8796	26.21	0.9777
3.5957	3.0712	2.1180	2.3047	2.7209	1.9451	19.71	0.9829
3.7913	3.4908	2.2939	2.5069	2.8091	1.9831	13.63	0.9875
3.9035	3.7678	2.4414	2.6244	2.9258	2.0403	6.50	0.9936
4.0156	4.0330	2.5462	2.7494	3.0271	2.0450	0.43	0.9998

Problem 10.7 A carbon-filled natural rubber is subjected to four test types: *uniaxial, biaxial, planar extension* and *volumetric tests*. Experimental data generated from these experiments are given in Table 10.7.

(a) Generate an `ABAQUS` model of a *tensile* test simulation using the dogbone specimen of Fig. 10.17.
(b) Using the uniaxial data of Table 10.7, carry out a finite element investigation based the following models:

  (i) Mooney-Rivlin
  (ii) Arruda-Boyce
  (iii) Ogden
  (iv) Polynomial
  (v) Yeoh

(c) Generate all the model constants for each of the above hyperelastic material models.
(d) Plot the resulting comparison stress-strain plots of the experimental data and all five hyperelasticity models.

Problem 10.8 A planar extension specimen of dimensions 80 × 50 mm^2 is shown in Fig. 10.22. The data from such an experiment [28], is given in Table 10.7.

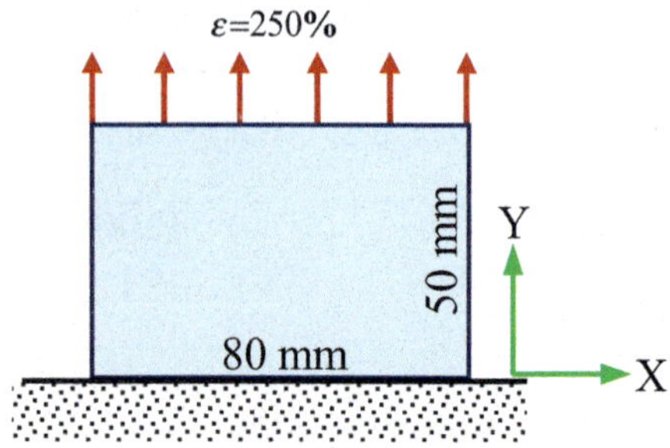

Fig. 10.22 An equibiaxial test specimen subjected to $x-$ and $y-$axes displacements

(a) Generate an `ABAQUS` model of the *planar extension* test simulation.
(b) Using the planar extension data of Table 10.7, carry out a finite element investigation based the following models:

  (i) Polynomial
  (ii) Arruda-Boyce
  (iii) Yeoh

(c) Plot the resulting comparison stress-strain plots of the experimental data and all three hyperelasticity models.

Problem 10.9 Use the biaxial test specimen of Fig. 10.21 to create a finite element virtual domain for investigating the three hyperelasticity material model in `ABAQUS`.

(a) Based on the biaxial data given in Table 10.7, determine the model constants for the following models: (a) Mooney-Rivlin; (b) Marlow; and, (c) Neo-Hookean material models.
(b) Plot the resulting comparison stress-strain plots of the experimental data and all three hyperelasticity models.

Problem 10.10 A steel plate of dimensions $300 \times 300 \times 10\,\text{mm}^3$, is subjected to blast loading from a trinitrotoluene (TNT) explosive located at a stand-off height of 120 mm from the plate. The plate is fixed securely on all four-sides and the explosive is located at the centre of the plate. Assume the steel material shows linear isotropic elasticity characterized by Young's Modulus, $E = 210\,\text{GPa}$, and Poisson Ratio, $\nu = 0.30$ (Fig. 10.23).

(a) Create the `ABAQUS` representation of the plate for use in the blast simulation.
(b) Undertake an FEM simulation of the blast response using 1 kg of TNT explosive. *Hint: Use the `CONWEP` option to simulate blast loading in `ABAQUS`.*
(c) If the steel material fails at a yield stress, $\sigma_y = 240\,\text{MPa}$, re-run the simulation assuming *perfect elasto-plastic* behaviour of the steel.
(d) In both the elastic and elasto-plastic cases, determine the displacement time plots of the top and bottom faces centre-deflection of the steel plate after the blast loading.

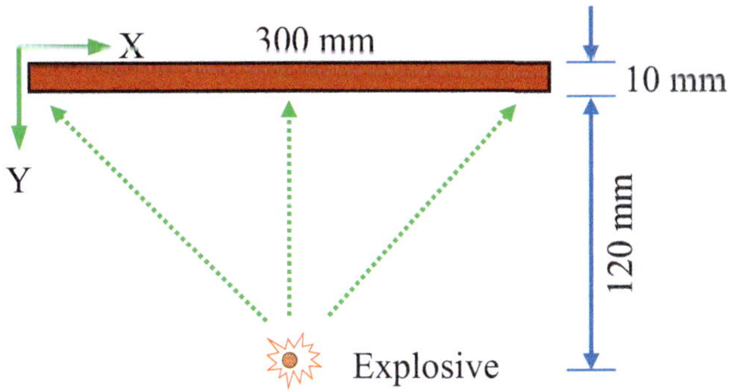

Fig. 10.23 A 2D schematic representation of a blast-loading arrangement for a metallic plane subjected to a TNT explosive blast

Fig. 10.24 A schematic representation of a steel-aluminium sandwich structure

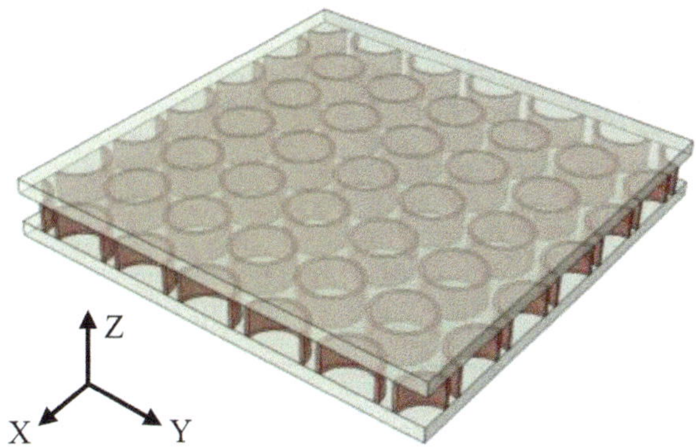

Problem 10.11 A steel-aluminum alloy sandwich structure is shown in Fig. 10.24. The steel plates have dimensions $300 \times 300 \times 10\,\text{mm}^3$. The aluminium core is made of annular cylindrical shape of inner, $R_i = 18\,\text{mm}$, and outer, $R_o = 20\,\text{mm}$, radii. The cylinder height is $H = 20\,\text{mm}$. The center-to-centre distance of the cylinders is 50 mm. The structure is subjected to a blast loading from 2.5 kg of trinitrotoluene (TNT) explosive located at a stand-off height of 120 mm from the plate.

The sandwich structure is fixed securely on all four-sides and the explosive is located at the centre of the plate. Assume that the material response of the steel plates is a classic metal plasticity and to be modelled here using a Johnson-Cook model with the properties given in Table 10.1. Also, consider the aluminium alloy to exhibit an elastoplastic behaviour with Young's Modulus, $E = 70\,\text{GPa}$, Poisson ratio, $\nu = 0.33$, and yield stress, $\sigma_y = 280\,\text{MPa}$.

(a) Create the `ABAQUS` representation of the sandwich structure plate for use in the blast simulation.
(b) Undertake an FEM simulation of the blast response of the structure. *Hint: Use the* `CONWEP` *option to simulate blast loading in* `ABAQUS`*.*

Table 10.8 Damage parameters of Johnson-Cook nonlinear plasticity model [18]

Damage variable	D_1	D_2	D_3	D_4	D_5
Values	0.05	3.44	−2.12	0.002	0.61

(c) Determine the displacement time plots of the top and bottom faces centre-deflection of the sandwich structure.

(d) If the steel plates are modelled using a Johnson-Cook damage feature,[18] with damage parameters given in Table 10.8, determine the new displacement-time plots of the top and bottom faces centre deflection of the sandwich structure.

References

1. Ali, A., Fouladi, M.H., Sahari, B.: A review of constitutive models for rubber-like materials. Am. J. Eng. Appl. Sci. **3**(1), 232–239 (2010)
2. Arruda, E.M., Boyce, M.C.: A three-dimensional constitutive model for the large stretch behavior of rubber elastic materials. J. Mech. Phys. Solids **41**(2), 389–412 (1993). http://dx.doi.org/10.1016/0022-5096(93)90013-6, http://www.sciencedirect.com/science/article/pii/0022509693900136
3. Beda, T.: Modeling hyperelastic behavior of rubber: a novel invariant-based and a review of constitutive models. J. Polym. Sci. B Polym. Phys. **45**(13), 1713–1732 (2007). http://dx.doi.org/10.1002/polb.20928
4. Beda, T.: An approach for hyperelastic model-building and parameters estimation a review of constitutive models. Eur. Polym. J. **50**, 97–108 (2014). https://doi.org/10.1016/j.eurpolymj.2013.10.006, http://www.sciencedirect.com/science/article/pii/S0014305713004989
5. Bergström, J.S., Boyce, M.C.: Deformation of elastomeric networks: relation between molecular level deformation and classical statistical mechanics models of rubber elasticity. Macromolecules **34**(3), 614–626 (2001). http://dx.doi.org/10.1021/ma0007942
6. Bischoff, J.E., Arruda, E.M., Grosh, K.: A new constitutive model for the compressibility of elastomers at finite deformations. Rubber Chem. Technol. **74**(4), 541–559 (2001). https://doi.org/10.5254/1.3544956
7. Boyce, M.C., Arruda, E.M.: Constitutive models of rubber elasticity: a review. Rubber Chem. Technol. **73**(3), 504–523 (2000). https://doi.org/10.5254/1.3547602
8. Buckley, C., Harding, J., Hou, J., Ruiz, C., Trojanowski, A.: Deformation of thermosetting resins at impact rates of strain. Part I: experimental study. J. Mech. Phys. Solids **49**(7), 1517–1538 (2001). http://dx.doi.org/10.1016/S0022-5096(00)00085-5, http://www.sciencedirect.com/science/article/pii/S0022509600000855
9. Burden, R., Faires, J., Burden, A.: Numerical Analysis. Cengage Learning (2015). https://books.google.co.uk/books?id=9DV-BAAAQBAJ
10. Chandrasekaran, H., M'Saoubi, R., Chazal, H.: Modelling of material flow stress in chip formation process from orthogonal milling and split hopkinson bar tests. Mach. Sci. Technol. **9**(1), 131–145 (2005). http://dx.doi.org/10.1081/MST-200051380

[18] In the *Johnson-Cook damage model*, the damage of an element is defined by the damage variable, $D = \dfrac{\Delta\varepsilon}{\Delta\varepsilon^f}$, where ε^f is the fracture strain. Fracture occurs when $D = 1.0$. The fracture strain is defined by $\varepsilon^f = [D_1 + D_2\exp(D_3\sigma^*)]\,[1 + D_4\ln(\dot{\varepsilon}^*)]\,[1 + D_5T^*]$. The variables $(\sigma^*, \dot{\varepsilon}^*, T^*)$ are constant values of stress, strain rate and temperature respectively [18] given $\sigma^* \le 1.5$.

11. Dunne, F., Petrinic, N.: Introduction to Computational Plasticity. Oxford Series on Materials Modeling. OUP, Oxford (2005). https://books.google.co.uk/books?id=jVxisjdpe8sC
12. Edwards, S., Vilgis, T.: The effect of entanglements in rubber elasticity. Polymer **27**(4), 483–492 (1986). http://dx.doi.org/10.1016/0032-3861(86)90231-4, http://www.sciencedirect.com/science/article/pii/0032386186902314
13. Fung, Y.: Elasticity of soft tissues in simple elongation. Am. J. Phys. **213**(6), 1532–1544 (1967). https://www.scopus.com/inward/record.uri?eid=2-s2.0-0014179220&partnerID=40&md5=3ffc65d20c6f8114c40f6ee211cfb2a0. Cited By 499
14. Gent, A.N.: A new constitutive relation for rubber. Rubber Chem. Technol. **69**(1), 59–61 (1996). http://dx.doi.org/10.5254/1.3538357
15. Hariharaputhiran, H., Saravanan, U.: A new set of biaxial and uniaxial experiments on vulcanized rubber and attempts at modeling it using classical hyperelastic models. Mech. Mater. **92**, 211–222 (2016). https://doi.org/10.1016/j.mechmat.2015.09.003. http://www.sciencedirect.com/science/article/pii/S0167663615001891
16. Hartmann, S.: Parameter estimation of hyperelasticity relations of generalized polynomial-type with constraint conditions. Int. J. Solids Struct. **38**(44–45), 7999–8018 (2001). https://doi.org/10.1016/S0020-7683(01)00018-X, http://www.sciencedirect.com/science/article/pii/S002076830100018X
17. Holzapfel, G.: Nonlinear Solid Mechanics: A Continuum Approach for Engineering. Wiley (2000). https://books.google.co.uk/books?id=_ZkeAQAAIAAJ
18. Johnson, G.R., Cook, W.H.: Fracture characteristics of three metals subjected to various strains, strain rates, temperatures and pressures. Eng. Fract. Mech. **21**(1), 31–48 (1985). http://dx.doi.org/10.1016/0013-7944(85)90052-9, http://www.sciencedirect.com/science/article/pii/0013794485900529
19. Jones, D.F., Treloar, L.R.G.: The properties of rubber in pure homogeneous strain. J. Phys. D. Appl. Phys. **8**(11), 1285 (1975). http://stacks.iop.org/0022-3727/8/i=11/a=007
20. Kiusalaas, J.: Numerical Methods in Engineering with Python. Cambridge University Press (2010). https://books.google.co.uk/books?id=9SG1r8EJawIC
21. Li, H., Buckley, C.: Evolution of strain localization in glassy polymers: a numerical study. Int. J. Solids Struct. **46**(7), 1607–1623 (2009). http://dx.doi.org/10.1016/j.ijsolstr.2008.12.002, http://www.sciencedirect.com/science/article/pii/S0020768308005039
22. McCrum, N., Buckley, C., Bucknall, C.: Principles of Polymer Engineering. Oxford Science Publications, Oxford University Press (1997). https://books.google.co.uk/books?id=UX-sAQAAQBAJ
23. Ogden, R.W.: Large deformation isotropic elasticity – on the correlation of theory and experiment for incompressible rubber-like solids. Proc. R. Soc. Lond. A Math. Phys. Eng. Sci. **326**(1567), 565–584 (1972). http://dx.doi.org/10.1098/rspa.1972.0026, http://rspa.royalsocietypublishing.org/content/326/1567/565
24. Ogden, R.W., Saccomandi, G., Sgura, I.: Fitting hyperelastic models to experimental data. Comput. Mech. **34**(6), 484–502 (2004). http://dx.doi.org/10.1007/s00466-004-0593-y
25. Peeters, F., Kussner, M.: Material law selection in the finite element simulation of rubber-like materials and its practical application in the industrial design process. In: Constitutive Models for Rubber, pp. 29–36 (1999)
26. Rivlin, R.S., Saunders, D.W.: Large elastic deformations of isotropic materials. VII. Experiments on the deformation of rubber. Philos. Trans. R. Soc. Lond. A Math. Phys. Eng. Sci. **243**(865), 251–288 (1951). http://dx.doi.org/10.1098/rsta.1951.0004, http://rsta.royalsocietypublishing.org/content/243/865/251
27. Sasso, M., Palmieri, G., Chiappini, G., Amodio, D.: Characterization of hyperelastic rubber-like materials by biaxial and uniaxial stretching tests based on optical methods. Polym. Test. **27**(8), 995–1004 (2008). http://dx.doi.org/10.1016/j.polymertesting.2008.09.001, http://www.sciencedirect.com/science/article/pii/S0142941808001529
28. Shahzad, M., Kamran, A., Siddiqui, M.Z., Farhan, M.: Mechanical characterization and fe modelling of a hyperelastic material. Mater. Res. **18**(5), 918–924 (2015)

29. Sweeney, J.: A comparison of three polymer network models in current use. Comput. Theor. Polym. Sci. **9**(1), 27–33 (1999). http://dx.doi.org/10.1016/S1089-3156(98)00050-6, http://www.sciencedirect.com/science/article/pii/S1089315698000506
30. Tan, J.Q., Zhan, M., Liu, S., Huang, T., Guo, J., Yang, H.: A modified Johnson-cook model for tensile flow behaviors of 7050-t7451 aluminum alloy at high strain rates. Mater. Sci. Eng. A **631**, 214–219 (2015). http://dx.doi.org/10.1016/j.msea.2015.02.010, http://www.sciencedirect.com/science/article/pii/S0921509315001276
31. Treloar, L.: Stress-strain data for vulcanized rubber under various types of deformation. Trans. Faraday Soc. **40**, 59–70 (1944)
32. Twizell, E., Ogden, R.: Non-linear optimization of the material constants in ogden's stress-deformation function for incompressible isotropic elastic materials. J. Aust. Math. Soc. Ser B. Appl. Math. **24**(04), 424–434 (1983)
33. Umbrello, D., MSaoubi, R., Outeiro, J.: The influence of Johnson-cook material constants on finite element simulation of machining of AISI 316l steel. Int. J. Mach. Tools Manuf. **47**(3), 462–470 (2007). http://dx.doi.org/10.1016/j.ijmachtools.2006.06.006, http://www.sciencedirect.com/science/article/pii/S0890695506001507
34. Wang, X., Shi, J.: Validation of Johnson-cook plasticity and damage model using impact experiment. Int. J. Impact Eng. **60**, 67–75 (2013). http://dx.doi.org/10.1016/j.ijimpeng.2013.04.010, http://www.sciencedirect.com/science/article/pii/S0734743X13000948
35. Ward, I., Sweeney, J.: An Introduction to the Mechanical Properties of Solid Polymers. Wiley (2005). https://books.google.co.uk/books?id=a9Qo8bWy5k0C
36. Whiteley, J.: Finite Element Methods: A Practical Guide. Mathematical Engineering. Springer International Publishing (2017). https://books.google.co.uk/books?id=MyMDDgAAQBAJ
37. Williams, M.L., Landel, R.F., Ferry, J.D.: The temperature dependence of relaxation mechanisms in amorphous polymers and other glass-forming liquids. J. Am. Chem. Soc. **77**(14), 3701–3707 (1955)
38. Wu, P., Giessen, E.V.D.: On improved network models for rubber elasticity and their applications to orientation hardening in glassy polymers. J. Mech. Phys. Solids **41**(3), 427–456 (1993). http://dx.doi.org/10.1016/0022-5096(93)90043-F, http://www.sciencedirect.com/science/article/pii/002250969390043F
39. Yeoh, O.: Some forms of the strain energy function for rubber. Rubber Chem. Technol. **66**(5), 754–771 (1993). https://www.scopus.com/inward/record.uri?eid=2-s2.0-0027701267&partnerID=40&md5=cdb49e8eec62e8dadf2087129b197281. Cited By 394
40. Yeoh, O.H., Fleming, P.D.: A new attempt to reconcile the statistical and phenomenological theories of rubber elasticity. J. Polym. Sci. B Polym. Phys. **35**(12), 1919–1931 (1997). http://dx.doi.org/10.1002/(SICI)1099-0488(19970915)35:12<1919::AID-POLB7>3.0.CO;2-K

Chapter 11
The Future of Finite Element Modelling

Abstract As powerful as finite element modelling (FEM) is, its future is expected to be both exciting and challenging. This chapter seeks to explore this future in order to identify strands of research that have to be carried out in order to sustain the impact of FEM within academic and industrial communities. This chapter starts by identifying the current challenges that limit the widespread adoption of this strategy of obtaining robust solutions to practical problems. After this, the discussion extends to an exploratory discourse on the future of FEM. In the next two decades, the landscape of FEM is expected to be significantly different from what pertains today. If this vision is to be achieved, the aspects highlighted in this chapter must be addressed. There is so much scope for improving the FEM process and this chapter makes the case for the sustained research around FEM so that this most essential design tool for engineers, and other allied experts, will continue to be sustained and advanced more and more even into the next century.

Keywords Next-generation FEM • Cloud-based FEM • Hybridized FEM • Limitations of FEM • Future FEM research

11.1 Introduction

This textbook is dedicated to understanding the principles of finite element modelling (FEM) and justifying the validity of results from the FEM framework. So far, the discussion has addressed these and in this last chapter, the focus switches to exploring the future of the FEM framework. As valuable a tool as the FEM solver can be, unfortunately, there are significant challenges facing the sustained development of FEM.

In Sect. 1.3.3, we presented the *limitations* of the finite element method. Also, in Sect. 1.3.4, we presented the current *adaptations* that can be, and have been, made to alleviate the impact of limitations of the FEM method. All these are laudable efforts and have largely enhanced the impact of FEM in addressing complex real life problems. Yet, there remain intractable problems that have to be tackled for the widespread application of FEM in many industries to be fully realized.

It is also important to consider what the future holds for the FEM method. One may not always get it right in making a prediction about the future, but in view

M. Okereke, S. Keates, *Finite Element Applications*, Springer Tracts in Mechanical Engineering, https://doi.org/10.1007/978-3-319-67125-3_11

of existing technologies and the current drive in the design community, some of the ideas posited here may prove to be correct. It is also the aim here to identify some of the research areas that can inspire the future FEM developer when more computing resources and capabilities become available. These, and many more, are the motivations for this chapter.

11.2 Chapter Objectives

At the end of this chapter, the reader should be able to:

(a) Understand the struggles of the FEM method.
(b) Appreciate the challenges and opportunities facing the FEM developer.
(c) Discuss the research directions that will inspire future uses of FEM.
(d) Suggest the features of future FEM frameworks.

11.3 What Have We Learned About the FEM Process?

We have learned a few things about FEM.

(a) ***FEM is a powerful design tool***: The FEM framework is a powerful tool that can be used to solve quite a lot of practical problems that in the past were difficult to solve. Analytical models exist and can be solved quickly. However, real life problems are not always amenable to such analytical formulations. For example, consider the study of the mechanics of the breakage of pieces wood into particles as they are travelling through a funnel-shaped silo before impacting onto a connected conveyor. There is no pre-defined analytical framework to study this, hence numerical schemes are best. The FEM tool is a framework that is designed to investigate this properly and generate robust results. It is this practical-leaning requirement of industrial problems that the FEM tool meets thus why we say that "*FEM is a powerful tool.*"
(b) ***FEM results are strongly dependent on mesh quality***: The principle of the FEM framework is the discretization of a virtual domain into '*finite elements,*' which represents the physical domain of the problem. The discretization process introduces errors as approximations have to be made when making a continuous, analogue original into a discrete digital model. These errors are passed into the FEM solver, which leads to an approximate solution made worse by the effect of discretization errors. As a result, the quality of the mesh, or in other words, the reduction of the discretization error, should be an important and essential consideration one has to take into consideration when assessing FEM results.

(c) ***FEM can simulate extensive loading conditions***: Another important lesson we have learnt from FEM is that it is a vital tool to solve quite an extensive range of loading conditions. The FEM method is essentially a numerical scheme that solves a boundary value problem. As a result, the user can specify quite a complex set of load conditions at the boundaries, which the FEM method can solve quickly. As a result, with FEM, we can assess static, impact rates, blast and ballistic loading scenarios to obtain realistic results. Even multiple types of combined loading conditions can be implemented on the boundary meshes. This again, drives the attraction of the FEM method within industry and academic research.

(d) ***FEM can be used to model nearly any material***: All FEM solvers have in-built material models that a user can easily use for a given FEM problem to study a given material response. The list of these in-built solvers continues to increase, however, it is always more limited than we might like. The possibility of including *user-defined material models* (see Sect. 10.9) has widened the range of materials that the FEM scheme can model. As a result, the FEM user can model, for example, steel and even, for example, heat-treated steel, simply by creating a unique user-defined material model for the heat-treated steel. This makes again the FEM process very attractive to both academic and industry research and development communities.

(e) ***FE solvers are not 'black boxes'***: Traditionally, the FEM uses in the community has been all about utilizing a chosen FEM solver to solve practical problems with little or no knowledge of the underlying mechanics of the problem. FEM solvers have always been seen as 'black boxes' that the user never really understands. This textbook (see Chap. 4) has shown that the information *under the hood* of a typical FEM solver can be understood. The presentation was taken from a practical implementation position. FEM solvers no longer need be black boxes. Nearly anyone can develop their own FEM solvers. In fact, there is a prevalence of 'in-house' FEM solvers designed for certain specialist functions, in research institutes and universities.

(f) ***FEM is limited but can be improved***: We have also presented the limitations of the FEM method and the adaptations that can be made to eliminate the known limitations. These limitations have led to new research teams developing variants of the FEM method and it is expected that this approach will continue in the future.

(g) ***FEM outputs must be validated***: FEM outputs are simply approximate solutions that depend on several factors highlighted in this book as *pillars of the FEM process*. As a result, FEM outputs have to be ranked along side existing *known information* about the material volume under consideration. This is the science of **validation**. Without validation, FEM outputs cannot really be relied upon. We have also learned that FEM outputs must always be checked rigorously before adopting them as design parameters.

11.4 The Hindrances to Widespread Adoption of FEM

In spite of the many benefits of the FEM process in both industrial and academic circles, there remain real hindrances to the widespread adoption of FEM. In this section, we will provide commentary on some of these hindrances with suggestions of how these can be tackled. Addressing these hindrances is essential for expanding the benefits of the FEM process for engineering designs, and the like.

11.4.1 Computing Cost of Simulations

The FEM process is a computational process which requires computational resources to be made available to undertake such an investigation. We already established that the problems solved using the FEM tool are quite challenging and not amenable to back-of-the-envelope calculations. The provision of computing resources is a requirement for choosing the FEM approach. The more challenging the problem is, the more the computing resources that is required for undertaking the simulations.

The FEM solvers that have been proven to be reliable design tools are usually locked behind a pay wall. As a result, users will often have to pay significant annual licenses to be able to use such frameworks. The open source FEM solvers are usually not as mature and easy to use as the commercial versions. They are also likely not be robustly validated for different design conditions. As a result, the validity of solutions from them is cast in doubt. Thus, most users (industries especially) will not use such *general public licence* FEM solvers for their operations.

Finally, there is a drive to make structural scale predictions of material behaviour based on input at microstructural or molecular length scales. Consequently, the FEM user will have to build representative volume elements (**RVEs**) of the problem that require detailed representation of the material behaviour from atomic, molecular, nanoscale or microscopic length scales. The predictions from such length scales have to be scaled up to structural scales during the validation stage. This requires extensive computing resources and high performance computing frameworks now have become the established frameworks for such investigations. These are often quite expensive too.

Until the challenge of the cost of simulations (software, computers, storage, etc.) are tackled, the ability of users to model nearly any possible materials, loading conditions, or length scales will always be limited. Advances are already being made here, but there are lots more that need to be done.

11.4.2 Insufficient Physics of the Problem

Another hindrance is the potentially limited understanding of the physics and mechanics of the problem under consideration. Already, existing in-built material models have been based on published information about the materials or systems under investigation. However, with improvements in material design and even innovations in structural designs, there is a growing set of design problems without a clear understanding of the underlying physics and mechanics.

Constitutive modellers like to describe their constitutive models as *'physical,' 'multi-physics,' 'representative,' 'real,'* models. This presupposes that the material model has been developed by considering the comprehensive physics of the problem, adopting the *multi-physics* dimensions and so have become *representative* of *real* materials. This is not always the case, as simplifying assumptions about the microstructure of the material, are often made. For example, modellers may focus on **isotropic** conditions when in reality most materials have **anisotropic** material behaviour. To address this challenge, there is a need to develop the correct constitutive formulations, even for such anisotropic systems. This is not always easy and continues to occupy material scientists and predictive model developers.

Use of non-physical models to capture real systems will introduce errors arising from the material response representations. This will influence the reliability of the FEM output. It is important to carry out more research in the theoretical physics of material models development. Until this is done, the paucity of knowledge will continue to hinder wider adoption of the FEM scheme.

11.4.3 Lack of Technical Know-How

The FEM process is technically challenging. It is taught and mastered at advanced undergraduate and post-graduate levels of study in many universities. If a company wants to incorporate the FEM process within their industrial processes, they will have to hire FEM experts or pay for extensive short courses to train staff in the FEM process. The high adoption cost, on especially human resources, is currently hindering small and medium-sized businesses (SMEs) from drawing from the benefits of the FEM process.

Even when companies have a work force that has the necessary technical knowledge of FEM, such skill sets are usually quite FEM solver specific. For example, the expert might be quite good in the use of ANSYS. The expert may therefore approach every problem using ANSYS, when in reality different solvers have different strengths. Locking oneself to one type of solver limits ones ability to leverage the benefits of other solvers. Thus, to create a computational modelling workforce with skills from different solvers, there is a need to recruit such staff or train staff on a wide set of FEM solvers. All these are expensive options and companies will not want to make such financial commitments without quantifying the impact of such skill set to their business. Hence, the lack of experts continues to limit the adoption of FEM in nearly every industry sector.

Also, companies will usually identify and choose a certain type of solver as their main analysis tool for all FEM work. This could be because the company validation processes are based around such solvers. They really do not have the option to try other solvers even when they might be differently adapted to delivering better results. Universities have many in-built FEM solvers that have been able to tackle some of the challenges of existing FEM frameworks. The technical know-how available in such universities is not typically available to such companies and also not demonstrated to the regulatory agencies to be approved. All these issues constitute hindrances to the adoption of FEM framework (i.e. innovative variants) in such companies.

11.4.4 Mandatory Integration of FEM Within Industry Processes

The use of FEM approaches within industry processes is generally seen as a good thing, but not something made mandatory for many industry processes. Of course, the cost and human resources challenges discussed above, are some of the main reasons why this is the case. However, having assessed the benefits of the FEM process, especially in dealing with complex real life design and manufacturing processes, it is absolutely essential that the FEM process be considered to be made mandatory for many industrial processes.

A typical industry sector that can benefit from such mandatory integration is in development of patient-specific healthcare solutions. For instance, FEM frameworks can be integrated within medical processes design like: respiratory lung motion analysis [41]; teeth restoration procedures [10]; 3D-printing technology for cranial reconstruction [27]; and, personalised medical product development [24].

Even when the FEM process is recommended to be integrated as a standard design tool for many industries, there is still the hindrance that many companies will not be able to achieve this due to the aforementioned considerations and challenges. The failure of such companies to abide by these policy demands will clearly affect the running of such companies and may even lead them to going out of business. This is a risk that many legislatures will not want to take. As a result, this continues to limit the widespread adoption of the FEM process within industry processes.

11.4.5 Verification, Validation and Accuracy of FEM Simulations

The last hindrance that will be addressed here lies in the area of validation of FEM outputs. This is a problem, because of the demand for numerically robust FEM solutions, informed by lower-level data informing structural-level response. We have to generate data at the atomic scale to validate atomic models, or at micrometer scales to validate micromechanical models, etc. Experiments at those scales can be challenging and so there is an absence of validation data at such length scales making it difficult to validate lower length scale FEM models.

The verification and accuracy of FEM simulations are also challenged by the absence of appropriate constitutive formulations at such length scales to allow the FEM outputs to be ranked against analytical tools. The consequence of this is limiting the accuracy of FEM solutions. It is important that this hindrance is addressed through innovative design of experiments at such length scales. Also, we should develop equipment that can help generate test data at lower length scales, or even structural or planetary length scales. When these are in place, the struggles around robust validation of FEM outputs would have been tackled adequately.

11.5 Research Directions for Next-Generation FEM Tools

There is so much scope to speculate about the future of the FEM process and the requirement of developing FEM tools to tackle current and future challenges of the FEM framework. It is the objective here to identify some of these research directions and identify what is required to make the research a success. The ideas presented here have been inspired by the work of Okereke and Akpoyomare [33].

11.5.1 Exploring Beyond the Design Space

At the core of the FEM process is the development of reliable prediction of known phenomenological behaviour of materials or structures. It is important that a set of experimental data exists that any numerical prediction must be compared with in order to verify the reliability of such numerical predictions. This is linked to the requirement of validation for the FEM scheme. That means, when experiments do not exist for such FEM predictions, there is an outstanding question as to how can such FEM simulations be judged. In other words, what happens for those constitutive responses that are difficult to replicate experimentally? If the design data does not exist for certain scenarios and it is crucial that understanding has to be developed for such scenarios, what will the engineer do?

The is the domain which we have described as *domain beyond the design space*, as conclusions drawn from them cannot be validated. The future FEM should address such challenges, because it is not always possible to have existing experimental data for all domains of the design space. The essence of FEM and all predictive tools is to be able to extrapolate model conclusions to those domains where we do not have the required experimental data.

According to Okereke and Akpoyomare [33],

> Future FEM frameworks must seek to liberate the design space within which the test materials are used. This implies developing a framework which is intelligent and intuitive in its ability to predict behaviours of the test material where traditional laboratory experiments are not viable. Such a framework will become a trusted design tool which will not only provide guidance on validated test cases, but also those cases where only a numerical solution is possible.

11.5.2 Cloud-Based FEM Frameworks

Computing cost will play a significant role in future FEM frameworks. Already, research institutions and companies are investing heavily in high-end computing capabilities to help with their FEM-related studies. The advance of **cloud-based computing** is helping organizations deal with the cost of computing without having to make investments in the hardware on site. The approach of having to rent computing power on a need-by-need basis is attractive to companies and helps with resource requirements planning.

The future FEM frameworks will have to be integrated within such cloud-based computing [1]. It should become second nature to FEM users to design a job and run it on-the-fly in the cloud just as it is currently done on the desktop. Already, many FEM solver companies have a presence in the cloud and offer different forms of cloud-based FEM computing services. However, the support and understanding of how these can be used meaningfully are quite limited.

It will, therefore, be a future expectation that the running of FEM simulations in the cloud will be quite advanced. Such cloud-based services might even be assessed using mobile apps as *mobile computing* continues to advance at great speed. Jobs that are currently run on servers can be run then on mobile apps and the framework will be as robust, if not better, than existing *desktop computing*.

Realizing the cloud computing potential will liberate the design advantages of integrating FEM within company processes. It is conceivable that the engineer can, whilst on site, run simulations on a mobile device app (connected to the cloud) and generate reports that can inform modifications or offer solutions to problems faced at work. The surgeon can in the course of a surgical operation, image the surgery site, and run patient-specific cloud-based FEM simulations to assess the suitability of, say, an implant on the skull at a given orientation. This is the future of FEM frameworks, and research to advance this cause should be encouraged as there abound multiple opportunities for improvement of quality of life whilst offering competitive economic advantage to companies using the frameworks.

11.5.3 Hybridized FEM Frameworks

This textbook has presented the formulations and complimentary pillars of the classic FEM framework. We have also identified the limitations of FEM and current adaptations of FEM frameworks (see Sects. 1.3.3 and 1.3.4). Future FEMs will not only be adapted to eliminate current limitations, but most certainly become what the authors call *hybridized FEM frameworks*.

The *hybrid FEM frameworks* will consist of innovative integration of the mathematics of classic FEM formulations described in this book with other numerical approaches. Some examples of these approaches include: the *Smoothed-particle hydrodynamics* (SPH) [6, 28], *Galerkin methods* [3, 13, 42], *Peridynamics* [14, 29, 30], *Isogeometric analysis (IGA)* [9, 12, 19], *Stochastic FEM (SFEM)* [2, 15, 35], *Adaptive FEM* [20, 32, 36] and others. There are clear advantages for these emerging approaches and although the core mathematics underlying their development might be different from classic FEM, it may be possible in the future to design common interface platforms where these multiple approaches can be integrated to generate an optimal solution to a problem. Figure 11.1 shows a schematic representation of the hybridized FEM frameworks.

It should be a research objective for developers to develop algorithms that allow for the interaction and the synergistic integration of these different solution methodologies into a novel solution package hybridized for a specific problem. The ideas proposed here are already being implemented by different research teams in the world. Some of these include the decade long work of Javier Bonet and colleagues at Swansea University [7, 25, 26]; ANSYS-Peridynamics research [34]; coupled finite element and *discrete element methods* (DEM) research [16, 23, 40] and others.

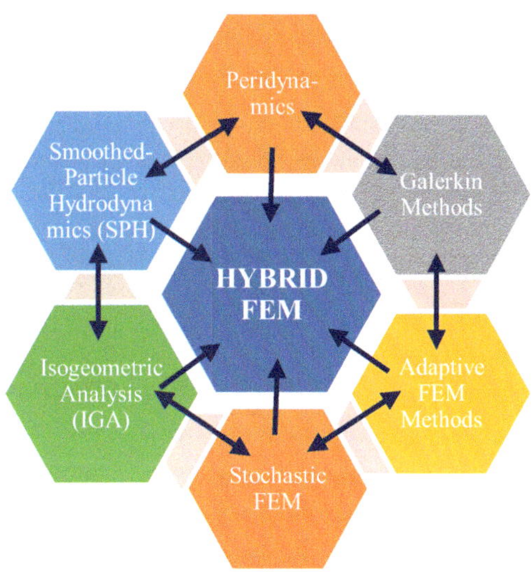

Fig. 11.1 A schematic representation of a hybridized FEM framework with the contributory interacting computational algorithms

The proposed hybridized FEMs go beyond the common approach used by current FEM developers in which they provide specific *Interfacing modules* (see Sect. 4.2.4) and results generated in one solution modality which do not really interact with another. It is currently more of solution silos without any interactions between them. Such an approach has proved beneficial to date, but the future hybridized FEM will take a different format. This will consist of in situ analysis where all these solution modalities interact and adapt as the problem is being solved. It is important that future developers gain a deep understanding of the fundamental theories of each solution modality and consequently explore how they can be harnessed into a cohesive dynamic solution tool.

11.5.4 Evolution of New Materials and Architectures

A McKinsey Global Institute report [31] identified *Advanced Materials* as the 10th most significant advance that will transform life, business and the global economy, with a potential economic impact of $0.2–$0.5 trillion in 2025. Some of the materials that are expected to contribute greatly to these are: graphene, carbon nanotubes, nanoparticles (e.g. nanoscale gold and silver), and other advanced and smart materials (piezoelectric materials, memory metals, and self-healing materials) [33].

In the context of FEM, all these materials have one thing in common, they pose significant constitutive modelling challenges driven by their heterogeneous composition, highly nonlinear behaviour and multiple length scale definition of material properties, etc. The challenge of future FEM frameworks will be to create a robust framework where these type of materials can be integrated as in-built material models or easy-to-code user defined materials.

As a result, the material response pillar of the FEM framework must be robust enough to deal with an ever increasing range of new materials and complex architectures. FEMs of the future should have enhanced robustness and adaptable designs such that emerging materials can be quickly integrated within the FEM framework and modelled with excellent detailed resolution of the material response. This is not currently possible in existing FEM frameworks, but this should be the motivation of material model developers, and when this is done, the FEM framework would become even more invaluable.

11.5.5 The Stochastic Microstructure Argument

Another challenge that FEM users of the future would have to face will be around the development of representative volume elements (**RVEs**) (see Sect. 5.3.2) of heterogeneous materials that are both *representative* of the test material as well as show *microstructures that are stochastically representative*. Today, most virtual domains are based on RVEs that are repetitive and do not often show the randomness of microstructure that real materials have. It is acknowledged in the community that it is very challenging to generate RVEs with truly stochastic microstructures. In the absence of such RVEs, the developers have resorted to using RVEs without such random representations.

As a result, a key question that faces developers of future FEM framework is: "*should one use a microstructure reconstruction (based on, say, tomographic images) or apply selected topological parameters to create regularized numerically-generated RVEs?*" [33]. This argument is what Okereke and Akpoyomare [33] described as the *stochastic microstructure argument.*

The use of tomographic images will always lead to realistic reconstructions of the microstructure and is therefore a preferred development option for future RVEs. However, the cost of generating them from expensive tomographic techniques, continues to make this an unattractive option for many companies. There is also the extra computational cost of working with quite complex microstructures that any gains from their use has to be compared against the cost of generating them.

Consequently, future developers of FEM frameworks have to tackle this argument by researching into less expensive techniques to generate the stochastic RVEs. Also, there is need for sustained research into developing advanced reconstruction techniques from tomographic images to virtual domains for the FEM process. The MIMICs software discussed in Sect. 1.4.1 is already achieving this expectation, but it is still an expensive option. Further work is required to make this a possibility.

Finally, there is the challenge of generating meshes for such complex microstructures. Dedicated meshing algorithms such as `Hypermesh` from the Altair Hyperworks Suite and ANSA (see Sect. 1.4.1) exist, but there is a need to continue to advance the meshing capabilities so that future FEM frameworks will have as standard robust meshing tools that can mesh any complex microstructure for any element type with minimal computational cost or user interference. All these will obviously lead to more computational cost, which will need higher computing power. Advances in quantum computing and high performance computing strategies will clearly play a role in this regard.

11.5.6 Structural Level Computational Mechanics Deficiency

Within FEM frameworks, the computational mechanics principles needed to describe microscale and microscale mechanics of materials is at an advanced stage. However, the structural-level computational mechanics for many problems are still limited. Even though it is quite straightforward to define the mechanics of structural level problems, when such problems take input from the atomic, molecular or micro-level scales, to describe structural scale problems. There is a significant gap in understanding of how the transfer of information, across length scales, can be implemented effectively within a computational framework. Effectiveness here is

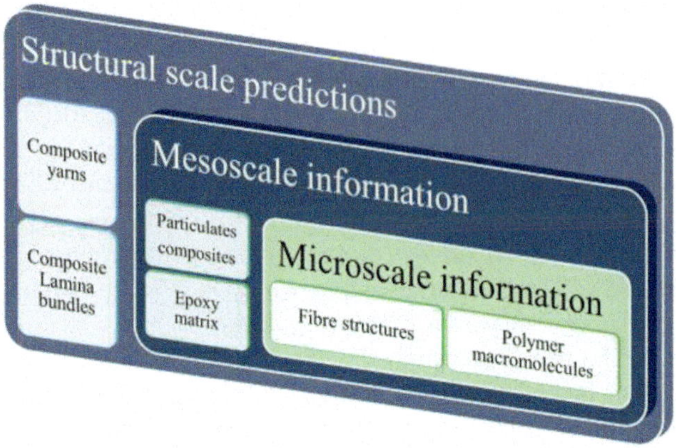

Fig. 11.2 A multiscale framework for composite material

quantified in two ways: *computationally inexpensive* and *fast convergence* to desired solution. This is what the authors here describe as the *structural-level computational mechanics deficiency.*

This challenge is very evident when the material responses of heterogeneous materials are considered. For example, a typical root section of a composite fan blade can be made from a three-dimensional reinforced textile composite. The composite is made up of weft, warp and binder yarns clearly defined at mesoscale length scales as shown in Fig. 11.2. Each yarn is made up of about 24,000 fibre strands, and all these are bound together using a matrix medium, which is typically an epoxy plastic. The architectural arrangements of the yarns lead to different types of textile composites such as orthogonal, layer-to-layer or tufted arrangements with clearly different mechanical properties.

When used in the fan blade, at structural scales of a few meters in length, the composite multiscale information has to be captured adequately for any mechanics at the structural scale to be reliably predicted from the fibre-level information. This is a real challenge within existing computational mechanics frameworks. This is a research area that future FEM frameworks will need to tackle in order to enhance predictions from such frameworks informed by micro-level data.

From the foregoing, the structural-level computational mechanics deficit is driven essentially by the computational cost required to predict structural response based on microscale inputs. Even though computing capabilities continue to improve, the state-of-the-art is still hindered by excessive demands on micro-processor capabilities and system memory of computing systems. Also, it is still computationally inefficient to achieve convergence of high mesh density problems: a necessity for structural level studies.

The research that should drive solutions to this hindrance will have to be focussed on innovative homogenization schemes that will translate microscale inputs into structural scale predictions with minimal computational and numerical implementation expense. Existing 'smearing' of structural responses highlights what is possible with this type of thinking, but it is severely limited as the 'smearing' exercise involves drastic erosion of detail of the microscale representation of the problem at the structural scale.

For example, if one modelled the textile composites at the microscale as consisting of 24,000 fibre strands, then at the structural scales, this detail is eroded as the structure is 'homogenized' into a homogeneous block of material. As a result any inter-fibre cracks at the yarn level cannot be replicated due to this extreme 'smearing'. Next-generation structural mechanics will have to account for such microscale details even at the structural scale.

In conclusion, Okereke and Akpoyomare [33] argued that the physics underpinning the bridging of scales between structural scale and microscale, without excessive smearing of the structural response, has not been properly understood. One of the feasible strategies that should be encouraged in future FEM frameworks was published by Edmans et al. [11]. The authors [33] proposed a nonlinear computational homogenization based on:

(a) *introducing a linear operator that maps the smooth part of a small-scale micro-displacement field to that of the large-scale; and,*
(b) *use of trace operator to impose boundary conditions on an RVE.*

Even though the Edmans' approach is still fraught with the same smearing losses discussed above, this innovative approach espouses sound mathematical principles to relate microscale data to the structural scale.

Improvements in the domain formulations across the scales and computational resolution of the problem should help establish improved structural scale predictions informed by microscale information. With advancements in this research area, one can envisage the application of even multiple silica particles interaction algorithms to describe computational solutions to seismic events such as earthquakes. This is the goal and when realized will make FEM highly effective as a solution tool for real problems of the future.

11.5.7 The Quest for High Fidelity Experimental Data

We have shown that the FEM framework gives approximate solutions. Validation plays a key role in establishing the reliability of the model predictions. Therefore, if validation data is not good enough, this will steer modelling development in the wrong direction. In fact, Blau et al. [5] argued that a key limitation to current FEM frameworks is the *paucity of high quality experimental data*. Also, Bentz et al. [4] stated that high quality experimental data are central to virtual testing and, consequently, the FEM process. If a poor set of experimental results is used in the FEM framework, then poor conclusions are likely to be drawn, in the famous cliche, *garbage in = garbage out*.

More prosaically, any simulation is only as good as the assumptions it contains and the quality of data used. Thus, the research challenge facing future uses of the FEM framework is the design and development of experimentation techniques to ensure always sound and reliable experimental data are always generated for the FEM scheme. Current data that feed into the FEM process are generated at the macroscale or structural length scale. These data are often used for validating models developed at microscale or even nanoscale. There is a dichotomy of usage here, which has been glossed over so far, with the common assumption that homogenization of lower length scale properties will make the resultant macroscale predictions comparable with such experimental data. This dichotomy of usage has to be explored in much depth in future.

The quest for high fidelity data is particularly important at microscale and sub-microscales. Such data will feed directly into modelling of, for example, *microelectromechanical systems* (MEMS). Hemker and Sharpe [18] showed that the flaw size within MEMS significantly influences the fracture strength and its evolution path, thus justifying the need to understand the microscale properties around the flaw site. This is very important in studies involving crack evolution and failure studies, as these originate from microscale defects within the material medium. The FEMs of the future will have to develop new techniques to investigate such length scale properties to generate robust data for the FEM framework. As a result, "*predictions at the microscale can be validated, establishing confidence that microscale models can be used as inputs for a higher scale analysis* [33]."

This research direction is already advancing at a rapid pace within the research community. In fact, there is a growing community of researchers working on innovative experiments to investigate and validate microscale models. Example experimentation techniques include: in situ nanoscale tensile tests [17]; Atomic Force Microscopy (AFM) based techniques for nanoscale measurements [38]; microscale fatigue experiments [39]; microscale plastic deformation experiment in micro-forming processes [8]; dynamic response of micro-machined beams [22]; lamellar level cortical bone load bearing characterization [21]; and finally an extensive set of tests for MEMS [37].

All these approaches describe an objective process of understanding the microscale behaviour of heterogeneous materials. These experiments are good starting points in the quest for high fidelity data and future FEM frameworks should be informed by outputs from such experiments for more reliable prediction. Research into similar experiments, not only at nanoscale or microscale, but even at molecular and atomic scales, have to be encouraged to ensure that high quality data are generated for the FEM process. This will significantly influence the impact of future FEM frameworks as design tool for both academic and industry sectors.

11.6 Conclusions

In this chapter, we have provided an overview of what has been learned so far with respect to the FEM framework. This established a baseline understanding of the state-of-the-art with respect to the use of FEM in academia and industry. Also, the FEM framework is fraught with several hindrances that are limiting its widespread use in a variety of sectors. The case has been made for what some of these hindrances are. These hindrances are identified to researchers explore studies to ensure that these hindrances are overcome. The full exploitation of the benefits of the FEM process cannot be achieved as long as these hindrances remain.

Finally, looking ahead into the future, this chapter has also presented some of the research directions that should occupy FEM researchers and enthusiasts to ensure in the next two decades, FEM will continue to be a design and research tool, that is capable of addressing the challenges of that generation. The options listed here are not exhaustive, but we have identified a few to begin the discussion about what the future FEM landscape will look like and what research needs to be carried out today to make sure that when we get there, we can actually recognize it as reward for the work we have done and continue to do.

At the end of this chapter, the reader should be able to:

(a) Give an overview of the FEM framework.
(b) Discuss the challenges and opportunities facing the FEM developer.
(c) Justify the research directions that will inspire future uses of FEM.
(d) Identify the unique features of future FEM frameworks.

References

1. Ari, I., Muhtaroglu, N.: Design and implementation of a cloud computing service for finite element analysis. Adv. Eng. Softw. **60**, 122–135 (2013). http://dx.doi.org/10.1016/j.advengsoft.2012.10.003, http://www.sciencedirect.com/science/article/pii/S096599781200141X. CIVIL-COMP: Parallel, Distributed, Grid and Cloud Computing
2. Arregui-Mena, J.D., Margetts, L., Mummery, P.M.: Practical application of the stochastic finite element method. Arch. Comput. Methods Eng. **23**(1), 171–190 (2016). http://dx.doi.org/10.1007/s11831-014-9139-3
3. Belytschko, T., Lu, Y.Y., Gu, L.: Element-free Galerkin methods. Int. J. Numer. Methods Eng. **37**(2), 229–256 (1994). http://dx.doi.org/10.1002/nme.1620370205
4. Bentz, D.P., Garboczi, E.J., Bullard, J.W., Ferraris, C., Martys, N., Stutzman, P.E.: Virtual testing of cement and concrete. In: Lamond, J.F., Pielert, J.H. (eds.) Significance of Tests and Properties of Concrete and Concrete-Making Materials. ASTM International, West Conshohocken (2006)
5. Blau, G., Lasinski, M., Orcun, S., Hsu, S.H., Caruthers, J., Delgass, N., Venkatasubramanian, V.: High fidelity mathematical model building with experimental data: a Bayesian approach.

Comput. Chem. Eng. **32**(4), 971–989 (2008). http://dx.doi.org/10.1016/j.compchemeng.2007.04.008. http://www.sciencedirect.com/science/article/pii/S0098135407000968. Festschrift devoted to Rex Reklaitis on his 65th Birthday

6. Bonet, J., Kulasegaram, S.: Correction and stabilization of smooth particle hydrodynamics methods with applications in metal forming simulations. Int. J. Numer. Methods Eng. 47(6), 1189–1214 (2000)
7. Bonet, J., Lok, T.S.: Variational and momentum preservation aspects of smooth particle hydrodynamic formulations. Comput. Methods Appl. Mech. Eng. **180**(1), 97–115 (1999). http://dx.doi.org/10.1016/S0045-7825(99)00051-1. http://www.sciencedirect.com/science/article/pii/S0045782599000511
8. Chan, W., Fu, M., Yang, B.: Experimental studies of the size effect affected microscale plastic deformation in micro upsetting process. Mater. Sci. Eng. A **534**, 374–383 (2012). http://dx.doi.org/10.1016/j.msea.2011.11.083, http://www.sciencedirect.com/science/article/pii/S0921509311013220
9. Cottrell, J., Hughes, T., Bazilevs, Y.: Isogeometric Analysis: Toward Integration of CAD and FEA. Wiley, Chichester (2009). https://books.google.co.uk/books?id=9Q9y0Xtz5E4C
10. de Paula Rodrigues, M., Soares, P.B.F., Valdivia, A.D.C.M., Pessoa, R.S., Verssimo, C., Versluis, A., Soares, C.J.: Patient-specific finite element analysis of fiber post and ferrule design. J. Endod. (2017). http://dx.doi.org/10.1016/j.joen.2017.04.024, http://www.sciencedirect.com/science/article/pii/S0099239917305307
11. Edmans, B., Alfano, G., Bahai, H.: Nonlinear multi-scale homogenization with different structural models at different scales. Int. J. Numer. Methods Eng. **94**(4), 355–373 (2013). http://dx.doi.org/10.1002/nme.4447
12. Farzam-Rad, S.A., Hassani, B., Karamodin, A.: Isogeometric analysis of functionally graded plates using a new quasi-3d shear deformation theory based on physical neutral surface. Compos. Part B Eng. **108**, 174–189 (2017). http://dx.doi.org/10.1016/j.compositesb.2016.09.029, http://www.sciencedirect.com/science/article/pii/S1359836816310174
13. Fernandez, P., Nguyen, N., Peraire, J.: The hybridized discontinuous Galerkin method for implicit large-eddy simulation of transitional turbulent flows. J. Comput. Phys. **336**, 308–329 (2017). http://dx.doi.org/10.1016/j.jcp.2017.02.015, http://www.sciencedirect.com/science/article/pii/S0021999117301080
14. Gerstle, W.: Introduction to Practical Peridynamics: Computational Solid Mechanics Without Stress and Strain. Frontier Research in Computation and Mechanics of Materials and Biology. World Scientific Publishing Company Pte Limited (2015). https://books.google.co.uk/books?id=WDc8DQAAQBAJ
15. Ghanem, R., Spanos, P.: Stochastic Finite Elements: A Spectral Approach. Civil, Mechanical and Other Engineering Series. Dover Publications (2003). https://books.google.co.uk/books?id=WzgKyTQQcAwC
16. Guo, N., Zhao, J.: A coupled FEM/DEM approach for hierarchical multiscale modelling of granular media. Int. J. Numer. Methods Eng. **99**(11), 789–818 (2014). http://dx.doi.org/10.1002/nme.4702
17. Haque, M.A., Saif, M.T.A.: In-situ tensile testing of nano-scale specimens in SEM and TEM. Exp. Mech. **42**(1), 123–128 (2002). http://dx.doi.org/10.1007/BF02411059
18. Hemker, K., Sharpe, W., Jr.: Microscale characterization of mechanical properties. Annu. Rev. Mater. Res. **37**, 93–126 (2007)
19. Hughes, T., Cottrell, J., Bazilevs, Y.: Isogeometric analysis: CAD, finite elements, nurbs, exact geometry and mesh refinement. Comput. Methods Appl. Mech. Eng. **194**(39), 4135–4195 (2005). http://dx.doi.org/10.1016/j.cma.2004.10.008, http://www.sciencedirect.com/science/article/pii/S0045782504005171
20. Ju, X., Mahnken, R.: Model adaptivity on effective elastic properties coupled with adaptive FEM. Comput. Methods Appl. Mech. Eng. **322**, 208–237 (2017). http://dx.doi.org/10.1016/j.cma.2017.04.013, http://www.sciencedirect.com/science/article/pii/S0045782516314554
21. Katsamenis, O.L., Chong, H.M., Andriotis, O.G., Thurner, P.J.: Load-bearing in cortical bone microstructure: selective stiffening and heterogeneous strain distribution at the lamellar level.

J. Mech. Behav. Biomed. Mater. **17**, 152–165 (2013). http://dx.doi.org/10.1016/j.jmbbm.2012.08.016. http://www.sciencedirect.com/science/article/pii/S1751616112002299

22. Kimberley, J., Lambros, J., Chasiotis, I., Pulskamp, J., Polcawich, R., Dubey, M.: Mechanics of energy transfer and failure of ductile microscale beams subjected to dynamic loading. J. Mech. Phys. Solids **58**(8), 1125–1138 (2010). http://dx.doi.org/10.1016/j.jmps.2010.04.005, http://www.sciencedirect.com/science/article/pii/S0022509610000748
23. Latham, J.P., Munjiza, A., Garcia, X., Xiang, J., Guises, R.: Three-dimensional particle shape acquisition and use of shape library for DEM and FEM/DEM simulation. Miner. Eng. **21**(11), 797–805 (2008). http://dx.doi.org/10.1016/j.mineng.2008.05.015, http://www.sciencedirect.com/science/article/pii/S0892687508001465. Discrete Element Methods (DEM) 07
24. Le, C., Okereke, M., Nguyen, V., Dao, V., Soe, S., Zlatov, N., Le, T., et al.: Personalised medical product development: methods, challenges and opportunities. Roman. Rev. Precis. Mech. Optic. Mechatron. **40**, 11–20 (2011)
25. Lee, C.H., Gil, A.J., Greto, G., Kulasegaram, S., Bonet, J.: A new jamesonschmidtturkel smooth particle hydrodynamics algorithm for large strain explicit fast dynamics. Comput. Methods Appl. Mech. Eng. **311**, 71–111 (2016). http://dx.doi.org/10.1016/j.cma.2016.07.033, http://www.sciencedirect.com/science/article/pii/S0045782516304182
26. Lee, C.H., Gil, A.J., Hassan, O.I., Bonet, J., Kulasegaram, S.: A variationally consistent streamline upwind petrovgalerkin smooth particle hydrodynamics algorithm for large strain solid dynamics. Comput. Methods Appl. Mech. Eng. **318**, 514–536 (2017). http://dx.doi.org/10.1016/j.cma.2017.02.002, http://www.sciencedirect.com/science/article/pii/S0045782516315201
27. Lim, J.Y., Kim, N., Park, J.C., Yoo, S.K., Shin, D.A., Shim, K.W.: Exploring for the optimal structural design for the 3D-printing technology for cranial reconstruction: a biomechanical and histological study comparison of solid vs. porous structure. Child's Nervous System (2017). https://doi.org/10.1007/s00381-017-3486-y
28. Liu, G.R., Liu, M.B.: Smoothed Particle Hydrodynamics: A Meshfree Particle Method. World Scientific, Singapore (2003)
29. Macek, R.W., Silling, S.A.: Peridynamics via finite element analysis. Finite Elem. Anal. Des. **43**(15), 1169–1178 (2007). http://dx.doi.org/10.1016/j.finel.2007.08.012, http://www.sciencedirect.com/science/article/pii/S0168874X07001035
30. Madenci, E., Oterkus, S.: Ordinary state-based peridynamics for plastic deformation according to von Mises yield criteria with isotropic hardening. J. Mech. Phys. Solids **86**, 192–219 (2016). http://dx.doi.org/10.1016/j.jmps.2015.09.016, http://www.sciencedirect.com/science/article/pii/S0022509615300909
31. Manyika, J., Chui, M., Bughin, J., Dobbs, R., Bisson, P., Marrs, A.: Disruptive Technologies: Advances That Will Transform Life, Business, and the Global Economy (McKinsey & Company) (2013) http://www.mckinsey.com/business-functions/digital-mckinsey/our-insights/disruptive-technologies, urldate = 10 July 2017
32. Morin, P., Nochetto, R.H., Siebert, K.G.: Data oscillation and convergence of adaptive FEM. SIAM J. Numer. Anal. **38**(2), 466–488 (2001). http://www.jstor.org/stable/3062065
33. Okereke, M., Akpoyomare, A., Bingley, M.: Virtual testing of advanced composites, cellular materials and biomaterials: a review. Compos. Part B Eng. **60**, 637–662 (2014). http://dx.doi.org/10.1016/j.compositesb.2014.01.007, http://www.sciencedirect.com/science/article/pii/S1359836814000109
34. Oterkus, S., Madenci, E., Oterkus, E., Hwang, Y., Bae, J., Han, S.: Hygro-thermo-mechanical analysis and failure prediction in electronic packages by using peridynamics. In: 2014 IEEE 64th Electronic Components and Technology Conference (ECTC), pp. 973–982. IEEE (2014)
35. Pivovarov, D., Steinmann, P.: On stochastic FEM based computational homogenization of magneto-active heterogeneous materials with random microstructure. Comput. Mech. **58**(6), 981–1002 (2016). http://dx.doi.org/10.1007/s00466-016-1329-4
36. Sheng, Z., Jirathearanat, S., Altan, T.: Adaptive FEM simulation for prediction of variable blank holder force in conical cup drawing. Int. J. Mach. Tools Manuf. **44**(5), 487–494 (2004). http://dx.doi.org/10.1016/j.ijmachtools.2003.11.001, http://www.sciencedirect.com/science/article/pii/S0890695503002980

37. Srikar, V.T., Spearing, S.M.: A critical review of microscale mechanical testing methods used in the design of microelectromechanical systems. Exp. Mech. **43**(3), 238–247 (2003). http://dx.doi.org/10.1007/BF02410522
38. Sundararajan, S., Bhushan, B.: Development of AFM-based techniques to measure mechanical properties of nanoscale structures. Sensors Actuators A Phys. **101**(3), 338–351 (2002). http://dx.doi.org/10.1016/S0924-4247(02)00268-6, http://www.sciencedirect.com/science/article/pii/S0924424702002686
39. Szczepanski, C., Jha, S., Shade, P., Wheeler, R., Larsen, J.: Demonstration of an in situ microscale fatigue testing technique on a titanium alloy. Int. J. Fatigue **57**, 131–139 (2013). http://dx.doi.org/10.1016/j.ijfatigue.2012.08.008, http://www.sciencedirect.com/science/article/pii/S0142112312002605. Fatigue and Microstructure: A special issue on recent advances
40. Wang, Z., Konietzky, H., Shen, R.: Coupled finite element and discrete element method for underground blast in faulted rock masses. Soil Dyn. Earthq. Eng. **29**(6), 939–945 (2009). http://dx.doi.org/10.1016/j.soildyn.2008.11.002, http://www.sciencedirect.com/science/article/pii/S0267726108002200
41. Werner, R., Ehrhardt, J., Schmidt, R., Handels, H.: Patient-specific finite element modeling of respiratory lung motion using 4D CT image data. Med. Phys. **36**(5), 1500–1511 (2009). http://dx.doi.org/10.1118/1.3101820
42. Zhang, L., Deng, Y., Liew, K.: An improved element-free Galerkin method for numerical modeling of the biological population problems. Eng. Anal. Bound. Elem. **40**, 181–188 (2014). http://dx.doi.org/10.1016/j.enganabound.2013.12.008, http://www.sciencedirect.com/science/article/pii/S0955799713002725

Glossary

Adjugate This is the transpose of **co-factor** matrix of a **square** matrix. It is also called the *adjoint* or *adjunct*.

Anisotropic If the material properties of a test material is different in every direction and plane of testing, the material is said to show anisotropy, and described as anisotropic material. Comprehensive description of the material model for such materials must consist of a **stiffness tensor** that has terms relating to every test direction and plane of analyses for such materials. It is a feature of a lot of materials with a heterogeneous microstructure. It is important that extensive experimental data is generated to support the model development effort for such materials. Also, computational science has become increasingly a standard for analysis and design of such anisotropic materials.

Arrhenius equation The classic formulation of the *Arrhenius Equation* gives the analytical relationship between temperature-dependence and reaction rates. It was developed by Svante Arrhenius in 1889. The original formulation of the Arrhenius equation is: $k = A\exp^{-E_a/(RT)}$, where k = rate constant, T = absolute temperature, A = pre-exponential factor, E_a = activation energy, and R = gas constant. It has since become adapted for other application areas, one of such is the temperature-dependence of viscoelastic materials. In this case, the temperature dependence of the relaxation times of polymer macromolecular segments is predicted using the Arrhenius equation.

Asymptotic near-tip field One of the methods of modelling cracks in fracture mechanics is called the *near-tip solution*. It involves understanding the crack growth around the crack tip by studying the stresses and deformations there. The values are defined in form of an asymptotic near-tip stress or deformation field. The values are not exact, but rather asymptotes of the actual stresses in the crack tip, hence their reference as *asymptotic near-tip field*. This method has been used successfully in fracture mechanics to define discontinuities arising from the effect of a crack in a material volume.

M. Okereke, S. Keates, *Finite Element Applications*, Springer Tracts in Mechanical Engineering, https://doi.org/10.1007/978-3-319-67125-3

Augmented matrix This type of matrix is formed when the columns of two matrices are appended to form a single matrix, usually for the purpose of undertaking an elementary row operation. For example consider the matrix: $Q = \begin{bmatrix} 5 & 8 & 15 \\ 25 & 10 & 12 \\ 8 & 14 & 3 \end{bmatrix}$, and $R = \begin{bmatrix} 5 \\ 8 \\ 14 \end{bmatrix}$, then, $S = (Q|R) = \begin{bmatrix} 5 & 8 & 15 & | & 5 \\ 25 & 10 & 12 & | & 8 \\ 8 & 14 & 3 & | & 14 \end{bmatrix}$ is the augumented matrix of Q and R matrices. A common application is during the *Gaussian elimination method* for solving linear equations.

Cloud-based computing This is also referred to, for short as, cloud computing. It is both a computing infrastructure and a software model, in which the user is granted access to computing resources hosted in a remote location, called *the cloud*. This framework allows access to shared pools of resources without geographical or time restrictions. This is converse to *desktop computing* in which the shared resources are restricted to a desktop and only specific users, in a given location and with access right to that desktop can assess the resources. The shared resources can include computer networks, servers, storage, applications and other services. Cloud computing is revolutionalizing the provisioning of computing resources to organizations and helps companies limit or even avoid up-front computing infrastructure usage costs. The enabling technology that makes cloud computing possible is a *virtualization* software that creates 'virtual devices' that perform computing tasks and can be managed as part of a shared resource pool.

Co-factor The co-factor of a matrix, F (see **determinant**), is obtained by multiplying the **minor** of a matrix by $(-1)^{i+j}$ where i is the i-th row and j refers to the j-th column. For example, given the matrix: $D = \begin{bmatrix} 2 & 3 & 5 \\ 5 & 10 & 5 \\ 0 & 6 & 4 \end{bmatrix}$, the co-factor of the first row and third column ($i = 1, j = 3$) becomes $C_{1,3} = (-1)^{(1+3)} M_{1,3} = M_{1,3}$ where $M_{1,3}$ is the minor of the matrix, D when the first row and third column are both deleted. A *co-factor* matrix is the matrix of all the co-factors associated with a given matrix.

Compliance tensor This is a matrix that represents the inverse of a **stiffness tensor**. It is the proportionality matrix that relates the strain at a material point to a stress at the same material point. For a uniaxial tensile test, the compliance matrix reduces to the inverse of the Young's Modulus.

Condition number This is a measure of the effect of changes in output value of a function with a slight change in its input values. It gives an indication of how sensitive model predictions can become due to slight errors in input to such models. When the condition number is high, then the system is said to be *ill-conditioned* while a low condition number problem is described as *well-conditioned*.

Constitutive behaviour This is a term used in engineering mechanics to describe the comprehensive material response of a material under the effect of mechanical loading. It is usually described based on a stress-strain diagram of the test material.

Convergence This is a technical term used in finite element studies to describe the speed at which the approximate solution to a problem is achieved. Rapid convergence is always desired and the objective of a good FE solution is to ensure that the solution converges quickly. It is therefore commonly said for a given numerical problem that the *solution converges rapidly*. Mesh size, element choice, boundary conditions, material formulation, etc., all play a part in improving or delaying the convergence to a desired solution.

Creep When a material is subjected to a constant load over a long time, the material is said to creep if it experiences a gradual elongation which continues to increase with time. The converse to creep is called stress relaxation, which refers to the reduction in stress experienced by a material when subjected to a constant elongation. Both are manifestations of viscoelastic behaviour.

Determinant This is a property of a square matrix and it is a single-valued number that defines a scaling parameter/value for that matrix. For a square G matrix, the determinant is represented in mathematics by: $\det(G)$ or $|G|$ or $\det G$. For example, given a 2×2 matrix G defined as: $G = \begin{bmatrix} g_{11} & g_{12} \\ g_{21} & g_{22} \end{bmatrix}$, the determinant becomes: $\det(G) = g_{11}g_{22} - g_{21}g_{12}$ and for a 3×3 matrix, $F = \begin{bmatrix} f_{11} & f_{12} & f_{13} \\ f_{21} & f_{22} & f_{23} \\ f_{31} & f_{32} & f_{33} \end{bmatrix}$, the determinant is: $|F| = f_{11}(f_{22}f_{33} - f_{32}f_{23}) - f_{12}(f_{21}f_{33}) - (f_{23}f_{31}) + f_{13}(f_{21}f_{32} - f_{22}f_{31})$.

Double precision The full name is *double precision floating point format*. It is an internal representation of numbers in digital computers in which the number takes 8 bytes (or 64 bits) in the computer memory. It offers an enhanced floating-point representation of numbers that is wide and dynamic in both range and precision. Many engineering codes have numeric values represented as double precision. In Fortran, is is written by adding a .D0 suffix to the numeric value.

Elasticity In Physics, this term is used to describe the ability of a material body to return to its original, undeformed shape when the load that caused its deformation is removed. The converse to elasticity is **plasticity**.

Elastoplastic This is a form of **plasticity** in which a material body under the effect of load, once it has attained its yield stress, begins to experience a constant-stress deformation, i.e. increasing strain without a corresponding increase in stress. The strain at which the elastoplasticity initiates is called the *plastic strain* and it is a central feature used in many material models of plasticity.

Equibiaxial In experimental mechanics, the term equibiaxial refers to a type of test in which the same load is applied in two axes usually perpendicular to one another. It should be contrasted with *uniaxial* where test is only in one direction only. Equibiaxial tests are useful in characterizing the behaviour of hyperelastic materials especially, subjected to a combined loading history. Structures such as tents, balloons, etc., experience a biaxial state of loading, hence such experiments are used to generate material properties for modelling reliably their material response.

Flow theory The full form of this is the *flow theory of plasticity*. It describes the mathematical formulation that establishes the transition from **elasticity** to **plasticity** behaviour. It is essential in classic plasticity literature to describe how the transition from linear elastic response to different forms of plasticity responses. It is usually based on considering energy dissipation that arises from a predominantly elastic response to a plastic deforming material system.

Fourier series This an approximation scheme for representing a periodic function, $g(x)$ as a sum of simple sine and cosine waves. It requires the decomposition of any periodic function into near infinite sets of oscillating functions. These oscillating functions are typically *sines* and *cosines*. If $g(x)$ is a square wave, the Fourier series representation can be: $g(x) = \frac{1}{2}a_0 + \sum_{n=1}^{\infty} a_n \cos(nx) + \sum_{n=1}^{\infty} b_n \sin(nx)$, where a_0, a_n and b_n are model constants for $n = 1, 2, 3, \cdots$. Fourier analysis is a dedicated field for using Fourier series to solve a large set of real life problems, particularly with differential equations.

Gauss' theorem This is also called *divergence theorem* or *Ostrogradsky's theorem*. It is used in vector calculus to relate the behaviour of a vector field inside a surface by observing or analyzing the flow (or flux) of a vector field through a surface. In other words, there exists a relationship between the behaviour observed at the surface and the internal volumetric space enclosed by the surface. It is the basis upon which the finite volume method works. It is widely used in engineering to describe systems involving flow across a surface..

Green's function This is another method for solving linear differential equation: $L[y] = f(x)$ where L is a *differential operator* and $f(x)$ is the *forcing function* or an *inhomogeneous term* defined in terms of x. The Green's function, $G(x, y)$ is an *integral kernel*, which can be represented by an inverse operator, L^{-1}. It is used in solving *boundary value problems* of *ordinary differential equations*. The Green's function has to satisfy the condition: $L_x G(x, y) = \delta(x - y)$, where the differential operator, L_x, has homogeneous boundary conditions: its solution corresponds to the data $\{\delta(x - y); 0\}$ and δ is the Dirac delta function.

Haar function Just like the Fourier series, which describes a function based on periodic functions of sine or cosine, the Haar function is a representation of a function in terms of a square-shaped functions such that they combine into a what is called a *wavelet family*, i.e. a family of wave-like oscillations. The Haar function is also a decomposition of a periodic function in terms of *orthonormal basis vectors* (unit vectors that are orthogonal to one another). The Haar function therefore consists of alternating values of 1's, 0's and -1's.

Higher order elements In Lagrangian element formulation, the shape functions can be described by linear polynomial expansions with the shape function having a linear dependence on the isoparametric natural coordinates. When this dependence is based on polynomial expansions higher than 1, then such element types are described as higher order elements. The higher order elements are useful in describing nonlinear discretization of the domains, for example, curved edges. They also offer improved features of element behaviour, for example, non-constant strain and stress across the element. However, they offer high computational cost to the FEM user and so should be used only when they are needed. The standard, linear, first order element types are used generically as the main element types in common commercial FEM solvers.

Hookean This is a type of material response in which the dependence of stress and strain is linear with the slope of the stress-strain plot being the modulus of elasticity, E. The name is taken from the Hooke's law, which describes this type of material response. Analytically, it is described by the equation: $\sigma = E\epsilon$, and it is the most common material response for many construction materials. For more details, refer to Sect. 10.4, especially the sections concerning *isotropic linear elasticity*.

Hooke's Law This is a law made popular by Robert Hooke (1635 - 1703), which established a relationship between the force, F, required to deform an elastic spring and the resulting displacement, x experienced by the spring. The law states that there is a linear relationship between the force, F, and the displacement, x. The constant of proportionality of the law is called the spring constant, k. The formula, according to Hooke's Law, becomes: $F = kx$. The law is the basis of linear elasticity formulation used in material science. Hence, such materials are said to exhibit a **Hookean** response.

Identity matrix This is a **square** matrix in which the diagonal terms of the matrix are all ones while the other off-diagonal terms are zeros. It is equivalent to a scalar number one. It is common in matrix manipulation and needed when a user wants to multiply or divide a matrix by one.

Isothermal This is a condition imposed on a test system, reaction or material in which temperature is constant. In other words, change in temperature is zero, i.e. $\Delta T = 0$. For this to happen, the system must have a mechanism that self-regulates its temperature thereby keeping it constant. For example the system could be connected to a *heat source/reservoir*. A thermostat can be used to enforce this condition. Material responses modelled under isothermal conditions have mechanical properties that do not change with temperature.

Isotropic In material science, isotropy is used to describe the uniformity of material properties in all directions of measurements. For example, a piece of polymer, is said to be isotropic, if the mechanical properties (e.g. Young's modulus) is the same in all directions namely $x-$, $y-$ and $z-$ axes. It is a common feature for materials that are homogeneous in microstructure.

Jacobian matrix This is the matrix formed from a set of all first-order derivatives of a mathematical function whose variables form a set of multidimensional vectors, otherwise called a vector-valued function. It is common to refer to this matrix simply as the Jacobian, for example the tangent stiffness matrix required for developing a UMAT. Also, the determinant of the matrix (when square) is called the Jacobian determinant, or jacobian for short. In continuum mechanics, for example, consider a deformation gradient, $\boldsymbol{F}$: a vector-valued function; its determinant, called *volume ratio*, $J = \det(\boldsymbol{F})$ is also called a Jacobian.

Lamé constants This is a set of constants that consist of Shear Modulus, G and Cross Modulus, λ. It is used to described the isotropic elasticity of a material. These constants are named after Gabriel Lamé (1795-1870). They appear in constitutive models of **isotropic isothermal** linear elasticity materials. Once they are known, the full set of material response of such isotropic linear elasticity materials are fully known. These constants can be related also to the Bulk Modulus, K, Young's Modulus, E, Poisson ratio, ν and even longitudinal or constrained modulus, M, of the material.

Macromolecular In polymer chemistry, addition of monomer units through a process called polymerization, leads to formation of polymers. A polymer is made up of large molecular units called a *macromolecule*. It consists of entangled mass of molecules that are held together by slip links or physical cross-links. The mechanics of such polymers is called *macromolecular mechanics* and it uses the network arrangement of the entangled mass to define the material response of such macromolecular structures. Examples of macromolecules can include such simple plastics as polyethylene (PE) and polypropylene (PP), as well as complex polymers like polytetrafluorethylene (PTFE). Even proteins and DNA can be modelled using macromolecular mechanics approaches.

Methods of joints and sections It is essential in *Mechanics of materials* (an undergraduate engineering mechanics course) that students compute the state of equilibrium of a pin-jointed truss structure. This is a common type of structural design that engineers are exposed to. This exercise requires calculating the effect of externally applied loads on the internal forces and displacements of the structural members (the trusses). The *method of joints* is one of the methods used to determine the internal forces on each member/truss. It requires solving a set of equilibrium equations for every joint in the structure, after disassembling the members. The *method of sections* is used to solve the equilibrium equations of a *sectioned off* rigid-body sub-set of the the structure.

Micromechanics Microscopic matter can be measured in order of microns (10^{-6} m). Micromechanics is a branch of engineering mechancis that considers the kinetics and kinematics of particles defined at the microscale. For example, in laminated composites, the fibre reinforcements have typical diameters in tens of microns, hence their material model representation is often described using micromechanics assumptions.

Minor Given a matrix: $D = \begin{bmatrix} 2 & 3 & 5 \\ 5 & 10 & 5 \\ 0 & 6 & 4 \end{bmatrix}$, the minor matrix of the second row and the second matrix, $M_{2,2}$ obtained by deleting the second row and the second column which results in: $M_{2,2} = \begin{bmatrix} 2 & 5 \\ 0 & 4 \end{bmatrix}$.

Nanomechanics The nanoscale is a length scale measured at 10^{-9} m. It is the scale at which sub-microscopic particles can be detected. For example, nanomaterials and nanocomposites have inclusions that are nanoscale in size and dispersed within a matrix medium. Nanomechanics is the branch of engineering mechanics that describes the interaction of kinetics and kinematics of such nanomaterial.

Orthotropic When the mechanical properties of a test material are different in three orthogonal planes of the material, that material is said to exhibit orthotropy and is described as an orthotropic material. An example of orthotropy is a multi-dimensional composite in which the properties in the $x-$, $y-$ and $z-$ axes are different. It is a common feature for materials that have composite and layered arrangements in microstructure.

Plasticity When a body is deformed under the effect of a load, if the material does not return to its undeformed state when the load is removed, it is said to exhibit plasticity and such deformation is called plastic deformation. The converse to plasticity is **elasticity**.

Poisson's ratio This is a mechanical property used in material science to describe the Poisson effect. This effect describes the behaviour of a material in which it tends to expand in a direction transverse to its main direction of compression. For a tensile test, the Poisson effect is manifest in the form of a contraction in a direction transverse to the main axis of stretching of the material. Mathematically, the Poisson effect is represented as the ratio of the transverse strain to the axial strain where the axial strain is the strain in the main axis of loading of the material. This ratio is described as the Poisson ratio and it is typically 0.5 and positive for perfectly incompressible material. Poisson ratio ranges from 0.0 to 0.5. Some materials, such as auxetic materials, polymeric foams, biological cells, show negative Poisson ratio. When such materials are stretched, they tend to get thicker.

Positive-definite This is a property of a real $n \times n$ matrix, A such that given a non-zero column vector, b, of n real numbers, the scalar of the matrix $b^T Ab$ exists and is positive.

Pseudorandom This is used to describe a sequence of numbers, digital output or statistical data that have no recognizable patterns of regularities (otherwise described as *statistical randomness*) even though the numbers were generated using a deterministic algorithm. Most computer-generated random variables are actually *pseudorandom*.

Rate-dependence Properties of materials can be said to show rate-dependence when such properties change with rate of loading. For example, if the Young's Modulus of a material is 20 GPa when generated at a slow loading rate, then the material is rate-dependent if this property does not remain constant and changes to, say, 50 GPa at fast or impact rates of loading. Polymers and biological tissues show rate-dependence and to provide a reliable prediction of their material response, one must account for such rate-dependence.

Relaxation time In viscoelasticity, this is the time it takes a polymer macromolecule, subjected to a large scale motion, to relax to a steady state following the application of a load. On a creep compliance versus log(time) plot, it is the time in between the relaxed and unrelaxed creep compliance. It is influenced by temperature, molecular weight, rate of loading and it is one of the most important parameter for describing the viscoelasticity of a material volume.

Serendipity The word "serendipity" is used to describe accidental and fortunate discovery or simply a happenstance. It was first used in c. 1754 by Horace Walpole in the book *The Three Princes of Serendip* (i.e. Sri Lanka). It is a book by Cristoforo Armeno in which the three princes make such 'fortunate discoveries.' The serendipity elements were 'accidentally discovered' by the originating authors and have since become the standard element type for many FEM solvers.

Singular If a **square** matrix does not have an inverse matrix, then it is described as *singular* or *degenerate*. This also implies that the **determinant** of such matrix is zero. However, if a matrix, R has an inverse (R^{-1}) that exists, then it is described as *non-singular* or *non-degenerate*.

Smoothness If a mathematical function, $g(x)$, has multiple derivatives that are continuous over the analysis domain, then it is described as a *smooth function*. The measure of the number of such derivatives is described by the *smoothness* property. In numerical methods, if the data can be sufficiently described by a smooth function, it will have a high smoothness property and, as a result, can converge to the desired solution quickly. Where this is not the case, the data is described as *non-smooth* and convergence will be delayed or might not even happen at all.

Square This is used in mathematics to describe a type of matrix in which the rows and columns are equal. They are regarded as $n \times n$ matrices for example, a 4×4 matrix which has 16 terms. It is a requirement for matrices that are *invertible* i.e. the inverse of the matrix exists. It is also essential condition for the **determinant** of a matrix to exist.

Stiffness tensor This is a matrix that relates the stress tensor, $\boldsymbol{\sigma}$ at a material point to the strain tensor, $\boldsymbol{\epsilon}$ at that material point. It is a collection of the moduli of elasticity in all normal directions and shear planes of testing for a given test material. It is a comprehensive representation of all the stiffnesses experienced within a material when subjected to a complex loading history. For a uniaxial tensile test, for an **isotropic** material, the stiffness tensor reduces to a scalar quantity: the Young's Modulus.

Strain energy The term is used to describe the energy stored by a system undergoing deformation. If the deforming material is a linear elastic material, the formula for strain energy is $U = \frac{1}{2}V\sigma\epsilon$, where V = volume of deforming materia; σ = stress experienced by deforming body; and, ϵ = strain.

Strain energy density function When the **strain energy**, U is divided by the volume of the deforming material, then this results in the strain energy density. If the material is not linear elastic, then the formulation of strain energy density has to be a function of other parameters. Some of these can include the deformation gradient, $\boldsymbol{F}$, strain tensor, $\boldsymbol{\epsilon}$, the stress invariants, I_i for $i = 1, 2, 3$. Development of constitutive models based on strain energy begins with establishing the correct formulation of the strain energy density function, for the material under investigation.

Symbolic computation This is a branch of computational mathematics about the development of algorithms and software for the manipulation of mathematical functions. Increasingly, they are becoming quite popular in computational mechanics for the derivation of partial derivatives of functions. Some of the common symbolic computing software are: Maple, Mathematica and, most recently, MATLAB™, which has a dedicated *symbolic mathematics toolbox* for undertaking symbolic computation.

Symmetric A matrix is described as *symmetric* if its transpose is equal to the same matrix. For example, given a matrix, A, it is described as symmetric if: $A = A'$, where A' is the transpose of A. Note that the symmetric matrix is always a **square** matrix. The terms on either side of diagonal terms of the square matrix are equivalent.

Taylor's series This is a numerical scheme in which a function, $g(x)$, is obtained as the *infinite sum* of the function's partial derivatives defined at a given point, x_0. Mathematically, the Taylor's series is represented as: $g(x_0) = g(a) + \frac{g'(a)}{1!}(x - x_0) + \frac{g''(a)}{2!}(x-x_0)^2 + \cdots \equiv \sum_{n=0}^{\infty} \frac{g^{(n)}(x_0)}{n!}(x-x_0)^n$. Here, $g^{(n)}(x_0)$ is the n-th partial derivative of the function, $g(x)$. The Taylor's series is used to obtain numerical approximations within the finite difference method.

Trace This is the sum of the diagonal terms of an $n \times n$ matrix with the diagonal read from the upper (top-right) term to the lower (bottom-right) term. The trace can only be computed for a square matrix. Given a square matrix, $\boldsymbol{D}$, the trace can be evaluated as: $\mathrm{tr}(\boldsymbol{D}) = d_{11} + d_{22} + \cdots + d_{nn} = \sum_{i=1}^{n} d_{ii}$, where d_{ii} is the i-th row and the i-th column diagonal term.

Transverse isotropic This is a special type of **orthotropic** material behaviour such that there exists a plane of symmetry in which all properties orthogonal to this plane have the same property. For example, a reinforced concrete will show this property. In the direction of reinforcement, the Young's Modulus, for example, will be high. However, all transverse directions to this reinforcement axis, will have the same Young's Modulus. For such materials, the elastic properties in the reinforcement and transverse directions have to be assembled for use within the material's **stiffness tensor**.

Uniaxial A type of loading on a test specimen in which the force and displacement are along the same axis (co-linear). For example, a typical tensile or compression test is a uniaxial experiment.

Unidirectional composite This term is used to describe a type of fibre reinforced composite in which the fibre reinforcement is alligned in the same directionl. As a result, in the direction of the fibre reinforcement, the composite shows excellent properties however, transverse to the fibre direction, the composite is very weak. A laminate is an assembly of unidirectional composites, called laminae, to form an assembly multidirectional composite.

User-defined material subroutine This is a set of programming instructions passed to a finite element solver to describe a new set of material response. It is usually the user that compiles the code that captures the new set of material behaviour that is not already existing with the finite element modelling solver.

Validation In numerical computing, this term relates to the exercise of checking the accuracy of computing results with respect to *known information* about the system under consideration. These *known information* are described as *validation data* and can be experimental data, material data, some analytical information about sub-domains of the problem, etc. It is an essential part of the FEM process and where it is non-existent, the validity of FEM conclusions will be cast in doubt.

Von Mises stress When a material body is subjected to combined multi-axial loading history, it is not always easy to identify a stress measure that can be used to compare with known material properties of the loaded material. Knowing the combined effect of the multiple stresses is essential for making design decisions involving the loaded material. The von Mises stress is one of many stress measures that gives a single value that the design engineer can compare with experimental data to decide whether the threshold of yielding of the test material has been exceeded in which case design fidelity compromised. At the threshold of yielding the von Mises stress must exceed the universal tensile strength (UTS) of the test material. It is a common stress measure in many FEM solvers, and contour plots of such solvers usually specify the stress measures in terms of the von Mises stress.

Young's Modulus This is the ratio of stress to strain within the linear elastic (or *proportionality*) limit of a material, measured from a stress-strain graph. It is a measure of the stiffness (or resistance to deformation) of a material volume under load. It is represented by *E*. Some textbooks refer to it as *elastic modulus* or *modulus of elasticity*. It is an important parameter for describing the linear elastic behaviour of a material.

Index

M. Okereke, S. Keates, *Finite Element Applications*, Springer Tracts in Mechanical Engineering, https://doi.org/10.1007/978-3-319-67125-3

The manufacturer's authorised representative in the EU is Springer Nature Customer Service Centre GmbH, Europaplatz 3, 69115 Heidelberg, Germany. If you have any concerns regarding our products, please contact ProductSafety@springernature.com

Printed and bound by CPI Group (UK) Ltd, Croydon, CR0 4YY
31/07/2026
02182892-0003